Axel Harnack, George Lambert Cathcart

An Introduction to the Study of the Elements of the Differential and Integral Calculus

Axel Harnack, George Lambert Cathcart

An Introduction to the Study of the Elements of the Differential and Integral Calculus

ISBN/EAN: 9783337811389

Printed in Europe, USA, Canada, Australia, Japan

Cover: Foto ©berggeist007 / pixelio.de

AN INTRODUCTION

TO THE STUDY OF THE

ELEMENTS

OF THE

DIFFERENTIAL AND INTEGRAL CALCULUS.

FROM THE GERMAN

OF THE LATE

AXEL HARNACK,

PROFESSOR OF MATHEMATICS AT THE POLYTECHNIKUM, DRESDEN.

WITH THE PERMISSION OF THE AUTHOR

WILLIAMS AND NORGATE,

14, HENRIETTA STREET, COVENT GARDEN, LONDON;

AND 20, SOUTH FREDERICK STREET, EDINBURGH.

1891.

LEIPZIG
PRINTED BY B. G. TEUBNER.

CONTENTS.

First Book. (Pp. 1—109.)

Real numbers and functions of real numbers.

Second Book. (Pp. 110—168.)

Complex numbers and functions of complex numbers.

Third Book. (Pp. 169—320.)

Integrals of functions of real variables.

Fourth Book. (Pp. 321—389.)

Integrals of complex functions. General properties of analytic functions.

First Book.

Real numbers and functions of real numbers.

The conceptions of space and of number are the subject matter of mathematical investigations. These investigations accordingly diverge into two main branches: Geometry and Analysis. It thus appears that Mathematics are of fundamental importance to all our knowledge of Nature: for our representations of space contain the simplest properties which are common to all things in the surrounding world; and accurate comparison or measurement of quantities leads always to concrete numbers of the units employed: in order to understand the result, we require a knowledge of numbers and of their combinations.

Nature in its phenomena is perpetually exhibiting change; the simplest changes we perceive externally are changes of place, motions. The representation of motion is necessarily combined with that of continuity, i. e. of an uninterrupted connexion in space and of an uninterrupted sequence in time. To describe thoroughly the phenomena of motion is to assign every circumstance in numbers of concrete units: so that if the series of numbers is also to enable us to describe motion, it must contain a continuous series of quantities. Thus the first problem of Analysis is: to develope the conception and the properties of the continuous series of numbers.

First Chapter.

Rational numbers.

1. The natural series of numbers, which arises by adding on a thing to others in counting, advances always by unity; each number is defined by the preceding number and by unity. This series of integers starting from unity can be continued on indefinitely. Now as each several number is a sum of repeatedly added units, such a sum of units can be composed of different given numbers. This arithmetical operation, merely a continued reckoning up of groups of units, is called Addition; it embraces all other operations, from it all others arise. The fundamental proposition for addition is: the sum of given numbers has alway the same value in whatever order the summands be reckoned up. The truth of this proposition cannot be deduced from simpler conceptions; it is immediately perceived from the intuition of an arbitrary but finite number of units, which form the sum. Addition is always possible, the sum is continually greater than any of the summands.

The problem of Subtraction follows from inversion of addition. What this requires is, given the sum and one summand to calculate the other summand (difference); that is, to reckon off from a given number (minuend) as many units as the other number (subtrahend) contains. This is only possible when the minuend is greater than the subtrahend. In case they are both equal, i.e. contain the same number of units, we denote the result by 0. The numerical conception 0 is therefore defined by the equation $a - a = 0$. Hence we obtain for calculating with 0 the equations: $a + 0 = a$, $a - 0 = a$.

2. To form a sum in which the several summands are equal to the same number a and the number of summands is b, is, to Multiply the number a by b. The result is called a multiple of a. But without distinguishing a and b they may be called factors and the result simply the product, since for multiplication of two or more numbers we have the fundamental proposition — which can be proved from the conception of summation: The product of given numbers has always the same value, in whatever way the factors may be interchanged or combined in groups.*) Multiplication is always possible.

The problem of Division follows from inversion of multiplication, when, given the product and one factor it is required to calculate the other factor; that is, to find that number (quotient) which multiplied by one of the given numbers (divisor) yields a product equal to the other (dividend). This is impossible unless the dividend is a multiple of the divisor; calling a the divisor, b the dividend, then if $a \leq b$ there is always an equation of the form: $b = a \cdot q + r$, where r (the remainder) must be a number of the series $0, 1, 2, \ldots a - 1$.

Two numbers can be multiples of a third number, this is then a common divisor of both. Two numbers, which have no common divisor besides unity, are relatively prime. A number is called absolutely prime which is not divisible by any other except unity. This distinction of divisible and prime numbers leads to the important theorem: Any number can be expressed only in a single manner as a product of absolutely prime numbers; but the investigation of the divisibility of a number rests on the following rule (employed by

---

*) The proofs of the theorems for rational numbers cannot be presented here: they are found in Baltzer, die Elemente der Mathematik, Vol. I.

Euclid): The greatest common divisor of two numbers a and b (where $b > a$) is found by forming by continued division the eqtations:

$$b = aq + r, \quad a = rq_1 + r_1, \quad r = r_1 q_2 + r_2 \quad \text{etc.}$$

The divisor of the last division, which leaves no remainder, is the greatest common divisor of a and b.

3. The first three arithmetical operations, performed on sums and differences lead to the following equations:

$$\text{(I.)} \quad \begin{cases} (a+b)+(c+d) = a+b+c+d \\ (a+b)-(c+d) = a+b-c-d \end{cases}$$
$$\begin{cases} (a+b)+(c-d) = a+b+c-d \\ (a+b)-(c-d) = a+b-c+d \end{cases}$$
$$\begin{cases} (a-b)+(c-d) = a-b+c-d \\ (a-b)-(c-d) = a-b-c+d \end{cases}$$

$$\text{(II.)} \quad (a+b)\,c = ac + bc, \quad (a-b)\,c = ac - bc$$

$$\text{(III.)} \quad \begin{aligned} (a+b)(c+d) &= ac+bc+ad+bd \\ (a+b)(c-d) &= ac+bc-ad-bd \\ (a-b)(c-d) &= ac-bc-ad+bd \end{aligned}$$

The differences on the left hand sides in these equations are assumed to be possible; some of them may be zero. A product which contains the factor 0, is therefore, as we learn from (II) and (III), itself equal to 0. It follows conversely from the same equations, that a product can never be 0 unless one of its factors have the value 0.

The similarity of the results in calculations with sums and differences, suggests the advantage of regarding from the outset the difference as a sum, for instance the difference $a - b$ as a sum of a actual units here to be reckoned up and b to be taken away, or as it is better expressed, of a Positive and b Negative units. The introduction of negative unity enables us to calculate also with differences in which the minuend is less than the subtrahend. Thus when $a < b$ the number $a - b$ expresses an excess of negative units, of so many, in fact, that $(b - a) + (a - b) = 0$.

In nature there are neither positive nor negative numbers in the abstract; there exist only things which can be counted. The distinction of positive or negative numbers — epithets which can only be understood in contrast to each other — has a meaning only for the process of adding and thence for all other arithmetical operations. But it is often of great advantage in applying calculation to physical problems, to distinguish the quantities we calculate with, in the sense of the positive and the negative unit.

Every subtraction is possible when we employ the negative unit, since there is now introduced an unlimited series of negative numbers besides the series of positive ones; zero separates the two. The remaining operations with negative numbers are given by the equations (I) (II) (III), since they are required to comply with the rules for differences in general. A special consequence is, that the product of two negative numbers is positive. The rules of signs for division by positive and negative numbers are also determined by the inversion of multiplication.

4. In order that the operation of division may be always possible, the positive or negative unit must be broken up into subordinate units; it is sufficient to introduce the numerical conception $+\frac{1}{b}$, where b can be any integer. For if the symbol $\frac{1}{b}$ be employed to denote that number which when multiplied by b produces 1, then a times this number will express the value of $\frac{a}{b}$. Here again it is to be remarked that in nature there exists no fractional number, but that this conception also has a sense only in reference to numerical combinations.

Any fraction can be replaced by another whose denominator is a multiple of the original denominator:

$$\frac{b}{a} = \frac{b a_1}{a a_1}.$$

Employing this transformation we derive the rules

$$\frac{b}{a} \pm \frac{b_1}{a_1} = \frac{b a_1 \pm b_1 a}{a a_1}.$$

We have further from the conception of multiplication

$$\frac{b}{a} c = \frac{bc}{a}.$$

Multiplication of a fraction $\frac{b}{a}$ by another $\frac{b_1}{a_1}$, in analogy with this last equation, is understood to be division of the fraction by a_1 and multiplication of this part by b_1; whence

$$\frac{b}{a} \cdot \frac{b_1}{a_1} = \frac{b b_1}{a a_1}.$$

This definition complies with the fundamental proposition of multiplication. By inversion we get the equation of division

$$\frac{b}{a} : \frac{b_1}{a_1} = \frac{b a_1}{a b_1}.$$

Thus we can now complete the equations for sums and differences

$$\text{(IV)} \qquad \frac{a \pm b}{c} = \frac{a}{c} \pm \frac{b}{c}, \quad \frac{a \pm b}{c \pm d} = \frac{a}{c \pm d} \pm \frac{b}{c \pm d}.$$

But there is one very important exception to these equations: the difference which occurs in the denominator must not be zero; a division is impossible if the divisor be zero. For if the dividend be not zero, there is no number, which when multiplied by zero, gives a product equal to it, since a product one of whose factors is zero vanishes; but if the dividend is likewise zero, the value of the quotient is completely indeterminate; no calculation can be performed with such a completely indeterminate number.

Integer and fractional numbers are both embraced under the expression Rational Numbers. Applying the first four rules of Arithmetic to them we always reproduce rational numbers. The series of rational numbers is unlimited: between any two, as a continued subdivision shows, we can always insert as many more rational numbers as we please, or to state the same thing in other words, a series of rational numbers which is arrived at by any finite number of subdivisions never forms a continuum.

— —

Second Chapter.

Radicals and irrational numbers in general.

5. Repeated multiplication gives rise to a fifth operation, that of raising to a power or Involution. $\left(\frac{a}{b}\right)^n$ means a product which consists of n (exponent) factors, each equal to $\frac{a}{b}$ (base). The values of the powers of a positive base are only positive, those of a negative base are positive or negative according as the exponent is an even or odd number. Involution is always possible within the range of rational numbers.

From one inversion of involution arises the problem of extracting roots, Evolution. Given the positive number $\frac{a}{b}$, where a and b are relatively prime, it is required to determine a positive number x, so that $x^n = \frac{a}{b}$ may be true for a given n. We assume the number $\frac{a}{b}$ positive, for, when it is negative and the exponent n even, as far as our conception of number yet reaches, no such root can be extracted, whereas when the exponent is odd, the nth root of the positive value of $\frac{a}{b}$ is to be taken, but with a negative sign. In like manner, it is to be remarked from the outset, that when the exponent is even, the root of a positive quantity can be given a positive or a negative sign. Accordingly, the purpose of the following considerations is only to show that the value itself can be determined.

If there be a fraction $x = \frac{p}{q}$ such that $\frac{p^n}{q^n} = \frac{a}{b}$ or $bp^n = aq^n$, since a and b may always be assumed relatively prime and since we know that there is only one way in which each number can be composed of prime factors, we see that this equation must break up into $a = p^n$, $b = q^n$. Thus a positive fraction is only equal to the nth power of a positive fraction, when its numerator and its denominator are each equal to the nth power of an integer; in particular, an integer can never be equal to the nth power of a fraction; for when $b = 1$, q must also $= 1$. To find out therefore whether $\frac{a}{b}$ is the nth power of a rational number, we must form the table of nth powers of integers and examine whether a and b appear in it.

Hence manifestly the roots of positive rational numbers cannot universally be extracted by means of rational numbers only, but evolution in like manner with the problems of subtraction and of division introduces a new conception into the theory of numbers. Employing Euclid's method of inclusion within limits, this conception is expounded as follows:

If σ be any rational number, then in the unlimited series of fractions

$$0, \frac{1}{\sigma}, \frac{2}{\sigma}, \frac{3}{\sigma}, \ldots \frac{m}{\sigma}, \ldots (m > \sigma),$$

there must be a consecutive pair $\alpha = \frac{\varrho}{\sigma}$ and $\beta = \frac{\varrho + 1}{\sigma}$, such that

$$\alpha^n < \frac{a}{b} < \beta^n.$$

The difference $\beta - \alpha$ is $\frac{1}{\sigma}$. Now if σ' be another number greater than σ, fractions with the greater denominator σ' can be inserted between the fractions α and β, this may be indicated by

$$\frac{\varrho}{\sigma}, \frac{\lambda}{\sigma'}, \frac{\lambda+1}{\sigma'}, \frac{\lambda+2}{\sigma'}, \ldots \frac{\mu}{\sigma'}, \frac{\varrho+1}{\sigma}.$$

In this series there must be two consecutive values α_1 and β_1 such that

$$\alpha_1^n < \frac{a}{b} < \beta_1^n,$$

where $\alpha_1 > \alpha$ or at least is equal to α, in case the required value occurs in the first interval, and $\beta_1 < \beta$ or at most is equal to β, in case the last interval should have to be taken. The difference $\beta_1 - \alpha_1$ is always less than $\frac{1}{\sigma}$, for this is the difference between the first and the last value of the series, intervening terms must have smaller differences. Continuing this procedure with a new denominator $\sigma'' > \sigma'$, we obtain two new fractions α_2 and β_2 having a difference smaller than $\frac{1}{\sigma'}$, and so on. This will result finally either in a number being found whose n^{th} power is equal to $\frac{a}{b}$; in which case the nature of $\frac{a}{b}$ is such, that it has a rational n^{th} root which can be expressed by one of the denominators $\sigma, \sigma', \sigma'', \ldots$, or, if not, the two series, that of the lower limits: $\alpha < \alpha_1 < \alpha_2 \ldots < \alpha_\mu \ldots < \alpha_{\mu+\nu} \ldots$ and that of the upper limits: $\beta > \beta_1 > \beta_2 \ldots > \beta_\mu \ldots > \beta_{\mu+\nu} \ldots$ go on indefinitely.

These two series have the following properties: Although the series of the α however far it is carried contains only increasing numbers, and that of the β only decreasing ones, and this is not affected if as is possible equalities occur among them, still each α is smaller than each β. Further, we can make the constantly positive differences $\beta_{\mu+\nu} - \alpha_{\mu+\nu}$, $\alpha_{\mu+\nu} - \alpha_\mu$, $\beta_\mu - \beta_{\mu+\nu}$ for each value of ν, smaller than any assigned value however small, merely by selecting only those terms in the series for which μ has a correspondingly great value. For if these differences are to be smaller than δ, we have only to determine the first value of μ for which $\beta_\mu - \alpha_\mu < \delta$; then we shall also have by the foregoing inequalities for every ν:

$$\beta_{\mu+\nu} - \alpha_{\mu+\nu} < \delta, \quad \alpha_{\mu+\nu} - \alpha_\mu < \delta, \quad \beta_\mu - \beta_{\mu+\nu} < \delta.$$

The numbers of the unlimited series of the α and likewise of the β have the property that their n^{th} powers are always coming nearer to the value $\frac{a}{b}$, so that they ultimately differ from it as little as we please. Hence it appears, that the numbers α and β themselves also approach more and more to one definite quantity, which, even though it does not exist among rational numbers, is yet called a numerical value, because it is connected with a rational number by a perfectly determinate arithmetical operation. We denote the quantity as the limiting value of the series α and β; it is written in the form $\sqrt[n]{\frac{a}{b}}$.

When the quantity is a rational number, its exposition as a limiting value depends solely on the choice of the denominators $\sigma, \sigma', \sigma''$ etc. All periodic decimals belong to this case; for instance, the value of

$$\text{limiting } 0\cdot 3; \quad 0\cdot 33; \quad 0\cdot 333; \quad 0\cdot 3333 \text{ etc.}$$

is $\frac{1}{3}$. Similarly, by the geometric progression, the limiting value of

$$1; \; 1 + \tfrac{1}{2}; \; 1 + \tfrac{1}{2} + (\tfrac{1}{2})^2; \; 1 + \tfrac{1}{2} + (\tfrac{1}{2})^2 + (\tfrac{1}{2})^3 \ldots,$$

is 2. By suitably choosing the denominator, therefore, a rational form can also be discovered for such values.

When on the other hand the quantity is not a rational number, — and of this we can make sure at the outset in the case of a radical by means of our opening proposition, — the only possible way of expressing the number, which is called an Irrational number, is as limiting value of a series. This exposition and definition embraces therefore besides rational quantities a new class, namely irrational radicals; but it will be subsequently seen to embrace still more than the roots of fractions. At the same time this process of evolution fixes our attention on what is properly to be understood henceforth by the calculation of a number which is to possess a given property. It cannot in general be required that this number shall be assigned in finite closed form, but rather we see that: *To calculate a number which shall have a definite property relatively to other given numbers, means, to find a series of rational numbers that can be unlimitedly continued which fulfil the property of the required number with ever increasing and ultimately* ***arbitrarily close*** *approximation.*

The question now arises, how such numbers are to be employed in calculation, how irrational numbers are added, multiplied and so on. This we proceed to show by the following investigations, based on the knowledge we have just attained somewhat generalised.

6. **Definition:** *By the General Conception of a Number is meant a series of positive or negative rational numbers that can be unlimitedly continued according to some rule:*

$$\alpha, \quad \alpha_1, \quad \alpha_2, \ldots \alpha_\mu, \ldots \alpha_{\mu+\nu}, \ldots,$$

having the properties, first, that its several numbers do not exceed a definite value in absolute amount (i. e. abstracting from sign) and second, that for any arbitrarily small prescribed number δ, some number α_μ can be found in the series, such that the difference $\alpha_{\mu+\nu} - \alpha_\mu$ between it and any succeeding number shall be in absolute amount smaller than δ. Of such a series we say, it **defines a number** which is called the **limiting value** of the series.

We do not adopt into the definition the property that from some certain place in the series the terms only increase or only decrease, as was the case with the series α and β above. But it is important to remark, that when this is the case we have the **theorem:**

If in the series

$$\alpha, \quad \alpha_1, \quad \alpha_2, \ldots \alpha_\mu, \ldots \alpha_{\mu+\nu}, \ldots$$

the terms only increase (or only decrease), and a superior limit can be assigned, which the terms do not exceed (or in the other case, an inferior one below which they do not fall), then such a series has always a limiting value, i.e., **both** properties stated in the definition hold good.

For, were no place to be found in the series of increasing numbers $\alpha_\mu, \alpha_{\mu+1}, \ldots \alpha_{\mu+\nu}, \ldots$ from which onwards the difference $\alpha_{\mu+\nu} - \alpha_\mu$ remains less than an arbitrary number δ, but rather, however large we take μ, numbers $\alpha_{\mu+\nu}, \alpha_{\mu+\nu'}, \alpha_{\mu+\nu''} \ldots$ were always to be found which satisfy the inequalities

$$\begin{array}{llll}
\alpha_{\mu+\nu} - \alpha_\mu > \delta & \text{and so} & \alpha_{\mu+\nu} > \alpha_\mu + \delta \\
\alpha_{\mu+\nu'} - \alpha_{\mu+\nu} > \delta & \text{and so} & \alpha_{\mu+\nu'} > \alpha_\mu + 2\delta \\
\alpha_{\mu+\nu''} - \alpha_{\mu+\nu'} > \delta & \text{and so} & \alpha_{\mu+\nu''} > \alpha_\mu + 3\delta \\
\text{— — —} & \text{— — —} & \text{— — —}
\end{array}$$

then we could form a series of **arbitrarily many** inequalities of this kind. Accordingly, as the factor of δ in the last inequality can be arbitrarily increased, the amounts of the numbers α must increase beyond all limits, which is contrary to hypothesis.

A series whose terms sink numerically below any assignable value is said to have the limiting value zero; thus we have the important fact, that zero also is included in this general conception of number.

A series whose limiting value is not zero, if it is to have both the properties we have stated, must have its terms after some definite place either only positive or only negative. For if α_μ and $\alpha_{\mu+\nu}$ differ in sign, the absolute amount of $\alpha_{\mu+\nu}-\alpha_\mu$ is not less than the greater of the two numbers, and so it is only possible for it to become smaller than δ when the series of the α has zero for limit.

7. We calculate with the numbers thus defined, by applying the operations of arithmetic to the terms of the series expressing them, since the limiting value can always be replaced with arbitrary approximation by a term of the series and thus the requirement of calculation just laid down is complied with. Suppose the given series are

$$\alpha,\ \alpha_1,\ \alpha_2,\ \ldots\ \alpha_\mu,\ \ldots \quad \text{and} \quad \beta,\ \beta_1,\ \beta_2,\ \ldots\ \beta_\mu\ \ldots;$$

each representing a number, we can embody this in the symbolical expressions

$$\mathsf{A} = \text{Lim}\,(\alpha_\mu),\quad \mathsf{B} = \text{Lim}\,(\beta_\mu).$$

Applying addition or subtraction it follows that in either case

$$\alpha \pm \beta,\quad \alpha_1 \pm \beta_1,\quad \alpha_2 \pm \beta_2,\ \ldots\ \alpha_\mu \pm \beta_\mu\ \ldots$$

form a new series possessing both the properties necessary for expressing a number. For when the absolute amount of the difference of $\alpha_{\mu+\nu}$ and α_μ, which we write briefly abs $[\alpha_{\mu+\nu}-\alpha_\mu]$, is $<\delta$ and abs $[\beta_{\mu+\nu}-\beta_\mu]<\delta$, the absolute amount of the difference of two terms of the new series is less than 2δ, for $\alpha_{\mu+\nu}$ and $\beta_{\mu+\nu}$ have at the most increased or diminished by the quantity δ in comparison with α_μ and β_μ. But the limiting value of this new series differs from the algebraic sum of the two given limiting values by less than an arbitrarily small assigned number, since the quantities α_μ and β_μ differ arbitrarily little from these limiting values; i. e. the limiting value of the new series is equal to the algebraic sum of the given numbers A and B. We formulate this in the equation

(I) $$\text{Lim}\,(\alpha_\mu) \pm \text{Lim}\,(\beta_\mu) = \text{Lim}\,(\alpha_\mu \pm \beta_\mu).$$

Subtraction here furnishes the special theorem: Two series of numbers which express the same limiting value yield when subtracted a series of numbers whose limit is zero. But also conversely we define: Two series of numbers whose difference has the limiting value zero express the same number, or, two numbers are called equal when the difference of corresponding terms of their defining series has the limiting value zero. This is to be regarded as the definition of equality of two numbers. The original definition: Two rational numbers are equal if they contain the same number of units or fractions of unity, is subordinate to this definition, which holds good for number in general.

Similarly it follows on applying multiplication that

$$\alpha\beta,\quad \alpha_1\beta_1,\ \alpha_2\beta_2,\ \ldots\ \alpha_\mu\beta_\mu\ \ldots$$

form a new series which likewise possesses the required properties. For when the amounts of α_μ and β_μ are at the most increased or diminished by δ, the product $\alpha_{\mu+\nu}\beta_{\mu+\nu}$ is also at the most only altered by the amount $\alpha_\mu\delta + \beta_\mu\delta + \delta^2$ which we can make as small as we please by a suitable choice of μ, since there are superior limits which α and β never exceed, and δ is arbitrarily small. Thus

(II) $$\text{Lim}\,(\alpha_\mu)\cdot\text{Lim}\,(\beta_\mu) = \text{Lim}\,(\alpha_\mu\cdot\beta_\mu).$$

We find by division the new series

$$\frac{\alpha}{\beta},\quad \frac{\alpha_1}{\beta_1},\quad \frac{\alpha_2}{\beta_2},\ \ldots\ \frac{\alpha_\mu}{\beta_\mu}\ \ldots$$

Excluding the case that the limiting value of the series of divisors vanishes, and so that of the β falls below any assignable value, there is a superior value which these quotients are certain never to exceed; and if α_μ and β_μ are altered by δ, then the difference

$$\text{abs}\left[\frac{\alpha_\mu \pm \delta}{\beta_\mu \pm \delta} - \frac{\alpha_\mu}{\beta_\mu}\right] = \text{abs}\left[\frac{\delta\,(\beta_\mu \pm \alpha_\mu)}{\beta_\mu\,(\beta_\mu \pm \delta)}\right],$$

is a number whose amount is arbitrarily small. Therefore

(III) $$\frac{\text{Lim}\,(\alpha_\mu)}{\text{Lim}\,(\beta_\mu)} = \text{Lim}\left(\frac{\alpha_\mu}{\beta_\mu}\right).$$

Similarly from (II) we find for involution with a positive integer exponent

(IV) $$(\text{Lim}\,(\alpha_\mu))^n = \text{Lim}\,(\alpha_\mu{}^n),$$

and thence for extracting the root of a positive number, i.e. of a number whose defining series from a certain place contains only positive terms:

(V) $$\sqrt[n]{\text{Lim}\,(\alpha_\mu)} = \text{Lim}\,(\sqrt[n]{\alpha_\mu}).$$

As before, these last two equations amount merely to the statements, that a rational or irrational number can be *quam proxime* raised to a power, or have a root extracted, by performing these respective operations upon a number in the defining series *quam proxime* equal to it in value.

It is convenient to base our introduction of powers with integer negative exponents, as Newton did, on the equation

$$\frac{A^m}{A^n} = A^{m-n} = \frac{1}{A^{n-m}},$$

whether $m \gtrless n$, from which we see that. A^0 is 1 for all values of A; as well as our introduction of powers with fractional exponents on the equation

$$(A^n)^m = A^{nm}$$

which is valid for integer exponents. For since $\sqrt[n]{A}$ is the number whose n^{th} power is A, we can by this rule put $\sqrt[n]{A} = A^{\frac{1}{n}}$; then

$$\left(A^{\frac{1}{n}}\right)^n = A^{\frac{n}{n}} = A^1, \quad A^{\frac{m}{n}} = \sqrt[n]{A^m}.$$

But at present we restrict this definition of the fractional exponent to a positive base, in order that the calculation may remain possible without any exception.

The following consideration, of importance in itself, is premised to the definition of a power with an irrational exponent:

Any number, which is expressed by a series of irrational numbers, can also be expressed by a series of rational numbers. For if

$$A_1, A_2, A_3, \ldots A_\mu, \ldots A_{\mu+\nu}, \ldots$$

be the series of irrational quantities with the necessary properties, suppose the quantities A to be defined by any arithmetical operation, ex: gr. as roots, let us conceive two series of rational numbers

$$\beta_1, \beta_2, \beta_3, \ldots \beta_\mu, \ldots \beta_{\mu+\nu}, \ldots$$
$$\alpha_1, \alpha_2, \alpha_3, \ldots \alpha_\mu, \ldots \alpha_{\mu+\nu}, \ldots$$

of which the series of the β is chosen arbitrarily but so as to have the limiting value zero, whereas each number α_μ is assumed so that abs $[A_\mu - \alpha_\mu] < \beta_\mu$. Then the series α represents the same limiting value as the first series. This is a statement of a general process for forming from a series any other with the same limiting value.

Accordingly if E be an arbitrary irrational exponent defined by the series

$$\varepsilon, \varepsilon_1, \varepsilon_2, \ldots \varepsilon_\mu \ldots,$$

the power is to be understood as the limiting value of the series

$$A^\varepsilon, A^{\varepsilon_1}, A^{\varepsilon_2}, \ldots A^{\varepsilon_\mu}, \ldots$$

whose terms possess both the required properties.

For, $$A^{\varepsilon_{\mu+\nu}} - A^{\varepsilon_\mu} = A^{\varepsilon_\mu}(A^\delta - 1).$$

But since $A^0 = 1$, the absolute amount of this quantity can be made arbitrarily small, as can be shown as follows. Conceive δ to be a positive rational fraction with numerator 1 and denominator an arbitrarily great positive number M, then, if $A > 1$,

$$A^\delta - 1 = \eta, \quad A = (1 + \eta)^M.$$

Multiplying $1+\eta$ continually on itself to M times produces a number which is certainly greater than $1+M\eta$; therefore

$$A > 1 + M\eta, \quad \eta < \frac{A-1}{M}.$$

But if $A < 1$, let us put

$$1 - A^{\delta} = \eta, \quad A^{-\delta} = \frac{1}{1-\eta} = 1 + \xi;$$

then

$$A = \frac{1}{(1+\xi)^M} < \frac{1}{1+M\xi}, \text{ therefore } \xi < \frac{\frac{1}{A} - 1}{M}.$$

Now if we consider, that by the procedure just explained the series

$$A^{\epsilon_1}, \; A^{\epsilon_2}, \; \ldots \; A^{\epsilon_\mu}, \; \ldots$$

also can be replaced by the series

$$\alpha_1^{\epsilon_1}, \; \alpha_2^{\epsilon_2}, \; \ldots \; \alpha_\mu^{\epsilon_\mu}, \; \ldots$$

when $\mathrm{Lim}\,(\alpha_\mu) = A$, the result of this investigation can be written in the form:

(VI) $$(\mathrm{Lim}\,\alpha_\mu)^{\mathrm{Lim}\,(\epsilon_\mu)} = \mathrm{Lim}\,(\alpha_\mu^{\epsilon_\mu}).$$

8. Inversion of involution presents a second problem: that of the logarithm. Two positive numbers A and B are given, each defined as limiting value of a series; it is required to determine a number x having the property that $B^x = A$. Here x is called the logarithm (exponent) of the number A (number) with regard to the base B, and is written

$$x = {}^{B}\log A.$$

It can be shown, in the first place that only one number x has this property — for we cannot at the same time have $(B \gtrless 1)$

$$B^x = A = B^{x'} \text{ and so } \frac{B^x}{B^{x'}} = B^{x-x'} = 1$$

without having $x - x' = 0$, — and moreover in the second place that this number x can be expressed as limiting value of a series. In fact, forming the series of terms

$$\ldots B^{-2}, \; B^{-1}, \; 1, \; B^1, \; B^2, \; \ldots$$

there will be among them two values, such that, if $B > 1$

$$B^{\lambda} < A < B^{\lambda+1}.$$

If we interpolate rational fractions between λ and $\lambda+1$ we have:

$$B^{\lambda+\frac{\lambda'}{\sigma}} < A < B^{\lambda+\frac{\lambda'+1}{\sigma}}.$$

As we increase the values of the denominator we obtain two series by which x is defined.

The logarithm of 1 with regard to any positive number as base is zero. In the sequel we assume as known the rules for calculating with logarithms.

Numbers positive and negative, rational and irrational, are all comprehended together under the name real numbers. However often we repeat the first operations of arithmetic on real numbers we reproduce numbers of the same kind. But an essential exception must be made regarding the last two operations. The values of negative numbers with an even exponent are positive, but on the other hand we are not at present in a position to assign what is the value of a negative number with its exponent a fraction having an even denominator or with its exponent irrational; since such fractions can occur in the defining series. Similarly in order to avoid having to point out further exceptions, we restricted ourselves in defining the logarithm to positive values of the base and of the number. Thus a gap still remains here which only the introduction of complex numbers will remove, but it will do so completely.

Further, we have so far made no attempt to give the most convenient methods for actual calculation of any power or logarithm, we have merely demonstrated its possibility and determinateness.*)

*) The conception of irrational number is treated in detail in Lipschitz, Lehrbuch der Analysis Vol. I. As to the general conception of number see: Cantor, Zur Theorie der trigonometrischen Reihen, Math. Ann. Vol. V; Heine, die Elemente der Functionenlehre, J. f. Math. Vol 74; and Dedekind, Stetigkeit und irrationale Zahl. (Braunschweig 1872).

Third Chapter.

Conception of a variable quantity and calculation with variables, specially with infinite quantities.

9. The totality of rational and irrational numbers forms the continuous series of numbers. To represent a continuum as a whole we necessarily require an intuitive image. We naturally take the right line, the simplest representation of a continuum in space.

Having assumed a unit of length, the points of a right line are determined by assigning their distances from a fixed zero-point, with the + or − sign, according as the point considered is right or left of that origin. The distance of each point can be expressed by a rational or irrational number. For, what is characteristic of a point is that it bounds a definite length from the origin; this length measured by the assumed unit is a number which can be expressed either in a finite form, or as is taught in Euclid's elements, with arbitrary approximation by continued subdivision of the unit; and this is the very idea of obtaining a series defining the distance. Thus to each point of the continuum belongs one and only one number, our general conception of number is never in default. And conversely: to each number belongs one and also only one point, since each number determines a length to be constructed, each length an end point. Thus according as a point moves upon a right line, its distance from the origin travels through the continuous series of numbers. The information we have now acquired can also be thus stated: Rational numbers enable us to express as approximately as can be desired every value of the continuous series of numbers.

In order to contemplate the conception of all numbers possible within an interval, that is, between two fixed values, we introduce into analysis the representation of variability apart from the image of a definite motion on the right line. This representation was first employed in the most general manner by Newton, the way for it having been prepared by Geometry, specially from the time of Descartes.*) *A quantity is said to be variable, when it is able to assume different numerical values.* As in purely arithmetical investigations we no longer consider what are the things given in number, so in the conception

*) (1596–1650).

of "variable quantity" we have also to free ourselves entirely from considering what this quantity represents. The distance of a movable point, the temperature, the tension of vapor, in a word, everything measurable in nature can enter into calculations as variable quantity. *A quantity is said to be continuous or continuously variable within an interval i.e. between two numbers, when there is no numerical value between two numbers x_0 and $x_0 + \delta$ however close together which it does not assume as it changes from x_0 to $x_0 + \delta$.* Thus the statement, a quantity changes continuously from the value a to the value b, amounts to this, the quantity travels through all numbers between a and b and there is no break in the sequence of the numbers.

But now since when we want to fix the variable, we are not always able to assign a value which it assumes in a closed form but only as the limiting value of a series, we often use a phrase drawing attention to this change towards a determinate value whether rational or irrational. We say: the continuous variable approximates infinitely to a determinate value b or converges to b, when the series of numbers through which it passes has this limiting value; that is, for any number δ however small, some place must be assignable in the series from which onwards all values of the continuous variable differ from b by less than δ. We shall always denote by the word "infinite", a continuous change towards a determinate limiting value.

In particular a variable becomes "infinitely small", when as it varies continuously there is no condition which prevents its absolute amount becoming less than any assignable number, i.e. when it has the limiting value zero.

The variable is called "infinitely great" and its limiting value written with the sign $\pm \infty$, when as it varies continuously there is no condition which prevents its absolute amount becoming greater than any assignable number. From the continuous series of numbers through which such a variable ultimately passes, discontinuous series of numbers can be selected according to some law, (for instance if we only pay attention to integers,) but they no longer satisfy the two properties already recognised as necessary for series defining numbers. Nevertheless certain calculations can be effected with these series of numbers, on which account we introduce them as a distinct numerical conception by the following more accurately stated definition:

A variable becomes in a determinate manner positively or negatively infinitely great, when the series of numbers through which it travels, has the properties, first that the values after a certain place are only positive or only negative, and, second that a place can be found in the series after which all values are greater in amount than an assigned number however great.

This second property must await closer discussion in subsequent examples; it does not preclude the terms of the series still oscillating on from any place, that is, becoming at one time greater and at another smaller. But when in the series of numbers only arbitrarily large values occur, yet the other property is not satisfied by all the terms, the variable becomes indeterminately infinite.

We denote variables by the last letters of the alphabet x, y, z; and quantities with fixed or constant values by the first letters a, b, c.

10. The principal laws for calculating with variables result from the rules established in Chapter II for limiting values; for, the numbers which there approximate discontinuously to the limiting value, occur among the other values in the continuous variation. Here we repeat the equations, written now with a reference to continuous variability, in order to append some special theorems to them:

(I) $$\text{Lim}\,(x \pm y) = \text{Lim}\,(x) \pm \text{Lim}\,(y).$$

This equation asserts: The limiting value to which the sum of two continuous variables tends, is equal to the sum of the limiting values to which the summands tend, and conversely. This proposition can be extended to several summands.

Special theorem: The sum or difference of two infinitely small quantities is itself an infinitely small quantity, that is, one converging to zero.

(II) $$\text{Lim}\,(x \cdot y) = \text{Lim}\,(x) \cdot \text{Lim}\,(y).$$

Special theorem: The product of a finite and of an infinitely small or the product of two infinitely small quantities is itself an infinitely small quantity, that is, one which converges to zero.

(III) $$\text{Lim}\left(\frac{x}{y}\right) = \frac{\text{Lim}\,(x)}{\text{Lim}\,(y)}.$$

If m denote a constant and x a variable which assumes only positive values, then

(IV) $$\text{Lim}\,(x^m) = (\text{Lim}\,x)^m.$$

If b be a positive constant, x an arbitrary variable, then:

(V) $$\text{Lim}\,(b^x) = b^{\text{Lim}\,(x)}.$$

Powers with a variable exponent are called "Exponentials" in contrast to powers with a constant exponent.

In equation (III) the condition is assumed, that Lim (y) is not zero. But the introduction of the conception of continuous variability furnishes an expedient which renders it possible henceforth to take account in calculations, of quotients even with an infinitely small divisor. For, zero being no longer defined only by the difference $a - a$, but as the limiting value of a series of numbers, the quotient $\frac{\text{Lim}\,(x)}{\text{Lim}\,(y)}$ can also still retain a determinate value when, for arbitrarily small values of y, the quotient $x:y$ expresses a number approximating to a determinate limiting value, finite or infinite. But this is only possible when the law of formation of the series $x:y$ is given, i. e. when to each number within the series x the number in the series y is known, by which the former is to be divided. When this is so, by the quotient $\frac{\text{Lim}\,(x)}{\text{Lim}\,(y)}$ we shall understand the value $\text{Lim}\left(\frac{x}{y}\right)$*). If x have a determinate finite limiting value, the series $x:y$ will consist of numbers, which increase in absolute amount beyond any limit. But if the limiting value of x itself be zero, then the series $x:y$ can have either a finite, or an infinitely small, or even an infinitely great or yet no definite limiting value at all, and this is to be decided only in each individual case by forming the series.

For equations (I), (II) the law holds also, that in order that the left side may be formed without ambiguity, the connexion between the places of the series x and those of the series y must be given. In this case the left sides can then still express determinate limiting values when the limits of one or of both series for x and y transcend any finite amount.**)

---

*) A simple example may make this clear to the beginner. Suppose the series given, in which y is to travel through all numbers from 1 to 0; let the series consist of the numbers $3y - y^2$, so that therefore for $y = 1$ $x = 2$, for $y = \frac{1}{2}$ $x = \frac{5}{4}$, for $y = \frac{1}{4}$ $x = \frac{11}{16}$, $\cdots$ $y = 0$ $x = 0$. If now by $\frac{\text{Lim}\,x}{\text{Lim}\,y}$ we understand the value $\text{Lim}\,\frac{x}{y}$, it will be expressed for every value of y by $\frac{3y - y^2}{y} = 3 - y$, and thus although y, and with it x, have the limiting value 0, the series of the quotients has the limiting value 3.

**) The last half of this chapter will become more intelligible when the investigations which immediately follow shall have furnished us with materials for definite examples

Fourth Chapter.

Conception and notation of functions of a variablo.

11. If the value of a variable y is determined by the value of a variable x in such a way that to each value of x within a certain interval one or more values of y can be calculated or assigned in any prescribed manner, then y is called a Function*) of the continuous variable x within this interval; y is also called the dependent and x the independent variable, or the argument of the function. If there be one value of the function for each value of the variable it is called a single-valued (one-valued or unique) function, if more values a many-valued (or ambiguous) function.

This dependence may be perfectly casual, the independent variable serving merely to indicate the position in a Table, the function being that which is found there: in the problems which deal with measurable quantities in Physics, the functions with which we are always concerned are those in which the dependence of the variables is ascertained by Observation. The dependence may be expressed by an equation between x and y, which may or may not have for all values of x a sense already defined. For instance, by the relation

$$y^2 = (x - 1)(3 - x) \text{ or } y = \pm\sqrt{(x-1)(3-x)}$$

a two-valued function of x is defined by reason of the double sign, but it can be calculated in real numbers only for the values, $1 \leq x \leq 3$.

When the relation between x and y is given in the form of the general equation $f(x, y) = 0$ the function is called implicit: but when the equation is in a form solved for y, $y = f(x)$, we call y an explicit function of x. The functional connexion between x and y may also arise by means of a third variable and the function be expressed by a parameter: $x = f(t)$, $y = \varphi(t)$; to each value of t belongs a value of x and of y; in this way values of x and y become connected.

It is usual to divide functions into algebraic and transcendental. When the equation by which y is defined can be brought to

*) The term "function" was introduced by John Bernoulli (1667—1748) (opera omnia t. II p. 241).

the form $f(x, y) = 0$ where f is integer and rational in x and y, we call y an algebraic function of x. Accordingly the most general type of an algebraic function with two variables is

$$A_0 y^n + A_1 y^{n-1} + A_2 y^{n-2} + \cdots + A_{n-1} y + A_n = 0,$$

where the quantities A are polynomials of any degree in x of the form

$$A = a_0 x^m + a_1 x^{m-1} \cdots + a_m,\ m \text{ and } n \text{ being positive integers.}$$

Here the only arithmetical operations performed on the variables are those of the first four species repeated a finite number of times and including integer powers.

All other functions are transcendental. To these the operations of arithmetic contribute the power with an irrational exponent, but chief of all the exponential function, the simplest type of which is $y = a^x$, and the logarithm $y = {}^b\log x$; we can only calculate these last two functions, so far as our conception of number yet reaches, for positive values of a and b, and further the second only for positive values of x.

12. Not less important are the trigonometric or as they are sometimes more precisely called goniometric functions which we are familiar with from the elements of geometry. These are geometrically defined as the ratios of lengths which depend on an angle. In order to indicate that angular quantity is always a pure number, we describe the magnitude of an angle, not by degrees, but by the ratio of the length of the circular arc to that of the generating radius which belongs to it. The length of the circular arc belonging to an angle, as well as that of the whole circumference, is proportional to the radius with which it is described. Hence the circumference may be denoted by $2\pi r$, where π is a number to which the Geometry of Euclid shows we can approximate as closely as we please by inscribing and circumscribing polygons. Accordingly, to each angle there will belong a number which determines as part or multiple of 2π the arc as part or multiple of the circumference. By geometric investigation we conclude the following properties of the functions:

$$y = \sin x \text{ and } y = \cos x.$$

1) as x goes from 0 to $\frac{1}{2}\pi$, $\sin x$ goes from 0 to 1, $\cos x$ from 1 to 0
" x " " $\frac{1}{2}\pi$ " π, $\sin x$ " " 1 " 0, $\cos x$ " 0 " -1
" x " " π " $\frac{3}{2}\pi$, $\sin x$ " " 0 " -1, $\cos x$ " -1 " 0
" x " " $\frac{3}{2}\pi$ " 2π, $\sin x$ " " -1 " 0, $\cos x$ " 0 " 1.

This is one of the cases in which it is good to adopt a distinction of signs in the geometric interpretation.

2) $(\sin x)^2 + (\cos x)^2 = 1$.

3) If the size of the angle be extended by turning the leg further, it follows that for any positive integer k:

$$\sin(x + 2k\pi) = \sin x; \quad \cos(x + 2k\pi) = \cos x.$$

4) If the rotation be effected in an opposite sense, the angle is to be denoted by a negative number; therefore we have then:

$$\sin(-x) = \sin(2\pi - x) = -\sin x$$
$$\cos(-x) = \cos(2\pi - x) = \cos x,$$

and generally:

$$\sin(x \pm 2k\pi) = \sin x; \quad \cos(x \pm 2k\pi) = \cos x.$$

Both functions, which always lie between the values $+1$ and -1, have the property of reproducing their values, whenever the independent variable is increased or diminished by an integer multiple of 2π. Such a function is called periodic and 2π its period.

5) It is proved geometrically that:

$$\sin(x \pm x_1) = \sin x \cos x_1 \pm \sin x_1 \cos x$$
$$\cos(x \pm x_1) = \cos x \cos x_1 \mp \sin x \sin x_1,$$

from which follow

$$\sin x - \sin x_1 = 2 \sin \tfrac{1}{2}(x - x_1) \cos \tfrac{1}{2}(x + x_1)$$
$$\cos x - \cos x_1 = -2 \sin \tfrac{1}{2}(x - x_1) \sin \tfrac{1}{2}(x + x_1).$$

6) The area of the circular sector $BCAM$, whose angle lies between 0 and $\frac{1}{2}\pi$, is greater than the area of the triangle ABM and less than the sum of the triangles ADM and BDM.

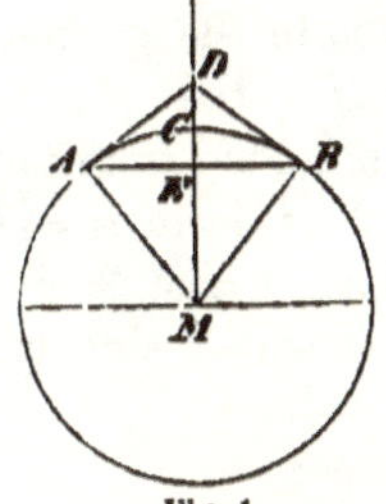

Fig. 1.

Let the radius $AM = 1$, the angle $AMC = x$, then $AE = \sin x$, $ME = \cos x$; further:

$$AD : AM = AE : ME,$$

and so $AD = \frac{\sin x}{\cos x}$; accordingly for the areas mentioned we have the inequality:

$$\frac{\sin x}{\cos x} > x > \sin x \cos x \quad \text{or} \quad \frac{1}{\cos x} > \frac{x}{\sin x} > \cos x.$$

This inequality holds, however small x is assumed; it holds also if x be made negative. The quotient $\frac{x}{\sin x}$ is of such a nature, that its numerator and denominator can become infinitely small, while the quotient itself converges to a determinate finite value; since for $x = 0$ $\cos x$ is $= 1$. Thus the value $\frac{1}{\cos x}$ which forms a superior limit, coincides with the lower limit $\cos x$ for $x = 0$. Therefore also the included value will be 1, that is

$$\text{Lim} \frac{x}{\sin x} = 1 \text{ for } x = 0.$$

The quotient $\frac{\sin x}{x}$ travels through the series of the reciprocal values and has likewise 1 for limit.

From these two functions two others are formed by division, which are also immediately expressed as ratios of lengths by geometry:

$$\tan x = \frac{\sin x}{\cos x}, \quad \cot x = \frac{\cos x}{\sin x};$$

$\tan x$ is 0 for $x = 0, \pm\pi, \pm 2\pi$ etc., $\cot x$ is 0 for $x = \pm\frac{1}{2}\pi, \pm\frac{3}{2}\pi, \pm\frac{5}{2}\pi$ etc.; $\tan x$ increases beyond any assignable limit and becomes $\pm\infty$ in a determinate manner when x converges to the values $\pm\frac{1}{2}\pi, \pm\frac{3}{2}\pi, \pm\frac{5}{2}\pi$ etc., so does likewise $\cot x$ when x converges to the values $0, \pm\pi, \pm 2\pi$ etc. Both functions have the period π.

We have so far not given any convenient method of calculating goniometric functions; only inasmuch as the determinateness of the problem, to find for any given angle its sine etc., is known geometrically, we ventured to introduce the conception of these functions and a symbolic notation for them.

13. From these four functions can be derived their inverses the circular functions. If for any function we conceive the values of y as given first, then the values of x appear as dependent; x is then called the inverse function of y. In a table of logarithms we can consider the logarithm as function of the number, but also inversely the number as function of the logarithm.

The nature of a function may however be such that its inverse function is defined only for isolated values of y. If we consider, for instance, the function $y = G(x)$, using $G(x)$ to denote the greatest integer in x, the inverse function is defined only for the integer values $y = 0, 1, 2, \ldots$; to each of these values then belong infinitely many different values of x, namely to $y = 0$ all values from $x = 0$ to $x = 1$ exclusive, to $y = 1$ all values of x between 1 and 2, etc.

When $y = \sin x$, the inverse denotes the angle x whose sine has the given value y: it is written:

$$x = \sin^{-1} y$$

which is read, the angle whose sine is y, or the arc whose sine is y. Similarly from the others $x = \cos^{-1} y$, $x = \tan^{-1} y$, $x = \cot^{-1} y$. Now since sine and cosine assume only values between -1 and $+1$, $\sin^{-1}$ and $\cos^{-1}$ are defined only for the values of y within this interval. Further, since different angles belong to the same value of either sine, cosine, tangent or cotangent, and so different numbers x, circular functions are many-valued. In order to be able to consider them as determinate numerical values in calculation, we adopt a convention which at the same time presents them as continuous functions: $\mathrm{Sin}^{-1} y$ denotes that number between $-\frac{1}{2}\pi$ and $\frac{1}{2}\pi$ whose sine is y.

$$(-1 \leqq y \leqq 1).$$

$\mathrm{Cos}^{-1} y$ denotes that number between 0 and π whose cosine is y.

$$(-1 \leq y < 1).$$

$\mathrm{Tan}^{-1} y$ denotes that number between $-\frac{1}{2}\pi$ and $\frac{1}{2}\pi$ whose tangent is y.

$$(-\infty \leq y \leq +\infty).$$

$\mathrm{Cot}^{-1} y$ denotes that number between 0 and π whose cotangent is y.

$$(-\infty \leq y \leq +\infty).$$

14. Algebraic functions and the four simple kinds of transcendentals: Exponential and Trigonometric functions with their inverses: Logarithmic and Circular functions, form in their innumerable combinations the proximate object of analytical investigation. Its course which is aimed at the actual calculation of the functions and at the knowledge of their properties, will reveal a close relation between exponential and trigonometric functions, and therefore also between the logarithm and the circular function. Further it will appear, that the operations of the Differential Calculus, when applied to these functions do not give rise to new ones, but that on the other hand, the Integral Calculus does teach us to exhibit and calculate new functions.

Fifth Chapter.

Geometrical representation of a function; its continuity and its differential quotient.

15. By the help of the Cartesian system of coordinates — most simply by a rectangular system — we present to ourselves an image of the entire course of a function, as a succession of points, whose number can be arbitrarily increased and whose uninterrupted connexion can be considered generally as a curve, when we lay off each value of x as a length along the axis of abscissæ, and each corresponding y as a perpendicular ordinate at the extremity of x. The extremity of this ordinate is the point which corresponds to the system of values x, y. But since points even though infinitely numerous never generate a curve, but it must always be made up of lines between points, as matter of fact the only way we obtain an image of the function i by constructing arbitrarily many separate points corresponding to diff rent systems of values, and connecting these points by right lines. Th approximate image of the function, a polygon with arbitrarily man angles, will present a certain general view of the whole course of tl function, which will be more correct the more we increase the numb of points constructed; but in its individual small parts this represe tation will never be quite exact. Specially, when the function ge metrically represented is very much twisted near any point, i. e. when presents polygons with salient and reentrant angles starting in and ou every image we thus procure will exhibit its course very imperfectly and will undergo very considerable changes as more points are em ployed in the construction, so that the true image of the function, in regard to all its properties at each point, cannot be fixed in this manner.

But when we say of a function: it can be exactly represented at a point by a curve, we are enunciating a definite property for this point; for we are assuming, that while the number of angles of the polygon is arbitrarily increased, the directions of the sides of the polygon proceeding from this point as an angle converge to fixed limiting positions. We shall in the present Chapter formulate the condition for this analytically.

Our first enquiry is: what properties does the course of the dependent variable y exhibit while the independent variable x assumes all possible values? There are three things of immediate importance in the examination of any proposed function.

First: Is its course everywhere continuous or not?

Second: What singularities occur among the values which the function assumes?

Third: What values does it take, when the independent variable becomes infinitely great?

16. Let the explicit function $y = f(x)$ be defined as a one-valued function for a determinate interval from $x = a$ to $x = b$, i. e. let one and only one determinate value of the function without any exception belong to each value of x. We call the function **continuous** on both sides of any point x in this interval, when there are no sudden changes in its values as we move from x to either side, that is, as we form the values of the function belonging to values arbitrarily little greater or less than x. In a form applicable to calculation we state what is required thus: *It must be possible for this value of x, to find a finite number h, which only converges to zero with $\delta = 0$, such that the absolute amount of the difference $f(x \pm \Theta h) - f(x)$ of the values of the function is less than an arbitrarily small assigned number δ, Θ denoting a variable between the limits 1 and 0.* r, there is thus fixed about this point x, a region $\pm h$ in which the lues of the function differ by less than the arbitrarily small quantity from the value at x, and this excludes a sudden change of the uction at that point. The same condition can be stated in other ords: *The value of the function for a determinate x must come as the niting value both from $f(x + h)$ and likewise from $f(x - h)$, when h comes infinitely small, i. e. converges continuously to zero, provided is limiting value is completely determinate**); for if so, by the fundaental property of a series defining a limiting value there will be a lue of h for which abs $[f(x \pm h) - f(x)] < \delta$, and this absolute nount will remain smaller than δ as h converges to zero.

Thus for instance

$$\sin(x \pm h) - \sin x = 2 \sin(\pm \tfrac{1}{2}h) \cos(x \pm \tfrac{1}{2}h).$$

ere h can be determined so as to make the first factor on the right de arbitrarily small, for the sine is smaller than its argument, the cond is finite for every x; accordingly by a suitable choice of h the roduct can always be made less than an assigned number δ, whence it sults that the sine is always a continuous function (see p. 30).

*) $\mathrm{Lim}\left(\cos \frac{1}{x+h}\right) = \mathrm{Lim}\left(\cos \frac{1}{x-h}\right)$, yet this function is not continuous r $x = 0$

With the foregoing considerations we have attained the conception of the Region or Neighbourhood of a point. By it we mean an arbitrarily small but still always finite interval at both sides of the value x. If a function satisfy the condition of continuity at each point of this interval, it is said to be continuous in this interval.

When a function is everywhere continuous in the interval from a to b, and has the values $f(x_1)$ and $f(x_2)$ for two points x_1 and x_2 of this interval, then, as x goes through the range of values from x_1 to x_2, the function also assumes each value which lies between $f(x_1)$ and $f(x_2)$ at least once. In other words: the continuous function does not overleap any value intermediate between two values which it assumes.

We can reduce the proof of this proposition to that of a simpler case. If m be a value which lies between $f(x_1)$ and $f(x_2)$, $f(x_1)$ being greater than $f(x_2)$, then the function $\varphi(x) = f(x) - m$ is positive for $x = x_1$ and negative for $x = x_2$. In order therefore to prove that $f(x)$ is equal to m for some point between x_1 and x_2, we have to show that somewhere between x_1 and x_2, $\varphi(x)$ is equal to 0. Therefore we have to prove the theorem: If a continuous function $\varphi(x)$ is positive for $x = x_1$ and negative for $x = x_2$, there is in the interval between x_1 and x_2 at least one point at which it is equal to 0.

Let the length of the interval $x_2 - x_1$ be l. Construct its middle point $x_1 + \frac{1}{2}l = x_3$. If at it the value of the function be zero, the theorem is already proved. If it be positive, let us take into consideration the interval from x_3 to x_2 within which the function changes its sign, but if negative, the interval from x_1 to x_3. Whichever case presents itself, the interval to be considered has the length $\frac{1}{2}l$, and at its commencement which we may call α_1 and which is either x_1 or x_3, the function is positive, at its termination β_1 which is either x_3 or x_2, it is negative. Now let us again halve this interval and consider the first or the second half, according as the function is negative or positive for the new middle point. We have then an interval of the length $\frac{1}{4}l$, in which the function changes sign; at its beginning α_2 the function is positive, at its end β_2 negative. As we continue this process of halving, we either reach a point at which the value of the function is zero, and then the theorem is proved; or we can mark off intervals without limit, whose lengths decrease ever more and more and converge to zero, for these lengths are

$$\frac{1}{2}l, \ \frac{1}{4}l, \ \frac{1}{8}l \cdots \text{etc.}$$

The initial points as well as the terminal points of these intervals converge to a determinate limit and it is the same for both. For the initial values $\alpha_1, \alpha_2, \alpha_3 \ldots$ never decrease, they are ever on the increase, though several in succession may remain equal; and all of them are less than β_1. From this it follows (§ 6) that they have a determinate limit. The final values β never increase and yet always remain greater than α_1; whence it follows that they likewise have a limit. Both limits are equal, because the difference $\beta_n - \alpha_n$ converges to zero. Accordingly a value X, between x_1 and x_2, is defined as well by the series $\alpha_1, \alpha_2, \ldots \alpha_n \ldots$ as by the series $\beta_1, \beta_2, \ldots \beta_n \ldots$ Now since $\varphi(\alpha)$ is always positive and $\varphi(\beta)$ always negative, and since we can take n so large that $X - \alpha_n$ and $\beta_n - X$ may be less than an arbitrarily small quantity h, we must also have by the definition of continuity

$$\text{abs}\,[\varphi(\alpha_n) - \varphi(X)] < \delta \quad \text{and} \quad \text{abs}\,[\varphi(\beta_n) - \varphi(X)] < \delta$$

if δ denote an arbitrarily small quantity. But if the value of $\varphi(X)$ were a positive quantity a, different from zero, we should have

$$\varphi(X) - \varphi(\beta_n) > a \quad \text{since } \varphi(\beta_n) < 0;$$

and if it were a negative quantity $-a$, different from zero, we should have

$$\varphi(\alpha_n) - \varphi(X) > a \quad \text{since} \quad \varphi(\alpha_n) > 0.$$

In either case the condition of continuity would not be fulfilled; we must therefore have $\varphi(X) = 0$.*)

Since a continuous function in the interval between a and b can never overleap a value lying between any two of its values, it is possible, unless the function is constant, to discover at every point x a finite region $+h$, such that abs $[f(x+h) - f(x)]$ is equal to a certain prescribed quantity $\frac{1}{3}\delta$ while abs $[f(x + \Theta h) - f(x)] < \frac{1}{3}\delta$. Now in the interval from a to b in which the function is to be continuous, suppose such a region h found first for the point a and laid off from that point on the side towards b, let $a + h = x_1$, then there is for the point x_1 likewise such an interval h_1, let $x_1 + h_1 = x_2$; proceed now to find the interval h_2 for the point x_2 and thus advance to a point x_3 and so on. The ultimate result of this process must be, either, that we arrive at the point b, or, that we include it in an interval, by a finite number of finite assignable intervals. For if, by reason of the intervals h ultimately sinking below any assignable limit, this process should go on infinitely and therefore should converge on some value x between a and b or on the value b, the condition of continuity could not be satisfied for this point x, since it requires that a finite interval can be assigned in which abs $[f(x-h) - f(x)] < \frac{1}{3}\delta$.

The smallest of the finite number of values h thus found, let us

---

*) A. Harnack: Lehrbuch d. Differential- und Integralrechnung von J.-A. Serret Vol. I p. 20—1. 1884.

call it h', is an interval for each point from a to b, within which the absolute difference between values of the function is certainly smaller than δ. For, to an arbitrary point x occurring, ex. gr. in the $x_1 x_2$ interval, belongs $x + h'$ which is at most in the $x_2 x_3$ interval; now we have

$$\text{abs}\,[f(x_1) - f(x)] < \tfrac{1}{3}\delta, \quad \text{abs}\,[f(x_2) - f(x_1)] = \tfrac{1}{3}\delta.$$
$$\text{abs}\,[f(x_2) - f(x+h')] < \tfrac{1}{3}\delta, \quad \text{therefore abs}\,[f(x+h') - f(x)] < \delta.$$

We have thus learned, that for each continuous function one finite value h can be assigned, which is sufficient for every x in the entire interval from a to b in order that it may satisfy the inequality

$$f(x \pm \Theta h) - f(x) < \delta,$$

δ being given arbitrarily. In consequence of this property Heine (see note p. 14) has called every continuous function uniformly continuous in its interval.

We can also state the result of these investigations thus: Any continuous function accomplishes any assignable finite change in its value only within an assignable finite interval; whereas with a discontinuous function a finite change takes place in an arbitrarily small interval.

17. Points at which the criterion of continuity is not satisfied are called points of discontinuity. Thus, for instance, the function

$$y = a \cdot \frac{e^{\frac{1}{x-a}} - 1}{e^{\frac{1}{x-a}} + 1}, \quad (e > 1)$$

is discontinuous for $x = a$, as we find in calculating its value by putting $x = a + h$ or $x = a - h$, and making h converge to zero. In the former case it is at once

$$a \frac{e^{\frac{1}{h}} - 1}{e^{\frac{1}{h}} + 1}.$$

For $h = 0$ this is a quotient whose numerator and denominator increase in a determinate manner beyond any assignable limit; yet it approximates to a fixed finite value. For as it can always be put equal to

$$a \frac{1 - e^{-\frac{1}{h}}}{1 + e^{-\frac{1}{h}}},$$

we see that ultimately the values of the second factor differ arbitrarily little from unity; because as h decreases, $e^{-\frac{1}{h}}$ is always a decreasing fraction. Determined thus, the value of the function for $x = \alpha$ is a. But if we put $x = \alpha - h$, then

$$y = a\frac{e^{-\frac{1}{h}} - 1}{e^{-\frac{1}{h}} + 1},$$

hence for $h = 0$ the limiting value is $-a$. Thus the function proceeds up to the value $-a$, when x beginning from a lower value increases to α, but there the function suddenly leaps to the value $+a$ for the same $x = \alpha$, henceforth it decreases continuously as x increases; accordingly, for $x = \alpha$ the function ceases to be one-valued.

The function already mentioned in § 13: $y = G(x)$ where by $G(x)$ is meant the greatest integer contained in x, is when investigated for positive values up from $x = 0$, a function everywhere one-valued, which is discontinuous at the points $x = 1, 2, 3, \ldots,$ the values of the function suddenly changing from 0 to 1, from 1 to 2, and so on. At all other points, however near they lie to the points of discontinuity, the function is continuous; we can even say, that for each distinct value of x, we can choose h so as to make $G(x + \Theta h) - G(x) < \delta$, only the value of h falls below any assignable limit when x comes arbitrarily near a point of discontinuity: the continuity ceases to be uniform; but we cannot make $G(x - \Theta h) - G(x) < \delta$ when x is one of the points of discontinuity, whence we see that the first inequality alone is not a sufficient condition of continuity.

But further, those points are also styled points of discontinuity, at which the value of the function itself exceeds any assignable limit, or at which it is quite indeterminate, because at such points also the condition of continuity as above formulated cannot be satisfied. For any given function it is then important to investigate, how it behaves in the neighbourhood of such a point.

Thus, for instance, $y = \frac{1}{x - a}$ will have a negative value for $x < a$, whose amount becomes greater the more x approaches the value a; as soon as x has become a little greater than a, the amount becomes positive and arbitrarily great; here therefore there is a change from $-\infty$ to $+\infty$, or

$$f(a + h) = \frac{1}{h} \quad \text{and} \quad f(a - h) = -\frac{1}{h}$$

present as h decreases, numerical series which become determinately (§ 9) positively and negatively infinite. The same thing can be seen, from the geometrical definition, to be the case with $\tan x$ when $x = \frac{1}{2}\pi$. On the other hand, the function $y = \left(\frac{1}{x - a}\right)^2$ is at both sides of a positively arbitrarily great. Continuity however is maintained for every value of x which differs from a by an arbitrarily small yet finite quantity.

The functions

$$y = \frac{1}{x-a}\sin\left(\frac{1}{x-a}\right) \quad \text{and} \quad y = \frac{1}{x-a}\left(\sin\left(\frac{1}{x-a}\right)\right)^2$$

behave quite indeterminately; for whereas in each case the first factor becomes arbitrarily great as x approximates to a, the second oscillates between -1 and $+1$ in the first case, between 0 and $+1$ in the other, so that from no point however close to a does

$$f(a \pm h) = \frac{1}{\pm h}\sin\left(\pm\frac{1}{h}\right) \quad \text{or} \quad \frac{1}{\pm h}\left(\sin\left(\pm\frac{1}{h}\right)\right)^2,$$

present a series with a determinate finite or infinite limiting value as h becomes infinitely small; neither the properties required in § 6 nor those in § 9 are here fulfilled. The oscillations of the sine function follow each other more rapidly the nearer x comes to a; if we assume $x = a + h$ and ask: by how much h must be diminished in order that the number under the sine may change by 2π, it follows, putting

$$\frac{1}{h} + 2\pi = \frac{1}{h'},$$

that

$$h' = \frac{h}{1 + 2\pi h},$$

therefore

$$h - h' = \frac{2\pi h^2}{1 + 2\pi h}.$$

Thus the difference, the interval in which the sine travels all through its values, is less than $2\pi h^2$, so that the number of oscillations of the sine in an arbitrarily small region near a is far beyond any assignable limit; for such a point, at which its argument becomes infinite, the sine (and likewise the cosine, tangent and cotangent) has no determinate value; therefore also it cannot be called continuous at this point, although it is continuous at every point however near to it.

If a continuous function do not become infinite for any value of x in an interval from a to b, then there is at least one value of x in the interval or coinciding with either limit a or b, for which it assumes a greatest value G assignable algebraically (that is, taking account of sign) and likewise one at which it assumes an algebraically least value g. This proposition is not self-evident. In case of a discontinuous function also, provided it remains finite, an upper (and a lower) limit can be assigned, beyond which the values of the function do not pass; moreover such limit can be fixed so that the values of the function come arbitrarily near it at some point, and yet the value itself never be actually reached. If we consider, for instance, the function $y = x$ in the interval from 0 to 1, but attribute to it at $x = 1$ the value $y = 0$ instead of $y = 1$, then it is a discontinuous function whose upper limit is the value 1, which it however does not assume in the interval from 0 to 1, both limits included.

For a continuous function the proposition is proved as follows. If we break up the interval a to b into n parts δ each $= \frac{b-a}{n}$, then, either, the function assumes the value of the upper limit G at one of the points of division, or the value G is upper limit at least for one of these intervals; that is, the values of the function come arbitrarily near this quantity; let this interval reach from x_μ to $x_{\mu+1}$. Choose $\delta' < \delta$ and subdivide this interval into parts δ'. By this we either hit upon the point with the maximum value, or G is the upper limit in an interval $x_{\mu'}$ to $x_{\mu'+1}$. Now if we divide this interval further into portions $\delta'' < \delta'$, and so on, we either arrive by a finite process at the point at which the maximum G is the value, or, we obtain an unlimited series of increasing quantities x_μ, $x_{\mu'}$, $x_{\mu''}$ . . ., which defines a point X. The value of the continuous function f at this point X cannot differ from the value G, since it is the property of a continuous function, that its value at each point can be derived as limiting value of neighbouring ones (§ 16);

$$f(x_\mu), \quad f(x_{\mu'}), \quad f(x_{\mu''}) \quad . \ . \ .$$

is the series by which $f(X)$ is defined. But the values of this series approach arbitrarily to the value G; consequently $f(X)$ also cannot differ from G by a finite quantity, thus $f(X) = G$.

In like manner we find a point at which the value of the function coincides with the lower limit g.

18. When a function is not restricted to a finite interval of x, it is always of importance to examine its behaviour when x becomes infinitely great, that is, to investigate what is the limiting value of the series of values of the function when formed for arbitrarily great values of x. The limiting value is either determinately finite, or determinately infinite, or indeterminate, according to the properties exhibited by the series.

The following are Examples:

(α) $\quad \text{Lim}\left(a + \frac{b}{x}\right)_{\text{for } x = \pm\infty} = a, \quad \text{Lim}\left(\frac{a}{x^m}\right)_{x=\pm\infty} = 0 \quad (m > 0).$

(β) $\quad \text{Lim}\,(x^m)_{x=+\infty} = +\infty\,(m > 0), \quad \text{Lim}\,{}^b\log\,(x)_{x=+\infty} = \infty\,(b > 1).$

(γ) $\begin{cases} \text{Lim}\,(\sin x)_{\text{for } x=\pm\infty} & \text{is indeterminate but finite,} \\ \text{Lim}\,(x \sin x)_{\text{for } x=\pm\infty} & \text{is unlimitedly indeterminate.} \end{cases}$

19. Inasmuch as the continuity of a function on both sides of a point precludes the function from becoming infinitely great in the neighbourhood of this point, the following general theorems regarding the property of continuity hold good:

a) The algebraic sum of two or more continuous functions is itself a continuous function.

For if $\varphi(x)$ and $\psi(x)$ be continuous, it is always possible to discover an interval h, such that

$$\text{abs}\,[\varphi(x \pm h) - \varphi(x)] < \tfrac{1}{2}\delta, \quad \text{abs}\,[\psi(x \pm h) - \psi(x)] < \tfrac{1}{2}\delta.$$

From this it follows that

$$\text{abs}\,[\{\varphi(x \pm h) \pm \psi(x \pm h)\} - \{\varphi(x) \pm \psi(x)\}] < \delta.$$

b) The product of two or more continuous functions is itself a continuous function.

For we have: $\text{abs}\,[\varphi(x \pm h) \cdot \psi(x \pm h) - \varphi(x) \cdot \psi(x)]$

$$= \text{abs}\,[\varphi(x \pm h)\{\psi(x \pm h) - \psi(x)\} + \psi(x)\{\varphi(x \pm h) - \varphi(x)\}].$$

If the interval h be so determined that

$$\text{abs}\,[\psi(x \pm h) - \psi(x)] < \varepsilon, \quad \text{abs}\,[\varphi(x \pm h) - \varphi(x)] < \varepsilon,$$

then we have:

$$\text{abs}\,[\varphi(x \pm h)\,\psi(x \pm h) - \varphi(x)\,\psi(x)] < \varepsilon \,.\, \text{abs}\,[\varphi(x \pm h) + \psi(x)].$$

But since φ and ψ have determinate upper limits in the neighbourhood of the point under consideration, by assuming ε suitably and determining h to correspond, this expression on the right can always be made smaller than any prescribed small number.

c) The quotient of two continuous functions is itself a continuous function, except at points at which the denominator vanishes.

For we have:

$$\text{abs}\left[\frac{\varphi(x \pm h)}{\psi(x \pm h)} - \frac{\varphi(x)}{\psi(x)}\right]$$

$$= \text{abs}\left[\frac{\psi(x)\{\varphi(x \pm h) - \varphi(x)\} - \varphi(x)\{\psi(x \pm h) - \psi(x)\}}{\psi(x)\,\psi(x \pm h)}\right].$$

A similar consideration to the last, shows that the numerator of this expression can be made as small as we please, while the denominator remains finite.

But if in the fraction $\frac{\varphi(x)}{\psi(x)}$ the denominator converges to zero for $x = a$, the numerator remaining finite, the value of the quotient becomes discontinuous, being either determinately infinite or indeterminate; though for every value of x however near this it is continuous.

If the numerator also converges to zero, then by the value of $\frac{\varphi(x)}{\psi(x)}$ for $x = a$ is to be understood the limiting value of the series obtained when x travels through a continuous series of numbers having a for its limit. Whether the quotient then acquires a determinate value and whether it is a continuous function at this point also, i. e. whether it arrives at the same value by proceeding from $x > a$ as from $x < a$, must be decided by particular methods. The most important case of this kind will be investigated in the following paragraphs: its solution gives rise to a general method.

d) If $u = \varphi(x)$ be a continuous function of x, and $y = f(u)$ a continuous function of u, then y is also a continuous function of x. For in order that

$$\text{abs}\,[f(\varphi(x+h)) - f(\varphi(x))] \text{ may be } < \delta,$$

the change of x must be so determined, that

$$\text{abs}\,[\varphi(x+h) - \varphi(x)] < \varepsilon,$$

ε denoting the quantity for which the condition

$$\text{abs}\,[f(u+\varepsilon) - f(u)] < \delta$$

is fulfilled for the continuous function f. But by hypothesis such a value h can be assigned.

20. Having ascertained the remarkable points in the course of a function, we proceed to investigate a measure for the way in which the function changes its values when the independent variable is increased or diminished. The leading idea in the Differential Calculus is the establishment of this measure with full mathematical precision.

Consider two values of the function, belonging to two different values of the argument x and $x + \Delta x$, Δx denoting here the increment of x must not be mistaken for a product and can be chosen $>$ or < 0; let the corresponding values of the function be denoted by y and $y + \Delta y$, then Δy can be calculated from the equation:

$$\text{(I)} \qquad \Delta y = f(x + \Delta x) - f(x).$$

The difference on the right assigns the magnitude of the change of y when the independent variable alters from the value x to the value $x + \Delta x$.

Now this change of y has to be compared with that of x. Keeping the value of Δx unchanged, the intensity of the change in the value of the function becomes greater when the value of Δy is increased. On the other hand, for a given value of Δy the intensity of change would be increased if Δy were produced by a smaller Δx. Thus *the quotient of differences*

$$\text{(II)} \qquad \frac{\Delta y}{\Delta x} = \frac{f(x+\Delta x) - f(x)}{\Delta x}, \; (\Delta x \gtrless 0)$$

gives a measure for the average (mean) intensity of the change in the interval from x to $x + \Delta x$. If the function changed uniformly in the interval, that is, if equal values of Δy always belonged to equal values of Δx, then if the intensity, or rate, of change were that given in equation (II), through the entire interval Δx it would give rise to the increase Δy in the function. The quotient of differences has the following properties: when f is a continuous function of x, the quotient of differences is for any finite value of Δx a continuous function of x; but on the same hypothesis it is also, secondly, a continuous function of Δx, as long as we restrict ourselves to finite but arbitrarily small values of Δx (see last Section).

When we now endeavour to determine a measure not for an arbitrary interval but for one point of the function, we have to make the interval Δx converge to zero. The quotient on the right then becomes infinitely great, unless its numerator $f(x + \Delta x) - f(x)$ also turn out to be a series of numbers having zero for a limit (§ 10). But assuming that this is the case, let us consider a point in whose immediate neighbourhood the unique function is continuous. What conditions must be fulfilled in order that for continually diminishing values of Δx, the quotient

$$\frac{f(x + \Delta x) - f(x)}{\Delta x}$$

may present a continuous sequence of numbers tending to a determinate limiting value: zero, finite or infinitely great? We can again give expression to the fact that we are passing through the interval Δx in the positive or negative sense, by considering Δx as a fixed arbitrarily small but finite value, and introducing a number Θ which moves continuously between the limits -1 and $+1$; then our present object is, to find whether

$$\frac{f(x + \Theta\Delta x) - f(x)}{\Theta\Delta x}$$

has a limiting value when Θ converges from -1 to zero, or from $+1$ to zero. The limiting values, arising in the two cases, can be different; in the first we call the quotient of differences regressive, in the second, progressive. Upon the latter we fix our attention:

The particular value zero will occur as limiting value, provided, for each number δ however small, a number Δx can be found such that the absolute amount of this quotient for every value of Θ is less than δ. Here the numerator can change sign arbitrarily, i. e. $f(x + \Theta\Delta x)$ can be at one time greater, at another less than $f(x)$, or in other words: Provided zero is the limiting value of the quotient of differences, the function $f(x)$ can, in the neighbourhood of the point x, oscillate arbitrarily many times about the axis of x.*)

*) The fluctuations (differences of ordinates), we remark only in passing, are infinitely small of an order higher than the first (Chap. VII). The function $x^2 \sin \frac{1}{x}$ has ex. gr. for $x = 0$ the differential quotient 0, for we have

$$\text{Lim}\left\{\frac{f(0 + h) - f(0)}{h}\right\} = \text{Lim}\left\{h \sin \frac{1}{h}\right\} = 0.$$

But if the limiting value is to be finite or determinately infinite, then since when $+1 \geq \Theta > 0$ the denominator of the quotient is only positive, the numerator must also always have one and the same sign for the same sign of Θ, that is,

$$f(x + \Theta\Delta x) \text{ is either only } > \text{ or only } < f(x).$$

The function only increases or only decreases in comparison with its value at x, while x increases; it no longer oscillates about this value in an arbitrarily small interval. This is a necessary condition. But the condition necessary and sufficient in order that the series of quotients of differences may have a finite limiting value, by § 6 is that

$$\text{abs}\left[\frac{f(x+\Delta x) - f(x)}{\Delta x} - \frac{f(x+\Theta\Delta x) - f(x)}{\Theta\Delta x}\right] < \delta, \quad (1 > \Theta \geq 0)$$

where δ signifies an arbitrary number.

This inequality can be interpreted as follows:

Calling the numerical difference between the quotients of differences belonging to the values Δx and $\Theta\Delta x$ which depends on Θ, a fluctuation of this quotient in the interval Δx, then this inequality asserts: The necessary and sufficient condition for the existence of a determinate finite limiting value consists in this, that for each number δ however small, a finite interval Δx can be ascertained, in which all fluctuations of the quotient of differences are less than δ.

But this amounts to saying, that in the neighbourhood of any point at which the quotient of differences of the continuous function has a determinate finite limiting value, the quotient of differences is not only a continuous function of Δx for every finite value of Δx, but retains this property also for $\Delta x = 0$.

If the quotient of differences has the property of only increasing or of only decreasing in the interval Δx, while Θ converges to zero, in other words: if at any point an interval Δx can be found in which the quotient of differences has neither maxima nor minima, then as in the Theorem in § 6 (cf. § 9), it has a limiting value either finite or determinately infinite. The possibility of no determinate limiting value arises accordingly only in case the quotient of differences at a point assumes in however small an interval infinitely many maxima and minima, whose differences cannot be arbitrarily diminished. In this case we no longer say that the quotient of differences is continuous inclusive of the value $\Delta x = 0$.

We can illustrate numerator and denominator of the above inequality geometrically as follows:

Let us represent the function $y = f(x)$ under the figure of a polygon with any number of angles, interpreting the values of x and y as lengths in a rectangular system of Cartesian coordinates. Then we perceive:

I. Δy is the difference of altitude of two points P and P_1, belonging to the values x and $x + \Delta x$.

II. The quotient of differences is equal to the trigonometric tangent of the angle, made by the chord PP_1 with the axis of abscissæ; it measures the mean intensity of rising.

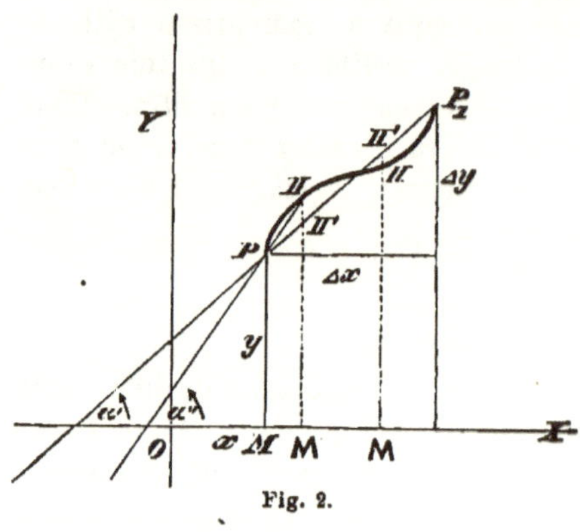

Fig. 2.

The equation of the chord, i. e. the right line passing through the points P and P_1, if ξ and η signify the coordinates of any point on it, is

$$\frac{\eta - f(x)}{\xi - x} = \frac{f(x + \Delta x) - f(x)}{\Delta x}.$$

The point on this line with the abscissa $\xi = x + \Theta \Delta x$ has the ordinate

$$\eta = f(x) + \Theta \{f(x + \Delta x) - f(x)\}.$$

On the other hand, to the abscissa $x + \Theta \Delta x$ belongs the vertex of the polygon

$$\eta = f(x + \Theta \Delta x).$$

Accordingly

$$\Theta \{ f(x + \Delta x) - f(x)\} - \{ f(x + \Theta \Delta x) - f(x)\}$$

is equal to the difference $M\Pi' - M\Pi = \Pi\Pi'$. Let us call $\frac{\Pi\Pi'}{MM}$ the measure of the deviation of the value of the function from the right line at this point in the interval Δx, then the above inequality, which includes the first mentioned condition, teaches us:

The necessary and sufficient condition that the quotient of differences may have a determinate finite limiting value, consists in it being possible to find an interval Δx, in which the deviations from the right line are in absolute amount less than any number δ however small.*)

Denoting by α' the angle which the chord PP_1 makes with the axis of abscissæ and likewise by α'' that which $P\Pi$ makes, we obtain another interpretation of the inequality: for

$$\tan \alpha' - \tan \alpha'' = \frac{f(x + \Delta x) - f(x)}{\Delta x} - \frac{f(x + \Theta \Delta x) - f(x)}{\Theta \Delta x} < \delta,$$

that is, it must be possible to find an interval Δx, in which the differences of the angles α', α'' become arbitrarily small. The sides of the polygon then approach to a fixed limiting position and the

*) Here the deviations can continually change sign, i. e. a curve can in any interval however small have infinitely many oscillations relatively to its tangent.

function at this point can be imagined as a curve, i. e. a geometrical figure having a definite direction at the point, so that its course at this point is represented as approximately as we can wish by a determinate right line. Therefore in geometrical language the condition that there should be a limiting value, is no other than the condition that a function should admit of being represented by a curve. (§ 15.)

On the other hand the limiting value becomes infinitely great in a determinate manner, when the measure of the deviation from the right line PP_1 increases always positively or always negatively beyond any finite amount or $\tan\alpha'$ becomes infinite (§ 24a). But it becomes completely indeterminate when the deviation is neither zero nor becomes infinite in a determinate manner, but oscillates between different limits.

This limiting value, defined by the equation:

(III) $\frac{dy}{dx} = \lim\limits_{\Theta=0} \frac{f(x+\Theta\Delta x)-f(x)}{\Theta\Delta x}$ or more briefly: $\lim\limits_{\Delta x=0} \frac{f(x+\Delta x)-f(x)}{\Delta x}$,

called the Progressive Differential Quotient of the function at the point x, affords a measure for the change of the function at the point, when x increases.

Likewise the Regressive Differential Quotient:

(IIIa) $\frac{dy}{dx} = \lim\limits_{\Theta=0} \frac{f(x-\Theta\Delta x)-f(x)}{-\Theta\Delta x}$ or more briefly: $\lim\limits_{\Delta x=0} \frac{f(x-\Delta x)-f(x)}{-\Delta x}$

affords a measure for the change of the function at that point when x decreases.

Instead of equations (III), provided $\frac{dy}{dx}$ is finite, we can also write

$$\frac{\Delta y}{\Delta x} = \frac{dy}{dx} + \delta,$$

where δ is precisely that difference between the terms of the series formed of the quotients of differences and the limiting value of that series which converges to zero with Δx. Here it is to be remarked, that since the quotient of differences is a continuous function of Δx, a finite value of Δx can always be found which shall satisfy this equation, however small but finite be the value of δ; but for a given δ this has not to be the same Δx for all values of x in an interval (see next paragraph). The equation

$$\Delta y = \Delta x \cdot \frac{dy}{dx} + \Delta x \cdot \delta$$

then asserts: the smaller we choose Δx, the more accurately is the corresponding change of the function equal to the product of the differential quotient by Δx, so that for the point itself this precisely denotes the measure of the increase. It is also evident, as stated at the outset, that the existence of determinate finite values for the progressive and regressive differential quotients, involves in itself always the condition of the continuity of the function at this point.

21. According as the progressive differential quotient is positive or negative, the function increases or decreases for increasing values of x at this point; and according as the regressive differential quotient is positive or negative, the function decreases or increases when the values of x decrease. But the distinction of the progressive and the regressive differential quotient for a continuous function is unnecessary in most of the cases we shall have to consider, for we have the theorem:

If in the neighbourhood on both sides of a point at which $f(x)$ is continuous, an interval Δx can be found for every value of x, such that the differences of the quotients of differences of this interval, formed for all values between 0 *and Δx, remain in absolute amount smaller than an arbitrarily small number δ, then the progressive differential quotient is a continuous function of x and the value of the regressive one is identical with it.*

Taking an arbitrarily small but finite quantity h, lay off the interval $\pm h$ on the two sides of a point x; this represents the neighbourhood on the two sides. The condition then asserts, that the difference

$$\text{abs}\left[\frac{f(x \pm h + \Delta x) - f(x \pm h)}{\Delta x} - \frac{f(x \pm \eta h + \Theta \Delta x) - f(x \pm \eta h)}{\Theta \Delta x}\right]$$

remains smaller than δ, while Θ and η assume all values from 0 to 1. This condition can be put into the words: The quotient of differences is a continuous function of both variables h and Δx; or it is a uniformly continuous function of h and Δx. (For the reason of this phrase, see Chap. IX.)

Let us denote the progressive differential quotient of $f(x)$ at the point x by $f_1(x)$, then, by hypothesis, Δx can always be chosen so small that

$$(1) \qquad f_1(x \pm h) = \frac{f(x \pm h + \Delta x) - f(x \pm h)}{\Delta x} \pm (< \delta).$$

In like manner in consequence of the hypothesis of our theorem we have for the differential quotient at the point x

$$(2) \qquad f_1(x) = \frac{f(x \pm h + \Delta x) - f(x \pm h)}{\Delta x} \pm (< \delta)$$

and it is of importance that in the entire interval the same value of Δx is sufficient for a given value of δ. Accordingly the amount of the difference $f_1(x \pm h) - f_1(x)$ is also less than 2δ, i. e. the progressive differential quotient is continuous in the neighbourhood of the point x.

To prove the second part of the statement, let us denote the regressive differential quotient at the point x by $f_2(x)$, then it is defined by the equation:

$$(3) \qquad f_2(x) = \frac{f(x - \Theta\Delta x) - f(x)}{-\Theta\Delta x} + \varepsilon,$$

ε converging to zero along with Θ. But if in equation (1) we make $\Delta x = h$ and then put for h in $f_1(x - h)$ the value $\Theta\Delta x$, we find:

$$(4) \qquad f_1(x - \Theta\Delta x) = \frac{f(x) - f(x - \Theta\Delta x)}{\Theta\Delta x} \pm (< \delta).$$

From this we see, that the quotient

$$\frac{f(x - \Theta\Delta x) - f(x)}{-\Theta\Delta x}$$

tends to a certain limit whose value is equal to $f_1(x)$, consequently $f_2(x) = f_1(x)$, as was to be proved.*)

The theorem of the equality of the progressive with the regressive differential quotient still continues true at the points at which $f_1(x)$ becomes positively or negatively infinite; provided both $f_1(x - \eta h)$ and $f_1(x + \eta h)$ become infinitely great for $\eta = 0$ in the same sense. For, if, in whatever way η and Θ converge to zero, both quotients

$$\frac{f(x - \eta h + \Theta\Delta x) - f(x - \eta h)}{\Theta\Delta x} \text{ and } \frac{f(x + \eta h + \Theta\Delta x) - f(x + \eta h)}{\Theta\Delta x}$$

increase beyond any finite amount in the same sense, then putting $\eta h = \Theta\Delta x$ in the first quotient and $\eta = 0$ in the second, the quotients

$$\frac{f(x) - f(x - \Theta\Delta x)}{\Theta\Delta x} \text{ and } \frac{f(x + \Theta\Delta x) - f(x)}{\Theta\Delta x}$$

also become determinately infinite in the same sense. Therefore the regressive differential quotient is identical with the progressive at the point x.

When progressive and regressive differential quotients of a function $f(x)$ are thus identical, the function which expresses them is the first differential coefficient of the function, it has been named by Lagrange (1736—1813) the first derived of f and is frequently denoted by $f'(x)$; for brevity we shall sometimes call the first derived function the first derivate.**)

Therefore $\frac{dy}{dx} = \frac{df(x)}{dx} = f'(x)$ are only different notations for one quantity, which is to be calculated from the formula:

$$\text{Lim} \frac{f(x \pm \Delta x) - f(x)}{\pm \Delta x} \quad \text{for } \Delta x = 0.$$

---

*) It will be seen § 100 that no further hypothesis is necessary in order to establish the identity of $f_2(x)$ and $f_1(x)$ than the continuity both of the progressive differential quotient and of the function $f(x)$.

**) The original mentions that the German name „Ableitung von f" is due to Crelle. After much of my work was printed off, the word "derivate" was suggested by my friend Dr. Atkinson, and I have ventured to use it since.

G. L. C.

The process of forming the differential quotient of y or its first differential coefficient, or the first derived of f with respect to x is briefly called differentiation.

22. It is however not only at each point that the values of the first derived of a function give us information respecting the manner in which the function changes whether positively or negatively. We are now going to demonstrate that the change of value of the function, even in an interval of finite extent, can be measured by the values of the first derivate. With this purpose in view we first prove the Lemma: *When a unique function, whose progressive differential quotient coincides with its regressive at each point within an interval from $x = a$ to $x = b$, has equal values at the extremities of this interval, there must be in the interval at least one point at which the first derivate vanishes.**) For, either, the function has everywhere the same value, in which case it is constant and its differential quotient everywhere zero; or, the function attains in at least one point within the interval its greatest or its least value (§ 17). It may even undergo repeated alternations of increase and decrease, one such it must have in any case. If x_1 be such a point, then in its immediate neighbourhood $f(x_1 - h) - f(x_1)$ will have the same sign as $f(x_1 + h) - f(x_1)$. Consequently the quotients

$$\frac{f(x_1 - h) - f(x_1)}{-h} \quad \text{and} \quad \frac{f(x_1 + h) - f(x_1)}{h}$$

differ in sign, however small we choose the value of h. Now these two quotients have the same limiting value, since by hypothesis the progressive differential quotient and the regressive are identical; but a positive and a negative series of numbers can have the same limit, only when this limit is zero, therefore at this point $f'(x_1) = 0$.

*) Such a function can be exhibited geometrically by tracing a curve of the form fig. 3:

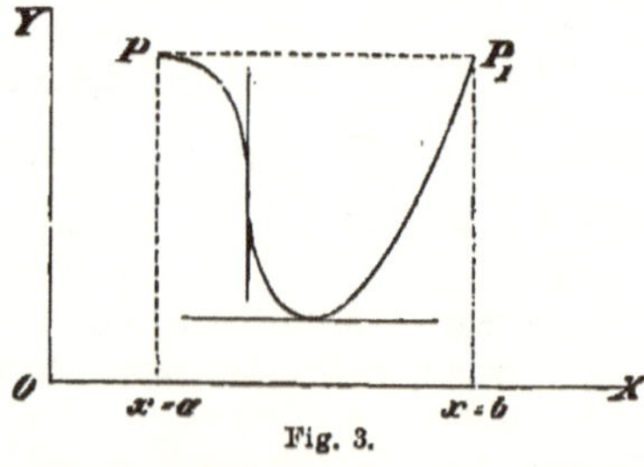

Fig. 3.

it can also have points at which the tangent is parallel to the axis of ordinates. The theorem asserts, what is manifest geometrically: that when the ordinates of the extremities are equal, there is between them at least one point with the tangent parallel to the axis of abscissæ.

Note. This proof does not assume the continuity of $f'(x)$, so that $f'(x)$ may become infinite in the interval. Moreover it does not require $f'(x)$ to have everywhere a determinate value; the only assumption it makes is that

$$\operatorname{Lim} \frac{f(x+h) - 2f(x) + f(x-h)}{h} = 0$$

at each point of the interval.

This Lemma furnishes the proof of the following proposition which is called the Theorem of the Mean Value:

If $f(x)$ be a unique function in the interval from a to b, whose progressive and regressive differential quotients are everywhere in the interval identical and determinate, then a value x_1 can always be found between a and b such that we shall have the quotient of differences

$$\frac{f(b) - f(a)}{b - a} = f'(x_1).\text{*)}$$

For if we denote the value of the quotient of differences by K, so that:

$$\{f(b) - Kb\} - \{f(a) - Ka\} = 0,$$

and form the function:

$$\varphi(x) = \{f(b) - Kb\} - \{f(x) - Kx\},$$

then this alike with $f(x)$ will be continuous, it will likewise have everywhere identical progressive and regressive differential quotients, and furthermore it will have the same value 0 both for $x = a$ and for $x = b$. Hence there must in the interval be a value x_1 which will make $\varphi'(x_1) = 0$.

But we have:

$$\varphi'(x_1) = K - f'(x_1) = 0, \text{ that is: } K = \frac{f(b) - f(a)}{b - a} = f'(x_1). \quad \text{Q. E. D.**)}$$

---

*) Geometrically:

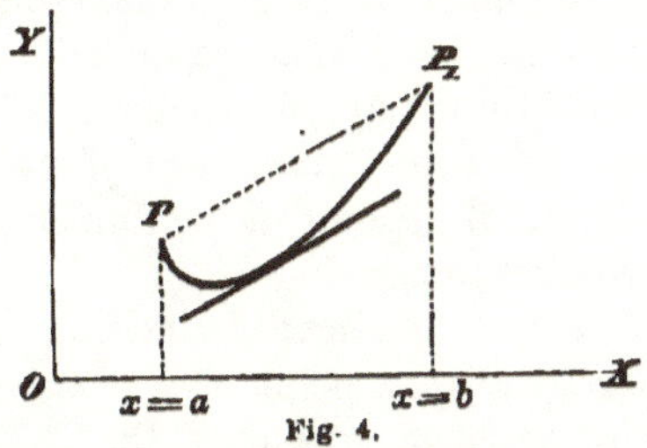

Fig. 4.

There is an intermediate point at which the tangent is parallel to the line joining the extremities. Here also the case may occur that the derivate becomes infinite: that is, that the tangent at a point is parallel to the axis of ordinates.

**) This proof of the proposition, which is also called the theorem of Rolle (1652—1719), is due to Serret (1819—85): Cours de calcul différentiel et intégral, 2e éd., t. I, p. 17 seq.

We can always express a value x_1 situated between a and b by the formula: $x_1 = a + \Theta(b - a)$, where Θ denotes a positive proper fraction, so that the equation of the Theorem of the Mean Value can also be written:

$$\frac{f(b) - f(a)}{b - a} = f'(a + \Theta(b - a)). \quad 0 < \Theta < 1.$$

Scholium: *If the progressive and regressive differential quotients vanish everywhere in the interval, the function is continuous in this interval and its value constant.* For then, if x denote any value in the interval

$$\frac{f(x) - f(a)}{x - a} = f'(a + \Theta(x - a)) = 0,$$

that is, $f(x) = f(a)$. (See also Integral Calculus. Bk. III. Chap. I.)

23. We have already indicated above, that the differential quotient, alike with the quotient of differences, admits of a simple geometric meaning. For when, as Δx goes on decreasing, the quotient $\frac{\Delta y}{\Delta x}$ tends to a determinate limiting value, and therefore (Fig. 2 § 20) the right line PP_1 approximates to a certain limiting position, this limiting line is called the tangent of the figure represented by the function. We must regard this as the definition of the tangent to a continuous series of points defined by an equation: Limiting position of the secant drawn through two arbitrarily near points. We can accordingly deduce the following proposition III from II:

III. The differential quotient is equal to the trigonometric tangent of the angle, which the touching line (tangent) at the point P of the curve forms with the axis of abscissæ; it measures the inclination of the curve at this point to the axis of abscissæ.

It will not be superfluous to remark, that a continuous series of points, which we call a curve, can be defined in two different ways: Either geometrically by the mechanical contrivance of a motion (as the circle by rotation of a fixed length) or analytically by a functional equation between the coordinates. In both cases proof must be produced that there is such a thing as a tangent, and it is only when this is forthcoming that the figure may in strictness be called a curve. In case of the geometric definition, kinematics proves there is a tangent, in case of the analytic, the proof is contained in the fact that the function admits of differentiation.

When the differential quotient is calculated, the problem of constructing the tangent at any point of any curve whose equation is given, is solved. This problem originated the Differential Calculus, of which Leibnitz (1646—1716) first published the principles in the very notations still employed, in the year 1684, in an essay of a few pages: "Nova methodus pro maximis et minimis, itemque tangentibus, quae nec fractas nec irrationales quantitates moratur, et singulare pro illis calculi genus", which appeared in the periodical: Acta Eruditorum, at Leipzig. Independently of him, Newton*) had already for years in working mechanical problems been developing the same method of calculation, as he has repeatedly mentioned and intimated in his letters from about 1670 until he ultimately published it in 1687 in his great work "Philosophiae naturalis principia mathematica" as quite an indispensable resource for investigating continuously measurable phenomena**) Here Newton introduced the conception of a variable, considering the independent variable as measure of the time. At the very outset he etablishes the theorem for calculation with continuous variables: "Quantities which in a given time continually approach to equality and which before the end of that time can come nearer to each other than any assigned quantity, are finally equal to each another", which is only a different statement of our fundamental proposition of § 5. Taking the distance described by a movable point as the dependent variable, the quotient of differences assigns the mean velocity with which a finite length is described, while the differential quotient measures the actual velocity at each point.

24. Geometrical corollaries and illustrations.

a) If the differential quotient for a finite value of x and of y is determinately infinite, then the tangent of the curve at this point is parallel to the axis of ordinates.

b) At points at which the progressive and regressive differential quotients differ, the direction of the tangent changes discontinuously; the curve forms an angle.

c) At points at which the function undergoes a break, provided it is continuous towards one side, it can also possess a differential quotient towards this side.

d) *If the function $f(x)$ be determinately infinite for a finite value*

---

*) The inscription on Newton's monument in Westminster Abbey runs: H. S. E. ISAACUS NEWTONUS, EQUES AURATUS, QUI ANIMI VI PROPE DIVINA PLANETARUM MOTUS FIGURAS, COMETARUM SEMITAS, OCEANIQUE AESTUS, SUA MATHESI FACEM PRAEFERENTE, PRIMUS DEMONSTRAVIT. RADIORUM LUCIS DISSIMILITUDINES, COLORUMQUE INDE NASCENTIUM PROPRIETATES, QUAS NEMO ANTE VEL SUSPICATUS ERAT, PERVESTIGAVIT. NATURAE, ANTIQUITATIS, S. SCRIPTURAE SEDULUS, SAGAX, FIDUS INTERPRES DEI O. M. MAIESTATEM PHILOSOPHIA APERUIT, EVANGELII SIMPLICITATEM MORIBUS EXPRESSIT. SIBI GRATULENTUR MORTALES TALE TANTUMQUE EXTITISSE HUMANI GENERIS DECUS.

Natus XXV. Decemb. A. D. MDCXLII; obiit Martii XX MDCCXXVI. (N. S. 1727.)

**) The treatise: Methodus fluxionum et serierum infinitarum, cum ejusdem applicatione ad curvarum geometriam, first appeared in 1736 after his death.

a of x, then the limiting value of $f'(x)$ is also determinately infinite for $x = a$, on the hypothesis that there is any determinate limiting value of $f'(x)$.

For, by the Theorem of the Mean Value, we have the equation:

$$f(a - \delta + h) - f(a - \delta) = h f'(a - \delta + \Theta h)$$

in which we first suppose h smaller than δ. But if we let h become arbitrarily nearly equal to δ, the left side becomes arbitrarily great, hence there are points in the interval from $a - \delta$ to a at which $f'(x)$ also becomes arbitrarily great; if then $f'(x)$ have any definite limit, this limit can only be infinitely great.

If the function $f(x)$ be determinately infinitely great for $x = \infty$, then for every finite value of h,

$$\operatorname*{Lim}_{x=\infty} \frac{f(x+h) - f(x)}{h} = \operatorname*{Lim}_{x=\infty} \frac{f(x)}{x}$$

*provided the first quotient have any determinate finite limiting value for any finite value of h.**).

For the proof we assume $h > 0$, and let $f(x)$ become positively or negatively infinite. Calling the limiting value of the left side K, x can be chosen so great that

$$K - \varepsilon < \frac{f(x+h) - f(x)}{h} < K + \varepsilon,$$

$$\text{or: } Kh - \varepsilon h < f(x+h) - f(x) < Kh + \varepsilon h;$$

where ε denotes a quantity, which is arbitrarily small by our choice of x. Putting for x the values $x, x+h, x+2h, \ldots x + \overline{n-1}h$, and adding the successive inequalities, we find:

$$nKh - n\varepsilon h < f(x + nh) - f(x) < nKh + n\varepsilon h, \quad \text{or,}$$

$$K - \varepsilon < \frac{f(x+nh) - f(x)}{nh} < K + \varepsilon.$$

In order that the argument of the function may become infinite in the most general manner, let us put $x = x_0 + ph$, where $0 < x_0 < h$, and p is a positive integer, then as soon as we have fixed p large enough, we have for all values of x_0,

$$K - \varepsilon < \frac{f(x_0 + \overline{p+n}h) - f(x_0 + ph)}{nh} < K + \varepsilon.$$

If we call $x_0 + \overline{p+n}h = \xi$, $nh = \xi - x_0 - ph$; thus we have

$$K - \varepsilon < \frac{f(\xi) - f(x_0 + ph)}{\xi - x_0 - ph} < K + \varepsilon, \quad \text{or}$$

$$K - \varepsilon < \frac{f(\xi) - f(x_0 + ph)}{\xi} \cdot \frac{\xi}{\xi - x_0 - ph} < K + \varepsilon.$$

*) Cauchy: Cours d'Analyse algébr. Cap. 2. The proposition has been extended by Stolz: Über die Grenzwerthe der Quotienten. Math. Ann. Vol. XIV. XV.

Now if we make n increase arbitrarily, ξ also increases without limit, thus as x_0 can be fixed arbitrarily, the argument ξ can assume any arbitrarily great value. Since zero is the limit of the quotient $\frac{f(x_0+ph)}{\xi}$, and 1 that of $\frac{\xi}{\xi - x_0 - ph}$, we have also

$$K - \varepsilon < \operatorname{Lim} \frac{f(\xi)}{\xi} < K + \varepsilon,$$

and because ε can from the outset be chosen arbitrarily small, we must have the limiting value $K = \operatorname{Lim}_{\xi=\infty} \frac{f(\xi)}{\xi}$.

The theorem also continues to hold, when the quotient

$$\frac{f(x+h)-f(x)}{h}$$

becomes determinately infinite for $x = \infty$. For if we can choose x so large, that the quotient continues larger than an arbitrary number K, it follows by the same process:

$$\frac{f(x+nh)-f(x)}{nh} > K$$

or

$$\frac{f(\xi)-f(x_0+ph)}{\xi} \frac{\xi}{\xi - x_0 - ph} > K,$$

that is

$$\operatorname{Lim} \frac{f(\xi)}{\xi} > K.$$

But we can replace the quotient $\frac{f(x+h)-f(x)}{h}$ (§ 22) by the value of the derived $f'(x+\Theta h)$, if $f(x)$ is a continuous function for all finite values of x and its derived is also everywhere determinate.

Therefore the inequality $K - \varepsilon < f'(x+\Theta h) < K + \varepsilon$ holds true for arbitrarily great values of x, but from this it follows: *If the function $f'(x)$ has any determinate limit for arbitrarily increasing values of x, this limit must likewise be equal to K.*

Examples:

(1) $f(x) = \log(x)$. $\operatorname{Lim} \frac{\log(x+h)-\log(x)}{h} = \frac{1}{h} \operatorname{Lim} \log\left(1+\frac{h}{x}\right) = 0$ for $x = \infty$. Therefore we have also $\operatorname{Lim} \frac{\log(x)}{x} = 0$. In next Chapter it will be shown, that the derived of the logarithm is $\frac{1}{x}$, so that in fact also $\operatorname{Lim} f'(x) = 0$.

(2) $f(x) = a^x$. $\operatorname{Lim} \frac{a^{x+h}-a^x}{h} = \frac{a^h-1}{h} \operatorname{Lim} a^x = \infty \quad (a > 1)$; consequently we have also $\operatorname{Lim} \frac{a^x}{x} = \infty$ for $x = \infty$.

(3) $f(x) = x + \frac{1}{x}\sin(x^2)$ is determinately infinite for $x = \infty$.

$$\operatorname{Lim} \frac{f(x+h)-f(x)}{h} = \frac{h + \frac{1}{x+h}\sin(x+h)^2 - \frac{1}{x}\sin(x^2)}{h} = 1.$$ Therefore

we have also $\operatorname{Lim} \frac{f(x)}{x} = 1$. But on the other hand $f'(x) = 1 - \frac{1}{x^2} \sin(x^2) + 2\cos(x^2)$ is for $x = \infty$ indeterminate.

e) If $f(x)$ has a finite determinate value for $x = \infty$, then for every finite value of h we have $\operatorname{Lim}_{x=\infty} \left\{\frac{f(x+h) - f(x)}{h}\right\} = 0$. This is also to be regarded as value of the differential quotient at $x = \infty$; it coincides with the value $\operatorname{Lim}\{f'(x)\}$ provided $f'(x)$ changes continuously to any determinate limiting value for $x = \infty$ as x increases arbitrarily.

f) The differential quotient can be indeterminate at all points of a continuous function, if ex. gr. the difference $f(x + \Delta x) - f(x)$ everywhere in an arbitrarily small interval Δx undergo change of sign without the amount of the quotient of differences $\frac{f(x+\Delta x) - f(x)}{\Delta x}$ converging to zero; this is a case, in which the function cannot be fixed under the figure of a curve, on the basis of the formula in accordance with which it is to be calculated, because the interpolation of more and more points ultimately displaces the angles of the polygon immeasurably, whereas it displaces the sides measurably.

g) In representing to ourselves how motion goes on in nature we presuppose no discontinuities, neither in regard to the places which the moving body occupies, nor in regard to the direction and magnitude of its motion. In some phenomena however (under the influence of blows) the changes are effected so rapidly, that we look upon the process as a discontinuous one.

The Continuity of a function was first precisely defined by Cauchy (1789—1857) (Analyse Algébr. 1821), to him we owe the foundation of the differential calculus generally in the form in which it is here developed. Riemann (1826—1866) directed attention to continuous functions which have within an arbitrarily small interval infinitely many points at which the progressive and the regressive differential quotients are indeterminate. Weierstrass was the first who gave an example of a continuous function having in no point a determinate value of the differential quotient either progressive or regressive (communicated in a paper of Du Bois-Reymond Journ. f. Math. Vol. 79). Here the function appears as limit of a series of functions, whose values ultimately differ arbitrarily little, while the same is not true for the values of the differential quotients which on the contrary vary between arbitrarily great positive and arbitrarily great negative values.

Sixth Chapter.

Differentiation of the simplest functions.

25. We shall first treat the functions defined in Chapter IV which are styled Elementary functions.

I. The algebraic, whose simplest type is $y = x^m$, and whose most general is the implicit function: $A_0 y^n + A_1 y^{n-1} + \dots A_{n-1} y + A_n = 0$ in which $A_0, A_1, \dots A_n$ are polynomials of any degree in x.

II. The transcendental, namely:

a) the exponential function $y = a^x$ and the trigonometric: $\sin x$, $\cos x$, $\tan x$, $\cot x$;

b) the logarithm $y = {}^a\log x$ and the circular functions: $\sin^{-1} x$, $\cos^{-1} x$, $\tan^{-1} x$, $\cot^{-1} x$.

The immediate aim of our investigation is: from the properties of these functions to obtain convenient methods of calculating them; for, with the exception of the case $y = x^m$ for m a positive or negative integer, in which the calculation is accomplished by carrying out an m-fold multiplication, we have been hitherto put off with the process of inclusion within limits or geometrical considerations. The most comprehensive problem is presented by the implicit algebraic function, but its treatment must be preceded by further general considerations.

26. For the formation of first differential quotients whether with positive or negative values of Δx, the following General Rules are required.

1) The differential quotient of a constant is equal to zero.

2) The differential quotient of a sum of functions is equal to the sum of the differential quotients of the summands. For if

$$y = f_1(x) + f_2(x) + f_3(x) + \dots + f_n(x),$$

then

$$y + \Delta y = f_1(x + \Delta x) + f_2(x + \Delta x) + \dots + f_n(x + \Delta x),$$

therefore:

$$\frac{\Delta y}{\Delta x} = \frac{f_1(x+\Delta x) - f_1(x)}{\Delta x} + \frac{f_2(x+\Delta x) - f_2(x)}{\Delta x} + \dots + \frac{f_n(x+\Delta x) - f_n(x)}{\Delta x}.$$

Hence it follows by Proposition I § 10 that:

$$\frac{dy}{dx} = f_1'(x) + f_2'(x) + \dots + f_n'(x) \quad \text{Q. E. D.}$$

Scholium: Two functions, which differ only by an additive constant, have the values of the differential quotient the same.

3) If y be the product of two functions: $y = \varphi(x) \cdot \psi(x)$

then $$\frac{\Delta y}{\Delta x} = \frac{\varphi(x+\Delta x) \cdot \psi(x+\Delta x) - \varphi(x) \cdot \psi(x)}{\Delta x}.$$

The expression on the right can be replaced for every finite Δx by
$$\varphi(x+\Delta x)\frac{\psi(x+\Delta x)-\psi(x)}{\Delta x} + \psi(x)\frac{\varphi(x+\Delta x)-\varphi(x)}{\Delta x}.$$

If φ and ψ be continuous and their differential quotients have determinate values at the point x, then, by Theorem II § 10 when Δx converges to zero:
$$\frac{dy}{dx} = \frac{d\{\varphi(x)\cdot\psi(x)\}}{dx} = \varphi(x)\,\psi'(x) + \psi(x)\,\varphi'(x).$$

This law can be extended to any definite number of factors.

Scholium: If a signify a constant, when $y = af(x)$ we have
$$\frac{dy}{dx} = \frac{d\,af(x)}{dx} = a \cdot f'(x).$$

4) If $y = \frac{\varphi(x)}{\psi(x)}$, then at a point where $\psi(x)$ does not vanish,
$$\Delta y = \frac{\varphi(x+\Delta x)}{\psi(x+\Delta x)} - \frac{\varphi(x)}{\psi(x)} = \frac{\psi(x)\,\varphi(x+\Delta x) - \varphi(x)\,\psi(x+\Delta x)}{\psi(x)\,\psi(x+\Delta x)}$$

or:
$$\frac{\Delta y}{\Delta x} = \left(\psi(x)\frac{\varphi(x+\Delta x)-\varphi(x)}{\Delta x} - \varphi(x)\frac{\psi(x+\Delta x)-\psi(x)}{\Delta x}\right) : \psi(x)\,\psi(x+\Delta x)$$

By Theorem III § 10 we have therefore:
$$\frac{dy}{dx} = \frac{\psi(x)\cdot\varphi'(x) - \varphi(x)\cdot\psi'(x)}{(\psi(x))^2}.$$

Scholium: If $y = \frac{1}{\psi(x)}$, then $\frac{dy}{dx} = -\frac{\psi'(x)}{(\psi(x))^2}$.

5) It is convenient to consider more composite functions in the form: $y = f(u)$, where u itself denotes a function of x.

For instance $y = (ax+b)^m$ may be treated under the form: $y = u^m$, where $u = ax + b$; or $y = \sin(x^2)$ as $y = \sin(u)$, where $u = x^2$.

In such a case when x increases by Δx it first makes u change; let the amount of the change be Δu, then: $\Delta y = f(u+\Delta u) - f(u)$

therefore $$\frac{\Delta y}{\Delta x} = \frac{f(u+\Delta u)-f(u)}{\Delta x} = \frac{f(u+\Delta u)-f(u)}{\Delta u}\cdot\frac{\Delta u}{\Delta x}.$$

If u be a continuous function of x, and f a continuous function of u, then when Δx becomes zero, Δu also converges to zero (§ 19 d). The limiting value of the first factor on the right is, as its form shows, simply the first derived of the function f taken with respect to u as independent variable, while that of the second is the differential quotient of u with respect to x. Consequently we have:

$$\frac{dy}{dx} = \frac{df(u)}{du} \cdot \frac{du}{dx}.$$

27. Explicit rational algebraic functions.

1) $y = x^m$.

α) m a positive integer. $-\infty < x < +\infty$.

For $x = \pm\infty$, the absolute value of the function also becomes infinitely great; for every finite value of x, the function is continuous and has a differential quotient finite in value; for, multiplying out we find

$$\frac{(x \pm \Delta x)^m - x^m}{\pm \Delta x} = \frac{\pm m x^{m-1} \Delta x + C_2 (\pm \Delta x)^2 + C_3 (\pm \Delta x)^3 \cdots + C_m (\pm \Delta x)^m}{\pm \Delta x}.$$

The coefficients C are finite, depending on x and m, not on Δx; their further determination does not here concern us. Accordingly:

$$\operatorname{Lim} \frac{(x \pm \Delta x)^m - x^m}{\pm \Delta x} = m x^{m-1} + C_2 \operatorname{Lim} (\pm \Delta x) + C_3 \operatorname{Lim} (\pm \Delta x)^2 \cdots$$
$$\cdots + C_m \operatorname{Lim} (\pm \Delta x)^{m-1},$$

therefore for $\Delta x = 0$: $\frac{dy}{dx} = \frac{d(x^m)}{dx} = m x^{m-1}$, in particular: $\frac{d(x)}{dx} = 1$. The value $\operatorname{Lim} (m x^{m-1})$ for $x = \infty$ is to be regarded as differential quotient at the point $x = \infty$.

β) m a negative integer. $-\infty < x < +\infty$.

$$y = x^m = \frac{1}{x^\mu} \quad (\mu = -m > 0).$$

For $x = 0$, the absolute value of the function is infinite. By § 26 Rule 4):

$$\frac{dy}{dx} = \frac{d \frac{1}{(x^\mu)}}{dx} = -\frac{\mu x^{\mu-1}}{x^{2\mu}} = -\frac{\mu}{x^{\mu+1}} = m x^{m-1}.$$

The differential quotient likewise is ∞ for $x = 0$, while for $x = \pm\infty$ both the function y and its differential quotient converge to zero.

$$2) \; y = a_0 x^m + a_1 x^{m-1} + a_2 x^{m-2} \cdots + a_{m-1} x + a_m = \mathsf{A},$$

m integer > 0. For every finite value of x we have by Rules 2) and 3):

$$\frac{dy}{dx} = m a_0 x^{m-1} + (m-1) a_1 x^{m-2} + \cdots a_{m-1} = \frac{d\mathsf{A}}{dx},$$

independent by Rule 1) of the additive constant a_m.

If $y = (ax + b)^m = u^m$, $u = ax + b$, we calculate the differential quotient by Rule 5) without having to expand the binomial:

$$\frac{dy}{dx} = \frac{dy}{du} \cdot \frac{du}{dx} = m u^{m-1} a = m a (ax + b)^{m-1}.$$

3) The fractional rational function:

$$y = \frac{a_0 x^m + a_1 x^{m-1} + \cdots + a_{m-1} x + a_m}{b_0 x^n + b_1 x^{n-1} + \cdots + b_{n-1} x + b_n} = \frac{\mathsf{A}}{\mathsf{B}},$$

m and n integer > 0. For every finite value of x for which B does not vanish we have:

$$\frac{dy}{dx} = \frac{B\frac{dA}{dx} - A\frac{dB}{dx}}{B^2}.$$

Application of differentiation to the deduction of the Binomial Theorem for positive integer exponents:

If we put:

(1) $$(1+x)^m = 1 + C_1 x + C_2 x^2 + C_3 x^3 + \cdots C_m x^m$$

and endeavour to determine each C of this equation, we obtain by differentiation:

(2) $$m(1+x)^{m-1} = C_1 + 2C_2 x + 3C_3 x^2 + \cdots mC_m x^{m-1}.$$

Multiplying both sides by $1+x$ and arranging by powers of x, we get:

(3) $$m(1+x)^m = C_1 + x(2C_2 + C_1) + x^2(3C_3 + 2C_2) + \cdots + x^{m-1}(mC_m + \overline{m-1}C_{m-1}) + x^m mC_m.$$

Comparing the right sides of equations (1) and (3) we find:

$$m = C_1,\ mC_1 = 2C_2 + C_1,\ mC_2 = 3C_3 + 2C_2, \cdots \cdots mC_{m-1} = mC_m + (m-1)C_{m-1},$$

or:

$$C_1 = m,\ C_2 = \frac{m(m-1)}{2},\ C_3 = \frac{m(m-1)(m-2)}{2 \cdot 3}, \cdots$$

$C_k = \frac{m(m-1)(m-2)\cdots(m-k+1)}{\underline{|k}} = m_k$, as we shall frequently denote the Binomial Coefficients, adopting also the symbol $\underline{|k} = 1 \cdot 2 \cdot 3 \cdots k$, in words: factorial k.

28. The Exponential function.

$$y = a^x. \quad (a > 0,\ -\infty < x < +\infty).$$

When $a > 1$, for $x = +\infty$ the function becomes positively infinite, but for $x = -\infty$ it is 0; when $a < 1$, for $x = +\infty$, $y = 0$, for $x = -\infty$ the function becomes positively infinite. For all the values of x, y has a positive value and has a differential quotient. We have

$$\frac{dy}{dx} = \text{Lim}\,\frac{a^{x+\Delta x} - a^x}{\Delta x} = a^x\,\text{Lim}\,\frac{a^{\Delta x} - 1}{\Delta x} \qquad (\Delta x \lessgtr 0),$$

and it has to be shown that the multiplier of a^x approximates to a determinate finite limiting value when Δx converges to zero. It fulfils the condition that its numerator and denominator have zero as limit. To facilitate calculation let us put:

$$a^{\Delta x} - 1 = \delta,\ \Delta x = {}^a\log(1+\delta),$$

then we have $\frac{a^{\Delta x} - 1}{\Delta x} = \frac{\delta}{{}^a\log(1+\delta)} = \frac{1}{{}^a\log\left\{(1+\delta)^{\frac{1}{\delta}}\right\}}$;

δ is positively arbitrarily small when either $a > 1$ and $\Delta x > 0$, or $a < 1$ and $\Delta x < 0$, in other cases negatively arbitrarily small. Writing $\frac{1}{\delta} = m$ the expression becomes $= \dfrac{1}{{}^a\log\left\{\left(1+\frac{1}{m}\right)^m\right\}}$. Now we have to show, that $\left(1+\frac{1}{m}\right)^m$ has a determinate finite limiting value, when m passes through the continuous series of numbers or through any discontinuous series of numbers whose limiting value exceeds any finite amount. At the same time we have to conduct the investigation in a way which shall present a convenient method of calculating this value.

Let us first make m pass through the series of positive integers, then by the binomial theorem as just proved, we have always:

$$\left(1+\frac{1}{m}\right)^m = 1 + m\cdot\frac{1}{m} + \frac{m(m-1)}{1\cdot 2}\left(\frac{1}{m}\right)^2 + \frac{m(m-1)(m-2)}{1\cdot 2\cdot 3}\left(\frac{1}{m}\right)^3 + \cdots + R$$

$$= 1 + 1 + \left(1-\frac{1}{m}\right)\frac{1}{1\cdot 2} + \left(1-\frac{1}{m}\right)\left(1-\frac{2}{m}\right)\frac{1}{1\cdot 2\cdot 3} + \cdots + R$$

$$= \Sigma_n + R.$$

Where by Σ_n we signify the sum of the first n terms, by R the sum of the last terms reckoned on from the $(n+1)$, so that therefore:

$$R = \left(1-\frac{1}{m}\right)\left(1-\frac{2}{m}\right)\cdots\left(1-\frac{n-1}{m}\right)\frac{1}{\underline{|n}}\cdot S$$

S embracing the $m+1-n$ terms

$$S = 1 + \left(1-\frac{n}{m}\right)\frac{1}{n+1} + \left(1-\frac{n}{m}\right)\left(1-\frac{n+1}{m}\right)\frac{1}{(n+1)(n+2)} + \cdots$$

$$+ \left(1-\frac{n}{m}\right)\cdots\left(1-\frac{m-1}{m}\right)\frac{1}{(n+1)\cdots m}.$$

If m be a considerable number, we can approximate arbitrarily to the value of $\left(1+\frac{1}{m}\right)^m$ by merely summing the first n terms of the series, n being a number much smaller than m.

For, the differences $1-\frac{1}{m}$, $1-\frac{2}{m}$, etc. in the expressions for R and S being positive proper fractions, the remainder R is certainly smaller than the value obtained when we put in it unity for each of these differences; and *à fortiori*

$$R < \frac{1}{\underline{|n}}\left(1 + \frac{1}{n+1} + \frac{1}{(n+1)^2} + \cdots + \frac{1}{(n+1)^{m-n}}\right) =$$

$$= \frac{1}{\underline{|n}}\cdot\frac{1-\left(\frac{1}{n+1}\right)^{m+1-n}}{1-\frac{1}{n+1}};$$

still more is: $\quad R < \dfrac{1}{\underline{|n}}\cdot\dfrac{1}{1-\frac{1}{n+1}} = \dfrac{1}{\underline{|n}}\cdot\dfrac{n+1}{n}.$

Therefore we have for each positive integer m:

$$\left(1+\frac{1}{m}\right)^m = \Sigma_n + \left(< \frac{1}{\underline{|n}} \frac{n+1}{n}\right), \quad (n < m+1).$$

Now retaining the value of n, and increasing m continually, the fractions $\frac{1}{m}, \frac{2}{m}, \ldots \frac{n-2}{m}$ which occur in Σ_n will approach always nearer to zero, i. e. Σ_n approximates to the limiting value

$$1 + \frac{1}{1} + \frac{1}{\underline{|2}} + \frac{1}{\underline{|3}} + \cdots + \frac{1}{\underline{|n-1}}.$$

The error incurred by equating to this sum the value for $m = \infty$ of $\left(1+\frac{1}{m}\right)^m$ is positive and less than $\frac{1}{\underline{|n}} \cdot \frac{n+1}{n}$, therefore we can choose n so as to make it arbitrarily small. We thus obtain for m arbitrarily great:

$$\left(1+\frac{1}{m}\right)^m = 1 + \frac{1}{1} + \frac{1}{\underline{|2}} + \frac{1}{\underline{|3}} + \frac{1}{\underline{|4}} + \frac{1}{\underline{|5}} + \cdots \text{ in inf.*)}$$

that is, the more terms of this sum we add up, the nearer we approximate to a determinate value which is denoted by e; we find

$$e = 2{,}7182818284\ldots$$

The number e is irrational, i. e. is expressed completely neither by a decimal fraction with a finite number of places, nor by a periodic decimal. The proof of this is simple: If we had $e = \frac{a}{b}$, where a and b are integers, we should find by multiplying the series by $\underline{|b}$:

$$a\,\underline{|b-1} = 2\underline{|b} + \frac{\underline{|b}}{\underline{|2}} + \frac{\underline{|b}}{\underline{|3}} + \cdots + 1 + \frac{1}{b+1} + \frac{1}{(b+1)(b+2)} + \cdots$$

or, bringing all integer values on the right over to the left, and then denoting the integer number on the left by G:

$$G = \frac{1}{b+1} + \frac{1}{(b+1)(b+2)} + \cdots < \frac{1}{b+1} + \frac{1}{(b+1)^2} + \cdots$$

This equation is impossible, for the value of the right side is smaller than $\frac{1}{b}$, therefore is a proper fraction.**)

If m be not an integer, but its value is included between the numbers n and $n+1$:

$$n + \alpha = m = n + 1 - \beta,$$

then

$$1 + \frac{1}{n} > 1 + \frac{1}{m} > 1 + \frac{1}{n+1},$$

therefore

---

*) Euler: Introductio in analysin infinit. I § 115.

**) Hermite has proved further, that the number e cannot be a root of an algebraic equation of any degree with rational coefficients. Sur la fonction exponentielle. Paris 1874.

$$\left(1+\frac{1}{n}\right)^m > \left(1+\frac{1}{m}\right)^m > \left(1+\frac{1}{n+1}\right)^m,$$

or

$$\left(1+\frac{1}{n}\right)^n \cdot \left(1+\frac{1}{n}\right)^\alpha > \left(1+\frac{1}{m}\right)^m > \left(1+\frac{1}{n+1}\right)^{n+1} : \left(1+\frac{1}{n+1}\right)^\beta.$$

Here both $\left(1+\frac{1}{n}\right)^n$ and $\left(1+\frac{1}{n+1}\right)^{n+1}$ converge to the value e, while as n increases arbitrarily $\left(1+\frac{1}{n}\right)^\alpha$ and $\left(1+\frac{1}{n+1}\right)^\beta$ have unity for their limit; thus superior and inferior limit approximate to the value e, therefore for the present m we also have

$$\text{Lim}\left(1+\frac{1}{m}\right)^m = e.$$

Lastly if m is negative, let us put $m = -\mu$; then

$$\left(1+\frac{1}{m}\right)^m = \left(1-\frac{1}{\mu}\right)^{-\mu} = \left(\frac{\mu}{\mu-1}\right)^\mu = \left(1+\frac{1}{\mu-1}\right)^\mu$$
$$= \left(1+\frac{1}{\mu-1}\right)^{\mu-1} \cdot \left(1+\frac{1}{\mu-1}\right).$$

Therefore

$$\text{Lim}\left(1+\frac{1}{m}\right)^m = \text{Lim}\left(1+\frac{1}{\mu-1}\right)^{\mu-1} \cdot \text{Lim}\left(1+\frac{1}{\mu-1}\right) = e \cdot 1.$$

Accordingly in whatever way Δx may converge to zero we have

$$\frac{dy}{dx} = a^x \cdot \text{Lim}\,\frac{a^{\Delta x}-1}{\Delta x} = a^x \cdot \text{Lim}\,\frac{1}{{}^a\log\left\{\left(1+\frac{1}{m}\right)^m\right\}}$$
$$= a^x \cdot \frac{1}{{}^a\log e} = a^x \cdot {}^e\log a.$$

Hence we see that the exponential function whose base is e, has the property of reproducing itself unaltered by differentiation; we have

$$\frac{d\,(e^x)}{dx} = e^x, \text{ since } {}^e\log e = 1.$$

The irrational number e is called the base of the natural system of logarithms; the logarithm relative to this base is briefly denoted by l.

29. The trigonometric functions.

$$\alpha)\ y = \sin x. \quad \beta)\ y = \cos x.$$

Although we have as yet defined these functions only geometrically for all finite values of x, the propositions in § 12 enable us to assign the derived function of each:

$$\text{for } \alpha)\ \frac{\Delta y}{\Delta x} = \frac{\sin(x+\Delta x) - \sin x}{\Delta x} = \frac{2\sin\frac{1}{2}\Delta x\cos(x+\frac{1}{2}\Delta x)}{\Delta x}$$

$$\frac{dy}{dx} = \text{Lim}\,\frac{\sin(\frac{1}{2}\Delta x)}{\frac{1}{2}\Delta x}\ \text{Lim}\cos(x+\tfrac{1}{2}\Delta x) = \cos x.$$

for β) $\frac{\Delta y}{\Delta x} = \frac{\cos(x+\Delta x) - \cos x}{\Delta x} = -\frac{2\sin\frac{1}{2}\Delta x \sin(x+\frac{1}{2}\Delta x)}{\Delta x}$

$$\frac{dy}{dx} = -\operatorname{Lim}\frac{\sin(\frac{1}{2}\Delta x)}{\frac{1}{2}\Delta x}\operatorname{Lim}\sin(x+\tfrac{1}{2}\Delta x) = -\sin x.$$

γ) $y = \operatorname{tang} x = \frac{\sin x}{\cos x}$, $\frac{dy}{dx} = \frac{\cos x \frac{d\sin x}{dx} - \sin x \frac{d\cos x}{dx}}{(\cos x)^2}$

$$\frac{dy}{dx} = \frac{1}{(\cos x)^2}.$$

δ) $y = \cot x = \frac{\cos x}{\sin x}$, $\frac{dy}{dx} = \frac{\sin x \frac{d\cos x}{dx} - \cos x \frac{d\sin x}{dx}}{(\sin x)^2}$

$$\frac{dy}{dx} = -\frac{1}{(\sin x)^2}.$$

All these four as well as their derived functions are quite indeterminate for infinite values of x. For all arguments for which the last two functions are infinite their derived functions are also infinite.

30. Inverse functions: the logarithm and the circular functions.

General rule: If $x = f(y)$, and $\frac{dx}{dy} = f'(y)$ is calculated, then if $y = \varphi(x)$ express the inverse function (§ 13), it follows that $\frac{dy}{dx} = \varphi'(x) = \frac{1}{f'(y)}$. For we have: $\frac{dy}{dx} = \operatorname{Lim}\frac{\Delta y}{\Delta x} = \operatorname{Lim}\left(1 : \frac{\Delta x}{\Delta y}\right)$. We calculate therefore $\varphi'(x)$ for any value of x, by substituting in the expression $\frac{1}{f'(y)}$ the corresponding value of y; points at which $f'(y) = 0$, demand special attention.

1. $y = {}^a\log x$. Inverse function: $x = a^y$. $(a > 0)$.

$$\frac{dy}{dx} = \frac{1}{a^y\, l\, a} = \frac{1}{x}\cdot\frac{1}{l\, a} = \frac{1}{x}\cdot{}^a\log e.$$

2. α) $y = \sin^{-1}x$. Inverse function: $x = \sin y$.

$$\frac{dy}{dx} = \frac{1}{\cos y} = \frac{1}{\sqrt{1-x^2}}.$$

The square root is positive because by our convention (§ 13)

$$-\frac{\pi}{2} \leqq y \leqq +\frac{\pi}{2} \text{ therefore } \cos y \geqq 0.$$

β) $y = \cos^{-1}x$. Inverse function: $x = \cos y$.

$$\frac{dy}{dx} = -\frac{1}{\sin y} = \frac{1}{-\sqrt{1-x^2}}.$$

The square root is positive because

$$0 \leqq y < \pi \text{ therefore } \sin y \geqq 0.$$

γ) $y = \tan^{-1} x$. Inverse function: $x = \tan y$.

$$\frac{dy}{dx} = (\cos y)^2 = \frac{1}{1+x^2}.$$

δ) $y = \cot^{-1} x$. Inverse function $x = \cot y$.

$$\frac{dy}{dx} = -(\sin y)^2 = -\frac{1}{1+x^2}.$$

In the first two of these functions the values ± 1, at which the definition of the functions ceases, form special points; in the last two, x goes from $-\infty$ to $+\infty$ and the functions as well as their differential quotients are finite even at these limits.

The logarithm and the circular functions are transcendental; but their differential quotients are algebraic.

31. But lastly we can also differentiate the explicit irrational function: $y = x^m$, where m means any real number, but x is positive, and the root is always taken positively. For, taking the natural logarithm of both sides of the equation $y = x^m$ we have $l(y) = m\,l(x)$. If we differentiate this equation, remembering that y on the left side is a function of x, it follows by Rule 5) § 26 that

$$\frac{1}{y}\frac{dy}{dx} = \frac{m}{x}, \text{ therefore: } \frac{dy}{dx} = \frac{m}{x}y = mx^{m-1}.$$

We have accordingly for every value of m the equation:

$$\frac{dy}{dx} = \frac{d(x^m)}{dx} = mx^{m-1}, \quad (y > 0).$$

It is to be noticed, that when $0 < m < 1$ the function is finite for $x = 0$, and infinitely great for $x = +\infty$, whereas its differential quotient is infinitely great at the former point and at the latter finite and equal to zero. For the Implicit Algebraic Function see Chap. X.

For all functions dealt with in this Chapter it is indifferent whether Δx is chosen positive or negative; i. e. all these functions have at each point the value of the progressive differential quotient equal to that of the regressive; each has a derived function or derivate.

Seventh Chapter.

Successive differentiation of explicit functions. Different orders of infinitely small quantities.

32. The first derivate of a function or its first differential quotient, as the calculations of last Chapter show, is itself again a function of the variable. For the linear function $y = ax + b$, the progressive and regressive differential quotient $\frac{dy}{dx} = a$ is constant, and this is the only continuous function with a constant differential quotient, as we shall prove in Chap. I of the Integral Calculus. Accordingly under similar hypotheses further functions can be derived by the same rules from each new derived function and their calculation always results from what precedes. Let $y = f(x)$ denote the original function, further let

$$\frac{dy}{dx} = \text{Lim}\,\frac{f(x + \Delta x) - f(x)}{\Delta x} = f'(x),$$

uniquely and determinately for each value of x, denote its first derivate, then provided the function $f'(x)$ is continuous and $\frac{f'(x+\Delta x) - f'(x)}{\Delta x}$ approximates to a determinate limiting value for $\Delta x = 0$, we obtain,

the second derivate: $f''(x) = \text{Lim}\,\frac{f'(x + \Delta x) - f'(x)}{\Delta x}$ for $\Delta x = 0$,

the third derivate: $f'''(x) = \text{Lim}\,\frac{f''(x + \Delta x) - f''(x)}{\Delta x}$ for $\Delta x = 0$, etc..

These higher derivates as well as the first can be immediately defined by means of the original function. For we have for $\Delta x \gtrless 0$

$$f'(x) = \text{Lim}\,\frac{f(x + \Delta x) - f(x)}{\Delta x},$$

$$f'(x + h) = \text{Lim}\,\frac{f(x + h + \Delta x) - f(x + h)}{\Delta x},$$

therefore:

$$f''(x) = \text{Lim}_{h=0}\,\text{Lim}_{\Delta x=0}\,\frac{f(x + h + \Delta x) - f(x + h) - f(x + \Delta x) + f(x)}{h\Delta x}.$$

But this double limit can be determined more simply. We saw § 22 by the Theorem of the Mean Value, that for a continuous function $\varphi(x)$, the quotient of differences $\frac{\varphi(x + \Delta x) - \varphi(x)}{\Delta x}$ can be always put equal to the value of the derivate formed for a point $x + \Theta\Delta x$ within the interval from x to $x + \Delta x$, where Θ denotes a number between 0 and 1, on the hypothesis that there exists for the entire interval a derived function (progressive identical with regressive). This being the case with the continuous function $\varphi(x) = f(x+h) - f(x)$, we can put:

$$\frac{f(x+h+\Delta x)-f(x+\Delta x)-f(x+h)+f(x)}{\Delta x} = f'(x+h+\Theta\Delta x)-f'(x+\Theta\Delta x),$$

thus

$$\frac{f(x+h+\Delta x)-f(x+\Delta x)-f(x+h)+f(x)}{h\Delta x} = \frac{f'(x+h+\Theta\Delta x)-f'(x+\Theta\Delta x)}{h}.$$

If we first make Δx, and then h converge to zero, we obtain the value $f''(x)$. Now we are going to show, that h may always be assumed equal to Δx in the quotient on the left, and thus h and Δx be made to vanish simultaneously. From substituting $h = \Delta x$, it follows that:

$$\frac{f(x+2\Delta x)-2f(x+\Delta x)+f(x)}{(\Delta x)^2} = \frac{f'(x+\Delta x+\Theta\Delta x)-f'(x+\Theta\Delta x)}{\Delta x}.$$

Let us give the right side the form:

$$\frac{f'(x+\Delta x+\Theta\Delta x)-f'(x)}{\Delta x(1+\Theta)}(1+\Theta) - \frac{f'(x+\Theta\Delta x)-f'(x)}{\Theta\Delta x}\cdot\Theta.$$

Now we can choose Δx so small that

$$\frac{f'(x+\Delta x(1+\Theta))-f'(x)}{\Delta x(1+\Theta)} \quad\text{and}\quad \frac{f'(x+\Theta\Delta x)-f'(x)}{\Theta\Delta x}$$

shall each differ from the value $f''(x)$ by less than the arbitrarily small quantity δ. Accordingly:

$$\frac{f(x+2\Delta x)-2f(x+\Delta x)+f(x)}{(\Delta x)^2} = f''(x) + (< 3\delta\Theta),$$

hence the new definition is: $f''(x) = \lim\limits_{\Delta x=0} \dfrac{f(x+2\Delta x)-2f(x+\Delta x)+f(x)}{(\Delta x)^2}$.

In like manner, provided $f''(x)$ remains the same when taken progressively and regressively and has a determinate derivate $f'''(x)$, we obtain for $f'''(x)$ the definition by means of the original function:

$$f'''(x) = \lim_{\Delta x=0} \frac{f(x+3\Delta x)-3f(x+2\Delta x)+3f(x+\Delta x)-f(x)}{(\Delta x)^3},$$

because: $f'''(x) =$

$$\lim_{h=0}\lim_{\Delta x=0} \frac{\dfrac{f(x+h+2\Delta x)-2f(x+h+\Delta x)+f(x+h)}{\Delta x^2} - \dfrac{f(x+2\Delta x)-2f(x+\Delta x)+f(x)}{\Delta x^2}}{h}.$$

For, by the Theorem of the Mean Value this last quotient is equal to

$$\frac{f'(x+2\Delta x+\Theta h)-2f'(x+\Delta x+\Theta h)+f'(x+\Theta h)}{(\Delta x)^2}$$

and also equal to

$$\frac{f''(x+\Delta x+\eta\Delta x+\Theta h)-f''(x+\eta\Delta x+\Theta h)}{\Delta x}, \quad (0 < \eta < 1).$$

We perceive as before, that h may be assumed $= \Delta x$ and then both made simultaneously converge to zero.

There is no difficulty in establishing the general equation of this kind, by showing that when it holds for n it is also true for $n + 1$.

33. These new expressions are no doubt less suited for calculating the higher differential quotients than those first formed; but they exhibit them to us as limiting values of higher quotients of differences, which is of importance for the theory. Euler (1707—1783) in his work: Institutiones calculi differentialis, Petrop. 1755, gave the following convenient exposition of this. If we denote the values of the function $y = f(x)$ which belong to the arguments

$x,\ x + \Delta x,\ x + 2\Delta x, \ldots x + n\Delta x$, respectively by

$y,\quad y_1,\quad y_2,\quad \ldots\quad y_n,$

we get the series of first differences:

$$y_1 - y = \Delta y,\ y_2 - y_1 = \Delta y_1,\ y_3 - y_2 = \Delta y_2, \ldots y_n - y_{n-1} = \Delta y_{n-1}.$$

From these we form the series of second differences:

$$\Delta y_1 - \Delta y = \Delta^2 y,\ \Delta y_2 - \Delta y_1 = \Delta^2 y_1,\ \Delta y_3 - \Delta y_2 = \Delta^2 y_2, \ldots$$
$$\Delta y_{n-1} - \Delta y_{n-2} = \Delta^2 y_{n-2},$$

and so of third differences:

$$\Delta^2 y_1 - \Delta^2 y = \Delta^3 y,\ \ \Delta^2 y_2 - \Delta^2 y_1 = \Delta^3 y_1,\ \ \Delta^2 y_3 - \Delta^2 y_2 = \Delta^3 y_2, \ldots$$
$$\Delta^2 y_{n-2} - \Delta^2 y_{n-3} = \Delta^3 y_{n-3},$$

on to the nth difference:

$$\Delta^{n-1} y_1 - \Delta^{n-1} y = \Delta^n y.$$

If we propose to express the higher differences by the original values of the function we find:

$$\Delta y = y_1 - y,$$
$$\Delta^2 y = (y_2 - y_1) - (y_1 - y) = y_2 - 2y_1 + y = f(x + 2\Delta x) - 2f(x + \Delta x) + f(x),$$
$$\Delta^3 y = \{(y_3 - y_2) - (y_2 - y_1)\} - \{(y_2 - y_1) - (y_1 - y)\} = y_3 - 3y_2 + 3y_1 - y$$
$$= f(x + 3\Delta x) - 3f(x + 2\Delta x) + 3f(x + \Delta x) - f(x), \ldots$$

whence results, that $f''(x) = \text{Lim}\,\frac{\Delta^2 y}{\Delta x^2}$, $f'''(x) = \text{Lim}\,\frac{\Delta^3 y}{\Delta x^3}$, etc.. This accounts for the notation for the higher differential quotients

$$\frac{d^2 y}{dx^2},\ \frac{d^3 y}{dx^3}, \text{ etc.; in general: } \frac{d^n y}{dx^n} = \text{Lim}\,\frac{\Delta^n y}{\Delta x^n}.$$

We can also form higher differences of the values of the independent variable, but they are found to vanish. For, from the values

$$x_1 = x + \Delta x,\ x_2 = x + 2\Delta x,\ x_3 = x + 3\Delta x, \ldots x_n = x + n\Delta x$$

we find:

$$x_1 - x = \Delta x,\ x_2 - x_1 = \Delta x_1,\ x_3 - x_2 = \Delta x_2, \ldots x_n - x_{n-1} = \Delta x_{n-1},$$

therefore

$$\Delta x_1 - \Delta x = \Delta^2 x = 0, \ \Delta^2 x_1 - \Delta^2 x = \Delta^3 x = 0$$
$$\Delta x_2 - \Delta x_1 = \Delta^2 x_1 = 0, \ \Delta^2 x_2 - \Delta^2 x_1 = \Delta^3 x_1 = 0$$
$$\Delta x_3 - \Delta x_2 = \Delta^2 x_2 = 0, \ \ldots \ldots \quad \ldots$$
$$\ldots \ldots \ldots \ldots$$

Thus the higher differences of the independent variable vanish, since its values are supposed to increase by equal amounts.

The same is the case, when the function y increases in proportion to x, that is, when $y = ax + b$.

Since according to the above method of determination the differential quotients

$$\frac{dy}{dx} = f'(x), \quad \frac{d^2y}{dx^2} = f''(x), \quad \frac{d^3y}{dx^3} = f'''(x) \ \text{ etc.}$$

have the signification of actual fractions, we can pass over from them also to the equations between the differentials:

$$dy = f'(x) . dx, \ d^2y = f''(x) . dx^2, \ d^3y = f'''(x) . dx^3, \ldots d^n y = f^n(x) . dx^n.$$

Of course we have now on each side of such an equation a vanishing quantity, so that it appears not to contain anything more than the self-evident identity $0 = 0$. Nevertheless it has a determinate content when we recollect how it originated. For it then asserts: the nth difference $\Delta^n y$ at the point x is more nearly equal to the product of Δx^n by the determinate value $f^n(x)$, the smaller Δx is chosen; so that the limiting value of the quotient $\frac{\Delta^n y}{\Delta x^n}$, which we have denoted by $\frac{d^n y}{dx^n}$ is equal to $f^n(x)$. *Thus an equation between infinitely small quantities has a determinate content, if it can be interpreted as a relation between the limiting values of continuous variables.*

Now it is further to be remarked, that in the above equations the differential dx occurs in increasing powers, so that we are enabled to distinguish infinitely small quantities of different orders. If dx be called infinitely small of the first order, then dx^2 is infinitely small of the second, dx^3 of the third, dx^n of the nth order. The ratio of two infinitely smalls of the nth order and of the mth order ($n > m$) is itself infinitely small of the $(n - m)^{\text{th}}$ order: $dx^n : dx^m = dx^{n-m}$. The numerator may be said to converge to zero much more rapidly than the denominator; if $n = m$ then the quotient is finite, equal to 1.

34. This measure of becoming infinitely small can be stated generally: *Two quantities are infinitely small of the same order when their quotient retains a finite value.* The derived functions

$$f'(x), \ f''(x), \ f'''(x) \ \text{ etc.}$$

have in general for any x finite values; consequently the differentials

dy, d^2y, d^3y, etc., in like manner as the corresponding powers of dx, are infinitely small respectively of the orders 1, 2, 3, etc. If we wish to investigate the order in which a function, which is known to vanish for $x = a$, is infinitely small at this point, we have to form the quotient $\frac{f(x)}{(x-a)^r}$ and to determine for what value of r it remains finite. We shall only become possessed of a general method for calculating such $\frac{0}{0}$ quotients by the investigations of next Chapter. It may be that the order of becoming infinitely small has to be expressed by a fractional or even irrational number, as, to cite only the simplest, in case of $f(x) = x^n$ at the point $x = 0$, where n is any positive number whatever. It is possible even, though we shall only mention it here, that no number can be found, but only a limit for r, below which the quotient is zero, above which it is infinitely great. The simplest example of this kind is $f(x) = x^a . \log(x)$, $(a > 0)$, in which a forms such a boundary between the values r.*)

In the applications of the differential calculus to problems of Geometry and Mechanics two courses always present themselves: either we start from equations between quotients of differences and pass over from these to differential quotients; or we start from equations between differences and pass from these to differentials. The latter frequently corresponds better to the immediate intuition. In this case we can from the outset facilitate calculation by omitting in the equation between the quantities still conceived as finite, all the terms which in the transition to differentials, become infinitely small of higher order than some term which occurs in the same sum with them. If for instance $y = x^n$ where n is a positive integer, then

$$\Delta y = (x + \Delta x)^n - x^n = n x^{n-1} \Delta x + n_2 x^{n-2} \Delta x^2 + n_3 x^{n-3} \Delta x^3 + \cdots \Delta x^n;$$

here as all terms on the right side become infinitely small of higher order than the first term, the equation $\Delta y = n x^{n-1} \Delta x$, though not exact for finite values, yet expresses for infinitely small ones the correct value $dy = nx^{n-1} dx$. In elementary Stereometry an application of this remark occurs, in proving the theorem for the cubature of a body bounded by planes: that the volume of a thin slice bounded by parallel planes can be calculated as that of a prism, provided the number of the parallel planes becomes infinite. In fact, the volume of a slice differs from that of a prism with equal base, by a quantity which is infinitely small of the second order when the volume of the prism comes to be considered infinitely small of the first order, on arbitrarily continued subdivision diminishing their

*) Cauchy, Sur les diverses ordres des quantités infiniment petites. Exercices de mathématiques. Tome 1.

thicknesses; it is easy to prove this from the simplest case of the three-sided pyramid by directly calculating the difference as a function of the thickness. Thence follows that the limiting value of the sums of the prisms is identical with the limiting value of the sums of the truncated pyramids, and this facilitates calculation from the outset.

35. Forming the higher differential quotients of the explicit functions treated in last Chapter we find:

$$\text{I. } y = x^m,\ \frac{dy}{dx} = m x^{m-1},\quad \frac{d^2y}{dx^2} = m(m-1)x^{m-2}\ \ldots$$

$$\frac{d^n y}{dx^n} = m(m-1)(m-2)\ldots(m-n+1)x^{m-n}.$$

When m signifies a positive integer, $\frac{d^m y}{dx^m}$ is constant.

$$\text{II. 1) } y = a^x,\ \frac{dy}{dx} = a^x\, la,\ \frac{d^2y}{dx^2} = a^x(la)^2 \ldots \frac{d^n y}{dx^n} = a^x(la)^n;\ (a>0.)$$

In particular for $y = e^x$, $\frac{d^n y}{dx^n}$ is $= e^x$.

$$\begin{aligned} 2)\ y = \sin x,\ \frac{dy}{dx} &= \quad \cos x = \sin(x + \tfrac{1}{2}\pi), \\ \frac{d^2y}{dx^2} &= -\sin x = \cos(x + \tfrac{1}{2}\pi) = \sin(x+\pi), \\ \frac{d^3y}{dx^3} &= -\cos x = \cos(x+\pi) = \sin(x + \tfrac{3}{2}\pi), \\ &\ldots\ldots\ldots\ldots\ldots\ldots \\ \frac{d^n y}{dx^n} &= \quad \sin(x + \tfrac{1}{2}n\pi), \\ \frac{d^{n+1}y}{dx^{n+1}} &= \quad \cos(x + \tfrac{1}{2}n\pi) = \sin(x + \tfrac{1}{2}(n+1)\pi). \end{aligned}$$

$$3)\ y = \cos x,\ \frac{dy}{dx} = -\sin x = \cos(x + \tfrac{1}{2}\pi), \ldots \frac{d^n y}{dx^n} = \cos(x + \tfrac{1}{2}n\pi).$$

If y be a sum of functions:

$$y = f_1(x) + f_2(x) + f_3(x) \ldots + f_m(x)$$

then

$$\frac{d^n y}{dx^n} = \frac{d^n f_1(x)}{dx^n} + \frac{d^n f_2(x)}{dx^n} + \frac{d^n f_3(x)}{dx^n} + \ldots + \frac{d^n f_m(x)}{dx^n};$$

for example:

$$y = \frac{1}{1-x^2} = \frac{1}{2}\left(\frac{1}{1-x} + \frac{1}{1+x}\right),\ \therefore\ \frac{d^n y}{dx^n} = \frac{1}{2}\left\{\frac{d^n \frac{1}{1-x}}{dx^n} + \frac{d^n \frac{1}{1+x}}{dx^n}\right\}$$

or explicitly

$$\frac{d^n y}{dx^n} = \frac{\lfloor n}{2}\left[\frac{1}{(1-x)^{n+1}} + \frac{(-1)^n}{(1+x)^{n+1}}\right].$$

If y be the product of two functions:

$$y = \varphi(x) \cdot \psi(x) = \varphi \cdot \psi \quad \text{then} \quad \frac{dy}{dx} = \varphi' \cdot \psi + \varphi \cdot \psi'$$

$$\frac{d^2y}{dx^2} = \varphi''\psi + 2\varphi'\psi' + \varphi\psi'', \quad \frac{d^3y}{dx^3} = \varphi'''\psi + 3\varphi''\psi' + 3\varphi'\psi'' + \varphi\psi'''.$$

In general, if we denote $\frac{d^k\varphi}{dx^k}$ by $\varphi^{(k)}$ and the binominal coefficients (§ 27) by n_k, ($n_0 = 1$), we have the Rule:

$$\frac{d^n y}{dx^n} = \varphi^{(n)}\psi + n_1\varphi^{(n-1)}\psi^{(1)} + n_2\varphi^{(n-2)}\psi^{(2)} + \cdots$$
$$+ n_k\varphi^{(n-k)}\psi^{(k)} + \cdots \varphi\psi^{(n)} = \sum_{k=0}^{k=n} n_k\varphi^{(n-k)}\psi^{(k)}.$$

For, if we assume this formula proved for any value of n, differentiating,

$$\frac{d^{n+1}y}{dx^{n+1}} = \sum_{k=0}^{k=n} n_k\left[\varphi^{(n+1-k)}\psi^{(k)} + \varphi^{(n-k)}\psi^{(k+1)}\right]$$
$$= \sum_{k=0}^{k=n} n_k\varphi^{(n+1-k)}\psi^{(k)} + \sum_{k=0}^{k=n} n_k\varphi^{(n-k)}\psi^{(k+1)}$$

Writing apart the first term of the first sum and the last of the second:

$$\frac{d^{n+1}y}{dx^{n+1}} = \varphi^{(n+1)}\psi + \varphi\psi^{(n+1)} + \sum_{k=1}^{k=n} n_k\varphi^{(n+1-k)}\psi^{(k)} + \sum_{k=0}^{k=n-1} n_k\varphi^{(n-k)}\psi^{(k+1)},$$

we can evidently combine each pair of terms of the two sums so that

$$\frac{d^{n+1}y}{dx^{n+1}} = \varphi^{(n+1)}\psi + (n_1 + n_0)\varphi^{(n)}\psi^{(1)} + (n_2 + n_1)\varphi^{(n-1)}\psi^{(2)} + \cdots$$
$$+ (n_k + n_{k-1})\varphi^{(n+1-k)}\psi^{(k)} + \cdots + \varphi\psi^{(n+1)}.$$

But it is a property of the binominal coefficients that

$$n_k + n_{k-1} = (n+1)_k,$$

thus this sum can be written according to the above notation:

$$\frac{d^{n+1}y}{dx^{n+1}} = \sum_{k=0}^{k=n+1} (n+1)_k\varphi^{(n+1-k)}\psi^{(k)},$$

which proves that if the assumed law holds for any n it remains valid for the following number, and therefore for all that follow; but its validity is directly seen for $n = 2$ and $n = 3$.

According to this Rule we obtain the following exposition for:

4) $y = \tan x$. If we put $y.\cos x = \sin x$, then, $y^{(n)}$ denoting $\frac{d^n y}{dx^n}$,

$$y'\cos x + y\cos(x + \tfrac{1}{2}\pi) = \sin(x + \tfrac{1}{2}\pi)$$
$$y''\cos x + 2y'\cos(x + \tfrac{1}{2}\pi) + y\cos(x + \tfrac{2}{2}\pi) = \sin(x + \tfrac{2}{2}\pi)$$
$$y'''\cos x + 3y''\cos(x + \tfrac{1}{2}\pi) + 3y'\cos(x + \tfrac{2}{2}\pi) + y\cos(x + \tfrac{3}{2}\pi)$$
$$= \sin(x + \tfrac{3}{2}\pi)$$

. . . .

$$y^{(n)}\cos x + n_1 y^{(n-1)}\cos(x + \tfrac{1}{2}\pi) + n_2 y^{(n-2)}\cos(x + \tfrac{2}{2}\pi) + \cdots$$
$$+ n_k y^{(n-k)}\cos(x + \tfrac{1}{2}k\pi) + \cdots y\cos(x + \tfrac{1}{2}n\pi) = \sin(x + \tfrac{1}{2}n\pi).$$

The calculation of the nth derived function $y^{(n)}$ from the last equation requires that of all preceding derivates from the preceding equations; the formula established for $y^{(n)}$ is called on this account a **recurring** formula. All the derived functions are finite, except where $\cos x = 0$.

5) $$y = \cot x, \quad y \sin x = \cos x.$$

$$y^{(n)} \sin x + n_1 y^{(n-1)} \sin(x + \tfrac{1}{2}\pi) + n_2 y^{(n-2)} \sin(x + \tfrac{2}{2}\pi) + \cdots$$
$$+ n_k y^{(n-k)} \sin(x + \tfrac{1}{2}k\pi) \cdots + y \sin(x + \tfrac{1}{2}n\pi) = \cos(x + \tfrac{1}{2}n\pi).$$

III. 1) $y = {}^a\log x, \ \frac{dy}{dx} = \frac{1}{x}\,{}^a\log e, \ \frac{d^2y}{dx^2} = -\frac{1}{x^2}\,{}^a\log e, \ \frac{d^3y}{dx^3} = \frac{1 \cdot 2}{x^3}\,{}^a\log e,$

$$\frac{d^ny}{dx^n} = (-1)^{(n-1)} \frac{\lfloor n-1}{x^n}\,{}^a\log e. \quad (a > 0.)$$

2) $$y = \sin^{-1} x, \quad (-1 \leqq x \leqq +1, \ -\tfrac{1}{2}\pi \leqq y \leqq +\tfrac{1}{2}\pi).$$

From the equation: $y' = \frac{1}{\sqrt{1-x^2}}$ i.e. $y'\sqrt{1-x^2} = 1$, follows on further differentiation, the quantity on the right being constant:

$$y''\sqrt{1-x^2} - \frac{y'x}{\sqrt{1-x^2}} = 0 \quad \text{or} \quad y''(1-x^2) - y'x = 0.$$

Differentiating this equation n times by the Rule of the Product, we find

$$y^{(n+2)}(1-x^2) = (2n+1) y^{(n+1)} x + n^2 y^{(n)}.$$

This is likewise a recurring formula for the calculation of all the derived functions; they become infinite for the arguments $x^2 = 1$.

3) $$y = \cos^{-1} x = \tfrac{1}{2}\pi - \sin^{-1} x. \qquad \frac{d^ny}{dx^n} = -\frac{d^n \sin^{-1} x}{dx^n}.$$

4) $$y = \tan^{-1} x, \qquad (-\tfrac{1}{2}\pi \leqq y \leqq +\tfrac{1}{2}\pi).$$

From the equation $\frac{dy}{dx} = \frac{1}{1+x^2}$, or $y'(1+x^2) = 1$, follows:

$$y^{(n+1)}(1+x^2) + 2n_1 y^{(n)} x + 2n_2 y^{(n-1)} = 0, \quad \text{or:}$$
$$y^{(n+1)}(1+x^2) = -2nxy^{(n)} - n(n-1)y^{(n-1)}; \quad \text{and lastly}$$

5) for $y = \cot^{-1} x = \tfrac{1}{2}\pi - \tan^{-1} x$, we have $\frac{d^ny}{dx^n} = -\frac{d^n \tan^{-1} x}{dx^n}$.

36. For circular functions we have thus found only recurring formulas; such formulas we shall obtain for every compound function by applying the Rule of the Product. But we can also propose the problem: to calculate the nth derivate by an **independent** formula not requiring first the calculation of all preceding derived functions.*)

As an example of obtaining an independent expression, we can treat $y = \tan^{-1} x$ in the following particularly simple manner.

---

*) The propositions bearing on this are discussed in detail in Schlömilch: Compendium der höheren Analysis, Vol. II, and Hoppe: Theorie der höheren Differentialquotienten.

We have: $y' = \frac{1}{1+x^2} = (\cos y)^2 = \cos y \cdot \sin(y + \frac{1}{2}\pi)$, therefore:

$$y'' = y'\{-\sin y \sin(y+\tfrac{1}{2}\pi) + \cos y \cos(y+\tfrac{1}{2}\pi)\} = y'\cos(2y+\tfrac{1}{2}\pi)$$
$$= (\cos y)^2 \sin 2(y+\tfrac{1}{2}\pi),$$
$$y''' = y'\{-2\cos y \sin y \sin 2(y+\tfrac{1}{2}\pi) + 2(\cos y)^2 \cos 2(y+\tfrac{1}{2}\pi)\}$$
$$= 2y'\cos y \cos(3y+\tfrac{2}{2}\pi) = 2(\cos y)^3 \sin 3(y+\tfrac{1}{2}\pi).$$

By reasoning from n to $n+1$ it is proved, that in general:

$$y^{(n)} = \underline{|n-1}\,(\cos y)^n \sin n(y+\tfrac{1}{2}\pi), \quad (\underline{|0} = 1).$$

Eighth Chapter.

Calculation of functions by infinite series. General theorems concerning series of powers.

37. We now proceed to employ the successive derived functions of a given function in presenting the Theorem of the Mean Value in a form which constitutes the basis of the most important theorem of the Differential Calculus.

Let $f(x)$ be a unique function from a to b, let its derived functions $f'(x), f''(x), \ldots f^{(n-1)}(x)$ be everywhere in the same interval continuous and therefore also finite, while we assume no other property of the nth derived $f^{(n)}(x)$, but that it has the same value at each point when formed progressively as regressively. Our first enquiry, in conformity with § 22, is whether the quotient

$$\frac{f(b) - f(a) - (b-a) f'(a)}{(b-a)^2},$$

which again may be denoted by K, can be expressed by means of higher derived functions. From the equation

$$f(b) - f(a) - (b-a) f'(a) - K(b-a)^2 = 0$$

it results as in § 22 that

$$\varphi(x) = f(b) - f(x) - (b-x) f'(x) - (b-x)^2 K$$

is a continuous function with a determinate differential quotient, and that it vanishes for $x = a$ and for $x = b$. There must therefore be some value x_1, such that

$$\varphi'(x_1) = -f'(x_1) + f'(x_1) - (b-x_1) f''(x_1) + 2(b-x_1)K = 0,$$

that is

$$K = \tfrac{1}{2} f''(x_1).$$

Accordingly we have the equation:

$$f(b) = f(a) + (b-a) f'(a) + \frac{1}{\underline{|2}} (b-a)^2 f''(a + \Theta(b-a)),\ 0 < \Theta < 1.$$

If we proceed similarly and put

$$f(b) - f(a) - (b-a) f'(a) - \frac{1}{\underline{|2}} (b-a)^2 f''(a) - \frac{1}{\underline{|3}} (b-a)^3 K = 0,$$

the value of K is found by the equation:

$$K = \frac{f'''(x_1)}{3},$$

so that

$$f(b) = f(a) + (b-a)f'(a) + \frac{(b-a)^2}{\underline{|2}} f''(a) + \frac{(b-a)^3}{\underline{|3}} f'''(a + \Theta(b-a)).$$

We introduced the quantity $\underline{|2}$ in the denominator from the first, in order that the equation for K arising from differentiation might assume as simple a form as possible.

Let us now put somewhat more generally for any value of n:

$$f(b) = f(a) + (b-a)f'(a) + \frac{(b-a)^2}{\underline{|2}} f''(a) + \cdots \frac{(b-a)^{n-1}}{\underline{|n-1}} f^{n-1}(a) + \frac{(b-a)^p}{p} K,$$

where p is to signify any positive integer, and let us enquire whether K can be expressed by values of the nth derived. Once more, the function

$$\varphi(x) = f(b) - f(x) - (b-x)f'(x) - \frac{(b-x)^2}{\underline{|2}} f''(x) - \cdots$$
$$\cdots - \frac{(b-x)^{n-1}}{\underline{|n-1}} f^{n-1}(x) - \frac{(b-x)^p}{p} K$$

is continuous, everywhere finite, has a determinate differential quotient, and vanishes for $x = a$ and for $x = b$: so that we must have

$$\varphi'(x_1) = -\left\{\frac{(b-x_1)^{n-1}}{\underline{|n-1}} f^n(x_1) - (b-x_1)^{p-1} K\right\} = 0$$

or, as x_1 must be different from b:

$$K = \frac{(b-x_1)^{n-p}}{\underline{|n-1}} f^n(x_1) = \frac{(b-a)^{n-p} \cdot (1-\Theta)^{n-p}}{\underline{|n-1}} f^n(a + \Theta(b-a)).$$

Accordingly

$$f(b) = f(a) + (b-a)f'(a) + \frac{(b-a)^2}{\underline{|2}} f''(a) + \cdots \frac{(b-a)^{n-1}}{\underline{|n-1}} f^{n-1}(a)$$
$$+ \frac{(b-a)^n \cdot (1-\Theta)^{n-p}}{p\,\underline{|n-1}} f^n(a + \Theta(b-a)).$$

The last term assumes particularly simple forms when p is put $= n$, or $= 1$, we have

$$\text{I. } f(b) = f(a) + (b-a)f'(a) + \frac{(b-a)^2}{\underline{|2}} f''(a) + \cdots \frac{(b-a)^{n-1}}{\underline{|n-1}} f^{n-1}(a)$$
$$+ \frac{(b-a)^n}{\underline{|n}} f^n(a + \Theta(b-a)), \text{ or}$$

$$\text{II. } f(b) = f(a) + (b-a)f'(a) + \frac{(b-a)^2}{\underline{|2}} f''(a) + \cdots \frac{(b-a)^{n-1}}{\underline{|n-1}} f^{n-1}(a)$$
$$+ \frac{(b-a)^n (1-\Theta)^{n-1}}{\underline{|n-1}} f^n(a + \Theta(b-a)).$$

Θ does not signify the same value in both equations, moreover all that immediately concerns us is to know, that in each case there is a value for Θ between 0 and 1.*)

38. By the formulas just established the problem of actually calculating the values of a function for a given interval of its argument x is solved. Previously, with the exception of the process of inclusion within limits, we had no means of doing this, even for the elementary functions: x^n (n arbitrary), a^x, ${}^a\log x$ (a positive), the trigonometric and circular functions; and yet in its absence we could perceive their uniqueness and continuity, and assign all their derived functions in terms of the same symbols of calculation.

The only arithmetical operations we have in our power actually to carry out are the two — summation and multiplication — performing them a finite number of times on a definite set of rational numbers, positive or negative, integer or fractional; irrational numbers must be replaced by their nearest rational approximations.**) Accordingly the only function whose calculation we can deem completed is the rational algebraic function:

$$y = f(x) = \frac{a_0x^n + a_1x^{n-1} + \cdots a_{n-1}x + a_n}{b_0x^m + b_1x^{m-1} + \cdots b_{m-1}x + b_m}.$$

To calculate any other function for an arbitrary value of x, is to assign a method according to which continued summations or multiplications have to be carried out, the result of which exhibits the required value with greater approximation the more frequently the operations are carried out as prescribed by the method.

The elementary functions of x must therefore admit of expression in the form either, of sums whose summands can be powers of the argument x, or, of products whose factors can likewise contain the argument in powers. When they are calculated in this manner, they can themselves be used in the calculation of more complicated functions. The number of such summands or of such factors will of course, in analogy with the exposition of an irrational number, be infinite, for otherwise every function could be brought to the rational algebraic form; but the arrangement of them will be such, that even a finite summation or multiplication is enough to generate a value whose difference from the required value of the function is demonstrably

---

*) The first formula was developed by Lagrange: Théorie des fonctions analytiques, 1797; the alteration contained in the second was given by Cauchy: Exercices de mathématiques, T. I. p. 29. Subsequently, still more general forms for the last term were devised by Schlömilch according to the method we have here followed.

**) Subtraction is summation with negative, division is multiplication with fractional numbers.

less than an arbitrarily small quantity. The calculation of the number e affords us an example of this. The formula of such an infinite sum or of such an infinite product is called a convergent one.*)

39. The property of convergence of any infinite series is expressed analytically as follows:

Let $u_1 + u_2 + u_3 + \cdots u_n + u_{n+1} + u_{n+2} + \cdots$

be the terms of the infinite series, which can be continued unlimitedly according to some law, the sums obtained by adding up, first n terms, then $n+1$, $n+2$, $\cdots n+k$ terms:

$$S_n = u_1 + u_2 + \cdots + u_n$$
$$S_{n+1} = u_1 + u_2 + \cdots + u_n + u_{n+1}$$
$$\cdots\cdots\cdots\cdots\cdots\cdots\cdots\cdots\cdots$$
$$S_{n+k} = u_1 + u_2 + \cdots + u_n + u_{n+1} + \cdots + u_{n+k}$$

must form a succession of numbers with a determinate finite limiting value S. This requires: first, that none of these sums, therefore also none of the terms u, increase beyond any finite amount, and second, that for any number δ however small, a place n be assignable, such that the amount of the difference $S_{n+k} - S_n$ for every value of k, shall be less than δ. But this difference is nothing else than the sum of k terms following the nth; *accordingly it must be possible to choose n so that for every value of k,*

$$\text{abs}\,[u_{n+1} + u_{n+2} \cdots + u_{n+k}] \textit{ shall be } < \delta.$$

Now let us denote by R_n the difference between the finite limiting value S and the sum S_n, then this quantity $S_{n+k} - S_n$ may also be written as $R_n - R_{n+k}$, similarly $S_{n+k} - S_{n+1}$ as $R_{n+1} - R_{n+k}$, etc., whence, provided the choice of n makes abs $[R_n - R_{n+k}]$ less than δ for every value of k, $R_{n+1} - R_{n+k}$, $R_{n+2} - R_{n+k}$, etc. are also certain to remain less in amount than 2δ for every value of k. For we have

$$\text{abs}\,[R_{n+1} - R_{n+k}] = \text{abs}\,[(S_{n+k} - S_n) - (S_{n+1} - S_n)]$$
$$\text{abs}\,[R_{n+2} - R_{n+k}] = \text{abs}\,[(S_{n+k} - S_n) - (S_{n+2} - S_n)], \text{ etc.}$$

And conversely, when R_n, $R_{n+1}, \cdots$ are smaller than δ, the differences

$$R_n - R_{n+k} = u_{n+1} + u_{n+2} \cdots + u_{n+k} = S_{n+k} - S_n$$

are also less than 2δ for every value of k.

If then we call the limit of the sum R_n of all terms from the $(n+1)^{th}$,

*) It is important for the beginner to realise clearly this requisite of calculability of a function; and so the essential difference between a rational function and all other denominations of functions $\sqrt{}$, log, sin, cos. The latter are only to be regarded as symbols by which the dependence of one number on another is expressed, whose properties are no doubt known, so that the nature of the dependence is completely defined ex. gr. by inversion of an arithmetical operation, or by geometric definitions, but for whose calculation we have as yet no fixed law.

the remainder of the series after n places, we can formulate the condition of convergence also as follows:

The necessary and sufficient condition for the convergence of an infinite series consists in this, that for any number δ however small, a place n can be found in it, such that its remainders R_n, R_{n+1}, R_{n+2}, $\cdots$ are always smaller in amount than δ.

This cannot possibly be fulfilled unless the amounts of the terms in the infinite series ultimately decrease and have zero as limit, but this condition alone is not sufficient for its convergence. The value of an infinite series which does not converge, will either be quite indeterminate, when the series of sums S_n oscillates between arbitrary values, or it will be determinately infinite positively or negatively. In both cases the series is said to be divergent.

40. Now the forms developed in § 37 express the simple functions as infinite series of powers. For, supposing the value of the function and of all its derived functions be known for the argument a, and that the value of the function for any other x is required, then in consequence of these equations we have, putting x for b:

$$f(x)=f(a)+(x-a)f'(a)+\frac{(x-a)^2}{\underline{|2}}f''(a)+\cdots\frac{(x-a)^{n-1}}{\underline{|n-1}}f^{n-1}(a)+R$$

$$R=\frac{(x-a)^n}{\underline{|n}}f^n(a+\Theta(x-a)),\ \text{or}=\frac{(x-a)^n(1-\Theta)^{n-1}}{\underline{|n-1}}f^n(a+\Theta(x-a)).$$

On the right side accordingly all terms are known, except the last, in which the unknown fraction Θ occurs. But if we can prove that this last term R, formed for arbitrarily increasing values of n, passes through a series of numbers having zero as limit, then neglecting this last term, we shall obtain the value of $f(x)$ with arbitrary approximation by summing as many terms as we please of the infinite series:

$$f(x)=f(a)+(x-a)f'(a)+\frac{(x-a)^2}{\underline{|2}}f''(a)+\frac{(x-a)^3}{\underline{|3}}f'''(a)+\cdots\text{in }\infty.$$

This is Taylor's series named after its discoverer*); it teaches: *If we know the value of a function and of all its derived functions for a single argument, we can calculate the value of the function for every other argument $x \lesseqgtr a$, if in the interval from a to x the function and as many of its derived functions as may be formed are continuous without becoming infinite, and if Lim (R) vanishes for $n=\infty$.*

41. The examination of the first hypothesis is apparently complicated, requiring the finiteness and continuity of all the derived

*) Taylor (1685–1731) in his chief work: Methodus incrementorum directa et inversa, 1715, established this series but without taking account of the remainder term. Mac Laurin (1698–1746) in his Treatise of Fluxions, 1742, developed the series for the special value $a=0$.

functions in the interval from a to x to be investigated; however it is simplified by the Theorem:

If the nth derived of a function formed progressively and likewise regressively, be within a finite interval everywhere determinate and only finite, not necessarily continuous, then all the derived functions preceding this one, as well as the function itself, must be continuous in this interval and cannot become infinite.

For if z be a value intermediate to a and x, the equations:

$$\frac{f^{n-1}(z+h)-f^{n-1}(z)}{h} = f^n(z) + \delta, \quad \frac{f^{n-1}(z-h)-f^{n-1}(z)}{-h} = \varphi^n(z) + \delta'$$

define the values of the progressive derived $f^n(z)$, and of the regressive $\varphi^n(z)$. From these by subtraction:

$$f^{n-1}(z+h) - f^{n-1}(z-h) = h\left(f^n(z) + \varphi^n(z)\right) + h(\delta + \delta').$$

If now f^n and φ^n are everywhere finite, we have at each point in the interval, $\text{Lim}\,[f^{n-1}(z+h) - f^{n-1}(z-h)] = 0$, for $h = 0$; so that the function is continuous at both sides of this point. Moreover f^{n-1} remains everywhere finite: for if M be the greatest value which f^n takes between $z = x_0$ and $x_0 + h$, and m the least, then in the entire interval from x_0 to $x_0 + h$ the expression

$$f^{n-1}(z) - f^{n-1}(x_0) - (z - x_0)\,m \text{ is} > 0,$$

because it vanishes for $z = x_0$ and its derived $f^n(z) - m$ remains always > 0; on the other hand

$$f^{n-1}(z) - f^{n-1}(x_0) - (z - x_0)\,M \text{ is} < 0 \text{ because } f^n(z) - M < 0.$$

Each value of z therefore will have its own proper fraction Θ, so that

$$f^{n-1}(z) - f^{n-1}(x_0) = (z - x_0)\,[m + \Theta(M - m)],$$

that is, to each z belongs a finite value of $f^{n-1}(z)$. A like method of proof being applied, the same thing follows for each preceding derived function as well as for $f(z)$ itself.

In determining the limiting value of R the following Theorem is at times of use:

If the values of the n derived functions up to $n = \infty$ remain finite in an interval, then Lim R vanishes.

For in the product

$$\frac{(x-a)^n}{\underline{|n-1}} = (x-a)\cdot\frac{x-a}{1}\cdot\frac{x-a}{2}\cdots\frac{x-a}{n-1} = (x-a)\,\varrho$$

we have

$$\varrho = \frac{x-a}{1}\cdot\frac{x-a}{2}\cdots\frac{x-a}{n-1}, \text{ therefore } \varrho^2 = \frac{(x-a)^2}{1\cdot n-1}\cdot\frac{(x-a)^2}{2\cdot n-2}\cdots\frac{(x-a)^2}{n-1\cdot 1},$$

therefore $\varrho^2 < \left(\frac{(x-a)^2}{n-1}\right)^{n-1}$, because $n - 1 \leq k(n-k)$, k being $\leq n - 1$,

hence $\varrho < \left(\frac{x-a}{\sqrt{n-1}}\right)^{n-1}$. Now if $x - a$ is finite, $\frac{x-a}{\sqrt{n-1}}$ is a fraction whose higher powers, when n increases, have zero as limit. In the present case therefore the first and second hypotheses are comprehended in one: A function whose n derived functions up to $n = \infty$ remain finite within an interval, can be calculated in this interval by a series of powers. But this proposition cannot be converted, because Lim R can vanish, without the n derived functions up to $n = \infty$ also being finite, as some of the following examples show.

42. Exponential series: $y = f(x) = e^x$.
For $x = 0$, $f(x)$ and all derived functions are known; in fact $f^n(x) = e^x$ therefore $f^n(0) = 1$. These are continuous functions for all finite values of x, and even for $n = \infty$ always remain finite. Accordingly Taylor's series converges and

$$e^x = 1 + \frac{x}{1} + \frac{x^2}{\underline{|2}} + \frac{x^3}{\underline{|3}} + \cdots \frac{x^k}{\underline{|k}} + \cdots \text{ in infin.}, \; -\infty < x < +\infty.^{*)}$$

If more generally $y = a^x$ $(a > 0)$, let us put $y = e^{x l a}$ and we have

$$a^x = 1 + \frac{x l a}{1} + \frac{(x l a)^2}{\underline{|2}} + \frac{(x l a)^3}{\underline{|3}} + \cdots \frac{(x l a)^k}{\underline{|k}} + \cdots \text{ in infin.}$$
$$-\infty < x < +\infty.$$

43. Trigonometric series: $y = f(x) = \sin x$.

$\frac{d^n y}{dx^n} = \sin(x + \frac{1}{2} n\pi)$ is finite for every finite argument. Accordingly Taylor's series converges, and as

$$f(0) = \sin(0) = 0, \; f''(0) = \sin(\tfrac{2}{2}\pi) = 0, \; f^{\text{IV}}(0) = \sin(\tfrac{4}{2}\pi) = 0,$$
$$f'(0) = \sin(\tfrac{1}{2}\pi) = 1, \; f'''(0) = \sin(\tfrac{3}{2}\pi) = -1, f^{\text{V}}(0) = \sin(\tfrac{5}{2}\pi) = 1,$$

we have:

$$\sin x = \frac{x}{1} - \frac{x^3}{\underline{|3}} + \frac{x^5}{\underline{|5}} - \frac{x^7}{\underline{|7}} \cdots (-1)^k \frac{x^{2k+1}}{\underline{|2k+1}} \cdots, \; -\infty < x < +\infty.$$

Again: $y = f(x) = \cos x$. $\quad \frac{d^n y}{dx^n} = \cos(x + \frac{1}{2} n\pi)$.

$$f(0) = \cos(0) = 1, \; f''(0) = \cos(\tfrac{2}{2}\pi) = -1, \; f^{\text{IV}}(0) = \cos(\tfrac{4}{2}\pi) = 1,$$
$$f'(0) = \cos(\tfrac{1}{2}\pi) = 0, \; f'''(0) = \cos(\tfrac{3}{2}\pi) = 0, \; f^{\text{V}}(0) = \cos(\tfrac{5}{2}\pi) = 0,$$
$$\cos x = 1 - \frac{x^2}{\underline{|2}} + \frac{x^4}{\underline{|4}} - \frac{x^6}{\underline{|6}} \cdots (-1)^k \frac{x^{2k}}{\underline{|2k}} \cdots, \; -\infty < x < +\infty.$$

The present series render it possible to calculate trigonometric tables for the sine and cosine of any number. When we wish to abstract quite from the geometric definition of sine and cosine, these series

*) The series itself was first established by Newton, as well as the series for sine and cosine; the number e, as already mentioned, was introduced by Euler as basis of exponential functions.

are to be considered as definitions of these functions, and from them all their properties already employed must directly follow.

44. To demonstrate this, we must show independently of previous considerations that the defining series converge and are continuous functions of x. For this purpose we prove the following

General theorems concerning series of powers.*)

I. *If the coefficients* $a_0, a_1 \ldots \ldots a_n \ldots$ *in any series of powers*

$$f(x) = a_0 + a_1 x + a_2 x^2 + \cdots a_n x^n + \cdots$$

are all of like sign, and for a definite positive value X *its terms after some certain one decrease and converge to zero, so that the quotient of a term by the preceding one is less than unity, and for* $n = \infty$ *is at most equal to unity, then the series converges for all positive values of* x *which are smaller than* X.

The quotient $\frac{a_{n+1}}{a_n} X$ after some definite place in the series being less than or at most equal to unity, if we take $x < X$, a proper fraction α can be assigned, such that

$$\frac{a_{n+1}\, x}{a_n} < \alpha, \quad \frac{a_{n+2}\, x}{a_{n+1}} < \alpha, \quad \frac{a_{n+3}\, x}{a_{n+2}} < \alpha \cdots,$$

therefore

$$a_{n+1}\, x < \alpha\, a_n, \quad a_{n+2}\, x^2 < \alpha^2\, a_n, \ \ldots\ a_{n+k}\, x^k < \alpha^k\, a_n \ldots,$$

hence

$$a_n x^n + a_{n+1}\, x^{n+1} \cdots + a_{n+k}\, x^{n+k} + \cdots < a_n x^n \{1 + \alpha + \alpha^2 \cdots + \alpha^k + \cdots\}$$

or:

$$a_n\, x^n + a_{n+1}\, x^{n+1} + \cdots a_{n+k}\, x^{n+k} + \cdots < a_n\, x^n \cdot \frac{1}{1-\alpha}\cdot$$

The quantity on the right is finite and positive and we can give n such a value as will make it less than any assignable quantity. Our statement is therefore proved.

For $\alpha = 1$ this reasoning applies no longer; so that we have always specially to investigate whether a series continues to converge, in case the quotient of two consecutive terms tends to the limiting value unity.

If on the other hand for a value x, the ratio of a term to the preceding one, from some initial place in the series, is always greater than 1, even though it may be $= 1$ for $n = \infty$, the series will have no determinate finite sum for this or any greater value of x, i. e. it will diverge, because the succeeding terms increase, and therefore also the remainder of the series does not converge to zero.

---

*) Abel (1802—1829): Recherches sur la série $1 + \frac{m}{1} x + \frac{m}{1} \frac{(m-1)}{2} x^2 +$ etc., Oeuvres I, p. 219 (1881). Crelle J. Vol. 1, p. 311 (1826).

Therefore in general the interval of convergence (convergency) of the series is given by the condition

$$\operatorname{Lim} \frac{a_{n+1}}{a_n} x < 1, \text{ or } x < \operatorname{Lim} \frac{a_n}{a_{n+1}} \quad (n = \infty).$$

II. The sum (difference) of two convergent series is itself a convergent series, whose terms consist of the sum (difference) of the terms of both.

If
$$f(x) = a_0 + a_1 x + a_2 x^2 + \cdots a_{n-1} x^{n-1} + R_n,$$
$$\varphi(x) = b_0 + b_1 x + b_2 x^2 + \cdots b_{n-1} x^{n-1} + R'_n,$$
be such, that a determinate n can be chosen, so that R as well as R' may become less than any arbitrarily small number, we have

$$f(x) \pm \varphi(x) = (a_0 \pm b_0) + (a_1 \pm b_1) x + \cdots (a_{n-1} \pm b_{n-1}) x^{n-1} + R_n \pm R'_n.$$

Now since we have $\operatorname{Lim} (R_n \pm R'_n) = 0$, for all values of x for which both series converge, we obtain for the algebraic sum the infinite series

$$f(x) \pm \varphi(x) = a_0 \pm b_0 + (a_1 \pm b_1) x + (a_2 \pm b_2) x^2 + \cdots$$

Still more generally if the series converge respectively for x and x', we have

$$f(x) \pm \varphi(x') = (a_0 \pm b_0) + (a_1 x \pm b_1 x') + (a_2 x^2 \pm b_2 x'^2) + \cdots$$

III. An infinite series, whose terms have different signs for some value of x, converges, if the limit of the sum of the positive terms be finite and also the limit of the sum of the negative terms be finite.

For by Theorem II such a series expresses the difference of the values of two convergent series. When this is the case, the series consisting of the same terms taken all with like sign converges and we shall see that it has the same limiting value even when the order of its terms is changed: such a series is said to be absolutely (unconditionally) convergent. But a series whose terms have different signs may converge without the sum of the positive and of the negative terms separately having finite limits, it is then said to be semiconvergent (conditionally convergent). A series converges unconditionally when the absolute value of the quotient of a term by the preceding one is less than unity for all values from some determinate n on to $n = \infty$. For then, even when all the terms are written with the same sign, the series fulfils the condition of convergence proved sufficient in Theorem I.

It is thus seen: that a conditional convergence can only arise by the ratio of a term to the preceding one being less in amount than unity, but becoming unity for $n = \infty$; and hence follows further: If a series of powers of x converges only conditionally for a determinate value X then it converges absolutely for every numerically smaller value of x; while it diverges for a greater value.

For, for a smaller value the quotient remains less than one, for a greater value it becomes greater than one; the terms of the series then do not decrease in amount but increase.

IV. *Every series of powers is a continuous function of the variable, within the interval in which it absolutely converges.*

Let $f(x)$ signify the value of the infinite series

$$a_0 + a_1 x + a_2 x^2 + \cdots a_n x^n + \cdots,$$

for which, since x is to be a value within its convergency, we have

$$\text{Lim}\left[\frac{a_{n+1}}{a_n} x\right] < 1,$$

it is required to show that $\text{Lim}\,[f(x \pm \delta) - f(x)] = 0$ for $\delta = 0$. Putting:

$$a_0 + a_1 x + a_2 x^2 + \cdots a_{n-1} x^{n-1} = \varphi(x)$$
$$a_n x^n + a_{n+1} x^{n+1} + a_{n+2} x^{n+2} + \cdots = \psi(x),$$

then as in Theorem I

$$\text{abs}\,\psi(x) \text{ is } < \text{abs}\,a_n x^n \cdot \frac{1}{1-\alpha}, \quad (0 < \alpha < 1).$$

Merely by the selection of a lower limit for n, we can thus make $\psi(x)$ as well as $\psi(x \pm \delta)$ and therefore also the amount of their difference $\psi(x \pm \delta) - \psi(x)$ less than a quantity ε however small, because the term $a_n x^n$ becomes arbitrarily small as n increases. When X denotes the greatest value of the interval for x, we must choose n so that $a_n < \frac{\varepsilon}{X^n}$.*) Accordingly

$$f(x \pm \delta) - f(x) = \varphi(x \pm \delta) - \varphi(x) + \varepsilon.$$

Now since $\varphi(x)$ denotes a rational integer function of x, which as already seen § 19 is continuous, the difference $f(x \pm \delta) - f(x)$ becomes smaller than an arbitrarily small quantity as δ decreases, i. e. $f(x)$ is a continuous function.

The Theorem also holds when the series converges at the limits of the interval of convergence for X, conditionally or unconditionally: that, for $\delta = 0$

$$\text{Lim}\, f(X - \delta) = f(X).$$

For we have here:

---

*) In consequence of this property, that for the same n, both $\psi(x)$ and $\psi(x \pm \delta)$ become less than ε, series of powers are said to be convergent in equal degree or uniformly. Abel was the first (*loc. cit.* Oeuvres I, p. 225) to point out, that continuity of the series does not of itself follow from the continuity of the terms of the series. Uniform convergence teaches also, that the infinite series in its entire convergency can be replaced *quam proxime* by the same rational integer function. The function expressed by the series of powers is styled therefore, after Weierstrass, one which has the character of a rational integer function.

$$\psi(X) = a_n X^n + a_{n+1} X^{n+1} + a_{n+2} X^{n+2} + \cdots$$

$$\psi(X - \delta) = a_n \left(\frac{X-\delta}{X}\right)^n X^n + a_{n+1} \left(\frac{X-\delta}{X}\right)^{n+1} X^{n+1} + \cdots$$

If the terms in $\psi(X)$ are all of like sign, ex. gr. positive, we see at once, that as $\frac{X-\delta}{X} < 1$, $\psi(X-\delta) < \left(\frac{X-\delta}{X}\right)^n \cdot \psi(X)$, so that a value can be given to n which will make $\psi(X-\delta)$ as well as $\psi(X)$, smaller than any arbitrarily small quantity ε.

But if the terms in $\psi(X)$ are different in sign, a special investigation is still required. This is based upon the following Lemma:

If $t_0, t_1, t_2, \ldots t_m, \ldots$ denote an infinite series of arbitrary quantities, and if the quantity

$$p_m = t_0 + t_1 + \cdots t_m$$

for all values of m be always algebraically less than a determinate quantity G, but greater than g, then if $\varepsilon_0, \varepsilon_1 \ldots$ denote decreasing positive quantities we have

$$g\,\varepsilon_0 < r = \varepsilon_0 t_0 + \varepsilon_1 t_1 + \cdots \varepsilon_m t_m < G\,\varepsilon_0.$$

Since

$$t_0 = p_0, \quad t_1 = p_1 - p_0, \quad t_2 = p_2 - p_1, \text{ etc.}$$

therefore

$$r = \varepsilon_0 p_0 + \varepsilon_1 (p_1 - p_0) + \varepsilon_2 (p_2 - p_1) + \cdots + \varepsilon_m (p_m - p_{m-1}),$$

or

$$r = p_0 (\varepsilon_0 - \varepsilon_1) + p_1 (\varepsilon_1 - \varepsilon_2) + \cdots p_{m-1} (\varepsilon_{m-1} - \varepsilon_m) + p_m \varepsilon_m.$$

As the differences $\varepsilon_0 - \varepsilon_1$, $\varepsilon_1 - \varepsilon_2, \ldots$ are positive, the value of this expression is less than

$$G(\varepsilon_0 - \varepsilon_1 + \varepsilon_1 - \varepsilon_2 \cdots - \varepsilon_m + \varepsilon_m) = G\,\varepsilon_0,$$

on the other hand it is greater than

$$g(\varepsilon_0 - \varepsilon_1 + \varepsilon_1 - \varepsilon_2 \cdots - \varepsilon_m + \varepsilon_m) = g\,\varepsilon_0.$$

Applied to the present case, in which $\left(\frac{X-\delta}{X}\right)$, $\left(\frac{X-\delta}{X}\right)^2 \cdots$ denote a series of decreasing positive quantities, it results from this Lemma that the amount of

$$\psi(X-\delta) \text{ is less than } \left(\frac{X-\delta}{X}\right)^n \cdot M,$$

where M represents the greatest numerical value in the series

$$a_n X^n,\ a_n X^n + a_{n+1} X^{n+1}, \ldots, a_n X^n + a_{n+1} X^{n+1} + \cdots + a_{n+k} X^{n+k}, \text{ etc.}$$

Since the series $f(X)$ converges, a place n can be found in it from which onwards the value of M is less than an arbitrarily small quantity ε, whence what we stated follows.

V. *The infinite series of powers is differentiated, by forming the series of differential quotients of its several terms.*

The series

$$a_1 + 2a_2 x + 3a_3 x^2 + \cdots n a_n x^{n-1} + \cdots$$

derived from

$$f(x) = a_0 + a_1 x + a_2 x^2 + \cdots a_n x^n + \cdots$$

certainly converges for all values of x which lie within the interval of convergence of the original series. For, the interval of the derived series is, according to the criterion, determined by

$$\text{abs Lim}\,\frac{(n+1)\,a_{n+1}\,x}{n\,a_n} < 1, \text{ or abs } x < \text{abs Lim}\,\frac{a_n}{a_{n+1}}\cdot\frac{n}{n+1}.$$

Now since $\text{Lim}\,\frac{n}{n+1} = \text{Lim}\left(1 - \frac{1}{n+1}\right)$ becomes $= 1$, for $n = \infty$, it follows that

$$\text{abs } x < \text{abs Lim}\,\frac{a_n}{a_{n+1}}.$$

Now in order to determine the differential quotient of the continuous function $f(x)$, let us first form the quotient of differences, doing so ex. gr. regressively, in order when possible to take account also of the upper limit of its convergency:

$$\frac{f(x-\Delta x) - f(x)}{-\Delta x} = \frac{\varphi(x-\Delta x) - \varphi(x)}{-\Delta x} + \frac{\psi(x-\Delta x) - \psi(x)}{-\Delta x}.$$

For any finite Δx however small, this continuous expression in Δx has a determinate finite value.

If we denote the infinite series $a_1 + 2a_2 x + \cdots + n a_n x^{n-1} + \cdots$ by $\chi(x)$, its remainder by $R_n(x)$, this equation takes the form

$$\frac{f(x-\Delta x) - f(x)}{-\Delta x} = \chi(x - \Theta\Delta x) - R_n(x - \Theta\Delta x) + \frac{\psi(x-\Delta x) - \psi(x)}{-\Delta x}.$$

Retaining the value of Δx, when we increase n arbitrarily the value of Θ changes on the right side. But as the remainder of a series of powers has the property, that after some determinate n, $R_n(x)$ becomes arbitrarily small for all values between x and $x - \Delta x$, then because as n increases the last quotient can also be made arbitrarily small, it follows that

$$\frac{f(x-\Delta x) - f(x)}{-\Delta x} = a_1 + 2a_2(x - \Theta\Delta x) + 3a_3(x - \Theta\Delta x)^2 + \cdots$$

For because the continuous function $\chi(x)$ comes arbitrarily near the quotient of differences in the interval from x to $x - \Delta x$, there must also be a point (compare § 17) where the two are equal. Now the differential quotient $f'(x)$ arises from the quotient of differences by continuous transition for $\Delta x = 0$. But as long as what is on the right side converges, it is by Theorem IV a continuous function of the variable $x - \Theta\Delta x$, therefore we have for $\Delta x = 0$:

$$f'(x) = a_1 + 2a_2 x + 3a_3 x^2 + \cdots (n-1) a_{n-1} x^{n-2} + \cdots,$$

as was to be proved. The regressive differential quotient of the infinite series of powers is a continuous function of x. For the progressive differential quotient we obtain in the same way the same series.

45. Applying these Theorems to the trigonometric series

$$\frac{x}{1} - \frac{x^3}{\underline{|3}} + \frac{x^5}{\underline{|5}} - \frac{x^7}{\underline{|7}} + \cdots (-1)^n \cdot \frac{x^{2n+1}}{\underline{|2n+1}} + \cdots = \sin x,$$

$$1 - \frac{x^2}{\underline{|2}} + \frac{x^4}{\underline{|4}} - \frac{x^6}{\underline{|6}} + \cdots (-1)^n \cdot \frac{x^{2n}}{\underline{|2n}} + \cdots = \cos x,$$

we perceive that each of them converges absolutely for all finite values of x, for we have for $n = \infty$:

$$\text{Lim}\left[\frac{x^{2n+1}}{\underline{|2n+1}} : \frac{x^{2n-1}}{\underline{|2n-1}}\right] = \text{Lim}\left[\frac{x^2}{2n \cdot (2n+1)}\right] = 0,$$

$$\text{Lim}\left[\frac{x^{2n}}{\underline{|2n}} : \frac{x^{2n-2}}{\underline{|2n-2}}\right] = \text{Lim}\left[\frac{x^2}{(2n-1) \cdot 2n}\right] = 0.$$

Accordingly the functions expressed by the series are continuous for all finite values of x. Further we have

1) $$\frac{d \sin x}{dx} = 1 - \frac{x^2}{\underline{|2}} + \frac{x^4}{\underline{|4}} \cdots (-1)^n \cdot \frac{x^{2n}}{\underline{|2n}} \cdots, \quad \text{i.e.} = \cos x.$$

$$\frac{d \cos x}{dx} = -\frac{x}{1} + \frac{x^3}{\underline{|3}} - \frac{x^5}{\underline{|5}} \cdots (-1)^n \cdot \frac{x^{2n-1}}{\underline{|2n-1}} \ldots, \text{i.e.} = -\sin x.$$

Next it follows from the series that

2) $\cos(-x) = \cos x$, $\sin(-x) = -\sin x$, $\cos(0) = 1$, $\sin(0) = 0$.

Now, since equations 1) teach that all the derived functions even for $n = \infty$ remain finite and continuous, we may develope $\cos(x+y)$ according to Taylor's series in powers of y, and thus obtain:

$$\cos(x+y) = \cos x + y\frac{d\cos x}{dx} + \frac{y^2}{\underline{|2}}\frac{d^2\cos x}{dx^2} + \frac{y^3}{\underline{|3}}\frac{d^3\cos x}{dx^3} + \frac{y^4}{\underline{|4}}\frac{d^4\cos x}{dx^4} + \cdots$$

$$= \cos x - y\sin x - \frac{y^2}{\underline{|2}}\cos x + \frac{y^3}{\underline{|3}}\sin x + \frac{y^4}{\underline{|4}}\cos x - \cdots$$

$$= \cos x\left(1 - \frac{y^2}{\underline{|2}} + \frac{y^4}{\underline{|4}} - \cdots\right) - \sin x\left(y - \frac{y^3}{\underline{|3}} + \frac{y^5}{\underline{|5}} - \cdots\right),$$

i. e.

3) $$\cos(x+y) = \cos x \cos y - \sin x \sin y.$$

In like manner we find:

$$\sin(x+y) = \sin x \cos y + \cos x \sin y.$$

Thus the theorem of addition, on which our previous calculations were based, is proved independently of geometrical considerations.

From 3) we get, putting $-y$ for y, in consequence of 2):

$$\cos(x-y) = \cos x \cos y + \sin x \sin y.$$

If we assume $x = y$, the proposition follows

4) $$1 = (\cos x)^2 + (\sin x)^2.$$

It remains to show the periodicity of both functions. With this in view we remark that the nature of the terms of both series is such, that $\sin x$ and $\cos x$ remain positive when x increases from 0 to 1, whence results, that within this interval $\sin x$ is a function increasing from 0, $\cos x$ a function decreasing from 1, for, the differential quotient of the first function is positive, that of the second negative. Now since taking account of the initial terms:

$$\sin 1 > 1 - \frac{1}{\underline{|3}} + \frac{1}{\underline{|5}} - \frac{1}{\underline{|7}}, \quad \cos 1 < 1 - \frac{1}{\underline{|2}} + \frac{1}{\underline{|4}} - \frac{1}{\underline{|6}} + \frac{1}{\underline{|8}},$$

it follows that

$$\sin 1 - \cos 1 > \frac{7}{24} + \left(\frac{1}{\underline{|5}} - \frac{1}{\underline{|7}}\right) + \left(\frac{1}{\underline{|6}} - \frac{1}{\underline{|8}}\right),$$

therefore $\sin 1 > \cos 1$. Hence there must be a value between 0 and 1 for which $\sin x = \cos x$. If we call this value $\frac{1}{4}\pi$ (< 1), we get by equation 4)

$$\sin \tfrac{1}{4}\pi = \cos \tfrac{1}{4}\pi = + \tfrac{1}{2}\sqrt{2}.$$

Also since

$$\cos 2x = (\cos x)^2 - (\sin x)^2; \quad \sin 2x = 2 \sin x \cos x,$$

we have

$$\cos \tfrac{1}{2}\pi = 0, \quad \sin \tfrac{1}{2}\pi = 1;$$

further

$$\cos \pi = -1, \ \sin \pi = 0; \quad \cos 2\pi = 1, \ \sin 2\pi = 0.$$

Consequently

$$\begin{aligned} \cos(x + \tfrac{1}{2}\pi) &= -\sin x, & \sin(x + \tfrac{1}{2}\pi) &= \cos x, \\ \cos(x + \pi) &= -\cos x, & \sin(x + \pi) &= -\sin x, \\ \cos(x + 2\pi) &= \cos x, & \sin(x + 2\pi) &= \sin x. \end{aligned}$$

The number 2π is the period. The course of the functions between $\frac{1}{4}\pi$ and $\frac{1}{2}\pi$ can be determined as follows.

We have: $$\cos(x + \tfrac{1}{4}\pi) = (\cos x - \sin x)\,\tfrac{1}{2}\sqrt{2},$$
$$\sin(x + \tfrac{1}{4}\pi) = (\cos x + \sin x)\,\tfrac{1}{2}\sqrt{2}.$$

As long as $x < \frac{1}{4}\pi$, the difference $\cos x - \sin x$ is always positive, therefore in the interval $x = \frac{1}{4}\pi$ to $x = \frac{1}{2}\pi$, $\cos x$ is a function continuously decreasing from the value $\frac{1}{2}\sqrt{2}$ to the value zero, $\sin x$ a function continuously increasing from the value $\frac{1}{2}\sqrt{2}$ up to unity. The number π, here defined purely arithmetically, we shall calculate when we come to circular series. The essential properties of the functions have thus been obtained immediately from the series defining them. Henceforth we shall always understand by $\sin x$, $\cos x$ only a symbolical representation of their respective series; $\sin^{-1} x$, $\cos^{-1} x$ are then defined as the inverse functions.

46. Binomial series.

$y=f(x)=(1+x)^m$. $x>-1$. m having any value; y always positive. For $x=0$ the value of y and of all derived functions is known; we have

$$f^n(x)=m(m-1)(m-2)\cdots(m-n+1)(1+x)^{m-n}.$$

These derived functions are continuous as long as $x>-1$; therefore

$$(1+x)^m=1+mx+\frac{m(m-1)}{\lfloor 2}x^2+\frac{m(m-1)(m-2)}{\lfloor 3}x^3+\cdots$$

$$\cdots+\frac{m(m-1)(m-2)\cdots(m-n+2)}{\lfloor n-1}x^{n-1}+R;$$

$R=m_n x^n(1+\Theta x)^{m-n}$, or $R=m_n n x^n(1-\Theta)^{n-1}(1+\Theta x)^{m-n}$;

$$m_n=\frac{m(m-1)(m-2)\cdots(m-n+1)}{\lfloor n}.$$

It is convenient to consider the second form of the remainder. Since $x>-1$, $1+\Theta x$ is a positive number for all values of Θ as required. Let us put

$$\frac{(1-\Theta)x}{1+\Theta x}=z,\quad (1+\Theta x)^{m-1}\cdot x=E,$$

then we have

$$R=m\cdot\frac{m-1}{1}\cdot\frac{m-2}{2}\cdots\frac{m-k}{k}\cdots\frac{m-(n-1)}{n-1}z^{n-1}E,$$

$$=m\cdot\frac{(m-1)z}{1}\cdot\frac{(m-2)z}{2}\cdots\frac{(m-k)z}{k}\cdots\frac{(m-(n-1))z}{n-1}\cdot E.$$

The factors E and m are finite. The product will certainly have its limit zero, when its factors begin somewhere to be proper fractions and remain proper fractions when n becomes ∞. For, if G be numerically the greatest of the fractions between $\frac{(m-k)z}{k}$ and $\frac{m-(n-1)}{n-1}z$, the product of these factors taken absolutely is less than G^{n-k}; but such a power has zero as limit. On the other hand the product will certainly increase beyond all limits, provided the factors once become greater than unity and remain so.

But now since as n increases, the amount of $\frac{m-(n-1)}{n-1}z$ approaches arbitrarily to that of z, the amount of z must be less than unity; therefore

for $x>0$, $\frac{(1-\Theta)x}{1+\Theta x}<1$, or $(1-\Theta)x<1+\Theta x$, i. e. $x<1$,

for $x<0$, $\frac{(1-\Theta)x}{1+\Theta x}>-1$, or $(1-\Theta)x>-1-\Theta x$, i. e. $x>-1$.

Result: If $-1<x<+1$, the positive function $(1+x)^m$ can be calculated for every m with arbitrary approximation from the infinite sum:

$$1 + \frac{m}{1}x + \frac{m(m-1)}{\underline{|2}}x^2 + \cdots \frac{m(m-1)\cdots(m-n+2)}{\underline{|n-1}}x^{n-1} + \cdots{}^{*)}.$$

The error which is incurred in breaking off the series at the nth term is at most equal to the maximum value of the remainder

$$m_n x^n (1+x)^{m-n} \text{ or } m_n x^n,$$

according as $x <$ or > 0.

It appears from what has been said that, except for positive integer values of m, for which the series is finite, the values of the terms of the series for $x > +1$ or < -1 increase beyond any limit, so that the series no longer converges.

The limiting cases: $x = +1$ or $= -1$ require a special consideration, not at present possible by means of the remainder, inasmuch as it should take account of the maximum value of $(1-\Theta)^{n-1}(1+\Theta x)^{m-n}$. It is plain at once, that if the series converge at all for these limits, the values it must express are respectively 2^m and 0^m; for, as long as the series converges, it is a continuous function of x and must therefore assume the same value as the continuous function $(1+x)^m$ with which it coincides for all values of x within these limits.

When $m > 0$ and $x = -1$, the series is of the form:

$$1 - \frac{m}{1} + \frac{m(m-1)}{\underline{|2}} - \frac{m(m-1)(m-2)}{\underline{|3}} + \cdots (-1)^n \frac{m(m-1)\cdots(m-n+1)}{\underline{|n}} \cdots$$

In this series the terms all take the same sign as soon as n becomes $> m$. But the sum of $2, 3, \cdots n+1$ terms is

$$-\frac{m-1}{1}, \frac{(m-1)(m-2)}{\underline{|2}}, -\frac{(m-1)(m-2)(m-3)}{\underline{|3}}, \cdots (-1)^n \frac{(m-1)(m-2)\cdots(m-n)}{\underline{|n}}.$$

Here each term is ultimately less in absolute amount then the one preceding; therefore we have a series of numbers all of one sign and each smaller than the one before it. This series of numbers has therefore a determinate limit and this limit is zero in consequence of the above remark.

For $x = +1$ we obtain the series

$$1 + \frac{m}{1} + \frac{m(m-1)}{\underline{|2}} + \frac{m(m-1)(m-2)}{\underline{|3}} + \cdots \frac{m(m-1)(m-2)\cdots(m-n+1)}{\underline{|n}} \cdots$$

These terms assume alternate signs when $n > m$ but yet the series converges absolutely, because according to what we have just seen, the series converges, even when we give all its terms the same sign. The series expresses the value 2^m.

When $m = -\mu < 0$, the series cannot converge for $x = -1$, for we have $(1-1)^{-\mu} = \infty$. Accordingly if the series converge at all for $x = +1$, it can do so only conditionally. We obtain:

---

*) Newton in the letters for Leibnitz of 13 June and 24 October 1676.

$$1 - \frac{\mu}{1} + \frac{\mu(\mu+1)}{\underline{|2}} - \frac{\mu(\mu+1)(\mu+2)}{\underline{|3}} + \cdots (-1)^n \cdot \frac{\mu(\mu+1)(\mu+2)\cdots(\mu+n-1)}{\underline{|n}} \cdots$$

This series, in which the terms alternate in sign, cannot converge if $\mu - 1 > 0$, $\mu > 1$, for then they increase incessantly in amount.

But if $\mu - 1 < 0$, the terms decrease incessantly in amount and consequently become zero as we saw above for R;

But a series, whose terms have alternate signs, decrease and have zero as limit, always converges.

For, if we denote the sum of the series from the nth term by

$$R_n = (u_n - u_{n+1}) + (u_{n+2} - u_{n+3}) \ldots$$

or

$$R_n = u_n - (u_{n+1} - u_{n+2}) - (u_{n+3} - u_{n+4}) \ldots$$

we perceive that R_n is positive but smaller than u_n. As n increases, the limit of R_n therefore, in like manner as that of u_n is zero.

Therefore the binomial series

$$(1 + x)^m = 1 + \frac{m}{1} x + \frac{m(m-1)x^2}{\underline{|2}} + \cdots$$

converges absolutely for all positive values of m even at the limits ± 1; on the other hand a negative m must be > -1 in order that the series may converge also for $x = +1$ and then it does so conditionally.

Although restricted in its convergence, the series can still be employed in extracting an arbitrary root of an arbitrary number; for, if a be the given base, $m = \frac{p}{q}$ a rational fraction, let us determine a number b^q as near as possible to a and put

$$a^{\frac{p}{q}} = \left(b^q - (b^q - a)\right)^{\frac{p}{q}} = b^p \left(1 - \frac{b^q - a}{b^q}\right)^{\frac{p}{q}},$$

then the binomial can be expanded.

47. Logarithmic series.*)

$$y = f(x) = l(1 + x), \text{ then } f^n(x) = \frac{(-1)^{n-1} \underline{|n-1}}{(1+x)^n}, \; x > -1,$$

$$l(1 + x) = x - \frac{x^2}{2} + \frac{x^3}{3} - \frac{x^4}{4} + \cdots \frac{(-1)^n x^{n-1}}{n-1} + R_n$$

$$R_n = \frac{(-1)^{n-1}}{n} \cdot \frac{x^n}{(1+\Theta x)^n}, \text{ or } (-1)^{n-1} \cdot \frac{(1-\Theta)^{n-1} x^n}{(1+\Theta x)^n}.$$

The remainder in the second form converges to zero when

$$\text{abs}\left[\frac{(1-\Theta)x}{1+\Theta x}\right] < 1, \text{ that is } -1 < x < +1.$$

---

*) Nic. Mercator (Logarithmotechnia 1668) and James Gregory (1636—1675) (Exercit. geometr. 1668); to the latter is due in the same work the series for $\tan^{-1} x$.

The first form of the remainder shows that the series still converges for $x = 1$.

Therefore

$$l(1+x) = x - \frac{x^2}{2} + \frac{x^3}{3} - \frac{x^4}{4} + \cdots (-1)^{n-1}\frac{x^n}{n}\cdots, \; -1 < x \leq +1.$$

In the particular case

$$l(2) = 1 - \tfrac{1}{2} + \tfrac{1}{3} - \tfrac{1}{4} + \tfrac{1}{5} - \cdots$$

of this last, we have an example of a merely conditionally convergent series; for neither the series

$$1 + \tfrac{1}{2} + \tfrac{1}{3} + \tfrac{1}{4} + \tfrac{1}{5} + \cdots,$$

nor the series

$$1 + \tfrac{1}{3} + \tfrac{1}{5} + \tfrac{1}{7}\cdots, \text{ nor } \tfrac{1}{2} + \tfrac{1}{4} + \tfrac{1}{6} + \tfrac{1}{8}\cdots,$$

are convergent, although their terms decrease and converge to zero, on the contrary, their sums increase beyond any finite amount; for

$$1 + \tfrac{1}{2} = \tfrac{3}{2}, \; \tfrac{1}{3} + \tfrac{1}{4} > \tfrac{1}{2}, \; \tfrac{1}{5} + \tfrac{1}{6} + \tfrac{1}{7} + \tfrac{1}{8} > \tfrac{1}{2}; \cdots$$

$$\frac{1}{2^n+1} + \frac{1}{2^n+2} + \cdots \frac{1}{2^{n+1}} > \tfrac{1}{2}, \text{ etc.*)}.$$

To obtain series useful for calculating the logarithm of **any** positive number, let us put $-x$ for x in the series just found, then if $x < 1$

$$l(1-x) = -x - \frac{x^2}{2} - \frac{x^3}{3} - \frac{x^4}{4}\cdots - \frac{x^{n-1}}{n-1} + R',$$

therefore

$$l\,\frac{1+x}{1-x} = 2\left(x + \frac{x^3}{3} + \frac{x^5}{5} + \cdots \frac{x^{2k+1}}{2k+1}\right) + R - R',$$

and as $R - R'$ converges to zero in the assumed interval, we have

---

*) The above divergent series $1 + \frac{1}{2} + \frac{1}{3} + \frac{1}{4}\cdots$ is called the **harmonic series**. It is important for subsequent applications to remark that the series

$$\frac{1}{1^\mu} + \frac{1}{2^\mu} + \frac{1}{3^\mu} + \frac{1}{4^\mu} + \frac{1}{5^\mu}\cdots$$

converges for all values of $\mu > 1$. For, grouping as above $\frac{1}{1^\mu} = \frac{1}{1^\mu}$,

$$\frac{1}{2^\mu} + \frac{1}{3^\mu} < 2\cdot\frac{1}{2^\mu} = \frac{1}{2^{\mu-1}},$$

$$\frac{1}{4^\mu} + \frac{1}{5^\mu} + \frac{1}{6^\mu} + \frac{1}{7^\mu} < 4\cdot\frac{1}{4^\mu} = \left(\frac{1}{2^{\mu-1}}\right)^2,$$

$$\frac{1}{8^\mu} + \frac{1}{9^\mu} + \cdots\cdots\cdots \frac{1}{15^\mu} < 8\cdot\frac{1}{8^\mu} = \left(\frac{1}{2^{\mu-1}}\right)^3,$$

. . . .

we see, that the sum of any number of terms of the series remains less than the sum of the same number of corresponding terms in the geometric progression whose ratio is the proper fraction $\frac{1}{2^{\mu-1}}$.

$$l\frac{1+x}{1-x} = 2\left(x + \frac{x^3}{3} + \frac{x^5}{5} + \cdots \frac{x^{2k+1}}{2k+1} + \cdots\right), \quad 0 < x < 1.$$

Now substitute:

$$\frac{1+x}{1-x} = \frac{z+a}{z} \text{ and so } x = \frac{a}{2z+a}, \quad a > 0,$$

then: $0 < z < \infty$

$$l(z+a) = l(z) + 2\left(\frac{a}{2z+a} + \tfrac{1}{3}\left(\frac{a}{2z+a}\right)^3 + \tfrac{1}{5}\left(\frac{a}{2z+a}\right)^5 + \cdots\right).$$

For instance, $z = 1$, $a = 1$:

$$l(2) = 2(\tfrac{1}{3} + \tfrac{1}{3}(\tfrac{1}{3})^3 + \tfrac{1}{5}(\tfrac{1}{3})^5 + \tfrac{1}{7}(\tfrac{1}{3})^7 + \cdots);$$

$z = 2$, $a = 1$:

$$l(3) = l(2) + 2(\tfrac{1}{5} + \tfrac{1}{3}(\tfrac{1}{5})^3 + \tfrac{1}{5}(\tfrac{1}{5})^5 + \cdots).$$

To pass from natural logarithms, with the base e, to common logarithms with the base 10, since ${}^{10}\log a = {}^{e}\log a : {}^{e}\log 10$ we have to calculate the number:

$$l(10) = l(2) + l(5) = 2.3025850929\ldots,$$

then we must multiply all values by $\frac{1}{l(10)} = 0.4342944819\ldots$

48. Circular series. $y = \tan^{-1}x$.

An independent expression of the nth derivate of the circular function $f(x) = \tan^{-1}x$ was given in § 36:

$$f^n(x) = \lfloor n-1 \cos^n y \sin n(y + \tfrac{1}{2}\pi) = \frac{\lfloor n-1}{(1+x^2)^{\frac{n}{2}}} \sin n(\tan^{-1}x + \tfrac{1}{2}\pi).$$

Now since for $x = 0$, y is also $= 0$, it follows that for this value:

$$f''(0) = 0, \ f^{\text{IV}}(0) = 0, \ f^{\text{VI}}(0) = 0 \ldots f^{2k}(0) = 0$$

$$f'(0) = 1, \ f'''(0) = -\lfloor 2, \ f^{\text{V}}(0) = \lfloor 4 \cdots f^{2k+1}(0) = (-1)^k \lfloor 2k.$$

The remainder R is by the first formula:

$$\frac{\lfloor n-1 \cdot x^n}{\lfloor n\,(1+\Theta^2x^2)^{\frac{n}{2}}} \sin n\left(\tan^{-1}\Theta x + \tfrac{1}{2}\pi\right) = \frac{1}{n}\left(\frac{x^2}{1+\Theta^2x^2}\right)^{\frac{n}{2}} \sin n\left(\tan^{-1}\Theta x + \tfrac{1}{2}\pi\right).$$

This first factor converges to zero, the third has a finite value. The middle factor does not become infinite for $n = \infty$ when the quotient within brackets is equal to or less than 1 for all values of Θ, i. e. when $x^2 \leq 1$. We have thus:

$$\tan^{-1}x = \frac{x}{1} - \frac{x^3}{3} + \frac{x^5}{5} \cdots (-1)^k \frac{x^{2k+1}}{2k+1} + \cdots, \quad -1 \leq x \leq +1.$$

The value of x being any proper fraction, this series presents the corresponding angle between $-\tfrac{1}{4}\pi$ and $+\tfrac{1}{4}\pi$.

The angle whose tangent has the value $+1$ is equal to $\tfrac{1}{4}\pi$; this number which is of fundamental importance for the periods of trigonometric functions can therefore be calculated by the formula:

$$\frac{\pi}{4} = 1 - \tfrac{1}{3} + \tfrac{1}{5} - \tfrac{1}{7} \cdots (-1)^k \frac{1}{2k+1} + \cdots,$$

but the convergence of this series is very slow, i.e. many terms must be summed in order to obtain a value at all approximate*); series that converge more rapidly can however be formed for calculating π. If x be a fraction small enough to give quickly an approximate value of $\varphi = \tan^{-1} x$ by the above series, let us form $\tan(m\varphi)$, where m is a positive integer to be chosen so that $m\varphi$ shall be nearly equal to $\frac{1}{4}\pi$, and $m\varphi - \frac{1}{4}\pi$ therefore be a very small angle; this will make

$$\tan(m\varphi - \tfrac{1}{4}\pi) = \frac{\tan m\varphi - 1}{1 + \tan m\varphi}$$

also a small fraction, so that $m\varphi - \frac{1}{4}\pi = \tan^{-1}\left(\frac{\tan m\varphi - 1}{1 + \tan m\varphi}\right)$ can be calculated from the series with rapid convergence. For $x = \frac{1}{5}$, $m = 4$:

$$\tan 2\varphi = \frac{2t\varphi}{1 - t^2\varphi} = \tfrac{5}{12}, \ \tan 4\varphi = \tfrac{120}{119}, \ \frac{\tan 4\varphi - 1}{1 + \tan 4\varphi} = \tfrac{1}{239},$$

therefore**)

$$\tfrac{1}{4}\pi = 4\tan^{-1}\tfrac{1}{5} - \tan^{-1}\tfrac{1}{239} = 4\{\tfrac{1}{5} - \tfrac{1}{3}\cdot(\tfrac{1}{5})^3 + \tfrac{1}{5}(\tfrac{1}{5})^5 - \tfrac{1}{7}(\tfrac{1}{5})^7 + \cdots\}$$
$$-\{\tfrac{1}{239} - \tfrac{1}{3}(\tfrac{1}{239})^3 + \tfrac{1}{5}(\tfrac{1}{239})^5 - \cdots\}$$
$$\pi = 3.1415926535\ldots$$

In order to calculate the angle corresponding to a tangent which is greater than 1, let us remember, that for

$$0 < \varphi < \tfrac{1}{2}\pi, \ \tan(\tfrac{1}{2}\pi - \varphi) = \frac{1}{\tan\varphi},$$

and for

$$-\tfrac{1}{2}\pi < \varphi < 0, \quad \tan(-\tfrac{1}{2}\pi - \varphi) = \frac{1}{\tan\varphi}.$$

Accordingly, if we put $\tan\varphi = x$, we have

$$\pm\tfrac{1}{2}\pi = \tan^{-1}x + \tan^{-1}\frac{1}{x}, \quad (x \gtrless 0),$$

therefore:

$$\tan^{-1}x = \pm\tfrac{1}{2}\pi - \tan^{-1}\frac{1}{x} = \pm\tfrac{1}{2}\pi - \left\{\frac{1}{x} - \frac{1}{3x^3} + \frac{1}{5x^5} - \frac{1}{7x^7} + \cdots\right\}.$$

We have thus obtained a series with ascending powers of $\frac{1}{x}$ or with descending powers of x.

---

*) When we add an odd number of terms in the above series we find a superior limit of the required value, an even number gives us an inferior limit. If ex. gr. the two limits are not to differ before the 11th place of decimals we must sum $\frac{1}{2}10^{10}$ places in the series.

**) First established by Machin, who in 1706 calculated π to 100 decimal places (vide Klügel: Math. Wörterbuch: Cyclotechnie). To obtain π accurately to 10 places, it is here enough to sum 15 terms of the first series and 8 of the second.

49. In the series for $\sin x$ and $\tan^{-1}x$ only odd powers of x occur, in that for $\cos x$ only even powers including zero. The first two functions are therefore characterised as odd, the other as even.

An odd function $f(x)$ can in general be defined by the property: that $f(x) = -f(-x)$; for an even function we have $f(x) = f(-x)$. Thence follows that an odd function, provided it is continuous for $x = 0$, must there vanish; it follows further by differentiation that all its derivates of odd order are even functions, on the other hand its even derivates are odd functions. Therefore these latter must also all vanish at the point zero. In the case of even functions on the other hand, all odd derivates are odd functions and vanish for $x = 0$.

50. The development of Taylor's series is based on the formation of the nth derivate. This marks the limit of its applicability; for, if for any function the general expression of this derivate be too complicated, the method loses in practicability. Thus, it is not hard to calculate in general from the recurring formula established for $\sin^{-1}x$ the values of the derived functions for $x = 0$, but that formula is not suited for forming the remainder.*) Therefore our first endeavour must be to decide as to the developability of a function and the convergence of the series of powers from the properties of the function itself exclusively, not also taking account as heretofore of the properties of all its derived functions. But then we shall recognise that a series of powers obtained in any way for a function in an interval must be identical with the series of Taylor, because $f(x)$ cannot be expressed by two different series of powers. The investigations require — for completeness — the extension of the domain of number and for this new conceptions must first be introduced by the theory of functions with more than one independent variable.

---

*) For the expansion of $\sin^{-1}x$ in a series, see Integral Calculus, Bk. III Chap. IV.

Functions of more than one independent variable.

51. When the value of a variable z is determined by the values of two independent variables x and y in such a way, that to each value of x in the interval from a to b and to each value of y in the interval from α to β belong one or more values of z, then z is said to be a function of the two independent variables x and y. Here we can also classify functions after the nature of their analytical expression into algebraic and transcendental, and the form in which the function is presented may be explicit: $z = f(x, y)$, or implicit: $f(x, y, z) = 0$, or again it may be brought about by two variable parameters:

$$x = \varphi(u, v),\ y = \psi(u, v),\ z = \chi(u, v).$$

The total course of the function is exhibited to intuition by the aid of a system of Cartesian coordinates in space — most simply by a rectangular one — each system of values x and y is represented by a point in the plane of xy, and from this the corresponding value of z is erected perpendicularly to the plane, towards one side or the other according as it is positive or negative. The extremity of this perpendicular represents the simultaneous system of values, x, y, z. The interval from $x = a$ to b, $y = \alpha$ to β determines in the plane xy a rectangle, the domain, for which the function is defined, and the points constructed lie above and below this. If x and y go through all values from $-\infty$ to $+\infty$, these points spread over the entire plane. A general view of the distribution of the points is arrived at, by beginning with a fixed value of one variable ex. gr. $x = a$, and giving to y different values between α and β; connecting the points thus constructed, a polygon arises in space, of which the right line $x = a$ is the projection in the plane xy. As the value x is altered, different polygons are obtained for these values of y; if we conceive the points for which y has the same value to be connected, there arises a net whose quadrangular meshes are more and more diminished by interpolating further points and such that it can have as its limit a determinate surface. This surface is accordingly the complete image of the function, its intersections with planes parallel to those of yz or zx are curves which form the limits of the polygons first constructed.

52. Considering the explicit function $z = f(x, y)$ let us assume it to be one-valued, and enquire when is it continuous in a domain for which it has determinate values. Conceive any point of the domain inclosed in a small rectangle having the lengths of its sides parallel to the axis of abscissæ $= 2h$ and of those parallel to the ordinate axis $= 2k$, so that $x \pm h$, $y \pm k$ are the coordinates of its four corners. Thus the coordinates of any point within this region or on its boundary are $x \pm \Theta h$, $y \pm \eta k$, $(0 \leq \Theta \leq 1,\ 0 \leq \eta \leq 1)$. If we denote by $f(x \pm \Theta h,\ y \pm \eta k)$ the corresponding value of the function, *then the function shall be called continuous at x, y, only when finite values of h and k can be found, for which the absolute amount of the difference: $f(x \pm \Theta h,\ y \pm \eta k) - f(x, y)$ is less than a prescribed arbitrarily small number δ for every value of the independent variables Θ and η.* For then and only then will every series of numbers obtained from $f(x \pm \Theta h, y \pm \eta k) - f(x, y)$ by making Θ and η converge to zero in any manner whatever, have zero as its limit. It is therefore necessary for continuity, that $f(x \pm \Theta h, y) - f(x, y)$ and likewise $f(x, y \pm \eta k) - f(x, y)$ become infinitely small, or in other words: that $f(x, y)$ be continuous as a function of the variable x alone or of the variable y alone; but yet this is not sufficient. Therefore to say $f(x, y)$ is a continuous function of both variables x and y, is different from saying that f is a continuous function of x as well as of y.

On the other hand we can replace the above definition by its equivalent: It must be possible to find at the point x, y, a finite value h and a finite value k, so that for all values equal to or less than h or k respectively, $f(x \pm \Theta h, y)$ shall be a continuous function of y alone, and $f(x, y \pm \eta k)$ a continuous function of x alone, in such a way that independently of Θh, we shall have

$$\text{abs}\,[f(x \pm \Theta h,\ y \pm \eta k) - f(x \pm \Theta h,\ y)] < \delta$$

for all values of η merely by the value chosen for k, and in like manner that independently of ηk, we shall have

$$\text{abs}\,[f(x \pm \Theta h,\ y \pm \eta k) - f(x, y \pm \eta k)] < \delta$$

for all values of Θ merely by the value chosen for h.

These conditions are enunciated in the words: $f(x, y)$ must be a uniformly continuous function of x as well as of y in the neighbourhood of the point x, y.

For according to this way of putting it, if we assume $\eta = 0$ in the second inequality, we have also for all values of Θ

$$\text{abs}\,[f(x \pm \Theta h, y) - f(x, y)] < \delta.$$

This inequality added to the first shows that

$$\text{abs}\,[f(x \pm \Theta h,\ y \pm \eta k) - f(x, y)]$$

becomes less than the arbitrarily small quantity 2δ for all values of Θ and η, merely by the values chosen for h and k.

This way of putting it is important, because it reduces the investigation of continuity of a function with two variables to that of uniform continuity in regard to each of them.

Examples:

1. The function $z = ax^\mu y^\nu$, where μ and ν are positive integers and a an arbitrary constant, is a continuous function of both variables.

For, the absolute amount of

$$(y \pm \eta k)^\nu - y^\nu = \{\nu_1(\pm \eta k) y^{\nu-1} + \nu_2(\pm \eta k)^2 y^{\nu-2} + \cdots (\pm \eta k)^\nu\}$$

is less than

$$N[\eta k + (\eta k)^2 + \cdots (\eta k)^\nu] = N \frac{(\eta k) - (\eta k)^{\nu+1}}{1-(\eta k)},$$

where N signifies the greatest among the coefficients within the above brackets. Assuming $\eta k < 1$, then the amount of the difference

$$(y \pm \eta k)^\nu - y^\nu \text{ is } < N\frac{\eta k}{1-\eta k}.$$

Consequently

$$a(x \pm \Theta h)^\mu [(y \pm \eta k)^\nu - y^\nu] < a(x \pm \Theta h)^\mu \cdot \frac{\eta k}{1-\eta k} \cdot N,$$

and if this is to be less than δ for all values of Θ and η, denoting by X the greatest absolute value which $(x \pm \Theta h)^\mu$ takes for all values of Θ, we have in order to determine k

$$\frac{\eta k}{1-\eta k} < \frac{\delta}{aXN}, \text{ i. e. } k < \frac{\delta}{\delta + aXN}.$$

A like consideration shows that the difference

$$a(y \pm \eta k)^\nu [(x \pm \Theta h)^\mu - x^\mu] \text{ is also } < \delta \text{ when } h < \frac{\delta}{\delta + aYM}.$$

The considerations in this example serve for the proof of the general theorem:

If $f(x, y) = \varphi(x) \cdot \psi(y)$ where φ and ψ are continuous functions of the variables x and y, then f is a uniformly continuous function of x as well as of y, i. e. a continuous function of both variables. For

$$\varphi(x \pm \Theta h)[\psi(y \pm \eta k) - \psi(y)]$$

can be made $< \delta$, exclusively by choice of k independently of Θh, and

$$\psi(y \pm \eta k)[\varphi(x \pm \Theta h) - \varphi(x)]$$

can be made $< \delta$, exclusively by choice of h independently of ηk.

2. $z = \frac{1}{xy}$ is discontinuous at all points of the right line $x = 0$ and of the right line $y = 0$. For, $\frac{1}{xy}$ is for all values of x a discontinuous function of y when $y = 0$, and for all values of y a discontinuous function of x when $x = 0$.

3. $z = \frac{y}{x}$ is discontinuous at all points of the right line $x = 0$ while y is finite, and is quite indeterminate at the point $x = 0$, $y = 0$.

4. The function $z = \sin(4\tan^{-1}\frac{y}{x})$ being defined for $x = 0$*) to be zero for all values of y including $y = 0$, is a discontinuous function in the point $x = 0$, $y = 0$, although for a constant y it is a continuous function of x and for a constant x a continuous function of y. But if we put $y = ax$ the function $\sin(4\tan^{-1}a)$ can take all possible values between the limits $+1$ and -1 as we approach the point $x = 0$, $y = 0$ in all possible directions, whereas it should there be zero: or, forming the differences for the neighbourhood of this point, no value can be chosen for k independently of Θh, nor for h independently of ηk which will make $\sin\left(4\tan^{-1}\frac{\pm\eta k}{\pm\Theta h}\right) < \delta$: the criterion of continuity therefore is not satisfied.

5. If we form the function $z = x^{\alpha} y^{-\beta}$, where α and β are positive and $\beta \leq \alpha$, and for all values of x replace it for $y = 0$ by the value $z = 0$, then it is a discontinuous function at $x = 0$, $y = 0$, although when we put $y = ax$ it is a continuous function: $z = x^{\alpha-\beta}a^{-\beta}$ of the variable x and so is continuous on every direction proceeding from the origin. For here too it is not possible to find a finite value of h independent of ηk, for which we have

$$(\pm\Theta h)^{\alpha}(\pm\eta k)^{-\beta} < \delta.$$

In a domain, in which the criterion of continuity holds without exception for every point, including its limits, $f(x, y)$ is a uniformly continuous function of both variables, i.e. a value can be assigned for h and one for k, which, whatever be the values of x and y, are sufficient to satisfy the inequality

$$f(x \pm \Theta h,\ y \pm \eta k) - f(x, y) < \delta, \quad (0 \leq \Theta \leq 1,\ 0 \leq \eta \leq 1).$$

For if it were assumed that such minima values could not be assigned to h and k, there should be points in the domain, in whose immediate neighbourhood the criterion of continuity could only be satisfied by h and k ultimately falling below any assignable value. That this is impossible is seen as follows: Suppose x_1, y_1 to be such a point, and determine h and k for a point $x_1 - \varepsilon$, $y_1 - \varepsilon'$ arbitrarily near it, so that

$$\text{abs}\,[f(x_1 - \varepsilon \pm \Theta h,\ y_1 - \varepsilon' \pm \eta k) - f(x_1 - \varepsilon,\ y_1 - \eta)] < \delta.$$

Now if the only way in which this inequality can continue to hold, when ε and ε' converge to zero, be by h and k also falling below any assignable value, then $-\varepsilon + \Theta h$ and $-\varepsilon' + \eta k$ will always remain less than zero, so that the point x_1, y_1 is never reached in this process.

*) Thomæ: Einleitung in die Theorie der bestimmten Integrale p. 31.

But this not being so, a finite region $\pm\Theta h$, $\pm\eta k$ can be assigned at the point x_1, y_1, for which

$$\text{abs}\,[f(x_1 \pm \Theta h,\ y_1 \pm \eta k) - f(x_1, y_1)] < \delta.$$

This finite region includes also the points $x_1 - \varepsilon$, $y_1 - \varepsilon'$ (for ε and ε' converge to zero, h and k have fixed finite values) and therefore the same assignable region is also sufficient for each of these points for satisfying the inequality, so that therefore what was assumed is in contradiction with the condition of continuity.

53. The first differential quotients of the function at a point, in whose neighbourhood it is continuous, can be formed in various ways: If we first leave y unaltered, while x increases or decreases by Δx, and denote the corresponding change of z by $\Delta_x z$, then the quotient of differences is

$$\frac{\Delta_x z}{\Delta x} = \frac{f(x \pm \Delta x, y) - f(x, y)}{\pm \Delta x}.$$

We assume that this approximates, when Δx converges to zero, to a determinate limiting value, as well for the $+$ as for the $-$ sign, but not necessarily the same for both. It is denoted after Jacobi by $\frac{\partial z}{\partial x}$ or $\frac{\partial f}{\partial x}$ and called the partial derived of z with regard to x progressive or regressive as the case may be, thus:

$$\frac{\partial f}{\partial x} = \frac{\partial z}{\partial x} = \text{Lim}\,\frac{f(x \pm \Delta x, y) - f(x, y)}{\pm \Delta x} \text{ for } \Delta x = 0.$$

The partial derived of z with regard to y secondly is got in the same way, x remaining unchanged:

$$\frac{\partial f}{\partial y} = \frac{\partial z}{\partial y} = \text{Lim}\,\frac{f(x, y \pm \Delta y) - f(x, y)}{\pm \Delta y} \text{ for } \Delta y = 0.$$

Obviously we have here also the proposition: If the progressive partial derived with regard to x or with regard to y be identical with the regressive one, the Theorem of the Mean Value holds:

$$f(x + h, y) - f(x, y) = hf'(x + \Theta h, y),$$
$$f(x, y + k) - f(x, y) = kf'(x, y + \eta k),$$

Θ and η will be respectively dependent on y and x.

But if x be changed by Δx and at the same time y by Δy, the ratio $\Delta y : \Delta x$ remaining quite arbitrary but finite, the increase is

$$\Delta z = f(x + \Delta x, y + \Delta y) - f(x, y).$$

Here also the question arises: What is the limiting value to which $\frac{\Delta z}{\Delta x}$ or $\frac{\Delta z}{\Delta y}$ tends, when Δx and Δy converge to zero in any manner in which their ratio always retains a finite limiting value $\frac{dy}{dx}$, the partial derivates $\frac{\partial f}{\partial x}$ and $\frac{\partial f}{d y}$ being supposed to have definite values in the neighbourhood of the point under consideration?

We have identically:

$$\frac{f(x+\Delta x, y+\Delta y)-f(x,y)}{\Delta x} = \frac{f(x+\Delta x, y+\Delta y)-f(x, y+\Delta y)}{\Delta x} - $$
$$+ \frac{f(x, y+\Delta y)-f(x, y)}{\Delta y} \cdot \frac{\Delta y}{\Delta x}.$$

If in the **first** quotient on the right we first make Δy vanish, it becomes $\frac{f(x+\Delta x, y)-f(x, y)}{\Delta x}$ and this passes over into $\frac{\partial f(x,y)}{\partial x}$ as Δx vanishes; but if reversing the order we first put $\Delta x = 0$, we get the expression $\frac{\partial f(x,y+\Delta y)}{\partial x}$, which will likewise become $\frac{\partial f(x,y)}{\partial x}$ for $\Delta y = 0$, only when it is a continuous function in regard to y. Now what value results when Δy and Δx converge to zero simultaneously in any manner? In order that $\frac{\partial f(x,y)}{\partial x}$ be again limiting value independently of the manner of convergence, the condition must be satisfied: that a Δx can be found and independently of it a Δy, such as will make the absolute amount of the difference

$$\text{abs}\left[\frac{f(x+\Delta x, y+\Delta y)-f(x,y+\Delta y)}{\Delta x} - \frac{f(x+\Theta\Delta x, y+\eta\Delta y)-f(x,y+\eta\Delta y)}{\Theta\Delta x}\right] < \delta,$$

where δ denotes an arbitrarily small quantity, while the proper fractions Θ and η assume all possible values. This inequality is expressed in the words: The quotient of differences must be a uniformly continuous function of Δx and of y.

This condition is necessary and sufficient — it cannot be replaced by any other. The differential quotient proceeds by continuous transition from the quotient of differences and we can therefore easily conclude, that from this requirement the continuity of the function $\frac{\partial f}{\partial x}$ in regard to y necessarily follows, without this therefore being fitted to replace the above. For, since the condition must be fulfilled for all values of Δy independently of the value Δx, it holds also for $\Delta x = 0$, i. e.,

$$\frac{\partial f(x, y+\Delta y)}{\partial x} - \frac{\partial f(x, y+\eta\Delta y)}{\partial x} < \delta. \qquad (0 \leq \eta \leq 1.)$$

If, putting $\Theta = 1$, $\eta = 0$, we write the above inequality in the form

$$\text{abs}\left[\frac{f(x+\Delta x, y+\Delta y)-f(x+\Delta x, y)}{\Delta y} - \frac{f(x, y+\Delta y)-f(x,y)}{\Delta y}\right]\frac{\Delta y}{\Delta x} < \delta,$$

since it holds for values of Δx and Δy however small, whose ratio has an arbitrary finite limiting value k, we see that the continuity of $\frac{\partial f}{\partial y}$ in regard to x is also involved in the above condition, for:

abs $\left[\frac{f(x+\Delta x, y+\Delta y)-f(x+\Delta x, y)}{\Delta y}-\frac{f(x, y+\Delta y)-f(x, y)}{\Delta y}\right]$ must be $< \delta k$,

therefore for $\Delta y = 0$, $\frac{\partial f(x+\Delta x, y)}{\partial y} - \frac{\partial f(x, y)}{\partial y}$ must also be $< \delta k$.

Accordingly the result is:

Provided we have at the point x, y, at which f is continuous, the quotient of differences:

$$\frac{f(x+\Delta x, y+\Delta y)-f(x, y+\Delta y)}{\Delta x}$$

a uniformly continuous function of Δx and y, then $\frac{\partial f}{\partial x}$ is a continuous function of the variable y, $\frac{\partial f}{\partial y}$ is a continuous function of the variable x, and for all values of $dy : dx$ the Total Differential Quotient with regard to x is equal to

$$\frac{dz}{dx} = \frac{\partial f}{\partial x} + \frac{\partial f}{\partial y} \cdot \frac{dy}{dx},$$

or in a more symmetrical statement: the Total Differential

$$dz = \frac{\partial f}{\partial x} dx + \frac{\partial f}{\partial y} dy,$$

i.e. is equal to the sum of the partial differentials.

The differential equation contrasts by its symmetry with the equation of differential quotients; but, since there are vanishing (infinitely small) quantities on both its sides, it derives a meaning only from the fact that a quotient equation can always be formed from it.

In most cases of calculation it is enough to replace the condition of this Theorem of the Total Differential by the narrower one: When the progressive and regressive differential quotient $\frac{\partial f}{\partial x}$ is a continuous function of both variables x and y in the neighbourhood of a point, and $\frac{\partial f}{\partial y}$ has a determinate value, we have also

$$dz = \frac{\partial f}{\partial x} dx + \frac{\partial f}{\partial y} dy.$$

For then we can replace the first quotient in the equation:

$$\frac{f(x+\Delta x, y+\Delta y)-f(x, y)}{\Delta x} = \frac{f(x+\Delta x, y+\Delta y)-f(x, y+\Delta y)}{\Delta x} + \frac{f(x, y+\Delta y)-f(x, y)}{\Delta y} \frac{\Delta y}{\Delta x},$$

by the mean value:

$$\frac{\partial f(x+\Theta \Delta x, y+\Delta y)}{\partial x},$$

which becomes $\frac{\partial f}{\partial x}$ for $\Delta x = 0$, $\Delta y = 0$.

Examples.

1. The function $z = \sqrt{x^2 + y^2}$ is unique and continuous even at the point $x = 0$, $y = 0$, but its first derived functions:

$$\frac{\partial z}{\partial x} = \frac{x}{\sqrt{x^2 + y^2}}, \quad \frac{\partial z}{\partial y} = \frac{y}{\sqrt{x^2 + y^2}}$$

have no determinate values at that point; therefore the Theorem of the Total Differential does not apply in this case, in the absence of further conventions respecting the partial derived functions.

2. If we replace in the function $z = (3x + 3) + y$ all values for $y = 0$ by the values $6x$, the function so formed is continuous in the neighbourhood of the point $x = 1$, $y = 0$; but $\frac{\partial f(1, 0)}{\partial x} = 6$, $\frac{\partial f(1, y)}{\partial x} = 3$ i. e. the partial differential quotient in regard to x is not a continuous function of y; in like manner $\frac{\partial f(1, 0)}{\partial y} = 1$, $\frac{\partial f(1 + \Delta x, 0)}{\partial y} = \mp\infty$; therefore although the function is continuous at the point, the Theorem of the Total Differential does not there hold for it.

3. An example, in which f is continuous in both variables and $\frac{\partial f}{\partial x}$ a continuous function of y and yet the Theorem of the Total Differential does not hold, is presented by the function: $z = x \sin\left(4 \tan^{-1} \frac{y}{x}\right)$ with the convention that for all values of y (including $y = 0$) whenever $x = 0$ we have also $z = 0$. This function is continuous in the neighbourhood of the point $x = 0$, $y = 0$. Now

$$\frac{\partial f(0, y)}{\partial x} = \operatorname{Lim} \frac{\Delta x \sin\left(4 \tan^{-1} \frac{y}{\Delta x}\right)}{\Delta x} = 0,$$

thus it is a continuous function of y. On the other hand we have

$$\frac{\partial f(x, 0)}{\partial y} = \operatorname{Lim} \frac{x \sin\left(4 \tan^{-1} \frac{\Delta y}{x}\right)}{\Delta y} = 4,$$

as long as x is different from 0; whereas $\frac{\partial f(0, 0)}{\partial y} = 0$. The Theorem of the Total Differential does not hold in this case.

The condition under which the Theorem of the Total Differential holds is the condition that the function can be represented by a surface. Just as we say of a function of one variable: it can be represented at a point by a curve, when the lines joining this point to neighbouring points converge to a fixed limiting position, so we say of a function of two variables: it can be represented at a point as a surface, when every plane, which can be drawn through that point and through any two other points belonging to the function, converges to the same fixed limiting position, in whatever manner these other two points close up to the original point. (It is a peculiarity in a surface when it behaves at a point as a cone does at its vertex;

in such a case there can no longer be any such thing as a fixed plane; the first partial derived functions are indeterminate.) Let the coordinates of the points be respectively x, y, z; $x+\Delta_1 x$, $y+\Delta_1 y$, $z+\Delta_1 z$; $x+\Delta_2 x$, $y+\Delta_2 y$, $z+\Delta_2 z$, so that

$$z=f(x,y),\quad z+\Delta_1 z=f(x+\Delta_1 x, y+\Delta_1 y),$$
$$z+\Delta_2 z=f(x+\Delta_2 x, y+\Delta_2 y).$$

The equation of a plane through these points, denoting its current coordinates by ξ, η, ζ, is:

$$\zeta - z = A(\xi - x) + B(\eta - y).$$

$$A=\frac{\Delta_1 z\Delta_2 y-\Delta_2 z\Delta_1 y}{\Delta_1 x\Delta_2 y-\Delta_2 x\Delta_1 y}=\frac{\frac{\Delta_1 z}{\Delta_1 x}\frac{\Delta_2 y}{\Delta_2 x}-\frac{\Delta_2 z}{\Delta_2 x}\frac{\Delta_1 y}{\Delta_1 x}}{\frac{\Delta_2 y}{\Delta_2 x}-\frac{\Delta_1 y}{\Delta_1 x}}$$

$$B=\frac{\Delta_2 z\Delta_1 x-\Delta_1 z\Delta_2 x}{\Delta_1 x\Delta_2 y-\Delta_2 x\Delta_1 y}=\frac{\frac{\Delta_2 z}{\Delta_2 x}-\frac{\Delta_1 z}{\Delta_1 x}}{\frac{\Delta_2 y}{\Delta_2 x}-\frac{\Delta_1 y}{\Delta_1 x}}.$$

Now if $\Delta_1 x$ and $\Delta_1 y$, as well as $\Delta_2 x$ and $\Delta_2 y$ converge to zero, while the limiting values of their ratios are denoted respectively by $\left(\frac{dy}{dx}\right)_1$, $\left(\frac{dy}{dx}\right)_2$, then, in case the Theorem of the Total Differential holds, we shall have:

$$\frac{\Delta_1 z}{\Delta_1 x}=\frac{\partial f}{\partial x}+\frac{\partial f}{\partial y}\left(\frac{dy}{dx}\right)_1,\quad \frac{\Delta_2 z}{\Delta_2 x}=\frac{\partial f}{\partial x}+\frac{\partial f}{\partial y}\left(\frac{dy}{dx}\right)_2,$$

therefore: $A=\frac{\partial f}{\partial x}$, $B=\frac{\partial f}{\partial y}$. And it follows conversely from these equations: if A and B in every limiting process take these values, then the Theorem of the Total Differential holds.

54. We now proceed to form the partial derived functions of higher order according to the rules for functions with one variable. By $\frac{\partial^2 f}{\partial x^2}$ we denote the function which is found by taking the derivate of $\frac{\partial f}{\partial x}$ with regard to x, and can again define it through the original function by means of the equation:

$$\frac{\partial^2 f}{\partial x^2}=\operatorname{Lim}\frac{f(x+2\Delta x,y)-2f(x+\Delta x,y)+f(x,y)}{\Delta x^2}$$

for $\Delta x=0$. Similarly:

$$\frac{\partial^2 f}{\partial y^2}=\frac{\partial\frac{\partial f}{\partial y}}{\partial y}=\operatorname*{Lim}_{\Delta y=0}\frac{f(x,y+2\Delta y)-2f(x,y+\Delta y)+f(x,y)}{\Delta y^2}.$$

By further differentiations the derivates $\frac{\partial^n f}{\partial x^n}$ and $\frac{\partial^n f}{\partial y^n}$ are obtained. But there can also be formed what are called mixed differential coefficients: when the function $\frac{\partial f}{\partial x}$ is differentiated with regard to y, a derivate is found which must be denoted by $\frac{\partial^2 f}{\partial y \partial x}$, and likewise when $\frac{\partial f}{\partial y}$ is differentiated with regard to x, $\frac{\partial^2 f}{\partial x \partial y}$. The two are defined through the original function by the equations

$$\frac{\partial^2 f}{\partial y \partial x} = \operatorname*{Lim}_{\Delta y = 0} \operatorname*{Lim}_{\Delta x = 0} \frac{\frac{f(x+\Delta x, y+\Delta y) - f(x, y+\Delta y)}{\Delta x} - \frac{f(x+\Delta x, y) - f(x, y)}{\Delta x}}{\Delta y}$$

$$\frac{\partial^2 f}{\partial x \partial y} = \operatorname*{Lim}_{\Delta x = 0} \operatorname*{Lim}_{\Delta y = 0} \frac{\frac{f(x+\Delta x, y+\Delta y) - f(x+\Delta x, y)}{\Delta y} - \frac{f(x, y+\Delta y) - f(x, y)}{\Delta y}}{\Delta x}.$$

The expressions on the right are identically equal to:

$$\frac{f(x+\Delta x, y+\Delta y) - f(x, y+\Delta y) - f(x+\Delta x, y) + f(x, y)}{\Delta x \Delta y} = \psi(\Delta x, \Delta y)$$

and differ from each other only in this, that in the first case the limiting value has to be formed by making Δx first converge to zero and then Δy, on the other hand in the second case in reversed order first Δy becomes zero then Δx; the question is, must these limiting values be identical? We assert: *They are identical always at a point, in whose neighbourhood* $\frac{\partial f}{\partial x}$ *and* $\frac{\partial^2 f}{\partial y \partial x}$ $\left(\textit{or again } \frac{\partial f}{\partial y} \textit{ and } \frac{\partial^2 f}{\partial x \partial y}\right)$ *are continuous functions of both variables* x *and* y; without implying that this is at the same time the necessary condition. For, in consequence of our hypotheses we can apply the Theorem of the Mean Value to the function

$$\Pi(x, y) = f(x, y+\Delta y) - f(x, y)$$

in which we first consider y and Δy as constant, and x as variable

$$\Pi(x+\Delta x, y) - \Pi(x, y) = \frac{\partial \Pi(x+\Theta \Delta x, y)}{\partial x} \cdot \Delta x,$$

or in full

$$\{f(x+\Delta x, y+\Delta y) - f(x+\Delta x, y)\} - \{f(x, y+\Delta y) - f(x, y)\}$$
$$= \Delta x \left\{\frac{\partial f(x+\Theta \Delta x, y+\Delta y)}{\partial x} - \frac{\partial f(x+\Theta \Delta x, y)}{\partial x}\right\}.$$

Accordingly

$$\psi(\Delta x, \Delta y) = \frac{\frac{\partial f(x+\Theta \Delta x, y+\Delta y)}{\partial x} - \frac{\partial f(x+\Theta \Delta x, y)}{\partial x}}{\Delta y} = \frac{\partial^2 f(x+\Theta \Delta x, y+\eta \Delta y)}{\partial y \partial x}.$$

For the theorem of the mean value, by hypothesis, likewise holds for the function $\frac{\partial f}{\partial x}$. If now $\frac{\partial^2 f}{\partial y \partial x}$ be a continuous function of both variables in the neighbourhood of x, y, $\frac{\partial^2 f(x+\Theta \Delta x, y+\eta \Delta y)}{\partial y \partial x}$ yields the same value, whether Δx first vanishes and then Δy, or first Δy and then Δx, i. e. ψ presents the same value independently of the order $\Delta x = 0$, $\Delta y = 0$. Therefore in this case the limiting values $\frac{\partial^2 f}{\partial y \partial x}$ and $\frac{\partial^2 f}{\partial x \partial y}$ are identical.

The only remark still to be made is that the value of $\frac{\partial^2 f}{\partial y\,\partial x}$ or ψ can also increase beyond all limits; though under the circumstances assumed, namely that the theorem of the mean value is to continue to hold, this can only occur by ψ becoming determinately infinite however we may approach this point; the limiting values then remain equal to one another, $+\infty$ or $-\infty$.

We can now further conclude, that under corresponding circumstances the order of differentiation is indifferent also for higher partial derived functions. For if

$$\frac{\partial^2 f}{\partial y\,\partial x} = \frac{\partial^2 f}{\partial x\,\partial y}$$

then it follows by differentiation ex. gr. with regard to x that:

$$\frac{\partial^3 f}{\partial x\,\partial y\,\partial x} = \frac{\partial^3 f}{\partial x^2\,\partial y}.$$

If then we put $\frac{\partial f}{\partial x} = p$ and the theorem just proved holds for the function p, that is to say, if not only $\frac{\partial p}{\partial x} = \frac{\partial^2 f}{\partial x^2}$ but also

$$\frac{\partial^2 p}{\partial y\,\partial x} = \frac{\partial^3 f}{\partial y\,\partial x\,\partial x}$$

be a continuous function of both variables, we have:

$$\frac{\partial^2 p}{\partial x\,\partial y} = \frac{\partial^2 p}{\partial y\,\partial x}, \text{ i. e. } \frac{\partial^3 f}{\partial x\partial y\partial x} = \frac{\partial^3 f}{\partial y\,\partial x^2} = \frac{\partial^3 f}{\partial x^2\,\partial y}. \quad \text{Q. E. D.}$$

55. By the help of the higher partial derived functions the higher total differential quotients are expressed as follows: In the function

$$\frac{dz}{dx} = \frac{\partial f(x,y)}{\partial x} + \frac{\partial f(x,y)}{\partial y}\cdot\frac{dy}{dx},$$

which depends on the two variables x and y, let x increase by Δx, y by Δy, then the limiting value of the quotient of differences $\frac{\Delta \frac{dz}{dx}}{\Delta x}$, which we shall have to denote by $\frac{d^2 z}{dx^2}$ when Δx vanishes, is to be calculated from the form:

$$\frac{d^2 z}{dx^2} = \text{Lim}\,\frac{\frac{\partial f(x+\Delta x, y+\Delta y)}{\partial x} - \frac{\partial f(x,y)}{\partial x}}{\Delta x}$$

$$+\frac{dy}{dx}\cdot\text{Lim}\left\{\frac{\frac{\partial f(x+\Delta x, y+\Delta y)}{\partial y} - \frac{\partial f(x,y)}{\partial y}}{\Delta x}\right\} + \frac{\partial f(x,y)}{\partial y}\cdot\text{Lim}\,\frac{\Delta\frac{dy}{dx}}{\Delta x}.$$

The first limiting value on the right is by the previous propositions the total derived function of $\frac{\partial f}{\partial x}$ with regard to x, therefore is equal to $\frac{\partial^2 f}{\partial x^2} + \frac{\partial^2 f}{\partial y\,\partial x}\cdot\frac{dy}{dx}$; likewise the second, the total derived of $\frac{\partial f}{\partial y}$ is equal to $\frac{\partial^2 f}{\partial x \partial y} + \frac{\partial^2 f}{\partial y^2} \cdot \frac{dy}{dx}$; the third limiting value is undetermined, just as $\frac{dy}{dx}$ itself is, as long as no law is assigned between the change of the variable x and that of the variable y; but if such a dependence exist, we shall have to denote Lim $\frac{\Delta \frac{dy}{dx}}{\Delta x}$ by $\frac{d^2 y}{dx^2}$, (ex. gr. when the increments of x are in a constant ratio to those of y, $dy : dx = k$, $d^2 y : dx^2 = 0$); therefore if $\frac{\partial^2 f}{\partial x \partial y} = \frac{\partial^2 f}{\partial y \partial x}$,

we have $\frac{d^2 z}{dx^2} = \frac{\partial^2 f}{\partial x^2} + 2 \frac{\partial^2 f}{\partial x \partial y} \cdot \frac{dy}{dx} + \frac{\partial^2 f}{\partial y^2} \left(\frac{dy}{dx}\right)^2 + \frac{\partial f}{\partial y} \cdot \frac{d^2 y}{dx^2}$.

It is to be remarked, that this expression is not symmetrically formed relatively to the differentials of x and y; this depends on the fact that we considered the variable x as independent, and formed the higher quotients of differences with regard to x (§ 33). It becomes symmetrical either, when as y varies in proportion to x, $\frac{dy}{dx}$ is to be considered constant, for then equal changes of y belong to equal changes of x so that as just stated $\Delta \frac{dy}{dx} = 0$,and consequently:

$$\frac{d^2 z}{dx^2} = \frac{\partial^2 f}{\partial x^2} + 2 \frac{\partial^2 f}{\partial x \partial y} \cdot \frac{dy}{dx} + \frac{\partial^2 f}{\partial y^2} \left(\frac{dy}{dx}\right)^2,$$

or, when the change of x as well as that of y is to be made depend on a third quantity.

This case we must consider more closely. If x as well as y be functions of the independent variable t whose change therefore conditions the change of value of x and of y, the differentials dx and dy are functions of t, multiplied by the differential dt; accordingly the differential quotient $\frac{dy}{dx}$ is a function of t. Writing $dx = \varphi(t) dt$, $dy = \psi(t) dt$, then $\frac{dy}{dx} = \frac{\psi(t)}{\varphi(t)}$; if t change and the differential quotient is to be determined, we have

$$\frac{d\left(\frac{dy}{dx}\right)}{dt} = \frac{\varphi(t)\psi'(t) - \varphi'(t)\psi(t)}{\varphi(t)^2}.$$

Evidently, in consequence of the equations $\frac{dx}{dt} = \varphi(t)$, $\frac{dy}{dt} = \psi(t)$ we can also write: $\frac{d^2 x}{dt^2} = \varphi'(t)$, $\frac{d^2 y}{dt^2} = \psi'(t)$, or $d^2 x = \varphi'(t) dt^2$, $d^2 y = \psi'(t) dt^2$; introducing these values into the above equation, it assumes the form

$$\frac{d\left(\frac{dy}{dx}\right)}{dt} = \frac{dx\, d^2 y - dy\, d^2 x}{dx^2 dt} \quad \text{or } d\left(\frac{dy}{dx}\right) = \frac{dx\, d^2 y - dy\, d^2 x}{dx^2},$$

i.e. if in a differential quotient, numerator and denominator are conceived to be functions of an independent variable, its differential relatively to this variable is formed according to the general rule for a quotient.

If the variable t coincide with x, we have $x = t$, hence

$$\varphi(t) = 1, \quad \varphi'(t) = 0,$$

therefore $d^2x = 0$ and we obtain the equation $d\left(\frac{dy}{dx}\right) = \frac{d^2y}{dx}$. If the variable t coincide with y, then $y = t$, $\psi(t) = 1$, $\psi'(t) = 0$, $d^2y = 0$ and we obtain $d\left(\frac{dy}{dx}\right) = -\frac{dy}{dx}\frac{d^2x}{dx}$. The same is the case if $\varphi(t)$ and $\psi(t)$ are constant, and the variables x and y therefore change in proportion to the independent variable. The second and all higher differentials of the independent variable therefore vanish.

If now in the equation $z = f(x, y)$ both x and y are to be conceived as dependent variables whose change is determined in any manner by a third variable t, the total first differential of z with regard to this variable is

$$\frac{dz}{dt} = \frac{\partial f}{\partial x}\frac{dx}{dt} + \frac{\partial f}{\partial y}\frac{dy}{dt} \quad \text{or } dz = \frac{\partial f}{\partial x}dx + \frac{\partial f}{\partial y}dy,$$

and now considering that both the partial derived functions and the differentials dx and dy depend on t, we find for the second differential

$$\frac{d^2z}{dt^2} = \frac{\partial^2 f}{\partial x^2}\left(\frac{dx}{dt}\right)^2 + \frac{\partial^2 f}{\partial y \partial x}\frac{dy}{dt}\frac{dx}{dt} + \frac{\partial^2 f}{\partial x \partial y}\frac{dx}{dt}\frac{dy}{dt} + \frac{\partial^2 f}{\partial y^2}\left(\frac{dy}{dt}\right)^2 + \frac{\partial f}{\partial x}\frac{d^2x}{dt^2} + \frac{\partial f}{\partial y}\frac{d^2y}{dt^2}$$

$$\text{or } d^2z = \frac{\partial^2 f}{\partial x^2}dx^2 + 2\frac{\partial^2 f}{\partial x \partial y}dx\,dy + \frac{\partial^2 f}{\partial y^2}dy^2 + \frac{\partial f}{\partial x}d^2x + \frac{\partial f}{\partial y}d^2y.$$

But it must be remembered as to this last equation, that it states a determinate relation between finite quantities only when x and y are given as functions of a quantity t, and both sides of the equation are divided by dt^2; that it has on the contrary no content at all, when nothing is determined as to the way in which x and y vary. We have accordingly the Rule.

In order to obtain the most general form of the second differential from $dz = \frac{\partial f}{\partial x}dx + \frac{\partial f}{\partial y}dy$, *we have to form the total differential of the terms on the right side, taking account of both differentials* d^2x *and* d^2y. *If* x *is to be taken as an independent variable, then* $d^2x = 0$; *if* y *also is to be considered as an independent variable,* d^2y *must also vanish.*

Example:

$$z = x^m y^n, \quad dz = mx^{m-1}y^n\,dx + nx^m y^{n-1}\,dy$$

$$d^2z = m(m-1)x^{m-2}y^n\,dx^2 + 2mnx^{m-1}y^{n-1}\,dx\,dy$$
$$+ n(n-1)x^m y^{n-2}\,dy^2 + mx^{m-1}y^n\,d^2x + nx^m y^{n-1}\,d^2y.$$

This Rule contains also the law of formation of further differentials:

$$d^3 z = \frac{\partial^3 f}{\partial x^3} dx^3 + 3\frac{\partial^3 f}{\partial x^2 \partial y} dx^2 dy + 3\frac{\partial^3 f}{\partial x \partial y^2} dx\, dy^2 + \frac{\partial^3 f}{\partial y^3} dy^3$$

$$+ 2\left\{\frac{\partial^2 f}{\partial x^2} dx\, d^2 x + \frac{\partial^2 f}{\partial x \partial y}(d^2 x dy + d^2 y dx) + \frac{\partial^2 f}{\partial y^2} dy\, d^2 y\right\} + \frac{\partial f}{\partial x} d^3 x + \frac{\partial f}{\partial y} d^3 y.$$

But if x and y are both to be taken as independent variables, this expression reduces to its first four terms. It is seen that these terms occur with binomial coefficients and that in general for independent variables

$$d^n z = \sum_{k=0}^{k=n} n_k \frac{\partial^n f}{\partial x^{n-k} \partial y^k} \cdot dx^{n-k} dy^k, \quad (n_0 = 1).$$

For, if we form the total differential of this equation, we find:

$$d^{n+1} z = dx \sum_{k=0}^{k=n} n_k \frac{\partial^{n+1} f}{\partial x^{n-k+1} \partial y^k} dx^{n-k} dy^k + dy \sum_{k=0}^{k=n} n_k \frac{\partial^{n+1} f}{\partial x^{n-k} \partial y^{k+1}} dx^{n-k} dy^k.$$

Except the first term of the first sum and the last of the second, each term occurs twice, only with a different binomial factor, so that

$$d^{n+1} z = \frac{\partial^{n+1} f}{\partial x^{n+1}} dx^{n+1} + (n_1 + n_0)\frac{\partial^{n+1} f}{\partial x^n \partial y} dx^n dy + \cdots$$

$$\cdots + (n_k + n_{k-1}) \frac{\partial^{n+1} f}{\partial x^{n+1-k} \partial y^k} dx^{n+1-k} dy^k + \cdots \frac{\partial^{n+1} f}{\partial y^{n+1}} dy^{n+1},$$

but since $n_k + n_{k-1} = (n+1)_k$, we have

$$d^{n+1} z = \sum_{k=0}^{k=n+1} (n+1)_k \frac{\partial^{n+1} f}{\partial x^{n+1-k} \partial y^k} dx^{n+1-k} dy^k, \quad \text{Q. E. D.}$$

There is no difficulty in extending these investigations to explicit functions with more than two independent variables, when once we have defined for them in a perfectly analogous manner what is meant by the continuity of the partial and total differential.*)

56. The knowledge of the partial derived of a function with more independent variables than one, leads likewise, as Lagrange has shown, to a calculation of the function by an infinite series of powers. To find the value of $z = f(x+h, y+k)$, when the values of the function and of all its partial derived functions at a point x, y are known, let us form the expression

1) $$F(t) = f(x + th, y + tk) = f(x', y').$$

This will be a continuous function of t for arbitrary values of h and k, only if f is a continuous function of the two variables within the region determined by h and k. Now if $F(t)$ can be expanded by Mac Laurin's series, that is, if

---

*) Theorems concerning functions with more than one variable were first systematically developed by Euler: Instit. calcul. diff. Pars I, 7.

2) $$F(t) = F(0) + \frac{t}{1} F'(0) + \frac{t^2}{\underline{|2}} F''(0) + \cdots \frac{t^n}{\underline{|n}} F^n(\Theta t),$$

we get for $t = 1$ the value:

3) $$F(1) = F(0) + \frac{1}{1} F'(0) + \frac{1}{\underline{|2}} F''(0) + \cdots \frac{1}{\underline{|n}} F^n(\Theta).$$

Now on the hypothesis that f and its partial derived functions are continuous in both variables, we have for every value of h and k the total derived of F with regard to t:

$$F'(t) = \frac{\partial f(x',y')}{\partial x'} \frac{dx'}{dt} + \frac{\partial f(x',y')}{\partial y'} \frac{dy'}{dt}$$

or because: $x' = x + ht$, $y' = y + kt$, therefore

$$\frac{\partial f}{\partial x'} = \frac{\partial f}{\partial x}, \quad \frac{\partial f}{\partial y'} = \frac{\partial f}{\partial y}, \quad \frac{dx'}{dt} = h, \quad \frac{dy'}{dt} = k,$$

$$F'(t) = h \frac{\partial f(x',y')}{\partial x} + k \frac{\partial f(x',y')}{\partial y} \text{ and } F'(0) = h \frac{\partial f(x,y)}{\partial x} + k \frac{\partial f(x,y)}{\partial y}.$$

Further we have

$$F''(t) = h^2 \frac{\partial^2 f(x',y')}{\partial x^2} + 2hk \frac{\partial^2 f(x',y')}{\partial x \partial y} + k^2 \frac{\partial^2 f(x',y')}{\partial y^2},$$

therefore

$$F''(0) = h^2 \frac{\partial^2 f(x,y)}{\partial x^2} + 2hk \frac{\partial^2 f(x,y)}{\partial x \partial y} + k^2 \frac{d^2 f(x,y)}{\partial y^2}$$

and so on, in general:

$$F^n(t) = \sum_{p=0}^{p=n} n_p h^{n-p} k^p \frac{\partial^n f(x',y')}{\partial x^{n-p} \partial y^p}.$$

If we substitute these values in equation 3) we find

$$F(1) = f(x+h, y+k) = f(x,y) + \left\{h \frac{\partial f}{\partial x} + k \frac{\partial f}{\partial y}\right\}$$

$$+ \frac{1}{2}\left\{h^2 \frac{\partial^2 f}{\partial x^2} + 2hk \frac{\partial^2 f}{\partial x \partial y} + k^2 \frac{\partial^2 f}{\partial y^2}\right\} + \cdots \frac{1}{\underline{|n}} \sum_{p=0}^{p=n} n_p h^{n-p} k^p \frac{\partial^n f(x+\Theta h, y+\Theta k)}{\partial x^{n-p} \partial y^p}.$$

This expression is the Theorem of the Mean Value in its most general form for a function with two variables. It leads to an infinite series proceeding by powers of h and k, whenever the remainder converges to zero as the values of n increase arbitrarily. A special case in which this occurs is when the partial derived functions have the property of remaining finite in the domain assigned by h and k when n becomes infinite. If they have not this property, the remainder may indeed still converge to zero, though, as the determination of the limit becomes difficult, we require other criteria to decide by.

Tenth Chapter.

Implicit Functions. Application of Taylor's series to evaluate quotients apparently indeterminate.

57. Our last investigations were shown to be applicable to the calculation of differential quotients of complicated functions of an independent variable when x and y depend on the quantity t; but they also admit of application to implicit functions.*)

In the case of an implicit function $f(x, y) = 0$, represented ex. gr. by the most general form of an algebraic function (§ 25), the value of the function hitherto named z is constantly equal to zero. By making it vanish we produce a dependence between the quantities x and y; for, if one or more values of y can be assigned for a determinate value of x so that $f = 0$, a change of the value of x will give rise to a perfectly determinate change of each of these values of y that satisfy the relation $f(x, y) = 0$.

Let us fix our attention on a determinate value of y and attempt to measure its change in relation to that of x.

When will y be a continuous function of x? It will be so when as the increase Δx becomes 0, the Δy belonging to it also converges to zero; i. e. when the equation $f(x + \Delta x, y + \Delta y) = 0$ is satisfied by a vanishing value of Δy simultaneously with a vanishing value of Δx.

Therefore in this case there are values arbitrarily near the point x, y, yet besides this point itself, for which the function of two variables $z = f(x, y)$ vanishes, and conversely the existence of such values within arbitrary proximity to the point x, y, coincides with the conception of continuity for y. If ex. gr. z be a unique and continuous function of both variables which can be shown to have positive as well as negative values in the neighbourhood of a point, it follows that there must be also a continuous series of values for which z vanishes.

When the change of value is continuous, the limiting value of the quotient $\frac{\Delta y}{\Delta x}$ can be enquired after.

---

*) Euler: Instit. calcul. diff. Pars I. 9.

If the function $z = f(x, y)$ when $z = 0$ have a total differential: $dz = \frac{\partial f}{\partial x} dx + \frac{\partial f}{\partial y} dy$, there is obviously a determinate sort of increase for x and y, for which z remains $= 0$, namely

$$\frac{\partial f}{\partial x} dx + \frac{\partial f}{\partial y} dy = 0, \text{ or } \frac{dy}{dx} = -\left(\frac{\partial f}{\partial x}\right) : \left(\frac{\partial f}{\partial y}\right).$$

Accordingly the value of the differential quotient of an implicit function can be determined, without requiring first to express it as an explicit one, by substituting in the expressions of the partial derived functions $\frac{\partial f}{\partial x}$ and $\frac{\partial f}{\partial y}$ the values of x and y at the point. The second total differential $d^2 z$ equated to zero gives the calculation of the second differential quotient $\frac{d^2 y}{dx^2}$, and so on:

$$0 = \frac{\partial^2 f}{\partial x^2} + 2 \frac{\partial^2 f}{\partial x \partial y} \cdot \frac{dy}{dx} + \frac{\partial^2 f}{\partial y^2} \left(\frac{dy}{dx}\right)^2 + \frac{\partial f}{\partial y} \cdot \frac{d^2 y}{dx^2},$$

$$0 = \frac{\partial^3 f}{\partial x^3} + 3 \frac{\partial^3 f}{\partial x^2 \partial y} \cdot \frac{dy}{dx} + 3 \frac{\partial^3 f}{\partial x \partial y^2} \left(\frac{dy}{dx}\right)^2 + \frac{\partial^3 f}{\partial y^3} \left(\frac{dy}{dx}\right)^3$$

$$+ 2 \frac{d^2 y}{dx^2} \left(\frac{\partial^2 f}{\partial x \partial y} + \frac{\partial^2 f}{\partial y^2} \cdot \frac{dy}{dx}\right) + \frac{\partial f}{\partial y} \cdot \frac{d^3 y}{dx^3}, \text{ etc.}$$

58. Application to the most general algebraic function of two variables.

The function $z = f(x, y) = A_0 y^n + A_1 y^{n-1} + \cdots A_{n-1} y + A_n$, where $A_0, A_1 \ldots A_n$ signify polynomials of any degree in x, which can likewise be arranged in powers of x:

$$z = B_0 x^m + B_1 x^{m-1} + \cdots + B_{m-1} x + B_m,$$

is a continuous function of both variables. For, written out in full it consists of a finite number of summands of the form: $a_{\mu\nu} x^\mu y^\nu$; each of these summands is, as was shown § 52, a continuous function of both variables, but a finite sum of continuous functions is itself a continuous function. We can easily prove this proposition in general. If for any values x, y, we have the function

$$z = f_1(x, y) + f_2(x, y) + \cdots + f_p(x, y),$$

and we form its value for $x \pm \Theta h$, $y \pm \eta k$, the difference between the new value and the old will remain less than δ, when we make h and k respectively equal to the smallest of the values that result for the individual summands, in order that the amount of

$$f_i(x \pm \Theta h, y \pm \eta k) - f_i(x, y)$$

may be less than $\frac{\delta}{p}$. Further, the algebraic function has a total differential of the form $dz = \frac{\partial f}{\partial x} dx + \frac{\partial f}{\partial y} dy$. For, its partial derived functions in x and y are themselves again algebraic and therefore continuous functions of both variables. Consequently, at any point for which $z = 0$, there is also a differential quotient identical when taken progressively and regressively, to be calculated from the equation:

$$\frac{dy}{dx} = -\frac{\frac{\partial f}{\partial x}}{\frac{\partial f}{\partial y}} = -\frac{mB_0x^{m-1} + (m-1)B_1x^{m-2} + \cdots + B_{m-1}}{nA_0y^{n-1} + (n-1)A_1y^{n-2} + \cdots + A_{n-1}},$$

and that becomes infinite at the points for which the denominator vanishes, where simultaneously $f = 0$ and $\frac{\partial f}{\partial y} = 0$.

The preliminary result is therefore: If in the algebraical equation $f(x, y) = 0$, y be considered as a function of x, this function has at each point a differential quotient; in other words: it can be carried on continuously from each point. Geometrically stated this is the theorem: an algebraical curve has at each point a tangent; it cannot break off at any point.

This theorem however undergoes modifications: for there can be points, at which numerator and denominator of the quotient $\frac{dy}{dx}$ vanish simultaneously; or simultaneously increase beyond all limits; these require a special investigation.

59. The Theorem of the Mean Value in its most general form (Taylor's series) shows directly whether a function can be carried on continuously for finite values of x and y. We found § 56 that the value of $f(x + h, y + k)$ can be calculated from the equation

$$f(x+h, y+k) = f(x,y) + \left\{h\frac{\partial f}{\partial x} + k\frac{\partial f}{\partial y}\right\} + \frac{1}{2}\left\{h^2\frac{\partial^2 f}{\partial x^2} + 2hk\frac{\partial^2 f}{\partial x \partial y} + k^2\frac{\partial^2 f}{\partial y^2}\right\} + \cdots$$

$$\cdots + \frac{1}{\underline{|n}}\sum_{p=0}^{p=n} n_p h^{n-p} k^p \cdot \frac{\partial^n f(x+\Theta h, y+\Theta k)}{\partial x^{n-p}\partial y^p}.$$

We commence from a point x_0, y_0, at which $f = 0$, and try to find another in its neighbourhood, i. e. for arbitrarily small values of h and k, at which $f(x_0 + h, y_0 + k)$ likewise vanishes. Denoting the values of the partial derived functions at x_0, y_0 by $()_0$, the values h and k must satisfy

$$0 = \left\{h\left(\frac{\partial f}{\partial x}\right)_0 + k\left(\frac{\partial f}{\partial y}\right)_0\right\} + \frac{1}{2}\left\{h^2\left(\frac{\partial^2 f}{\partial x^2}\right)_0 + 2hk\left(\frac{\partial^2 f}{\partial x \partial y}\right)_0 + k^2\left(\frac{\partial^2 f}{\partial y^2}\right)_0\right\} + \cdots$$

$$\cdots + \frac{1}{\underline{|n}}\sum_{p=0}^{p=n} n_p h^{n-p} k^p \left(\frac{\partial^n f}{\partial x^{n-p}\partial y^p}\right)_0.$$

Since arbitrarily small values of h and k are in question, we see that: unless $\frac{\partial f}{\partial x}$ and $\frac{\partial f}{\partial y}$ vanish simultaneously, terms containing higher powers of h and k are arbitrarily small compared with terms of the first dimension; thus the continuation of the implicit function is indicated in the direction of its differential quotient

$$\frac{k}{h} = -\frac{\frac{\partial f}{\partial x}}{\frac{\partial f}{\partial y}}.$$

But now if $\frac{\partial f}{\partial x}$ and $\frac{\partial f}{\partial y}$ simultaneously vanish, the first member of our equation disappears, and since for the arbitrarily small values of h and k for which we are enquiring, the 3rd, 4th etc. powers are arbitrarily small compared with the second, the limit of the ratio of k to h is to be extracted from the quadratic equation:

$$h^2\left(\frac{\partial^2 f}{\partial x^2}\right)_0 + 2hk\left(\frac{\partial^2 f}{\partial x \partial y}\right)_0 + k^2\left(\frac{\partial^2 f}{\partial y^2}\right)_0 = 0.$$

When this equation is written in the form

$$\left\{\frac{k}{h}\left(\frac{\partial^2 f}{\partial y^2}\right)_0 + \left(\frac{\partial^2 f}{\partial x \partial y}\right)_0\right\}^2 = \left\{\left(\frac{\partial^2 f}{\partial x \partial y}\right)_0\right\}^2 - \left(\frac{\partial^2 f}{\partial x^2}\right)_0\left(\frac{\partial^2 f}{\partial y^2}\right)_0,$$

it shows, that the ratio $\frac{k}{h}$ has two real values, different or equal, or no real value at all, according as

$$\left\{\left(\frac{\partial^2 f}{\partial x \partial y}\right)_0\right\}^2 - \left(\frac{\partial^2 f}{\partial x^2}\right)_0\left(\frac{\partial^2 f}{\partial y^2}\right)_0$$

is greater than, equal to or less than zero. In the last case the function $f(x, y) = 0$ cannot be continued from the point $x_0\, y_0$ in any direction by real values of x and y; while in the first, two different directions are found. Its curve has an isolated point or a double point with real branches, singularities that can also occur in algebraic curves. If $\left(\frac{\partial^2 f}{\partial x^2}\right)_0$, $\left(\frac{\partial^2 f}{\partial x \partial y}\right)_0$, $\left(\frac{\partial^2 f}{\partial y^2}\right)_0$, also vanish, we are led to a cubic equation which presents for the ratio of $k : h$ either three real values, different or equal, or else only one real value. Such singularities can rise higher, but the further discussions require theorems as to the number and nature of the solutions of equations of the nth degree. These remarks contain only the first germs of a problem which may be stated generally thus: The implicit algebraic function $f(x, y) = 0$ is given. For $x = x_0$, y takes the value y_0. It is required to express y as an explicit function of x by a convergent series of powers, subject to the condition, that the relation $f(x, y) = 0$ remain always satisfied, and that for $x = x_0$, $y = y_0$. But we must postpone the solution of this problem until later on, for it requires a considerable extension of our previous conceptions. In the first place we must be able to tell, how many values of y belong to a determinate x_0. This requires the investigation of complex solutions. We must then solve in general the question as to whether a function, in whatever way it may be defined, can be expanded in a series of powers. (Bk. II Ch. IV. Bk. IV Ch. III).

60. We return to the question as to the differential quotients at these singular points. We assert, the values of $k:h$ calculated from the above-mentioned quadratic or cubic equation present, when real, the different values of the differential quotient $\frac{dy}{dx}$ at such point. This proposition is manifest, for $k:h$ is a quotient of differences, and its limiting value defines the differential quotient. It can however be deduced yet otherwise from the original equation of definition:

$$\frac{dy}{dx} = -\frac{\partial f}{\partial x} : \frac{\partial f}{\partial y}.$$

We consider first the following simple but important case, (cf. § 19c): If in a quotient $\frac{\varphi(x)}{\psi(x)}$, numerator and denominator vanish simultaneously for a determinate value $x = a$, φ and ψ may be any continuous functions, not only algebraic, then (§ 10) this quotient has a meaning only in so far as it is the limit towards which the values at neighbouring points tend; thus

$$\frac{\varphi(a)}{\psi(a)} = \lim_{h=0} \frac{\varphi(a+h)}{\psi(a+h)}, \text{ or } \frac{\varphi(a)}{\psi(a)} = \lim_{h=0} \frac{\varphi(a-h)}{\psi(a-h)}.$$

If now the Theorem of the Mean Value in its first form hold for the functions φ and ψ, the following general rule is found for calculating the limiting value. Put

$$\varphi(a+h) - \varphi(a) = h\varphi'(a+\Theta h), \text{or, as } \varphi(a) = 0, \varphi(a+h) = h\varphi'(a+\Theta h),$$
$$\psi(a+h) - \psi(a) = h\psi'(a+\eta h), \text{ or, as } \psi(a) = 0, \psi(a+h) = h\psi'(a+\eta h),$$

then

$$\frac{\varphi(a+h)}{\psi(a+h)} = \frac{\varphi'(a+\Theta h)}{\psi'(a+\eta h)}, \text{ and so } \lim_{h=0} \frac{\varphi'(a+\Theta h)}{\psi'(a+\eta h)} = \frac{\varphi'(a)}{\psi'(a)},$$

i. e. the value of the quotient $\frac{\varphi(x)}{\psi(x)}$ at a point where numerator and denominator vanish simultaneously, is, on the hypothesis assigned, equal to the quotient of the first derived functions of φ and ψ at this point.

If $\varphi'(a)$ and $\psi'(a)$ likewise vanish, but if the theorem of the mean value is applicable in its extended form, we have

$$\varphi(a+h) = \frac{h^2}{\underline{|2}} \varphi''(a+\Theta h), \quad \psi(a+h) = \frac{h^2}{\underline{|2}} \psi''(a+\eta h),$$

therefore

$$\frac{\varphi(a)}{\psi(a)} = \lim_{h=0} \frac{\varphi(a+h)}{\psi(a+h)} = \lim_{h=0} \frac{\varphi''(a+\Theta h)}{\psi''(a+\eta h)} = \frac{\varphi''(a)}{\psi''(a)}.$$

This demonstration is not possible for the special case that a

is not a finite value, when numerator and denominator simultaneously vanish for $x = \infty$. This case shall be dealt with in § 62.

Now the same process holds good for a quotient of the form: $\frac{\varphi(x, y)}{\psi(x, y)}$, in which, as in case of the implicit function, y is a function of x. This quotient must likewise be derived as limiting value from neighbouring points, when for $x = a$, $y = b$, numerator and denominator vanish. We have

$$\frac{\varphi(a, b)}{\psi(a, b)} = \operatorname*{Lim}_{h=0,\, k=0} \frac{\varphi(a+h, b+k)}{\psi(a+h, b+k)},$$

and by the theorem of the mean value

$$\varphi(a+h, b+k) = h \frac{\partial \varphi(a+\Theta h, b+\Theta k)}{\partial a} + k \frac{\partial \varphi(a+\Theta h, b+\Theta k)}{\partial b},$$

$$\psi(a+h, b+k) = h \frac{\partial \psi(a+\eta h, b+\eta k)}{\partial a} + k \frac{\partial \psi(a+\eta h, b+\eta k)}{\partial b}.$$

Accordingly

$$\frac{\varphi(a, b)}{\psi(a, b)} = \frac{\frac{\partial \varphi}{\partial a} + \frac{\partial \varphi}{\partial b} \cdot \operatorname{Lim} \frac{k}{h}}{\frac{\partial \psi}{\partial a} + \frac{\partial \psi}{\partial b} \cdot \operatorname{Lim} \frac{k}{h}}.$$

If now in our case, in which $\varphi = -\frac{\partial f}{\partial x}$, $\psi = +\frac{\partial f}{\partial y}$, assuming that the implicit function $f(x, y) = 0$ can be continued, we determine the value of $\frac{dy}{dx} = \frac{0}{0}$ by the same rule, remembering that by the above proof $\operatorname{Lim} \frac{k}{h} = \frac{dy}{dx}$, we obtain the equation

$$\frac{dy}{dx} = -\frac{\frac{\partial^2 f}{\partial x^2} + \frac{\partial^2 f}{\partial x \partial y} \cdot \frac{dy}{dx}}{\frac{\partial^2 f}{\partial x \partial y} + \frac{\partial^2 f}{\partial y^2} \cdot \frac{dy}{dx}}, \text{ or } \frac{\partial^2 f}{\partial y^2}\left(\frac{dy}{dx}\right)^2 + 2 \frac{\partial^2 f}{\partial x \partial y} \cdot \frac{dy}{dx} + \frac{\partial^2 f}{\partial x^2} = 0,$$

in which the values of the partial derived functions are taken at the point under consideration. This equation coincides with that already found for h and k, and teaches that the quotient $\frac{dy}{dx}$, provided it is real, remains even in the singular points a continuous function of x, for it can be derived as limiting value from neighbouring points. But we may not take this second manner of calculating, as a proof that such a thing exists, for here we are taking for granted not only the existence of $\frac{dy}{dx}$ but also its continuity.

It is easily seen how this calculation adapts itself also to higher singularities, the value of $\frac{\varphi(a, b)}{\psi(a, b)}$ being developed on the hypothesis, that all first derived functions, then all second, etc. vanish. We do not here enter on the special case, that a and b are both infinite.

61. But in the second place, a quotient appears in an indeterminate form, when as numerator and denominator simultaneously approach a point, they increase beyond any limit; or in other words become simultaneously infinite at a point. If the quotient be an algebraic function, this can of course only occur when either or both of the quantities x and y become infinite. Here the simplest case is disposed of by the following theorem:

Suppose $f(x, y) = 0$ an algebraic function in which the highest order of the terms, i. e. the highest sum of the exponents of x and y is n, and let the terms of equal dimensions be grouped together, we can thus bring it into the form

$$x^n f_n\left(\frac{y}{x}\right) + x^{n-1} f_{n-1}\left(\frac{y}{x}\right) + \cdots + x^{n-k} f_{n-k}\left(\frac{y}{x}\right) + \cdots x f_1\left(\frac{y}{x}\right) + f_0 = 0,$$

where $f_{n-k}\left(\frac{y}{x}\right)$ signifies a polynomial of the $(n-k)^{\text{th}}$ order in the quotient $y : x$. Substituting for $y : x$ a definite value g, the equation

$$x^n f_n(g) + x^{n-1} f_{n-1}(g) + \cdots + x f_1(g) + f_0 = 0$$

determines those values of x to which a value y belongs such that $y : x = g$. The question: For what values of g does x become infinite? can be reduced, by putting $\frac{1}{z}$ for x, to the simpler one: When is

$$f_n(g) + z f_{n-1}(g) + \cdots + z^{n-1} f_1(g) + z^n f_0 = 0$$

satisfied by $z = 0$? The answer is: When $f_n(g) = 0$, only. This is in general an equation of the nth degree in g. Suppose we had found g a definite solution of it. In our present notation we have:

$$\frac{\partial f}{\partial x} = n x^{n-1} f_n\left(\frac{y}{x}\right) + (n-1) x^{n-2} f_{n-1}\left(\frac{y}{x}\right) + \cdots + f_1\left(\frac{y}{x}\right) - x^{n-2} y f_n'\left(\frac{y}{x}\right)$$

$$- x^{n-3} y f'_{n-1}\left(\frac{y}{x}\right) - \cdots - \frac{y}{x} f_1'\left(\frac{y}{x}\right),$$

$$\frac{\partial f}{\partial y} = x^{n-1} f_n'\left(\frac{y}{x}\right) + x^{n-2} f'_{n-1}\left(\frac{y}{x}\right) + \cdots f_1'\left(\frac{y}{x}\right).$$

Dividing numerator and denominator of the quotient $-\frac{\partial f}{\partial x} : \frac{\partial f}{\partial y}$ by x^{n-1} and putting $\frac{y}{x} = g$, $x = \infty$, all terms that retain x in their denominators will go out, and as $f_n(g) = 0$ there will only remain in the numerator the term $g f_n'(g)$, and in the denominator $f_n'(g)$. Accordingly, unless the special case $f_n'(g) = 0$ occur, we have for $x = \infty$,

$$\frac{dy}{dx} = g = \left(\frac{y}{x}\right).$$

Therefore the differential quotient of an algebraic function has a determinate value, even at points where x is infinite, if the ratio $y : x$ have any determinate value, even zero. If we went through the demonstration for the ratio $x : y$ in an analogous manner, we should find the theorem also holds for the points not already considered, at which y is infinite, but x finite, so that the ratio $y:x$ is infinite; here we have $\frac{dx}{dy}=0$.

62. We conclude these considerations with the problem: Let $\varphi(x)$ and $\psi(x)$ be any two functions, which both become infinite for $x=a$; it is required to find the limiting value of the quotient

$$\frac{\varphi(x)}{\psi(x)} \text{ for } x=a.$$

Putting $x=a+\frac{1}{z}$ consider φ and ψ as functions in z, then they become infinite for $z=\infty$, i. e. when z increases positively or negatively beyond all limits. Therefore the problem is reduced to this other, to calculate $\operatorname{Lim}\frac{\varphi(z)}{\psi(z)}$ for $z=\infty$ or $-\infty$, when $\operatorname{Lim}\varphi(z)=\infty$, $\operatorname{Lim}\psi(z)=\infty$. It was shown in § 24d, that provided $\frac{f(z+h)-f(z)}{h}$ has any determinate limiting value for $z=\pm\infty$, we have then

$$\operatorname{Lim}\frac{f(z)}{z}=\frac{f(z+h)-f(z)}{h}$$

for every value of h. If now the function f be continuous from a point $z=z_1$, and infinite only when $z=\infty$, if further its differential quotient be everywhere determinate, which we know cannot be the case unless from a certain point the function only increases or only decreases, and if the progressive differential quotient be identical with the regressive, then it follows that

$$\operatorname{Lim}\frac{f(z)}{z}=\operatorname{Lim}\frac{f(z+h)-f(z)}{h}=\operatorname{Lim}f'(z+\Theta h).$$

When these formulas apply to φ and ψ, we have

$$\operatorname{Lim}\frac{\varphi(z)}{z}=\operatorname{Lim}\varphi'(z+\Theta h),\quad \operatorname{Lim}\frac{\psi(z)}{z}=\operatorname{Lim}\psi'(z+\Theta' h),$$

and so by division:

$$\operatorname{Lim}\frac{\varphi(z)}{\psi(z)}=\operatorname{Lim}\frac{\varphi'(z+\Theta h)}{\psi'(z+\Theta' h)}\quad\text{for } z=\infty,$$

i. e. *Provided the derived functions φ' and ψ' have each a determinate value for $z=\infty$ the same when taken progressively as regressively, then their quotient also has; and the limiting value of the quotient of the functions φ and ψ, which become determinately infinite, is equal to the quotient of the derived functions.**)

Examples:

1) $$\left\{\frac{l(a+be^x)}{\sqrt{\alpha+\beta x^2}}\right\}_{x=\infty}=\frac{1}{\sqrt{\beta}}.$$

---

*) The hypotheses of the theorem can be further generalised, see
Rouquet: N. Annal. de mathém., 2. Sér., T. XVI.
Stolz: Math. Annal., Vol. XV.

For we have:

$$\varphi'(x) = \left\{\frac{be^x}{a+be^x}\right\}_{x=\infty} = 1,\quad \psi'(x) = \left\{\frac{\beta x}{\sqrt{\alpha+\beta x^2}}\right\}_{x=\infty} = \sqrt{\beta}.$$

2) If we put $\varphi(x) = x + \sin x$, $\psi(x) = x$, we have for $x = \infty$, $\text{Lim}\left\{\frac{\varphi(x)}{\psi(x)}\right\} = \text{Lim}\left\{\frac{x+\sin x}{x}\right\} = 1$, but $\text{Lim}\left\{\frac{\varphi'(x)}{\psi'(x)}\right\} = \left\{\frac{1+\cos x}{1}\right\}$ is indeterminate. Although the function $\varphi(x)$ increases continuously, and becomes determinately infinite, it does not satisfy either the condition that

$$\frac{\varphi(x+h)-\varphi(x)}{h}$$

or that $\varphi'(x)$ have a determinate value for $x = \infty$.

If $\varphi(x)$ and $\psi(x)$ both vanish for $x = \infty$ (see § 60), let us write: $\frac{1}{\varphi(x)} = \varphi_1(x)$, $\frac{1}{\psi(x)} = \psi_1(x)$; both functions φ_1 and ψ_1 become ∞ for $x = \infty$; so by the rule last found

$$\left\{\frac{\varphi(x)}{\psi(x)}\right\}_{x=\infty} = \left\{\frac{\psi_1(x)}{\varphi_1(x)}\right\}_{x=\infty} = \text{Lim}\left\{\frac{\psi_1'(x+\Theta h)}{\varphi_1'(x+\Theta' h)}\right\}_{x=\infty} = \text{Lim}\left\{\frac{\psi'(x+\Theta h)}{\varphi'(x+\Theta' h)}\left\{\frac{\varphi(x+\Theta' h)}{\psi(x+\Theta h)}\right\}^2\right\}_{x=\infty}.$$

Hence we have the equation:

$$\left\{\frac{\varphi(x)}{\psi(x)}\right\}_{x=\infty} = \left\{\frac{\varphi'(x+\Theta' h)}{\psi'(x+\Theta h)}\right\}_{x=\infty}.$$

But the problem of determining the value is not directly solved by this equation, since φ' and ψ' must also vanish for $x = \infty$ (§ 24c); it may however lead to a simplification.

Example:

$$\left\{\frac{l\left(1+\frac{a}{x}\right)}{l\left(1+\frac{1}{x}\right)}\right\}_{x=\infty} = \left\{\frac{-\frac{a}{x^2}\cdot\frac{1}{1+\frac{a}{x}}}{-\frac{1}{x^2}\cdot\frac{1}{1+\frac{1}{x}}}\right\}_{x=\infty} = a.$$

Second Book.

Complex Numbers and functions of Complex Numbers.

First Chapter.

The complex number and the Operations of Arithmetic.

63. In a manner similar to that in which Subtraction, Division and Evolution each required the conception of number to be extended so as to include respectively, negative, fractional and irrational numbers, the attempt to render all the seven operations of Arithmetic possible on real numbers without any exception, requires the adoption into analysis of a new conception, the complex number.

Regarding roots of positive quantities we have the propositions:

1. The root of a number is equal to the product of the roots of its factors.

2. If the exponent of the root be a product mn, the root can be reduced by taking the mth root of the nth root of the quantity or inversely.

These propositions being extended to the even roots of negative quantities, it will appear that the problem of evolution is solved in all cases, as soon as we adopt the square root of negative unity into the numerical system and define the arithmetical operations with it; for we have

$$\sqrt[2n]{-a} = \sqrt[2n]{a}\,\sqrt[n]{\sqrt{-1}}.$$

$\sqrt{-1}$ is called the imaginary unit and, after Gauss, briefly denoted by $\pm i$. In like manner as real positive and negative numbers arise from ± 1 by multiplication, division and involution, so positive and negative imaginary numbers are obtained from $\pm i$:

$$\pm(i+i) = \pm 2i,\ \pm(2i+i) = \pm 3i, \cdots \pm(ai+i) = \pm(a+1)i,$$

$$i - i = 0 \cdot i = 0.$$

The most general imaginary number is: $\pm ai$, where a signifies an arbitrary real number rational or irrational.*)

*) It being already known that all quadratic equations could not be solved by means of real quantities, the introduction of imaginary numbers became unavoidable, when it was found in the irreducible case of solving cubic equations

64. The contrasted epithets "real" and "imaginary" favor the erroneous impression, which indeed has impeded the systematic introduction of imaginary numbers into analysis, that numbers of the first kind possess a practical reality which those of the second have not. Considering the arithmetical operations merely, without application to physical quantities, fractions, irrational numbers, imaginaries, all form legitimate extensions of the conception of number, that are connected with the integer by determinate arithmetical operations. In the applications of these operations on the other hand everything depends on the kind of numbers introduced at the outset in framing a problem analytically. If ex. gr. in the case of discrete quantities, from the way in which the problem is proposed only integers are admissible, the proposed problem is seen to be impossible when the result is a fraction. Likewise in the result of a calculation referring to physical quantities, a negative number will have a meaning applicable to these quantities, only when from the first the quantities were distinguished in the sense of positive and negative. In analogy with this, even when the result of calculation is imaginary, its meaning is no longer unreal, when the actual quantities considered, are characterised not only by real, but also by imaginary numbers. The simplest example of a representation of intuitive quantities by imaginary numbers is the geometric interpretation, which we shall deal with as we go on. "As mathematical science strives towards doing away with exceptions to rules and towards contemplating different propositions from one point of view, it is often compelled to enlarge its conceptions or to establish new ones, and this nearly always denotes a progress in the science. A great example of this is the introduction of imaginary quantities into analysis." (v. Staudt, Beiträge zur Geometrie der Lage. Heft 1. Vorwort.)

65. It follows from the definition of the imaginary unit that:

$$i^2 = i \cdot i = (\sqrt{-1})^2 = -1.$$

Accordingly we understand by:

$$i^3 = i^2 \cdot i = -1 \cdot i = -i,$$
$$i^4 = i^3 \cdot i = -i \cdot i = -(i^2) = +1, \cdots$$

Hence follows by inversion:

$$\frac{1}{i} = -i, \quad \frac{1}{i^2} = -1, \quad \frac{1}{i^3} = i, \quad \frac{1}{i^4} = 1, \ldots$$

that a real solution could be expressed only by means of square roots of negative quantities (Bombelli 1579). Since that time imaginary numbers have never disappeared from analysis. The earliest who found their employment fruitful was Euler. But it was the works of Gauss (1777—1855) and of Cauchy which first manifested the importance of the complex number as a generalised conception of number.

on the circle with radius r. All numbers with the same amplitude belong to points on a right line proceeding from the origin.

We found for calculating $\cos\varphi$ and $\sin\varphi$ the convergent series:

$$\cos\varphi = 1 - \frac{\varphi^2}{\lfloor 2} + \frac{\varphi^4}{\lfloor 4} - \frac{\varphi^6}{\lfloor 6} + - \cdots R \qquad \text{Lim } R = 0,$$

$$\sin\varphi = \varphi - \frac{\varphi^3}{\lfloor 3} + \frac{\varphi^5}{\lfloor 5} - \frac{\varphi^7}{\lfloor 7} + - \cdots R' \qquad \text{Lim } R' = 0.$$

If we take their sum, having multiplied the latter by i, we get:

$$\left(1 - \frac{\varphi^2}{\lfloor 2} + \frac{\varphi^4}{\lfloor 4} - \frac{\varphi^6}{\lfloor 6} + \cdots + R\right) + i\left(\varphi - \frac{\varphi^3}{\lfloor 3} + \frac{\varphi^5}{\lfloor 5} - \frac{\varphi^7}{\lfloor 7} + \cdots + R'\right).$$

Employing the properties of the powers of i, this assumes the form:

$$1 + i\varphi + \frac{(i\varphi)^2}{\lfloor 2} + \frac{(i\varphi)^3}{\lfloor 3} + \frac{(i\varphi)^4}{\lfloor 4} + \frac{(i\varphi)^5}{\lfloor 5} + \cdots (R + iR').$$

The remainder of this series, a complex number, converges to zero for all values of φ; accordingly the convergent series:

$$1 + i\varphi + \frac{(i\varphi)^2}{\lfloor 2} + \frac{(i\varphi)^3}{\lfloor 3} + \frac{(i\varphi)^4}{\lfloor 4} + \frac{(i\varphi)^5}{\lfloor 5} + \cdots \text{ in inf.}$$

expresses the complex quantity: $\cos\varphi + i\sin\varphi$ as approximately as can be desired. But this series is got from the exponential series found in § 42, by writing $i\varphi$ for x; accordingly*) we denote it by the symbol $e^{i\varphi}$, and obtain in this notation the theorem:

Every complex number can be written in the form $re^{i\varphi}$, where $e^{i\varphi}$ stands for the infinite series just defined.

In consequence of this definition we have:

$$e^{\frac{1}{2}i\pi} = i, \quad e^{i\pi} = -1, \quad e^{\frac{3}{2}i\pi} = -i, \quad e^{i2\pi} = 1. \quad e^{\pm i2k\pi} = 1.$$

68. Complex numbers form a group complete in themselves; that is, every operation of arithmetic when applied to complex numbers presents without exception a result which can be expressed by a complex number. Before this can be shown, we must define what is now to be understood by the operations of arithmetic; the definitions must be framed so as to embrace those already given for real numbers.

69. Sum and difference (§ 65):

$$(a + ib) \pm (a' + ib') = (a \pm a') + i(b \pm b'),$$

or

$$r(\cos\varphi + i\sin\varphi) \pm r'(\cos\varphi' + i\sin\varphi')$$
$$= (r\cos\varphi \pm r'\cos\varphi') + i(r\sin\varphi \pm r'\sin\varphi').$$

In this second form we prove most easily the theorem: The modulus of

---

*) Euler: Introductio, I. Cap. VIII. These equations establish the connexion stated in § 14 to exist between exponential and trigonometrical functions:

$\cos\varphi + i\sin\varphi = e^{i\varphi}$, $\cos\varphi - i\sin\varphi = e^{-i\varphi}$, $\cos\varphi = \frac{1}{2}(e^{i\varphi} + e^{-i\varphi})$, $\sin\varphi = \frac{1}{2i}(e^{i\varphi} - e^{-i\varphi})$.

the sum or of the difference of two complex numbers may have any value not greater than the sum, and not less than the difference of the moduli of the summands. For, putting

$$(r\cos\varphi \pm r'\cos\varphi') + i(r\sin\varphi \pm r'\sin\varphi') = R(\cos\psi + i\sin\psi),$$

if we identify the real parts and also the imaginary parts

$$R\cos\psi = r\cos\varphi \pm r'\cos\varphi', \quad R\sin\psi = r\sin\varphi \pm r'\sin\varphi',$$

we find for the required modulus R that:

$$R^2 = r^2 + r'^2 \pm 2rr'\cos(\varphi - \varphi'); \text{ but, } -1 \leqq \cos(\varphi - \varphi') \leqq +1.$$

70. Multiplication.

If m be a positive integer, $(a + ib)\cdot m$ must be understood to mean that $a + ib$ is to be put m times as summand; from this follows

$$(a + ib)\cdot m = am + ibm;$$

in agreement with this we define in general:

$$(a+ib)(a'+ib') = a'(a+ib) + ib'(a+ib) = aa' + iba' + iab' + i^2bb'$$
$$= aa' - bb' + i(ba' + ab').$$

This keeps up the proposition of the interchangeability of factors.

In the second form we obtain:

$$r(\cos\varphi + i\sin\varphi)\,.\,r'(\cos\varphi' + i\sin\varphi')$$
$$= rr'\{\cos\varphi\cos\varphi' - \sin\varphi\sin\varphi' + i(\sin\varphi\cos\varphi' + \cos\varphi\sin\varphi')\}$$
$$= rr'\{\cos(\varphi + \varphi') + i\sin(\varphi + \varphi')\} = rr'_{\varphi+\varphi'}.$$

The modulus of the product is equal to the product of the moduli of its factors; the amplitude of the product is equal to the sum of their amplitudes.

The product vanishes only when one of its factors vanishes.

We have in general:

$$r_\varphi \cdot r'_{\varphi'} \cdots r^\nu_{\varphi^\nu} = (rr' \cdots r^\nu)_{\varphi+\varphi'+\ldots\varphi^\nu}.$$

In the third form the equation is written:

$$re^{i\varphi}\cdot r'e^{i\varphi'}\cdot r''e^{i\varphi''}\cdots r^\nu e^{i\varphi^\nu} = (rr'\cdots r^\nu)\,e^{i(\varphi+\varphi'+\ldots\varphi^\nu)};$$

showing that here also: Powers of the same base e with imaginary exponents are multiplied by adding the exponents.

71. Division is defined as inverse of multiplication: The meaning of $a + ib : a' + ib'$ is that a number should be determined, which, multiplied by $a' + ib'$ shall give a product equal to $a + ib$. As $\frac{a+ib}{a+ib} = 1$, the calculation of the quotient can be reduced to the multiplication of two complex numbers.

$$\frac{a+ib}{a'+ib'} = \frac{a+ib}{a'+ib'}\cdot\frac{a'-ib'}{a'-ib'} = \frac{(a+ib)(a'-ib')}{a'^2+b'^2} = \frac{aa'+bb'}{a'^2+b'^2} + i\,\frac{ba'-ab'}{a'^2+b'^2}.$$

Two numbers of the form $a' + ib'$ and $a' - ib'$ are said to be conjugate; their product, equal to the square of their common modulus, is called the Norm. (Gauss.)

Again: $\frac{r(\cos\varphi + i\sin\varphi)}{r'(\cos\varphi' + i\sin\varphi')} = \frac{r}{r'}\cdot(\cos\varphi + i\sin\varphi)(\cos\varphi' - i\sin\varphi')$

$$= \frac{r}{r'}\{\cos(\varphi - \varphi') + i\sin(\varphi - \varphi')\},$$

i. e., $r_\varphi : r'_{\varphi'} = \left(\frac{r}{r'}\right)_{\varphi-\varphi'}$, or in the third form, $re^{i\varphi} : r'e^{i\varphi'} = \frac{r}{r'}\cdot e^{i(\varphi-\varphi')}$.

The modulus of the quotient is equal to the quotient of the two moduli, its amplitude is equal to the difference of the amplitudes.

If the modulus of the divisor vanish, the quotient is infinitely great.

72. The Power with a real exponent.

a) The power with a positive integer exponent n is defined as n-fold multiplication of the base.

$$(a + ib)^n = a^n + na^{n-1}(ib) + n_2a^{n-2}(ib)^2 + \cdots + (ib)^n,$$

or:

$$\{r(\cos\varphi + i\sin\varphi)\}^n = r^n(\cos n\varphi + i\sin n\varphi).*)$$

In the third form we obtain: $(re^{i\varphi})^n = r^n e^{in\varphi}$.

b) When the exponent is a positive rational fraction $\frac{m}{n}$, m and n being relatively prime, the power is understood to be the nth root of the mth power of the base, or the mth power of the nth root of the base. We shall see by calculation that both values coincide.

To determine the nth root of a number means to find that number which raised to the nth power is equal to the given number. If $\sqrt[n]{a + ib} = u + iv$, then $(u + iv)^n$ must $= a + ib$.

But the values of u and v are assigned far more easily in the trigonometrical form.

When $\{r(\cos\varphi + i\sin\varphi)\}^{\frac{1}{n}} = \varrho(\cos\psi + i\sin\psi)$,

$r(\cos\varphi + i\sin\varphi)$ must $= \varrho^n\cdot(\cos n\psi + i\sin n\psi)$.

Thence: $r\cos\varphi = \varrho^n\cos n\psi,\quad r\sin\varphi = \varrho^n\sin n\psi,$

and so:

$r^2 = \varrho^{2n}$, i. e. $\varrho = +\sqrt[n]{r}$; and $\cos\varphi = \cos n\psi$, $\sin\varphi = \sin n\psi$.

The last two equations are only satisfied, if $n\psi = \varphi$ or differ from φ by integer multiples of 2π: $n\psi = \varphi \pm 2k\pi$, so $\psi = \frac{\varphi}{n} \pm \frac{2k\pi}{n}$. Now as k goes through all positive or negative integers, we obtain infinitely many values for ψ. But all of them that differ only by

*) De Moivre (1667—1756): Miscellanea analytica (1730).

multiples of 2π belong to the same number ϱ_ψ. There are therefore only n such numbers; they belong to the values $k = 0, 1, 2 \ldots n-1$. For, in the first place, among the different forms for ψ, all that have a negative sign for k can evidently be brought to be forms with a positive sign by addition of integer multiples of 2π; then again, for $k = (n-1) + k'$, we have $\frac{2k\pi}{n} = 2\pi + \frac{2(k'-1)\pi}{n}$.

Every complex number has therefore n distinct n^{th} roots; these are included in the form:

$$\sqrt[n]{r(\cos\varphi + i\sin\varphi)} = r^{\frac{1}{n}}\left(\cos\frac{\varphi+2k\pi}{n} + i\sin\frac{\varphi+2k\pi}{n}\right), \text{ i. e.}$$

$$\sqrt[n]{r_\varphi} = \left(\sqrt[n]{r}\right)_{\frac{\varphi+2k\pi}{n}}, \text{ or, } \sqrt[n]{re^{i\varphi}} = \sqrt[n]{r}\cdot e^{i\frac{\varphi+2k\pi}{n}}; \; (k = 0, 1, 2 \cdots n-1).$$

Every positive or negative real number also has n distinct roots; of these however in case of a positive number for which $\varphi = 0$, when n is odd, only one is real: $k = 0$; when n is even, two are real: $k = 0$, $k = \frac{1}{2}n$; in case of a negative number for which $\varphi = \pi$, when n is odd, one is real: $k = \frac{1}{2}(n-1)$; when n is even, they are all imaginary or complex.

$$\sqrt[n]{+1} = \cos\frac{2k\pi}{n} + i\sin\frac{2k\pi}{n}, \quad \sqrt[n]{-1} = \cos\frac{(2k+1)\pi}{n} + i\sin\frac{(2k+1)\pi}{n},$$

$$\sqrt[n]{+i} = \cos\frac{(4k+1)\pi}{2n} + i\sin\frac{(4k+1)\pi}{2n}, \quad \sqrt[n]{-i} = \cos\frac{(4k+3)\pi}{2n} + i\sin\frac{(4k+3)\pi}{2n},$$

$$(k = 0, 1, 2 \ldots n-1).$$

Accordingly by the first definition for an arbitrary fractional exponent we have:

$$(r(\cos\varphi + i\sin\varphi))^{\frac{m}{n}} = \sqrt[n]{r^m(\cos m\varphi + i\sin m\varphi)}$$

$$= r^{\frac{m}{n}}\left(\cos\frac{m\varphi+2k\pi}{n} + i\sin\frac{m\varphi+2k\pi}{n}\right),$$

$$(k = 0, 1, 2 \ldots n-1).$$

By the second:

$$(r(\cos\varphi + i\sin\varphi))^{\frac{m}{n}} = \left\{r^{\frac{1}{n}}\left(\cos\frac{\varphi+2k'\pi}{n} + i\sin\frac{\varphi+2k'\pi}{n}\right)\right\}^m$$

$$= r^{\frac{m}{n}}\left(\cos\frac{m\varphi+2k'm\pi}{n} + i\sin\frac{m\varphi+2k'm\pi}{n}\right),$$

$$(k' = 0, 1, 2 \ldots n-1).$$

But the last expressions on the right are identical in both equations; for, since m and n are relatively prime, each number 0, m, $2m, \ldots (n-1)m$, divided by n, leaves a different remainder; hence $\frac{2k'm\pi}{n}$ expresses, it may be in altered order, values differing from those in the upper line only by multiples of 2π. The value belonging to $k = 0$ is styled the simplest value among the roots.

c) The power with an irrational exponent is understood to be the limiting value of the series of numbers obtained by forming the powers whose exponents are the rational numbers of the series that defines the irrational number. If $\frac{m}{n}, \frac{m'}{n'}, \frac{m''}{n''} \dots$ be the series of defining numbers, $(r(\cos\varphi + i\sin\varphi))^\mu$ represents all those numbers whose modulus is the limiting value of the series $r^{\frac{m}{n}}, r^{\frac{m'}{n'}}, r^{\frac{m''}{n''}} \dots$, and whose amplitudes are the limiting values of the series:

$$\frac{m(\varphi+2k\pi)}{n}, \quad \frac{m'(\varphi+2k\pi)}{n'}, \quad \frac{m''(\varphi+2k\pi)}{n''} \dots$$

$$(k = 1, 2, \dots n, n+1, \dots n' \dots).$$

To each integer value of k belongs a different limiting value of the amplitude, so that therefore for an irrational exponent there are infinitely many numbers of the form:

$$(r(\cos\varphi + i\sin\varphi))^\mu = r^\mu\{\cos\mu(\varphi + 2k\pi) + i\sin\mu(\varphi + 2k\pi)\},$$

i. e., $\qquad (r_\varphi)^\mu = r^\mu{}_{\mu(\varphi+2k\pi)}$, or, $(re^{i\varphi})^\mu = r^\mu \cdot e^{i\mu(\varphi+2k\pi)}$

all with the same modulus r^μ.

d) A power with a negative exponent must as before signify the reciprocal value of the power with a positive exponent. We have

$$(r(\cos\varphi + i\sin\varphi))^{-\mu} = \frac{1}{(r(\cos\varphi+i\sin\varphi))^\mu} = \frac{1}{r^\mu\{\cos\mu(\varphi+2k\pi)+i\sin\mu(\varphi+2k\pi)\}}$$

$$= r^{-\mu}\cdot\{\cos\mu(\varphi + 2k\pi) - i\sin\mu(\varphi + 2k\pi)\}$$

for every μ; a result that can also be identified with the form of De Moivre's theorem, since we can write this equation as follows:

$$(r(\cos\varphi + i\sin\varphi))^{-\mu} = r^{-\mu}\cdot\{\cos(-\mu(\varphi+2k\pi)) + i\sin(-\mu(\varphi+2k\pi))\}.$$

73. The power with a complex exponent. (Exponential.) The symbol $e^{i\varphi}$, denoting a power with a real base and a purely imaginary exponent, was defined in § 67 as the sum of the infinite series

$$1 + \frac{i\varphi}{1} + \frac{(i\varphi)^2}{\underline{|2}} + \frac{(i\varphi)^3}{\underline{|3}} + \frac{(i\varphi)^4}{\underline{|4}} + \cdots$$

whose real and imaginary constituents considered apart, form convergent infinite series with the values $\cos\varphi$ and $i\sin\varphi$. In connexion with this we define the power with the base e and complex exponent $x + iy$ as the value of the product of the exponential expression e^x by the exponential expression e^{iy}; i. e. in a formula

$$e^{x+iy} = e^x \cdot e^{+iy} = e^x(\cos y + i\sin y)$$

$$= \left(1 + \frac{x}{1} + \frac{x^2}{\underline{|2}} + \frac{x^3}{\underline{|3}} + \cdots\right)\left(1 + \frac{iy}{1} + \frac{(iy)^2}{\underline{|2}} + \cdots\right).$$

We shall see in next Chapter how this product of two infinite series may be combined in a single infinite series. From the definition follows the fundamental property of the exponential (§ 70):

$$e^{x+iy} \cdot e^{x'+iy'} = e^{x} \cdot e^{iy} \cdot e^{x'} \cdot e^{iy'} = e^{x+x'} \cdot e^{i(y+y')} = e^{x+x'+i(y+y')},$$

that is, powers with a real base e and complex exponents are multiplied by adding the exponents.

Since $e^{\pm i2k\pi} = 1$, we have $e^{(x+iy)} \cdot e^{\pm i2k\pi} = e^{x+i(y\pm 2k\pi)} = e^{x+iy}$; the exponential remains unaltered when its exponent is increased or decreased by an integer multiple of $2i\pi$; it has the period $2i\pi$.

If n be a real number, we have

$$(e^{x+iy})^n = (e^{x} \cdot e^{iy})^n = e^{nx} \cdot e^{iny} = e^{(x+iy)n}.$$

In order that we may be able to reduce the most general exponential expression always to the base e, we proceed to define:

74. The Logarithm.

By the logarithm of a complex number $a + ib$ in regard to the base e we understand that number $x + iy$, which has the property that $e^{x+iy} = a + ib$; we denote it by $x + iy = l(a + ib)$.

In order to calculate the numbers x and y let us determine

$$a + ib = r(\cos\varphi + i\sin\varphi).$$

If then ϱ be the logarithm of the positive number $r = \sqrt{a^2 + b^2}$ to the base e, we have: $a + ib = e^{\varrho}(\cos\varphi + i\sin\varphi) = e^{x}(\cos y + i\sin y)$. Equating separately the real parts and the imaginary parts, we find $x = \varrho$, $y = \varphi \pm 2k\pi$, therefore $l(a + ib) = l(r) + i(\varphi \pm 2k\pi)$, or, employing the definition of φ (§ 67),*)

$$l(a + ib) = l(+\sqrt{a^2+b^2}) + i\tan^{-1}\frac{b}{a} \pm i2k\pi \qquad \text{when } a > 0,$$

$$l(a + ib) = l(+\sqrt{a^2+b^2}) + i\tan^{-1}\frac{b}{a} \pm i(2k + 1)\pi \quad \text{when } a < 0.$$

Every number has therefore infinitely many logarithms in regard to the base e; they differ by integer multiples of $2i\pi$. To $k = 0$ belongs the **simplest** value of the logarithm. A real positive number a has **one** real logarithm, while the real **negative** number a has complex logarithms only, that differ from those of the positive number by $i\pi$.

$$l(+1) = \pm i2k\pi. \quad l(-1) = \pm i(2k + 1)\pi.$$

The equation: $e^{x+iy} \cdot e^{x'+iy'} = e^{(x+x')+i(y+y')} = (a + ib)(a' + ib')$, shows that: $l(a + ib) + l(a' + ib') = l\{(a + ib)(a' + ib')\}$, an equation however that does not always hold between the **simplest** values of the logarithms.

75. Powers with complex base and complex exponent. (General exponential expressions.)

---

*) The connexion between the logarithm and circular functions is founded on these equations.

By the expression $(a+ib)^{a'+ib'}$ we understand

$$\{e^{l(a+ib)}\}^{a'+ib'}=e^{(a'+ib').l(a+ib)}.$$

Putting $$l(a+ib)=x+i(y\pm 2k\pi),$$

we find: $$(a+ib)^{a'+ib'}=\{e^{x+i(y\pm 2k\pi)}\}^{(a'+ib')},$$

and further: $$\{e^{x+i(y\pm 2k\pi)}\}^{(a'+ib')}=e^{xa'-(y\pm 2k\pi)b'+i(xb'+(y\pm 2k\pi)a)}$$

$$=e^{xa'-(y\pm 2k\pi)b'}\{\cos(xb'+(y\pm 2k\pi)a')+i\sin(xb'+(y\pm 2k\pi)a')\}.$$

Both the modulus of this number and its amplitude have in general infinitely many values, corresponding to all integer values of k; to $k=0$ belongs the simplest value. For $b'=0$ this equation includes the former definition of a power with a real exponent. If a and b both vanish, this definition fails; because for the logarithm the value x is infinite and the value y is completely indeterminate. By this most general definition also only the simplest value of $(e)^{x+iy}$ is equal to $e^{x+iy}=e^x\cdot(\cos y+i\sin y)$, its general value being:

$$(e)^{x+iy}=e^{x\mp 2k\pi y+i(y\pm 2k\pi x)}=e^{x\mp 2k\pi y}\{\cos(y\pm 2k\pi x)+i\sin(y\pm 2k\pi x)\}.$$

However it is usual to denote only the simplest value by the symbol e^{x+iy}. Inverting this definition we obtain further the definition of the logarithm for any base:

The logarithm of a number $a+ib$ in regard to the base $a'+ib'$ is that number $u+iv$, which has the property that:

$$(a'+ib')^{u+iv}=a+ib.$$

Putting $a+ib=e^{\xi+i(\eta\pm 2k\pi)}$, and $a'+ib'=e^{s+i(t\pm 2k'\pi)}$; u and v are to be calculated from the equations:

$$su-(t\pm 2k'\pi)v=\xi,\quad sv+(t\pm 2k'\pi)u=\eta\pm 2k\pi,$$

which determine them uniquely for each value of k and k'; to $k=0$, $k'=0$ belong the simplest value of the logarithm:

$$u=\frac{s\xi+t\eta}{s^2+t^2},\quad v=\frac{s\eta-t\xi}{s^2+t^2}.$$

This closes the circle of arithmetical operations; their results can always be assigned in complex numbers; it is specially to be observed that the power with a negative base and any exponent, as also the logarithm with a negative base or with a negative number are now adopted as numerical conceptions into analysis.

Second Chapter.

Complex series. Complex variable. Functions of a complex variable.

76. By the sum of an infinite series, whose terms are complex:

$$\Sigma(u + iv) = (u_0 + iv_0) + (u_1 + iv_1) + \cdots (u_n + iv_n) \cdots$$

is to be understood the complex number $U + iV$ whose real part U is equal to the sum of the infinite series $u_0 + u_1 + \cdots u_n \cdots$, and its imaginary part iV to the sum of the series $i(v_0 + v_1 + \cdots v_n \cdots)$. The complex series has therefore a determinate value and is said to be convergent, only when both U and V have determinate finite limiting values, that is to say, when both the series $u_0 + u_1 + \cdots u_n \cdots$ and $v_0 + v_1 + \cdots v_n \cdots$ converge.

Addition of two infinite series.

If 1) $p_0 + p_1 + p_2 + \cdots + p_n \cdots$ and 2) $q_0 + q_1 + q_2 \cdots + q_n + \cdots$ be two convergent infinite series having complex terms

$$(p_n = u_n + iv_n, \; q_n = u_n' + iv_n'),$$

and their sums respectively P and Q, the series

3) $$(p_0 + q_0) + (p_1 + q_1) + (p_2 + q_2) + \cdots (p_n + q_n) \cdots$$

is convergent and its sum is $P + Q$. For, putting:

$$P_n = p_0 + p_1 + \cdots p_n, \quad Q_n = q_0 + q_1 + \cdots q_n,$$

as n increases, P_n and Q_n approximate to the limits P and Q, hence

$$(p_0 + q_0) + (p_1 + q_1) + \cdots (p_n + q_n) = P_n + Q_n$$

the sum of the $n + 1$ first terms of series 3) has, as n increases, the limiting value $P + Q$.

77. The complex series is said to be absolutely convergent, when the sum of the positive terms of $u_0 + u_1 + u_2 \cdots$ and of $v_0 + v_1 + v_2 \cdots$, and likewise the sum of the negative terms of each, have finite limiting values, or as this property is described: when the series u and the series v are absolutely convergent.*) When each series consists only of terms with one and the same sign, the only conception we

*) The conception of absolute convergence was introduced by Cauchy; Dirichlet (1805—1859) noticed the contrast with infinite series in which the limit of the sum depends on the order of the terms: Abhandlungen der Berliner Akademie, 1837.

require of convergence is that of absolute or unconditional convergence; but when their terms have different signs, ex. gr. are alternately positive and negative, a special property is enunciated. (Cf. § 44, III.)

If a complex series converges absolutely, the series of its moduli

$$\sqrt{u_0^2+v_0^2}+\sqrt{u_1^2+v_1^2}+\sqrt{u_2^2+v_2^2}+\cdots\sqrt{u_n^2+v_n^2}+\cdots$$

also converges.

If we denote the terms $u+iv$ separated according to the signs of u and v so that

$$1)\quad \Sigma(u+iv)=\Sigma(u^{(1)}+iv^{(1)})+\Sigma(u^{(2)}-iv^{(2)})+\Sigma(-u^{(3)}+iv^{(3)}) + \Sigma(-u^{(4)}-iv^{(4)}),$$

then on the hypothesis of absolute convergence, each of the series:

$$2)\qquad \Sigma u^{(1)},\ \Sigma u^{(2)},\ \Sigma u^{(3)},\ \Sigma u^{(4)},\ \Sigma v^{(1)},\ \Sigma v^{(2)},\ \Sigma v^{(3)},\ \Sigma v^{(4)}$$

converges, and therefore by the theorem of addition the series:

$$3)\qquad \Sigma(u^{(1)}+v^{(1)}),\ \Sigma(u^{(2)}+v^{(2)}),\ \Sigma(u^{(3)}+v^{(3)}),\ \Sigma(u^{(4)}+v^{(4)}).$$

also converge. But now:

$$4)\qquad u^{(1)}+v^{(1)}>\sqrt{(u^{(1)})^2+(v^{(1)})^2},\ u^{(2)}+v^{(2)}>\sqrt{(u^{(2)})^2+(v^{(2)})^2},$$
$$u^{(3)}+v^{(3)}>\sqrt{(u^{(3)})^2+(v^{(3)})^2},\ u^{(4)}+v^{(4)}>\sqrt{(u^{(4)})^2+(v^{(4)})^2}.$$

Consequently:

$$5)\qquad \sum\sqrt{(u^{(1)})^2+(v^{(1)})^2},\ \sum\sqrt{(u^{(2)})^2+(v^{(2)})^2},\ \sum\sqrt{(u^{(3)})^2+(v^{(3)})^2},\ \sum\sqrt{(u^{(4)})^2+(v^{(4)})^2}$$

must have finite values, and therefore the sum of these four series $\sum\sqrt{u^2+v^2}$ is likewise a finite quantity.

The converse proposition is also true; for since the modulus of a sum is not greater than the sum of the moduli of its summands, when the sums 5) have finite limiting values, the moduli of the sums $\Sigma(u^{(1)}+iv^{(1)})$, $\Sigma(u^{(2)}-iv^{(2)})$, $\Sigma(-u^{(3)}+iv^{(3)})$, $\Sigma(-u^{(4)}-iv^{(4)})$ must also be finite; this requires that the sums 2) should converge.

*The necessary and sufficient condition for the absolute convergence of a complex series is therefore the convergence of the series of moduli belonging to it.**)

An absolutely convergent series tends to the same finite limiting value, and is therefore said to have the same sum, even when the arrangement of its terms is changed according to any law.

Let U_n+iV_n be the sum of the terms

$$(u_0+iv_0)+(u_1+iv_1)+\cdots(u_n+iv_n),$$

and $\lim\limits_{n=\infty}(U_n+iV_n)=U+iV$; further let:

*) Cauchy: Cours d'Analyse algébrique.

$$(u_0' + iv_0') + (u_1' + iv_1') + \cdots (u_p' + iv_p') = U_p' + iV_p'$$

denote the sum of the first $p + 1$ terms of an infinite series that is formed from the other only by a different arrangement of the terms, then however large we suppose n, we can always choose p so that all the terms which are contained in $U_n + iV_n$ shall occur in $U_p' + iV_p'$. This latter expression will also contain other terms but their indices are greater than n. Consequently:

$$(U_p' + iV_p') - (U_n + iV_n) = (u_q + iv_q) + (u_r + iv_r) \cdots + (u_z + iv_z)$$
$$(q, r, \cdots z > n).$$

If now we make n and therefore also p increase arbitrarily,

$$\operatorname{Lim}(u_q + u_r + \cdots u_z) \text{ and } \operatorname{Lim}(v_q + v_r + \cdots v_z)$$

will become smaller than an arbitrarily small quantity; because since the series of moduli converges exclusively by choice of n:

$$\operatorname{Lim}\left(\sqrt{u_q^2 + v_q^2} + \sqrt{u_r^2 + v_r^2} + \cdots \sqrt{u_z^2 + v_z^2}\right), (q, r, \cdots z > n)$$

becomes arbitrarily small; therefore we have:

$$\operatorname{Lim}(U_p + iV_p) = \operatorname{Lim}(U_n + iV_n) = U + iV.$$

The fundamental proposition of addition is thus proved to apply to a sum of infinitely many summands that converges absolutely.

On the other hand, an infinite series that converges only conditionally, changes its value when the arrangement of the terms is altered; it presents a different sum when the summands are reckoned up in a different order.

Let us take as an example the series with real terms cited § 47:

$$S = 1 - \tfrac{1}{2} + \tfrac{1}{3} - \tfrac{1}{4} + \tfrac{1}{5} - \tfrac{1}{6} + \cdots = \operatorname*{Lim}_{n=\infty} \sum_{m=1}^{m=n} \left(\frac{1}{2m-1} - \frac{1}{2m}\right);$$

with a different rule of arrangement of its terms let us form the series:

$$S' = 1 + \tfrac{1}{3} - \tfrac{1}{2} + \tfrac{1}{5} + \tfrac{1}{7} - \tfrac{1}{4} \cdots = \operatorname*{Lim}_{n=\infty} \sum_{m=1}^{m=n} \left(\frac{1}{4m-3} + \frac{1}{4m-1} - \frac{1}{2m}\right),$$

then, since we can also consider S under the form:

$$S = \operatorname*{Lim}_{n=\infty} \sum_{m=1}^{m=n} \left(\frac{1}{4m-3} - \frac{1}{4m-2} + \frac{1}{4m-1} - \frac{1}{4m}\right),$$

we shall have:

$$S' - S = \operatorname*{Lim}_{n=\infty} \sum_{m=1}^{m=n} \left(\frac{1}{4m-2} - \frac{1}{4m}\right) = \tfrac{1}{2} \operatorname*{Lim}_{n=\infty} \sum_{m=1}^{m=n} \left(\frac{1}{2m-1} - \frac{1}{2m}\right) = \tfrac{1}{2} S,$$

hence we find: $S' = \tfrac{3}{2} S$.

It is obvious that the terms of such a semiconvergent series can be arranged so that their sum shall amount to any required value C. For, Σa denoting the combination of all the positive terms, Σb that of all the negative, both these series increasing beyond any finite amount; let us form a series by first taking so many a terms that their sum is greater than C, then joining on so many b terms that the sum becomes less than C, and so on. As this alternating arrangement is continued, the deviation from C will never amount to more than the value of the term preceding the last change of sign; therefore since the quantities a and b converge to zero, the value of the sum of the series has the limit C.*)

78. Multiplication of two infinite series.

If the moduli of the terms of the two complex series (§ 76):

$$1)\quad p_0 + p_1 + p_2 + \cdots p_n \cdots \text{ and } 2)\ q_0 + q_1 + q_2 + \cdots q_n \cdots$$
$$(p_n = u_n + iv_n, \qquad q_n = u_n' + iv_n')$$

likewise form convergent series:

$$3)\quad \varrho_0 + \varrho_1 + \varrho_2 + \cdots \varrho_n + \cdots \quad 4)\ \varrho_0' + \varrho_1' + \varrho_2' + \cdots \varrho_n' + \cdots$$
$$(\varrho_n = \sqrt{u_n^2 + v_n^2}, \qquad \varrho_n' = \sqrt{u_n'^2 + v_n'^2})$$

then the series

$$5)\qquad p_0 q_0 + (p_0 q_1 + p_1 q_0) + (p_0 q_2 + p_1 q_1 + p_2 q_0) + \cdots$$
$$\cdots (p_0 q_n + p_1 q_{n-1} + \cdots p_k q_{n-k} + \cdots p_n q_0) + \cdots$$

is convergent and its sum is $P \,.\, Q$.

To prove this theorem**) we require the Lemma:

If R and R' be the sums of the two series 3) and 4) that consist only of positive terms, the series

$$6)\qquad \varrho_0\varrho_0' + (\varrho_0\varrho_1' + \varrho_1\varrho_0') + (\varrho_0\varrho_2' + \varrho_1\varrho_1' + \varrho_2\varrho_0') + \cdots$$
$$\cdots (\varrho_0\varrho_n' + \varrho_1\varrho_{n-1}' + \cdots \varrho_k\varrho_{n-k}' + \cdots \varrho_n\varrho_0') + \cdots$$

is convergent and its sum $= R \,.\, R'$.

For, denoting the sum of the first $n+1$ terms of series 3), 4) and 6) respectively by R_n, R_n', S_n, it is plain, that $S_n < R_n R_n'$; and if we call m the greatest integer in $\frac{1}{2} n$, that $S_n > R_m R_m'$. For, the product $(\varrho_0 + \varrho_1 + \cdots \varrho_n)(\varrho_0' + \varrho_1' + \cdots \varrho_n')$ contains more terms than occur in S_n, while all the terms of the product

$$(\varrho_0 + \varrho_1 + \cdots \varrho_m)(\varrho_0' + \varrho_1' + \cdots \varrho_m')$$

occur in S_n, and in addition to them other positive quantities. It follows from the inequality, which holds for every n however great,

$$R_n R_n' > S_n > R_m R_m',$$

that $\operatorname{Lim} S_n = \operatorname{Lim} R_n R_n' = \operatorname{Lim} R_m R_m'$, since as n is arbitrarily increased $\operatorname{Lim} R_n = \operatorname{Lim} R_m$, $\operatorname{Lim} R_n' = \operatorname{Lim} R_m'$. Forming now the difference:

---

*) Dirichlet: *loc. cit.*; Riemann: Ueber die Darstellbarkeit einer Function durch eine trigonometrische Reihe. Werke, p. 221.

**) Cauchy: Cours d'Analyse algébrique.

$$R_n . R_n' - S_n = (\rho_1 \rho_n' + \rho_2 \rho'_{n-1} + \cdots \rho_n \rho_1') + (\rho_2 \rho_n' + \rho_3 \rho'_{n-1} + \cdots \rho_n \rho_2')$$
$$+ (\rho_3 \rho_n' + \rho_4 \rho'_{n-1} + \cdots \rho_n \rho_3') + \cdots \rho_n \rho_n',$$

inasmuch as by the theorem just proved $\lim_{n=\infty} (R_n R_n' - S_n) = 0$, we find, that the sum on the right also has zero as its limit.

This result enables us to prove the theorem regarding the product in the following manner.

If S_n' be the sum of the first $n+1$ terms of series 5), we have:

$$P_n Q_n - S_n' = (p_1 q_n + p_2 q_{n-1} + \cdots p_n q_1)$$
$$+ (p_2 q_n + p_3 q_{n-1} + \cdots p_n q_2) + \cdots p_n q_n;$$

and it has to be shown, that the sum of the products on the right has zero as its limit when n increases. This will be the case only when the modulus of this complex expression converges to zero. Now as the modulus of a sum of complex numbers is not greater than the sum of the moduli of the summands, what we have stated is proved, since

$$(\rho_1 \rho_n' + \rho_2 \rho'_{n-1} + \cdots \rho_n \rho_1') + (\rho_2 \rho_n' + \rho_3 \rho'_{n-1} + \cdots \rho_n \rho_2') + \cdots \rho_n \rho_n'$$

converges to zero.

We apply this theorem to the exponential function as defined § 73:

$$e^{x+iy} = e^x(\cos y + i \sin y) = \left(1 + \frac{x}{1} + \frac{x^2}{\underline{|2}} + \frac{x^3}{\underline{|3}} + \cdots\right)\left(1 + \frac{iy}{1} + \frac{(iy)^2}{\underline{|2}} + \frac{(iy)^3}{\underline{|3}} + \cdots\right).$$

The product of the two infinite series is expressed by one infinite series:

$$1 + \frac{x+iy}{1} + \left(\frac{x^2}{\underline{|2}} + \frac{x}{1} \cdot \frac{iy}{1} + \frac{(iy)^2}{\underline{|2}}\right) + \left(\frac{x^3}{\underline{|3}} + \frac{x^2}{\underline{|2}} \cdot \frac{iy}{1} + \frac{x}{1} \frac{(iy)^2}{\underline{|2}} + \frac{(iy)^3}{\underline{|3}}\right)$$
$$+ \cdots \left(\frac{x^n}{\underline{|n}} + \frac{x^{n-1}}{\underline{|n-1}} \frac{iy}{1} + \frac{x^{n-2}}{\underline{|n-2}} \frac{(iy)^2}{\underline{|2}} + \cdots \frac{x^{n-k}}{\underline{|n-k}} \frac{(iy)^k}{\underline{|k}} + \cdots \frac{(iy)^n}{\underline{|n}}\right) + \cdots,$$

that can be contracted in the form:

$$1 + \frac{x+iy}{1} + \frac{(x+iy)^2}{\underline{|2}} + \frac{(x+iy)^3}{\underline{|3}} + \cdots \frac{(x+iy)^n}{\underline{|n}} + \cdots$$

This is the exponential series with a complex argument, it expresses the simplest value of the exponential function e^{x+iy}.

79. *A quantity is called a complex variable when it is able to assume different complex numerical values.*

Whilst any set of real numbers can always be figured by the points of a finite right line, a limited range of complex numbers is in general presented to intuition by a "domain" of two dimensions of the plane bounded by some curve. In each individual case it must be specially assigned whether the points of the boundary curve themselves belong to the domain or not. Such a domain can in particular cases reduce to a linear figure, a domain of one dimension: to the points of a portion of a curve or of a finite right line.

Thus, for example, all complex numbers whose moduli are less than r_1 and greater than r_2 form a domain exhibited geometrically by the plane ring bounded by the two concentric circles with radii r_1 and r_2 round the origin. But complex numbers whose modulus is equal to r_1 form a linear figure, namely the circumference of the circle whose radius is r_1.

A domain of two dimensions is said to be connected, when we can pass from any one point within it to any other without crossing its boundary.

A domain of one dimension is said to be connected, when we can pass from any point in the domain to any other point in it without leaving the domain.

A quantity is said to be unrestrictedly variable within a domain, when it is able to assume all numerical values belonging to this domain.

A complex quantity is said to be continuously variable, when all values which it assumes, belong always to a finite connected domain.

In particular a variable is called unrestrictedly continuous at a certain point, when it can assume all values which belong to a finite domain however small including that point. On the other hand, the variable is restrictedly continuous at this point, when the values it assumes near the point form a domain whose boundary passes through the point, or form a domain of only one dimension. It is discontinuous at this point when the point is isolated by itself, and so belongs to no domain.

A further circumstance has to be noticed here: If a real variable is said to alter continuously within an interval from a value a to a value b, this informs us what numerical values it assumes, or in geometric language, we know the path along which it travels. If a complex quantity change continuously from a complex value a to a complex value b, this tells nothing at all of its intervening values. The ways in which it changes are just as illimitable as the continuous lines which can be drawn joining one point of the plane to another.

Continuous change of a complex quantity $z = x + iy$ requires that both the real constituent x and the factor y of the imaginary, vary continuously. A complex variable becomes infinitely small, when its modulus becomes infinitely small, i. e. when both x and y have zero as limit A complex quantity becomes infinitely great, when its modulus becomes infinitely great, i. e. when either x or y or both together increase numerically beyond any finite amount.

To the infinitely great values of complex numbers correspond in the plane of the figure the infinitely distant points, and as a complex number $x + iy$ can become infinite, while the ratio $x : y$ assumes all possible values, there is in the plane on each direction through the origin an infinitely distant point. In so far the limiting conception of projective plane geometry, in accordance with which there is an infinitely remote point belonging to each direction, coincides with the system of complex numbers in regard to the infinite values.

On the other hand in analytical investigations the occurrence of an infinite quantity always requires special considerations; these are mostly independent of the ratio $x : y$ and can be reduced by a simple substitution to what takes place with a finite value of the variable. For, should the variable z increase beyond all bounds, the equation $z = \frac{1}{z'}$ correlates to such values only a single value of z', namely the value 0, leaving it still possible for the ratio of the vanishing real numbers x' and y' in $z' = x' + iy'$ to assume any value whatever. The infinite values of z are therefore in this manner concentrated into one point.

There are two ways of expressing geometrically this connexion. In the first we say directly, the plane is closed at infinity by a point. This statement does not coincide with any actual representation any more than does the statement, that the plane is bounded by a right line at infinity, for there can be no such thing as a representation of infinity; it is only a way of stating how certain limiting processes are completed. But by passing from the plane to the sphere, we can procure ourselves an intuitive image of this conception. Let us suppose a sphere of any radius placed on the plane at the origin and its highest point joined with the points of the plane, each joining line meets the sphere in a second point. Let us consider this point as the image of the complex number $x + iy$ which was originally represented by the point of the plane with the coordinates x and y. Here, then, all the points corresponding to arbitrarily remote points of the plane, converge on the sphere to a single point namely its highest point. In this way we can utilise a finite sphere in representing the system of numbers; we shall however in what follows abide by the plane.

For there is still a second possibility of giving an intuitive form to the way in which we contemplate the infinite by a point.

Let us lay out in the plane of z all numbers whose modulus does not exceed a certain arbitrary limit R; but let us lay out in another plane z', whose points are coordinated to those of the first by the equation $z' = \frac{1}{z}$, all numbers whose modulus is greater than R. Let us put $z = r(\cos\varphi + i\sin\varphi)$ and $z' = \varrho(\cos\psi + i\sin\psi)$, then $\varrho(\cos\psi + i\sin\psi) = \frac{1}{r}(\cos\varphi - i\sin\varphi)$, therefore $\varrho = \frac{1}{r}$, $\psi = -\varphi$.

To all points of the plane z, which are outside the circle $r = R$, correspond points in the plane z' within the circle $\varrho = \frac{1}{R}$. Circles round the origin $z = 0$ change into circles round the origin $z' = 0$, but these are described in the opposite direction; to the infinite in the plane z corresponds only one point namely $z' = 0$. It is this Circular Relation between two planes,*) called Transformation by reciprocal radii vectores or by Inversion, which we shall subsequently employ in reference to infinite values of z.

80. When the values of a complex variable $w = u + iv$ are so determined by the values of a complex variable $z = x + iy = r_\varphi$, that to each value of z within a determinate domain one or more values of w can be assigned by means of any finite or infinite number of arithmetical operations (§ 38) on z, w is said to be a function of the complex variable z.

Here also functions are distinguished into one-valued and many-valued, according to the number of values of w belonging to one value of z; into algebraic and transcendental, according to the form in which the variables occur; and into explicit and implicit according as the equation defining the function is solved for w or not.

The total course of a one-valued (monotropic) function is realised by help of two planes. To each value $x + iy = r_\varphi$ of the quantity z corresponds a point of plane A having the rectangular coordinates x and y, to each corresponding value $u + iv$ of w, a point of plane B having as rectangular coordinates u and v. If to each value of z belong a determinate value of w changing continuously with z, then to each point of plane A will correspond a point of plane B, to each line a line, to each connected area a connected area. If on the other hand, w changes discontinuously at some points, while z changes continuously, disconnected portions of plane B will correspond to a connected area of plane A. In a word, the dependence of the quantity w on z is geometrically represented as a Transformation of plane B upon plane A.**) Such a transformation, for instance, was already investigated in last Section by means of the equation: $w = \frac{1}{z}$.

81. Commencing with the case of an explicit function

$$w = u + iv = f(z) = f(x + iy) = f\{r(\cos\varphi + i\sin\varphi)\}$$

— the quantities u and v are functions of the real variables x and y,

---

*) Möbius (1790—1868): Abhandl. der sächs. Gesellsch. d. Wissensch., 1855. This paper on Circular Relationship (Kreisverwandtschaft) follows earlier notices of the same subject in his Gesammelte Werke, vol. II, p. 243.

**) Riemann: Grundlagen für eine allgemeine Theorie der Functionen. Werke, p. 5.

or of r and φ — let us ask what is the analytical characteristic that such a function is unrestrictedly continuous in a domain for which it has determinate values. The foregoing discussions indicate that it must be possible to find surrounding any point z at which w is to be continuous, a connected domain of two dimensions of finite extent however small (the Neighbourhood of the point z), to which corresponds a connected domain of w; i. e. the quantities u and v must vary continuously when the quantities x and y, or r and φ vary continuously; in other words: u and v must be continuous functions of both the real variables x and y, or r and φ (§ 52). When the quantities u and v are expressed as functions of r and φ it must however further be remarked, that should the function w be one-valued, an increase of φ by multiples of 2π must not alter the values of these functions.

Denoting the increment of z by $\Delta z = \Delta x + i\Delta y$ and putting:

$$w + \Delta w = (u + \Delta u) + i(v + \Delta v) = f(z + \Delta z),$$

then in whatever way Δx and Δy converge to zero, we must have

$$\operatorname{Lim} \Delta u = 0 \text{ and } \operatorname{Lim} \Delta v = 0.$$

These two conditions are combined in the single statement:

The function $w = f(z)$ is continuous at a point z, when this point can be included within a domain such that the modulus of the difference:

$$mod\,[\Delta w] = mod\,[f(z + \Delta z) - f(z)] = \sqrt{\Delta u^2 + \Delta v^2}$$

for every point $z + \Delta z$ in this domain, shall be less than any arbitrarily small prescribed number δ.

82. Turning now to the formation of definite examples of functions of a complex variable, before all things we restrict ourselves to such as admit of the calculation of one or more values of w by a given formula for each value of the argument z. The most general instrument our previous investigations have provided for this purpose is the series of powers, which embraces the explicit rational or irrational functions.

Of such series we have already become acquainted in the domain of real quantities with the exponential series, and its inverse, the logarithm; to these we can reduce trigonometric and circular functions (§ 67 and § 74).

Accordingly we propose to ourselves the task of studying in the complex domain: first the explicit rational and irrational functions, next the exponential function and its inverse the logarithm, and then in general the properties of functions expressed by series of powers. These problems form the basis of the general Theory of Functions; to them the following investigations always return, inasmuch as in their progress the methods for the complete solution are gradually attained.

By these methods we shall be ultimately enabled to carry out the development of implicit algebraic functions.

1. The power with a positive integer exponent: $w = z^m$ is a one-valued function and continuous for the entire plane, because $u = r^m \cos m\varphi$, $v = r^m \sin m\varphi$ are continuous functions of r and φ that do not alter as the amplitude φ is increased by multiples of 2π. Such a function having everywhere in the neighbourhood of any definite point z a finite value that changes continuously with z, we describe as behaving regularly in the neighbourhood of that point, or as having that point as a regular point.

To investigate how this function behaves for infinite values of z, let us put $z = \frac{1}{z'}$, then $w = \frac{1}{z'^m}$ and to the infinite corresponds the point $z' = 0$. At this point w becomes infinite, that is, its modulus increases determinately beyond all limits in whatever way the point is approached. The point infinity is therefore a singular point for this function. But inasmuch as the function $w = \frac{1}{z'^m}$ is such that its product by $z'^m = \left(\frac{1}{z}\right)^m$ at the point $z' = 0$ has the finite value 1, this singular point is called non-essential; the general definition being:*)

If a function $f(z)$ become infinite at a finite point $z = \alpha$ or at infinity $z = \infty$, this singular point is called non-essential, provided an integer m can be assigned, such that the product:

$$((z-\alpha)^m f(z)), \text{ or: } \left(\frac{1}{z}\right)^m f(z)$$

is equal to a finite quantity G, in whatever manner z is made converge to α, or to ∞, as the case may be; or more strictly: a domain must be assignable about the point, within which the above product shall differ from G by less than an arbitrarily small number δ. Moreover, when the value of $f(z) = \frac{G}{(z-\alpha)^m}$ for $z = \alpha$, or $= Gz^m$ for $z = \infty$, we say that for such a value of z the function becomes infinite in the order m, or, that in the point its infinitude is equal to m.

It follows hence: Every rational integer function of the nth degree in z is one-valued and continuous for the entire plane and has no singularity except one, which is non-essential, in the point infinity. For, the function $f(z) = a_0 + a_1 z + a_2 z^2 + \cdots a_n z^n$, in which $a_n \gtrless 0$, is for every finite z a sum of functions that are all continuous, and only for $z = \infty$ becomes infinite, as $f\left(\frac{1}{z'}\right) = a_0 + \frac{a_1}{z'} + \frac{a_2}{z'^2} + \cdots \frac{a_n}{z'^n}$ does for $z' = 0$; but for $z' = 0$, $z'^n f\left(\frac{1}{z'}\right) = \frac{1}{z^n} f(z) = a_0 z'^n + a_1 z'^{n-1} + \cdots a_n$ is equal to a_n. Thus in $z = \infty$ the infinitude of $f(z)$ is equal to n.

*) Weierstrass: Zur Theorie der eindeutigen analytischen Functionen. Abhandl. d. Akad. d. Wissensch., Berlin 1876. Reprinted in his: Abhandlungen aus der Functionenlehre. 1886.

2. The rational fractional function: $\frac{1}{(z-\alpha)^m}$ is one-valued and continuous in the entire plane, except at the non-essential singular point $z = \alpha$. The point infinity is not singular, since by the above substitution we have for it: $\frac{1}{(z-\alpha)^m} = \left\{\frac{z'^m}{(1-\alpha z')^m}\right\}_{z'=0}$.

The most general rational fractional function is of the form:

$$\frac{a_0 + a_1 z + a_2 z^2 + \cdots a_n z^n}{b_0 + b_1 z + b_2 z^2 + \cdots b_m z^m}.$$

It will be shown in next Chapter, that every rational integer function of the degree n may be broken up into n linear factors. If this result be assumed here, and this fraction therefore written in the form:

$$\frac{a_n}{b_m} \cdot \frac{(z-\alpha_1)(z-\alpha_2)\cdots(z-\alpha_n)}{(z-\beta_1)(z-\beta_2)\cdots(z-\beta_m)},$$

where some of the quantities α as well as some of the β may be equal, but each α will be supposed different from each β because otherwise factors could be cancelled, then obviously: the rational fractional function is one-valued and continuous in the entire finite plane except at the non-essential singular points $\beta_1, \beta_2 \ldots \beta_m$. The point infinity likewise is a non-essential singular point when $n > m$, but is a regular point when $n \leq m$. For, the function behaves for $z = \infty$ as:

$$\frac{a_n}{b_m} \cdot \frac{\left(\frac{1}{z'}-\alpha_1\right)\left(\frac{1}{z'}-\alpha_2\right)\cdots\left(\frac{1}{z'}-\alpha_n\right)}{\left(\frac{1}{z'}-\beta_1\right)\left(\frac{1}{z'}-\beta_2\right)\cdots\left(\frac{1}{z'}-\beta_m\right)} = \frac{a_n}{b_m} \cdot z'^{m-n} \frac{(1-\alpha_1 z')(1-\alpha_2 z')\cdots(1-\alpha_n z')}{(1-\beta_1 z')(1-\beta_2 z')\cdots(1-\beta_m z')}$$

behaves for $z' = 0$.

3. The simplest explicit irrational algebraic function: $w = (z' - a)^m$, m a rational fractional number $\frac{p}{q}$, reduces to the form z^m by the substitution $z' - a = z$. All propositions which we shall prove for the function z^m, can be easily transferred to the more general form $(z' - a)^m$ since we have only to observe that whatever holds for the point $z = 0$ relates to the point $z' = a$.

If $z = r(\cos\varphi + i\sin\varphi)$, then for $k = 0, 1, \ldots q-1$,

$$w = z^m = r^{\frac{p}{q}}\left(\cos\frac{p}{q}(\varphi + 2k\pi) + i\sin\frac{p}{q}(\varphi + 2k\pi)\right).$$

This exposition shows that the function can be calculated. For, both cosine and sine are series of powers. Each integer value of k determines at each point the corresponding value of a branch of the function; it is therefore a many-valued function, no longer single-valued as those in the previous examples. At the point $z = 0$ and at $z = \infty$ all branches have the same values, namely:

when $\frac{p}{q} > 0$, at $z = 0$ the value 0, at $z = \infty$ the value ∞;

when $\frac{p}{q} < 0$, at $z = 0$ the value ∞, at $z = \infty$ the value 0.

These two points are called **branching points** or **ramifications** of the function; it has moreover, in the first case the point $z = \infty$, and in the second the point $z = 0$, as an infinity point.

The next inquiry is, how can we group together the values belonging to a single branch of the function, so as to have each branch by itself in general a continuous function.

Draw from the origin out to infinity any curve that does not cross itself; the simplest that can be chosen is one of the axes of coordinates, ex. gr. the positive part of the axis of abscissæ.

To each point of this right line belong q values of the function

$$w = r^{\frac{p}{q}} \left(\cos \frac{p}{q} (\varphi + 2k\pi) + i \sin \frac{p}{q} (\varphi + 2k\pi)\right)$$

$$(k = 0, 1, \ldots q - 1).$$

Selecting one of these values for a point at an arbitrarily small finite distance from the origin, for instance the value belonging to $k = 0$, let us attribute to all other points of the curve those values that proceed from the assumed value by continuous change of r and φ, for which therefore k is likewise zero. In this manner the values $w = r^{\frac{p}{q}}$ are chosen for the positive axis of abscissæ. Now in order to construct the values of one branch of the function for other points of the plane, suppose concentric circles drawn round the origin with all possible values of r and attribute to their points those values of w that result on continuous change of φ, the circles being described in one and the same direction, ex. gr. from the positive axis of abscissæ to the positive axis of ordinates.

Along a circle with radius r corresponds in this way to the point:

$$\varphi = 0; \quad z = r, \quad w = r^{\frac{p}{q}},$$

$$\varphi = \frac{\pi}{2}; \quad z = ri, \quad w = r^{\frac{p}{q}} \left(\cos \frac{p}{q} \frac{\pi}{2} + i \sin \frac{p}{q} \frac{\pi}{2}\right),$$

$$\varphi = \pi; \quad z = -r, \quad w = r^{\frac{p}{q}} \left(\cos \frac{p}{q} \pi + i \sin \frac{p}{q} \pi\right),$$

$$\varphi = \frac{3\pi}{2}; \quad z = -ri, \quad w = r^{\frac{p}{q}} \left(\cos \frac{p}{q} \frac{3\pi}{2} + i \sin \frac{p}{q} \frac{3\pi}{2}\right),$$

$$\varphi = 2\pi; \quad z = r, \quad w = r^{\frac{p}{q}} \left(\cos \frac{p}{q} 2\pi + i \sin \frac{p}{q} 2\pi\right).$$

Thus w is a continuous function all along each circle, only its final values for $\varphi = 2\pi$ do not coincide with its initial values for

$\varphi = 0$. Therefore also its values that belong to points arbitrarily near the positive axis of abscissæ with positive ordinates, differ by finite quantities from the values possessed by points below that axis. A branch of the function constructed in this way is therefore discontinuous along the positive axis of abscissæ. It is usual also to state the matter geometrically thus: A branch of the function $z^{\frac{p}{q}}$ is continuous in the connected surface that consists of the infinite plane perforated from its zero point to its infinity point.

It is in fact easily seen that everywhere else the branch is not only unique, as follows from the construction, but also continuous. For, $z = \varrho(\cos\psi + i\sin\psi)$ being any point for which ψ differs finitely from 0 or 2π, let us surround it with a small circle of radius $\Delta\varrho$; the coordinates of points upon or within the circumference of this circle are

$$z + \Delta z = z + \varepsilon(\cos\Theta + i\sin\Theta) = r(\cos\varphi + i\sin\varphi)$$
$$(\varepsilon \leqq \Delta\varrho,\ 0 \leqq \Theta < 2\pi),$$

so that:

$$r\cos\varphi = \varrho\cos\psi + \varepsilon\cos\Theta, \quad r\sin\varphi = \varrho\sin\psi + \varepsilon\sin\Theta.$$

Hence follows that $r = \sqrt{\varrho^2 + \varepsilon^2 + 2\varrho\varepsilon\cos(\psi - \Theta)}$; therefore $\Delta\varrho$ can be chosen so as to make the difference abs $[r - \varrho]$ less than an arbitrarily small quantity δ, whence it follows further that we can also put $\varphi = \psi \pm \eta$, where η is arbitrarily small. If now ψ differ from 0 or 2π by a finite quantity, $\psi \pm \eta$ is always a positive number between zero and 2π, and the values of the function:

$$w + \Delta w = (\varrho \pm \delta)^{\frac{p}{q}}\left(\cos\frac{p}{q}(\psi \pm \eta) + i\sin\frac{p}{q}(\psi \pm \eta)\right)$$

differ, as the respective series show, arbitrarily little from those of:

$$w = \varrho^{\frac{p}{q}}\left(\cos\frac{p}{q}\psi + i\sin\frac{p}{q}\psi\right).$$

This method of rendering the function w unique and continuous, by drawing a section that must not be crossed by the argument z as it varies, was introduced by Cauchy. Riemann perfected it by a process which enables us to contemplate simultaneously all branches, and to render the function unique and continuous along all paths without restriction. This is effected, for a function admitting of q values, by making the variable z move upon q different plane leaves.

We shall first consider the simplest case, assuming $q = 2$. Besides the one plane perforated along the positive axis of x, to which we have coordinated the values of the function starting from the values $w = r^{\frac{p}{q}}$, we take a second plane for the motion of z. In this, conformably with the general equation:

$$w = r^{\frac{p}{q}}\left(\cos\frac{p}{q}(\varphi + 2k\pi) + i\sin\frac{p}{q}(\varphi + 2k\pi)\right)$$

for $q = 2$, we coordinate to the points of the positive axis of abscissæ the values belonging to $k = 1$:

$$w = r^{\frac{p}{2}}\{\cos(p\pi) + i\sin(p\pi)\}$$

and proceed with laying out the remaining values as before. The amplitudes thus assigned to points with arbitrarily small negative ordinates will, for $\varphi = 2\pi$, conduct continuously into the values:

$$w = r^{\frac{p}{2}}(\cos 2p\pi + i\sin 2p\pi) = r^{\frac{p}{2}},$$

that therefore again differ finitely from the initial values chosen for the second plane; but they coincide with the initial values of the first. Accordingly let two banks be distinguished along the positive axis of abscissæ in the first leaf; call one with positive values of the ordinates, $I^{(+)}$, the other with negative values, $I^{(-)}$; similarly in the second leaf the two banks $II^{(+)}$, $II^{(-)}$. Conceiving the planes superposed, join $I^{(+)}$ with $II^{(-)}$, $I^{(-)}$ with $II^{(+)}$, so that thus both planes or leaves cross along the entire positive axis of abscissæ, the branching section. Thus arises a connected two-leaved Riemann's surface that is called a winding surface of the first order. To each point of this surface corresponds one determinate value of the function, to each continuous curve on the surface, whether remaining in the same leaf or passing over into the other, correspond values of the function which change continuously. As each closed curve must cross the branching section either an even number of times or not at all, it leads to a final value which is identical with the initial one.*) When there are more than two leaves, ex. gr. when $q = 5$, we have corresponding to:

$I^{(+)}$ the values $r^{\frac{p}{5}}$,

$II^{(+)}$ " " $r^{\frac{p}{5}}\left(\cos\frac{2p\pi}{5} + i\sin\frac{2p\pi}{5}\right)$,

$III^{(+)}$ " " $r^{\frac{p}{5}}\left(\cos\frac{4p\pi}{5} + i\sin\frac{4p\pi}{5}\right)$,

$IV^{(+)}$ " " $r^{\frac{p}{5}}\left(\cos\frac{6p\pi}{5} + i\sin\frac{6p\pi}{5}\right)$,

$V^{(+)}$ " " $r^{\frac{p}{5}}\left(\cos\frac{8p\pi}{5} + i\sin\frac{8p\pi}{5}\right)$,

---

*) It is manifest, that instead of the positive axis of abscissæ any other curve, that does not cross itself, can be chosen as branching section; the two-leaved surface which arises, is in its totality always the same.

$$\text{I}^{(-)} \text{ the values } r^{\frac{p}{5}}\left(\cos\frac{2p\pi}{5} + i\sin\frac{2p\pi}{5}\right),$$

$$\text{II}^{(-)} \;,,\quad ,,\quad r^{\frac{p}{5}}\left(\cos\frac{4p\pi}{5} + i\sin\frac{4p\pi}{5}\right),$$

$$\text{III}^{(-)} \;,,\quad ,,\quad r^{\frac{p}{5}}\left(\cos\frac{6p\pi}{5} + i\sin\frac{6p\pi}{5}\right),$$

$$\text{IV}^{(-)} \;,,\quad ,,\quad r^{\frac{p}{5}}\left(\cos\frac{8p\pi}{5} + i\sin\frac{8p\pi}{5}\right),$$

$$\text{V}^{(-)} \;,,\quad ,,\quad r^{\frac{p}{5}}\left(\cos\frac{10p\pi}{5} + i\sin\frac{10p\pi}{5}\right) = r^{\frac{p}{5}}.$$

We have therefore to connect $\text{I}^{(-)}$ with $\text{II}^{(+)}$, $\text{II}^{(-)}$ with $\text{III}^{(+)}$, $\text{III}^{(-)}$ with $\text{IV}^{(+)}$, $\text{IV}^{(-)}$ with $\text{V}^{(+)}$, finally $\text{V}^{(-)}$ with $\text{I}^{(+)}$, so that there arises a connected five-leaved surface; every closed curve crosses the branching section either not at all or a number of times that is a multiple of 5.

If the exponent m be a real irrational number, the function is infinitely many-valued; a single branch is constructed in the same manner as before, moreover a Riemann's surface can also be formed but now it must consist of infinitely many leaves.

4. The exponential function: e^{x+iy} defined by its infinite series, which has the same meaning as $e^x(\cos y + i\sin y)$, is a one-valued and continuous function in the entire finite plane. But the point infinity is a singular point, it is moreover an essential singular point.

For, putting $z = \frac{1}{z'}$, and therefore:

$$e^z = e^{\frac{1}{z'}} = e^{\frac{1}{x'+iy'}} = e^{\frac{x'-iy'}{x'^2+y'^2}} = e^{\frac{x'}{x'^2+y'^2}}\left(\cos\frac{y'}{x'^2+y'^2} - i\sin\frac{y'}{x'^2+y'^2}\right),$$

when the ratio $y' : x'$ of the vanishing values of x' and y' has any arbitrary fixed limiting value k we have:

$$(e^z)_{z=\infty} = e^{\frac{1}{x'}\cdot\frac{1}{1+k^2}}\left(\cos\frac{k}{x'(1+k^2)} - i\sin\frac{k}{x'(1+k^2)}\right)_{x'=0}.$$

The modulus of this expression converges to $+\infty$ or to zero according as x' approximates positively or negatively to the value zero, while the functions cosine and sine oscillate between the limits -1 and $+1$. We can also make the values x' and y' converge to zero so that the modulus may tend to any arbitrary finite value e^k, by putting $\text{Lim}\, \frac{x'}{x'^2+y'^2} = k$, for $x' = 0$, $y' = 0$, and therefore $x' = y'^2 k$; *in a word: in the essential singular point, that for e^z is situated in the point $z = \infty$, or for $e^{\frac{1}{z'}}$ in the point $z = 0$, the function is completely indeterminate, it assumes every complex value without any restriction.*

Another property distinguishing the essential singular point from the non-essential, is, that with the latter an integer m can be assigned for which $\lim_{z=a} (z-\alpha)^m f(z) = G$. With the former this is not possible. For, putting here: $e^{\frac{1}{z}} = 1 + \frac{1}{1}\frac{1}{z} + \frac{1}{\underline{|2}}\frac{1}{z^2} + \frac{1}{\underline{|3}}\frac{1}{z^3} + \cdots$, we have:

$$z^m e^{\frac{1}{z}} = z^m + \frac{1}{1} z^{m-1} + \frac{1}{\underline{|2}} z^{m-2} + \cdots \frac{1}{\underline{|m}} + \frac{1}{\underline{|m+1}} \frac{1}{z} + \cdots;$$

and however great we may choose m, we cannot give it any finite value such that for $z = 0$ the right side shall remain finite.

A second property is illustrated by the exponential function: it is a periodic function; the period is $2i\pi$.

$$e^{z+2i\pi} = e^x \{\cos(y+2\pi) + i\sin(y+2\pi)\} = e^z.$$

If we divide the plane into infinite strips by right lines parallel to the axis of abscissæ at distances 2π, the function reproduces itself symmetrically in each of these strips.

5. **The logarithm** $u + iv$ of the number $x + iy = r_\varphi$ in regard to the base e is by the definition (§ 74) an infinitely many-valued function. But as long as the simplest value

$$u + iv = l(x+iy) = l(r) + i\varphi$$

is considered, it is a one-valued function. Only the points $r = 0$ and $r = \infty$ are branching points; at these the real constituent of the function increases beyond any limit, and the imaginary is completely indeterminate. Conceiving therefore a branching section laid from the zero point to the infinity point, as in the example of the irrational function, one branch of the infinitely many-valued function is continuous in this perforated plane.

It is important for a subsequent application to interpret further in the following manner the significance of the branching section for the different values of the logarithm. If we make the variable z describe a finite closed curve, beginning at a point A and returning to it, which curve neither crosses itself nor includes the origin, and so meets the positive part of the axis of abscissæ either in an even number of points or not at all, then as r and φ vary continuously, the value of $l(z)$ on the return to the point A is just the same as at first; for, z has

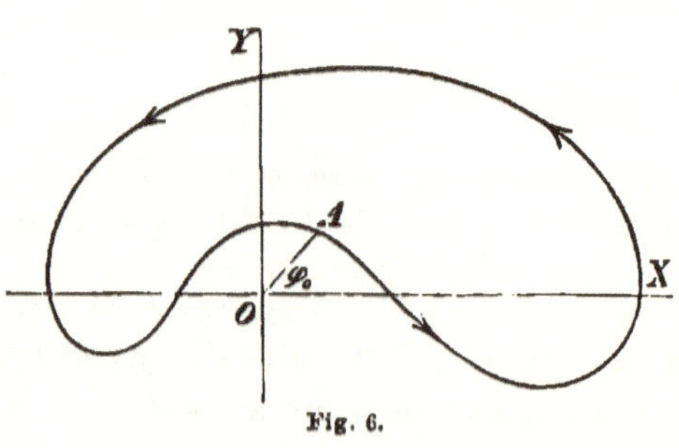

Fig. 6.

returned to the same leaf. In fig. 6, φ decreases from the value φ_0 to a determinate negative value, then increases passing through zero up to a value greater than π and then decreases to the value φ_0. Something similar to this will occur along every other such curve even when there are different pairs of intersections with the positive axis of abscissæ.

If on the other hand we make the argument z describe a finite closed curve, not crossing itself but including the origin $z = 0$ and so meeting the positive axis of abscissæ in an odd number of points, the value of $l(z)$, as r and φ vary continuously, will on the return to the point A come to be different from the initial one by $2i\pi$. In the adjoining figure let z travel from the point A so as to have the enclosed space on the left, then φ increases from φ_0 up to 2π, becomes then $> 2\pi$, decreases to a value between 2π and $\frac{3}{2}\pi$ and again increases, ultimately passing beyond 2π to the value $\varphi_0 + 2\pi$. Thus the value of the logarithm is $l(r) + i(\varphi_0 + 2\pi)$, while it was initially $l(r) + i\varphi_0$.

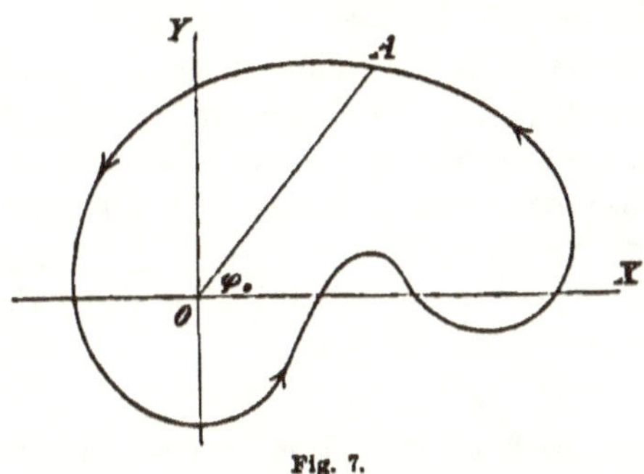

Fig. 7.

We can extend these considerations to curves that repeatedly go round the origin, crossing themselves in doing so, and formulate the following rule:

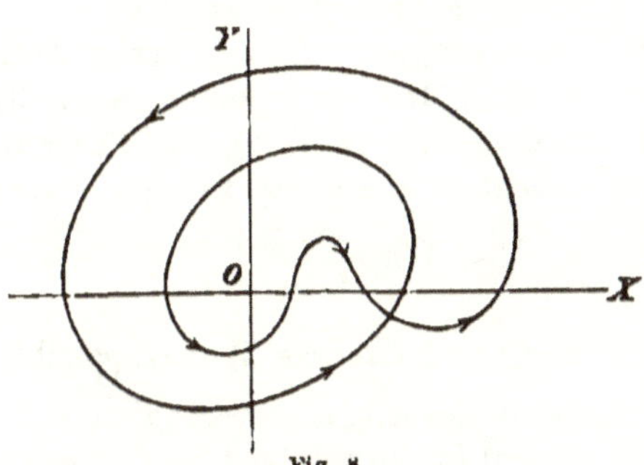

Fig. 8.

If in a determinately directed circuit the positive axis of abscissæ is crossed n times from below upwards, the value of $l(z)$ is increased by $2i\pi n$, for, each crossing shows that a circuit is completed. The path between two crossings only signifies that there has been no circuit, when, between the two, the amplitude φ had a retrograde motion, so that at an even or an odd number of points the curve has cut the positive axis of abscissæ from above downwards. If there be m such points, $2i\pi m$ is to be deducted; therefore the value of the logarithm changes by $2i\pi(n - m)$, when the numbers of the crossings are respectively n and m.

83. The complex series of powers in general.

The infinite series:

1) $a_0 + a_1(z' - \alpha) + a_2(z' - \alpha)^2 + \cdots a_n(z' - \alpha)^n + \cdots,$

in which the coefficients $a_0, a_1, \ldots a_n \ldots$ as well as α signify determinate complex numbers, can always be given the form:

2) $$a_0 + a_1 z + a_2 z^2 + \cdots a_n z^n + \cdots$$

by substituting z for $z' - \alpha$.

If we denote the modulus of z by r, and that of a_k by A_k, the proposed series converges absolutely for every value of z for which:

3) $$A_0 + A_1 r + A_2 r^2 + \cdots A_n r^n + \cdots$$

converges; and conversely the convergence of series 3) follows from the absolute convergence of series 2) (§ 77). The geometric statement of this connexion is: When a series of powers converges absolutely for a determinate point z, it likewise converges absolutely for all other points at the same distance from the origin $z = 0$.

But then the infinite series also converges absolutely at all points within this circle. For if r' be a value less than r:

4) $$A_0 + A_1 r' + A_2 r'^2 + \cdots A_n r'^n + \cdots$$

must be a convergent series, since each of its terms is positive and less than the corresponding term of series 3), consequently the sum of series 4) has a determinate value between zero and the sum of 3). Therefore the domain of convergence (convergency) of series 2) is always a circle round the origin $z = 0$; the convergency of series 1) is a circle round the point $z' = \alpha$. For, the value of the modulus of the difference determines all points z' that are equally distant from α.

By the radius of the circle of convergence of any series 2) is meant the greatest value of r up to which the corresponding series 3) converges. We are enabled to calculate this value by the Theorem (§ 44) that beyond a determinate place n in the series we must have:

$$\text{Lim}\,\frac{A_{n+1}R}{A_n} < 1, \quad \text{or} \quad R < \text{Lim}\,\frac{A_n}{A_{n+1}}.$$

When $R = \text{Lim}\,\frac{A_n}{A_{n+1}}$ the series 3) and likewise 2) may possibly converge; therefore in each case a special investigation is required as to whether the series is or is not convergent in the points of the bounding circle. But outside this circle it cannot converge, even conditionally, because then the moduli of the terms after a certain place in series 2) increase beyond any finite amount. When a series converges for a value of z even conditionally, it converges absolutely for all values having a smaller modulus. For, semiconvergence, in which the series of moduli ceases to converge, still requires that: $\text{Lim}\, R\,\frac{A_{n+1}}{A_n} < 1$

since the moduli of the terms cannot increase, and only for $n = \infty$ can we have $R \operatorname{Lim} \frac{A_{n+1}}{A_n} = 1$. If then we give R a smaller value, the property of absolute convergence is satisfied. Accordingly, series can be semiconvergent, if at all, only in points on the circle of convergence*).

Since each of its terms is unique, an infinite series of powers as long as it converges, is a one-valued function of the complex variable, that does not anywhere become infinite. *This function is continuous*, i. e. *when z and $z \pm \delta$ are complex values for which the series converges,*

$$Lim\ mod\,[f(z \pm \delta) - f(z)] = 0, \qquad for\ \delta = 0.$$

To prove this, we separate the terms of the series $f(z)$ into the groups:

$$\varphi(z) = a_0 + a_1 z + \cdots a_{n-1} z^{n-1},$$
$$\psi(z) = a_n z^n + a_{n+1} z^{n+1} + \cdots a_{n+k} z^{n+k} + \cdots;$$

inasmuch as from the nth term onwards, within the circle of convergence:

$$A_{n+1} R < A_n \alpha, \quad A_{n+2} R^2 < A_n \alpha^2, \cdots A_{n+k} R^k < A_n \alpha^k,$$

where α denotes a proper fraction, we must therefore have:

$$\mathrm{mod}\ \psi(z) < A_n R^n \frac{1}{1-\alpha}.$$

Accordingly merely by choosing a lower limit for n, we are able to make both mod $\psi(z)$ and mod $\psi(z \pm \delta)$, and therefore also:

$$\mathrm{mod}\ [\psi(z \pm \delta) - \psi(z)]$$

less than an arbitrarily small quantity ε. Now, since:

$$\mathrm{mod}\ [f(z \pm \delta) - f(z)] = \mathrm{mod}\ [\varphi(z \pm \delta) - \varphi(z) + \psi(z \pm \delta) - \psi(z)]$$
$$< \mathrm{mod}\ [\varphi(z \pm \delta) - \varphi(z)] + \varepsilon$$

---

*) Examples:

1. The binomial series:

$$1 + mz + \frac{m(m-1)}{1 \cdot 2} z^2 + \frac{m(m-1)(m-2)}{1 \cdot 2 \cdot 3} z^3 + \cdots$$

in which m is a real number and z complex, converges absolutely as long as mod $[z]$ is less than 1; it diverges if mod $[z] > 1$. For mod $[z] = 1$:

If $m > 0$, it also converges absolutely along the entire circle of convergence.

If $m < 0$ but > -1, the series is semiconvergent, with the exception of the point $z = -1$, in which it diverges.

If $m \leq -1$, the series diverges in all points of the circle of convergence. All these results can be deduced from § 46. In investigating the case $-1 < m < 0$ put $p_n = 1 + m_1 z + \cdots m_n z^n$, multiply both sides of the equation by $(1 + z)$ and consider the limiting value for $n = \infty$.

2. The logarithmic series:

$$z - \frac{z^2}{2} + \frac{z^3}{3} - \frac{z^4}{4} + \cdots$$

converges absolutely for mod $[z] < 1$, it diverges for mod $[z] > 1$. On the circle of convergence it is semiconvergent, except at the point $z = -1$ at which it diverges.

and $\varphi(z)$ is a rational integer function; having chosen a lower limit for n we can always determine δ so as to make the difference on the right less than any prescribed number, i. e.:

$$\text{Lim mod}\,[f(z \pm \delta) - f(z)] = 0.$$

It is proved as in § 44, IV that this theorem of uniform or equable convergence holds even on the limiting circle for a point in which the infinite series converges, by varying z along a radius and so putting:

$$\psi(z) = a_n z^n + a_{n+1} z^{n+1} + \cdots a_{n+k} z^{n+k} + \cdots$$

$$\psi(z - \delta) = a_n \left(\frac{z-\delta}{z}\right)^n z^n + a_{n+1}\left(\frac{z-\delta}{z}\right)^{n+1} z^{n+1} + \cdots$$

The amount of $\psi(z - \delta)$ is less than that of $\left(\frac{z-\delta}{z}\right)^n M$, where M is put for the greatest amount in the series of the complex numbers:

$$a_n z^n, \quad a_n z^n + a_{n+1} z^{n+1}, \quad a_n z^n + a_{n+1} z^{n+1} + a_{n+2} z^{n+2}, \text{ etc..}$$

For, choosing $z - \delta$ upon the radius to z, $\frac{z-\delta}{z} = \varrho$ is a positive real quantity less than 1. Thus:

$$\psi(z - \delta) = \varrho^n a_n z^n + \varrho^{n+1} a_{n+1} z^{n+1} + \varrho^{n+2} a_{n+2} z^{n+2} + \cdots,$$

and the powers of ϱ form a decreasing series that converges to zero.

But the Lemma of Abel in § 44, IV can be stated as follows for complex quantities.

If $t_0, t_1, \ldots t_m \ldots$ denote an infinite series of arbitrary complex quantities and if the amount of the quantity:

$$p_m = t_0 + t_1 + \cdots t_m$$

is for all values of m always less than G, then the amount of:

$$r = \varepsilon_0 t_0 + \varepsilon_1 t_1 + \cdots \varepsilon_m t_m \text{ is } < G \cdot \varepsilon_0,$$

when $\varepsilon_0, \varepsilon_1 \ldots$ denote real positive decreasing numbers. For, we have

$$r = p_0(\varepsilon_0 - \varepsilon_1) + p_1(\varepsilon_1 - \varepsilon_2) + \cdots p_{m-1}(\varepsilon_{m-1} - \varepsilon_m) + p_m \varepsilon_m$$

as before, and so,

$$\text{mod}\, r \leqq (\varepsilon_0 - \varepsilon_1)\,\text{mod}\, p_0 + (\varepsilon_1 - \varepsilon_2)\,\text{mod}\, p_1 + \cdots (\varepsilon_{m-1} - \varepsilon_m)\,\text{mod}\, p_{m-1} + \varepsilon_m\,\text{mod}\, p_m.$$

The numerical value on the right side is less than:

$$G(\varepsilon_0 - \varepsilon_1 + \varepsilon_1 - \varepsilon_2 + \cdots \varepsilon_{m-1} - \varepsilon_m + \varepsilon_m) = G \cdot \varepsilon_0.$$

84. The differential quotient of a function of a complex variable at a point in which the function is continuous is formed as follows. Supposing the complex variable $z = x + iy$ to receive the increment $\Delta z = \Delta x + i\Delta y = \Delta r e^{i\varphi}$, let us consider the quotient of differences

$$\frac{f(z + \Delta z) - f(z)}{\Delta z} \quad \text{or} \quad \frac{f(z + \Delta r e^{i\varphi}) - f(z)}{\Delta r e^{i\varphi}}.$$

The limiting values, to which the real and imaginary constituents of this quotient tend for $\Delta z = 0$, constitute the derived of the complex function.

We shall concern ourselves in the sequel only with functions for which, except in singular points, this limiting quantity is a function of $z = x + iy$ exclusively and is thus independent of the value φ or of the ratio $\Delta y : \Delta x$; such functions are called analytic functions. For an analytic function $f(z)$ therefore:

$$1)\qquad \operatorname{Lim} \frac{f(z + \Delta z) - f(z)}{\Delta z} = f'(z);$$

it has a derived function, not only as in real functions identical when taken progressively and regressively but the same in every direction. We shall show that every function expressed by an infinite series of powers is analytic within the convergency of this series, and that also conversely every function that is analytic within a domain can within this domain be expressed by infinite series of powers*).

Equation 1) can also be written thus:

$$2)\qquad \frac{df(z)}{dz} = f'(z) \quad \text{or:} \quad df(z) = f'(z) \, . \, dz = f'(z)\,(dx + i\,dy).$$

This last form is the equation for the Total Differential of the complex function.

If we make the complex variable z change only by the real part Δx or by the purely imaginary part $i\Delta y$, we obtain as limiting values of the quotient of differences the partial derived functions with regard to x or to y. But these likewise, in consequence of our hypothesis, satisfy the following equations:

$$3)\qquad \begin{aligned} \frac{\partial f(z)}{\partial x} &= \operatorname{Lim} \frac{f(z + \Delta x) - f(z)}{\Delta x} = f'(z), \\ \frac{\partial f(z)}{i\,\partial y} &= \operatorname{Lim} \frac{f(z + i\Delta y) - f(z)}{i\Delta y} = f'(z). \end{aligned}$$

Therefore the analytic function regarded as a function of the two variables x and y satisfies the equation:

$$4)\qquad \frac{\partial f}{\partial x} = \frac{1}{i}\frac{\partial f}{\partial y}, \quad \text{or} \quad \frac{\partial f}{\partial x} + i\,\frac{\partial f}{\partial y} = 0.$$

If we ask whether these equations are also sufficient conditions, that there may be at a point one derived function depending only on z for every direction; the answer is:

Provided there exist in the neighbourhood of the point, definite values of the partial derived functions $\frac{\partial f}{\partial x}$ and $\frac{1}{i}\frac{\partial f}{\partial y}$ that are always

*) Riemann styled functions on the hypothesis of their analytical property, simply, functions of a complex variable. Cauchy called functions that are analytic in a domain without exception, synectic. Briot and Bouquet call such functions *holomorphe*.

equal and moreover are continuous functions of the complex variable $x+iy$, we have also:

$$\frac{df(z)}{dz}=\frac{\partial f}{\partial x}=\frac{1}{i}\frac{\partial f}{\partial y}.$$

For, we have:

$$\frac{f(x+iy+\Delta x+i\Delta y)-f(x+iy)}{\Delta x+i\Delta y}=\frac{f(x+iy+\Delta x+i\Delta y)-f(x+iy+i\Delta y)}{\Delta x}\cdot\frac{\Delta x}{\Delta x+i\Delta y}$$
$$+\frac{f(x+iy+i\Delta y)-f(x+iy)}{i\Delta y}\cdot\frac{i\Delta y}{\Delta x+i\Delta y}.$$

Now Δy can be chosen so small that:

$$\frac{f(x+iy+i\Delta y)-f(x+iy)}{i\Delta y}=\frac{1}{i}\frac{\partial f(x+iy)}{\partial y}\pm\delta,$$

where δ denotes a quantity arbitrarily small in amount. Further we can choose the value of Δx so that:

$$\frac{f(x+iy+\Delta x+i\Delta y)-f(x+iy+i\Delta y)}{\Delta x}=\frac{\partial f(x+iy+\Theta\Delta x+i\Delta y)}{\partial x}$$
$$=\frac{\partial f(x+iy)}{\partial x}\pm\delta=\frac{1}{i}\frac{\partial f(x+iy)}{\partial y}\pm\delta.$$

From this it follows: values can be assigned to Δx and Δy such that for them and for all smaller values the above quotient of differences shall differ from the value $\frac{1}{i}\frac{\partial f}{\partial y}$ at most by the quantity

$$\frac{\delta\Delta x}{\Delta x+i\Delta y}+\frac{\delta i\Delta y}{\Delta x+i\Delta y},$$

the modulus of δ being arbitrarily small. Since the amount of the quotients by which δ is multiplied cannot increase beyond all limits, what we have stated is proved.

The real and imaginary constituents $u+iv$ into which the complex function $f(z)$ resolves, are functions of the two real variables x and y. But as the equation of condition 4) must be fulfilled, they are functions of two variables of a special kind: the functions u and v cannot be independent. In fact from:

$f(z)=u+iv$; we have: $\frac{\partial f}{\partial x}=\frac{\partial u}{\partial x}+i\frac{\partial v}{\partial x}$, $\frac{\partial f}{\partial y}=\frac{\partial u}{\partial y}+i\frac{\partial v}{\partial y}$,

then by equation 4) we find:

$$\left(\frac{\partial u}{\partial x}+i\frac{\partial v}{\partial x}\right)+i\left(\frac{\partial u}{\partial y}+i\frac{\partial v}{\partial y}\right)=0. \tag{5}$$

By separating the real and the imaginary, this equation resolves into:

$$\frac{\partial u}{\partial x}=\frac{\partial v}{\partial y},\quad \frac{\partial u}{\partial y}=-\frac{\partial v}{\partial x}. \tag{6}$$

The two constituents of an analytic function are therefore generally continuous functions with determinate differential quotients. For these functions the Theorem of the Total Differential holds. For, if we write:

$$\frac{df(z)}{dz} = \frac{du + i\,dv}{dx + i\,dy} = \frac{\frac{du}{dx} + i\frac{dv}{dx}}{1 + i\frac{dy}{dx}},$$

since $f'(z) = \frac{\partial u}{\partial x} + i\frac{\partial v}{\partial x}$, we obtain the equation:

$$\frac{du}{dx} + i\frac{dv}{dx} = \left(\frac{\partial u}{\partial x} + i\frac{\partial v}{\partial x}\right)\left(1 + i\frac{dy}{dx}\right),$$

from which and 6) we find for the total differential quotients of u and v:

$$\frac{du}{dx} = \frac{\partial u}{\partial x} - \frac{\partial v}{\partial x}\cdot\frac{dy}{dx} = \frac{\partial u}{\partial x} + \frac{\partial u}{\partial y}\cdot\frac{dy}{dx},$$

$$\frac{dv}{dx} = \frac{\partial v}{\partial x} + \frac{\partial u}{\partial x}\cdot\frac{dy}{dx} = \frac{\partial v}{\partial x} + \frac{\partial v}{\partial y}\cdot\frac{dy}{dx}.$$

Conversely, if the Theorem of the Total Differential be supposed to hold for u and v, equations 6) are sufficient conditions that the combination $u + iv = w$ be expressible by arithmetical operations on the single variable $z = x + iy$. For then, replacing x by $z - iy$ in the combination $u + iv$, this must become altogether independent of the variable y, i. e. its partial differential quotient with respect to y must vanish. (§ 26. 1.)

Denoting the result of substituting $x = z - iy$ in $u + iv$ by $(u) + i(v)$, we find

$$\frac{\partial (u)}{\partial y} = \frac{\partial u}{\partial x}(-i) + \frac{\partial u}{\partial y}, \quad \frac{\partial (v)}{\partial y} = \frac{\partial v}{\partial x}(-i) + \frac{\partial v}{\partial y},$$

when $z - iy$ is substituted for x in the derived functions on the right. Accordingly, combining these, we have:

$$\frac{\partial (u)}{\partial y} + i\frac{\partial (v)}{\partial y} = \left(\frac{\partial u}{\partial y} + \frac{\partial v}{\partial x}\right) - i\left(\frac{\partial u}{\partial x} - \frac{\partial v}{\partial y}\right).$$

In consequence of equations 6) the expressions in the brackets vanish; they are therefore sufficient conditions that w should depend on z only.

85. The property of an analytic function of a complex variable, that its first derived $f'(z)$ is independent of the ratio $\frac{dy}{dx}$, is important in the geometrical transformation upon plane A of plane B that represents the values of the function $w = f(z)$. If we consider in plane A a triangle $PP'P''$, whose vertices belong to the values z, $z + \Delta z$, $z + \Delta z'$, to these correspond in plane B three points $QQ'Q''$, whose values we may denote by w, $w + \Delta w$, $w + \Delta w'$. Transposing the system of coordinates in each plane so as to make the points P and Q the origins of the systems, and putting:

$$\Delta z = \Delta r\,.\,e^{i\varphi}, \quad \Delta z' = \Delta r'\,.\,e^{i\varphi'}, \quad \Delta w = \Delta\varrho\,.\,e^{i\psi}, \quad \Delta w' = \Delta\varrho'\,.\,e^{i\psi'},$$

the quantities introduced are in each plane the polar coordinates of the other two vertices of the triangle in regard to the origin. But by the analytic property, the quotients:

$$\frac{\Delta w}{\Delta z} = \frac{\Delta\varrho}{\Delta r}\cdot e^{i(\psi - \varphi)} \quad\text{and}\quad \frac{\Delta w'}{\Delta z'} = \frac{\Delta\varrho'}{\Delta r'}\cdot e^{i(\psi' - \varphi')}$$

have the same limiting value; therefore unless this value be 0 or ∞:

$$\operatorname{Lim}\frac{\Delta r'}{\Delta r} = \operatorname{Lim}\frac{\Delta\varrho'}{\Delta\varrho}, \quad \varphi' - \varphi = \psi' - \psi,$$

i. e. if the sides of the triangle $PP'P''$ and therefore also those of the triangle $QQ'Q''$ be infinitely small, the angles P, P', P'', and Q, Q', Q'', are respectively equal and their containing sides proportional. Therefore two corresponding infinitely small parts and so in general the smallest parts of the planes A and B are similar. To two curves that cross in plane A correspond in plane B two curves that cross at the same angle*). This likewise holds for the transformation of z and w upon two spheres.

86. Applications.

1. The positive integer power: $w = z^n$ has the derived function:

$$\frac{dw}{dz} = \operatorname{Lim}\frac{(z+\Delta z)^n - z^n}{\Delta z} = \operatorname{Lim}\frac{(z + \Delta r e^{i\varphi})^n - z^n}{\Delta r e^{i\varphi}} = nz^{n-1}.$$

For if we consider w as a function of x or of y, we have:

$$\frac{\partial w}{\partial x} = \frac{\partial(x+iy)^n}{\partial x} = n(x+iy)^{n-1}, \; \frac{\partial w}{\partial y} = \frac{\partial(x+iy)^n}{\partial y} = in(x+iy)^{n-1}.$$

From this follows, that the rational integer function:

$$w = a_0 + a_1 z + a_2 z^2 + \cdots a_n z^n,$$

has the derived function:

$$\frac{dw}{dz} = a_1 + 2a_2 z + \cdots n a_n z^{n-1}.$$

2. The rational fractional function has a finite first derived function except at the singular points, in which its denominator vanishes.

3. The explicit irrational function: $w = (z-a)^m$, which is unique and continuous in the plane perforated along a curve starting from a, has a derived function. We determine it, so as to exhibit its ambiguity, as follows:

If we put $z - a = r(\cos\varphi + i\sin\varphi)$ and therefore:

$$w = r^m(\cos m(\varphi + 2k\pi) + i\sin m(\varphi + 2k\pi)),$$

when we fix upon some one value of k, we have:

$$\begin{aligned} dz &= dr(\cos\varphi + i\sin\varphi) + r(-\sin\varphi + i\cos\varphi)\,d\varphi \\ &= (\cos\varphi + i\sin\varphi)(dr + ir\,d\varphi), \\ dw &= mr^{m-1}dr(\cos m(\varphi+2k\pi) + i\sin m(\varphi+2k\pi)) \\ &\quad + mr^m(-\sin m(\varphi+2k\pi) + i\cos m(\varphi+2k\pi))\,d\varphi \\ &= (\cos m(\varphi+2k\pi) + i\sin m(\varphi+2k\pi))\,mr^{m-1}(dr + ir\,d\varphi), \end{aligned}$$

*) This property of the function of a complex variable was noticed by Gauss in solving the important problem in mapping: "To represent the parts of one given surface upon another given surface so that the copy may be similar in its smallest parts to the original." See his answer to the prize problem proposed by the Royal Society of Sciences in Copenhagen for 1822. Werke, Vol. IV, p. 189.

accordingly:

$$\frac{dw}{dz} = m r^{m-1} \frac{(\cos m(\varphi + 2k\pi) + i \sin m(\varphi + 2k\pi))}{\cos \varphi + i \sin \varphi}$$

$$= m r^{m-1} \{\cos(\overline{m-1}\varphi + 2km\pi) + i \sin(\overline{m-1}\varphi + 2km\pi)\}.$$

Since $\cos(2km\pi) = \cos 2k(m-1)\pi$, $\sin(2km\pi) = \sin 2k(m-1)\pi$, this expression takes the form:

$$\frac{dw}{dz} = m r^{m-1} \{\cos(m-1)(\varphi + 2k\pi) + i \sin(m-1)(\varphi + 2k\pi)\} = m(z-a)^{m-1}.$$

The value of k originally chosen remains unchanged in the derivate, which is as many-valued a function as the original.

4. For the exponential function:

$$w = e^z = e^{x+iy} = e^x (\cos y + i \sin y)$$

we have:

$$\frac{\partial w}{\partial x} = e^x (\cos y + i \sin y), \quad \frac{\partial w}{\partial y} = e^x (-\sin y + i \cos y),$$

therefore it has the derivate: $\frac{dw}{dz} = e^x (\cos y + i \sin y) = e^z$.

5. The logarithm of the number $z = x + iy$ to the base e has the value:

$$w = l(x+iy) = l(+\sqrt{x^2+y^2}) + i \tan^{-1} \frac{y}{x} \pm i2k\pi \quad \text{when } x > 0,$$

$$w = l(x+iy) = l(+\sqrt{x^2+y^2}) + i \tan^{-1} \frac{y}{x} \pm i(2k+1)\pi \quad \text{when } x < 0.$$

A branch of the function is continuous in the plane perforated along the positive axis of x; for it we have:

$$\frac{\partial w}{\partial x} = \frac{x}{x^2+y^2} - \frac{iy}{x^2+y^2} = \frac{1}{x+iy},$$

$$\frac{\partial w}{\partial y} = \frac{y}{x^2+y^2} + \frac{ix}{x^2+y^2} = \frac{i}{x+iy},$$

therefore the derivate is: $\frac{dw}{dz} = \frac{1}{z}$.

6. The complex infinite series of ascending positive integer powers:

1) $$f(z) = a_0 + a_1 z + a_2 z^2 + \cdots a_n z^n + \cdots$$

is an analytic function within its circle of convergence.

Let $z + h$ be a complex value within this circle, then:

2) $$f(z+h) = a_0 + a_1(z+h) + a_2(z+h)^2 + \cdots a_n(z+h)^n + \cdots;$$

arranging this absolutely convergent infinite series by powers of h, the coefficients of these powers are infinite series, which we shall prove to be the successive derived functions of $f(z)$. With this in view let us provisionally denote the series:

3) $$a_1 + 2a_2 z + 3a_3 z^2 + \cdots n a_n z^{n-1} + \cdots \text{ by } f_1(z).$$

This series converges absolutely within the same circle for which the original series 1) is convergent; for, R its radius of convergence is determined from the inequality § 83:

$$R \operatorname{Lim} \frac{n+1}{n} \frac{A_{n+1}}{A_n} < 1, \text{ or } R < \operatorname{Lim} \frac{A_n}{A_{n+1}}, \text{ since } \operatorname{Lim}_{n=\infty} \frac{n+1}{n} = 1.$$

Similarly we obtain by continued differentiation of the several terms the following series that all converge within the same circle:

4) $2a_2 + 3 . 2a_3 z + \cdots n(n-1) a_n z^{n-2} \ldots = f_2(z)$

$3 . 2a_3 + 4 . 3 . 2a_4 z + \cdots n(n-1)(n-2) a_n z^{n-3} \ldots = f_3(z)$

. . . .

Introducing this notation we obtain for series 2) arranged by powers of h the value:

5) $$f(z+h) = f(z) + \frac{h}{1} f_1(z) + \frac{h^2}{\underline{|2}} f_2(z) + \cdots \frac{h^n}{\underline{|n}} f_n(z) + \cdots,$$

where $\operatorname{Lim} \frac{h^n}{\underline{|n}} f_n(z)$ is certainly zero, because the sum of the moduli of $\frac{h^n}{\underline{|n}} f_n(z)$ is smaller than the sum of the moduli of the absolutely convergent series of powers:

$$a_n(z+h)^n + a_{n+1}(z+h)^{n+1} + a_{n+2}(z+h)^{n+2} + \cdots \text{ etc..}$$

The convergency of this new series 5) is therefore a circle with its centre at the point $h = 0$, i. e. at the point z, and its radius H at least equal to: $R - \operatorname{mod}[z]$; for, as the circle, whose radius is equal to this, touches the inside of the original one, all its points lie within that circle and for them series 2) converges absolutely, therefore series 5) derived from it by arranging its terms differently also converges absolutely.

We have still to convince ourselves directly, that it is allowable to reason thus from the absolutely convergent series 2), for this might seem doubtful, since in the new arrangement of terms each coefficient requires the summation of an infinite series. Let us therefore examine whether n can be chosen so as to make the difference between the first $n+1$ terms of series 2) and of series 5) arbitrarily small, always assuming the absolute convergence of the former series. Putting:

$$f(z) = a_0 + a_1 z + a_2 z^2 + \cdots a_n z^n + \varrho_n,$$
$$f_1(z) = a_1 + 2a_2 z + \cdots n a_n z^{n-1} + \varrho_n',$$
$$f_2(z) = 2a_2 + 3.2 a_3 z + \cdots n(n-1) a_n z^{n-2} + \varrho_n'',$$

. . . .

$$f_n(z) = \underline{|n} . a_n + \varrho_n^{(n)},$$

the difference between the sums of their first $n+1$ terms is:

$$\varrho_n + \frac{h}{1} \varrho_n' + \frac{h^2}{\underline{|2}} \varrho_n'' + \cdots \frac{h^n}{\underline{|n}} \varrho_n^{(n)}.$$

Now since series 2) converges absolutely, we can always choose n, such that for every value of h: .

$$R_{n,k} = a_{n+1}(z+h)^{n+1} + a_{n+2}(z+h)^{n+2} + \cdots + a_{n+k}(z+h)^{n+k}$$

shall be smaller than δ, as long as $z+h$ lies in the convergency of series 1.), even when we put for each term its absolute value; when δ is prescribed the value of n is determined. But now k can always be chosen so great as to make the amount of:

$$R_{n,k} - \left(\varrho_n + \frac{h}{1}\varrho_n' + \frac{h^2}{\underline{|2}}\varrho_n'' + \cdots \frac{h^n}{\underline{|n}}\varrho_n^{(n)}\right)$$

smaller than any quantity however small. For, each of the series ϱ converges absolutely; therefore in each of them can be found a place k from which onwards the remainder of that series is constantly smaller than a determinate quantity δ'. Hence we have:

$$\operatorname{mod}\left[R_{n,k} - \left(\varrho_n + \frac{h}{1}\varrho_n' + \cdots \frac{h^n}{\underline{|n}}\varrho_n^{(n)}\right)\right] < \delta + \delta' . e^{\operatorname{mod}[h]},$$

therefore we have also:

$$\operatorname{mod}\left[\varrho_n + \frac{h}{1}\varrho_n' + \cdots \frac{h^n}{\underline{|n}}\varrho_n^{(n)}\right] < 2\delta + \delta' . e^{\operatorname{mod}[h]}.$$

This proves that the difference between the sums can be made arbitrarily small by choice of n alone, so that series 2) and 5) must be identical.

Each point within the circle of convergence of 1) *can therefore be taken as the centre of an expansion, and its convergency will be at least as great as the circle touching the inside of that original boundary circle.*

Now from series 5) it follows that:

$$\frac{f(z+h)-f(z)}{h} = f_1(z) + \frac{h}{\underline{|2}} f_2(z) + \cdots \frac{h^{n-1}}{\underline{|n}} f_n(z) + \cdots,$$

therefore: $$\operatorname{Lim}\frac{f(z+h)-f(z)}{h} = f_1(z),$$

i. e. *the first derived function of* $f(z)$ *is expressed by the infinite series* 3):

$$f'(z) = f_1(z) = a_1 + 2a_2 z + 3a_3 z^2 + \cdots n a_n z^{n-1} + \cdots \text{ etc.}.$$

Or, the complex series of powers is differentiated by forming the series of first derivates of its individual terms. This series converges within the same circle of convergence as the original series.

Further, by differentiating the series for $f'(z)$ it follows that:

$$f''(z) = f_2(z), \text{ similarly } f'''(z) = f_3(z), \text{ etc.}.$$

Accordingly we have series 5):

$$f(z+h) = f(z) + \frac{h}{1}f'(z) + \frac{h^2}{\underline{|2}}f''(z) + \cdots \frac{h^n}{\underline{|n}}f^n(z) + \cdots$$

identical with Taylor's expansion for the function of a complex variable given by an infinite series of powers.

Comparing this with 2) we find the meaning of the coefficients in the expansion; we have:

$$a_0 = f(0),\ a_1 = f'(0),\ a_2 = \frac{1}{\underline{|2}}f''(0), \cdots a_n = \frac{1}{\underline{|n}}f^n(0), \text{ etc.}.$$

Hence we see that whenever it is possible to express a unique function $f(z)$ by a series of powers of z, it can be possible only in a single way. For, the coefficients of this series must be equal to the values of the function and of its successive derivates at the point $z = 0$.

In fact, if two developments convergent within the same domain:

$$f(z) = a_0 + a_1 z + a_2 z^2 + \cdots a_n z^n + \cdots$$

$$f(z) = b_0 + b_1 z + b_2 z^2 + \cdots b_n z^n + \cdots$$

were found, we should have:

$$0 = (a_0 - b_0) + (a_1 - b_1) z + (a_2 - b_2) z^2 + \cdots (a_n - b_n) z^n + \cdots,$$

and since all the successive derivates of the constant zero vanish for all values of z, we must also have:

$$a_0 - b_0 = 0,\ a_1 - b_1 = 0,\ a_2 - b_2 = 0,\ \ldots\ a_n - b_n = 0, \text{ etc.}.$$

In the French Translation by E. Picard of the above Memoir of Weierstrass, which appeared under the revision of the Author in the Annales de l'École Normale, 2e Série, T. VIII, 1879, entitled: "Mémoire sur les fonctions analytiques uniformes", the opening statement reads as follows:

Among unique (uniformes) functions of a single variable, rational functions form a distinct class which we proceed to define by their characteristic property.

We shall say that a unique function $f(z)$ of the complex variable z is regular in the neighbourhood of a point a, when for all values of z comprised within a circle having its centre at a and a radius sufficiently small, the function can be developed in a series of the form: $a_0 + a_1(z-a) + a_2(z-a)^2 + \ldots$, the coefficients $a_0, a_1, a_2, \ldots$ being constants. In case the point a were at infinity we should replace $z - \infty$ by $\frac{1}{z}$.

Every point a' in whose neighbourhood the function $f(z)$ is not regular will be called a singular point of $f(z)$, and we shall distinguish two kinds of singular points: if a positive integer (m) power of $(z - a')$ can be found such that the product $(z - a')^m f(z)$ is regular in the neighbourhood of a' and does not vanish for $z = a'$, this point will be called a pole of the function; if not, we shall say that a' is an essentially singular point.

We may accordingly say that the function $f(z)$ has a determinate value for $z = a$, not only when it is regular in the neighbourhood of this point, but even when a is a pole, for in both cases we shall have for values of z sufficiently near a

$$f(z) = (z - a)^{-m} \{a_0 + a_1(z - a) + a_2(z - a)^2 + \ldots\},$$

where m is an integer and a_0 is not zero. When $m > 0$, the function will be infinitely great for any infinitely small value of $(z - a)$ and this is an indispensable condition in order that we may be able to write $f(a) = \infty$.

Moreover within the circle of convergence of the above series the function has no other singular point than a when m is positive and it has none at all when m is negative or zero. Consequently, if we can demonstrate that in the neighbourhood of a given point z_0, and at a distance less than any given quantity, there are singular points of $f(z)$ different from z_0, we shall be able to assert that z_0 is an essentially singular point of $f(z)$.

After these explanations, the class of rational functions of one variable z can now be defined as that which comprehends collectively all the unique functions of z that have only poles.

It is not only to illustrate this chapter that I have placed this Extract here, but further I wish to remark, that throughout the Translation the word "pôle" is used wherever the phrase, "ausserwesentliche singuläre Stelle" occurs in the original Memoir. In H. Laurent: Traité d'Analyse, T. III, p. 366 etc., these points are called "infinis ou pôles", and this alternative word I render by "infinite" of the function or "infinity point"; it occurs once in Picard's translation where the original has "Werthe, für die $f(z) = \infty$". The desire to avoid using the overtasked word "pole" has been my motive in employing for "ausserwesentlich" the word "non-essential" which I feel to be objectionable. It might be less so, if unique analytic functions only had to be considered; but when we have other analytic functions in view, it is unpleasant to have to remember, that essential and non-essential singular points do not exhaust the catalogue of singular points. Those epithets are exhaustive only of the points in which analytic functions become infinite.

G. L. C.

Third Chapter.

Tho vanishing values of a series of powers, specially those of the integer rational algebraic function.

87. The Theorem last proved in the previous Section for series of powers assists us in enquiring: *how many points are there within a circle of convergence, for which the function $f(z)$ vanishes?*

The function vanishes in a point z_0, when, taking this point as centre, it gives rise to an expansion in which the first term $f(z_0)$ is zero. When other succeeding terms also vanish, so that the expansion begins with the term $\frac{h^n}{\underline{|n}} f^n(z_0)$, the point is called a vanishing or zero point (nullity) of the order (nullitude) n; it must be counted as n vanishing points; the quotient $\frac{f(z_0+h)}{h^n}$ then remains finite at the point $h = 0$. We have first to prove:

The function $f(z)$ cannot be zero at infinitely many points within a finite circle of convergence, unless it be zero identically, i. e. everywhere in the circle, so that all the coefficients of the series vanish.

In fact, if there be infinitely many vanishing points, there is also a region of arbitrarily small extent which contains infinitely many of them. For if the entire domain be divided into an arbitrarily great finite number of parts, there must still be in at least one of these parts infinitely many vanishing points. Let z be a point in such a region; the expression

1) . $$f(z+h) = f(z) + \frac{h}{1} f'(z) + \frac{h^2}{\underline{|2}} f''(z) + \cdots$$

must become zero for a value h, whose modulus is arbitrarily small. Since the coefficients of $h, h^2 \ldots$ are finite, the amount of the terms multiplied by h is arbitrarily small; accordingly if the expression is to vanish, the amount of $f(z)$ must also be smaller than any finite quantity, i. e. since $f(z)$ is a determinate value, we must have:

$$f(z) = 0.$$

Now considering the product:

$$h\left(f'(z) + \frac{h}{\underline{|2}} f''(z) + \frac{h^2}{\underline{|3}} f'''(z) + \cdots\right);$$

there is an arbitrarily small but finite value h for which it is zero, therefore the bracketed factor vanishes. Hence follows as before:

$$f'(z) = 0.$$

In like manner from the vanishing of the product:

$$h^2\left(\frac{1}{\underline{|2}} f''(z) + \frac{h}{\underline{|3}} f'''(z) + \cdots\right)$$

it results that $f''(z)$ vanishes, and similarly it is found that:

$$f'''(z) = 0, \ldots f^{(n)}(z) = 0, \ldots \text{ etc.}.$$

Thus $f(z+h)$ and all its derivates vanish for all values of $z+h$ that lie within the circle of convergence of series 1). In this circle a point z' for which mod $z' <$ mod z, being chosen as centre, gives rise to an expansion the radius of whose convergency

$$H' = R - \text{mod } [z'] \text{ is } > H.$$

The coefficients in this new expansion all vanish, i. e. the function is zero everywhere in this greater circle also. Taking a point z'' within this circle as centre, we obtain a new circle, and this process can be continued till we reach a circle that includes the origin $z = 0$. Since for this point the function and all its derivates vanish, the same is true for all points of the original circle of convergence of the series:

$$f(z) = a_0 + a_1 z + a_2 z^2 + \cdots a_n z^n + \cdots;$$

that is, we have

$$f(0) = a_0 = 0, f'(0) = a_1 = 0, f''(0) = a_2 \underline{|2} = 0, \cdots f^n(0) = a_n \underline{|n} = 0, \text{ etc.}.$$

By means of this proposition we can generalise the Theorem proved at the close of last Chapter concerning the unique expression of a function by a series of powers. For, from it follows that:

When the values of two series of powers are the same even only in infinitely many points of a domain, the series are identical throughout the entire common part of their convergencies. For let these series be:

$$a_0 + a_1(z-\alpha) + a_2(z-\alpha)^2 + \cdots \text{ and } b_0 + b_1(z-\beta) + b_2(z-\beta)^2 + \cdots,$$

then their difference can be expanded for a point γ, within the domain common to their circles of convergence, in a series of powers

$$c_0 + c_1(z-\gamma) + c_2(z-\gamma)^2 + \cdots \text{ etc.}.$$

This series vanishes in infinitely many points, accordingly it is zero within its entire circle of convergence. But from this series we can attain to any other point lying at a finite distance however small inside the boundaries, by adopting a new point within its circle of convergence as centre of an expansion, that in like manner must vanish, and continuing this process.

88. Suppose a domain of convergence with the radius R is to be investigated; since there is only a finite number of vanishing points in each finite part of the plane, we can assume, that none of these points lies upon the bounding circle. Let the vanishing points within it be $z_1, z_2, \ldots z_\nu$, and their respective orders $\lambda_1, \lambda_2, \ldots \lambda_\nu$.

Calling the product:

$$(z - z_1)^{\lambda_1}(z - z_2)^{\lambda_2}\ldots(z - z_\nu)^{\lambda_\nu} = \Pi(z);$$

$f(z)$ is divisible by $\Pi(z)$ and the quotient is a series of powers that converges for the original domain and is not zero at any point in it. To see this, let us put $h = z - z_1$ in the above development, hence:

$$f(z) = f(z_1 + z - z_1) = f(z_1) + \frac{z - z_1}{1} f'(z_1) + \frac{(z - z_1)^2}{\underline{|2}} f''(z_1) + \cdots;$$

but since z_1 is a vanishing point of the order (nullitude) λ_1, we have:

$$f(z_1) = f'(z_1) = f''(z_1) = \cdots = f^{\lambda_1 - 1}(z_1) = 0,$$

therefore:

$$\frac{f(z)}{(z - z_1)^{\lambda_1}} = \frac{1}{\underline{|\lambda_1}} f^{\lambda_1}(z_1) + \frac{z - z_1}{\underline{|\lambda_1 + 1}} f^{\lambda_1 + 1}(z_1) + \cdots \text{ etc..}$$

This absolutely convergent series proceeding by powers of $z - z_1$ can be rearranged by powers of z and resumes the original circle of convergence. For, then again the circle of convergence of this series must include that for the development by powers of $z - z_1$. Since the new resulting series vanishes in the point z_2 in the order λ_2, it is divisible by $(z - z_2)^{\lambda_2}$; in this way we ultimately obtain:

$$\frac{f(z)}{\Pi(z)} = \varphi(z), \text{ or: } f(z) = \Pi(z)\,.\,\varphi(z),$$

where $\varphi(z)$ is a series of powers that is not zero at any point in the domain.

The propositions hitherto proved apply in particular to the integer rational algebraic function:

$$f(z) = a_0 + a_1 z + a_2 z^2 + \cdots + a_n z^n.$$

The convergency of this function, in which $a_0, \ldots a_n$ mean determinate finite complex values, is the entire plane, i. e. to each finite value of z belongs a determinate finite value of $f(z)$; this function cannot vanish for infinitely many values of z without vanishing identically; and further if $z_1, z_2 \ldots z_\nu$ be vanishing points, $f(z)$ is divisible by $\Pi(z)$, the quotient being again an integer rational algebraic function; in this case, if $\lambda_1 + \lambda_2 + \cdots \lambda_\nu = n$, $\varphi(z)$ is constant and equal to a_n. Therefore an integer rational algebraic function of the order n can certainly not have more than n vanishing points in the entire plane.

89. Forming the logarithm of $f(z)$ we have:

$$lf(z) = l\Pi(z) + l\varphi(z) = \lambda_1 l(z - z_1) + \lambda_2 l(z - z_2) + \cdots + \lambda_\nu l(z - z_\nu) + l\varphi(z).$$

For each of the logarithms on the right we take one of its infinitely many values, and let z describe the circumference of the bounding circle from any point A, keeping the inside of the circle on the left. Since the argument $z - z_1$ vanishes only once within the circle, namely for $z = z_1$, the value of the logarithm when z returns to the point A will differ by $2i\pi$ from its initial value, § 82, 5; the same will happen with $l(z - z_2), \ldots l(z - z_\nu)$; on the other hand $l\varphi(z)$ on its return to the point A will resume the value it had at first, because $\varphi(z)$ does not vanish at any point within the circle, but has for every point a determinate finite continuously changing value, so that no branching point of its logarithm is included. Accordingly we see that: *If there be ν vanishing points of the function $f(z)$ within the circle of convergence, the value of $lf(z)$ changes by $2i\pi\nu$ when the argument z describes the entire circumference;* and conversely: *If the logarithm of the function change by $2i\pi\nu$ when its argument z describes the circle of convergence, ν is the number of vanishing points of $f(z)$ within this circle.*

90. Applying this Theorem to the algebraic function, we can take the circle, for whose points the values of z are to be formed, with a radius so large that the amount of the term $a_n z^n$ shall far exceed all the rest, and accordingly the amount of:

$$\frac{a_0 + a_1 z + a_2 z^2 + \cdots a_{n-1} z^{n-1}}{a_n z^n}$$

be smaller than an arbitrarily small quantity δ; to attain this, we have only to take mod z greater than unity and then to determine that

$$\text{mod}\, z \text{ shall be also} > \frac{A_0 + A_1 + \cdots A_{n-1}}{A_n} \cdot \frac{1}{\delta},$$

each A denoting the modulus of the corresponding a; then let us put:

$$lf(z) = l(a_n z^n) + l\left(1 + \frac{a_0 + a_1 z + a_2 z^2 + \cdots a_{n-1} z^{n-1}}{a_n z^n}\right) = l(a_n z^n) + l(1 + \varepsilon);$$

writing ε for the complex quantity, whose modulus is smaller than δ.

Now $l(1 + \varepsilon)$, formed from a determinate point upon the bounding circle, differs everywhere inappreciably from $l(1) = \pm 2ki\pi$; thus if we begin with any value of the logarithm, ex. gr. the simplest, since when ε changes its value, the corresponding logarithm must vary continuously, it will always differ only inappreciably from the simplest value of $l(1)$ namely zero; therefore when z has returned to the original point, the value of $l(1 + \varepsilon)$ will not have increased by a multiple of $2i\pi$.

But while z describes the circle, $l(a_n z^n) = nl(a_n^{\frac{1}{n}} z)$ changes by $2i\pi \,.\, n$, since the point $z = 0$ is included within this circle. Accordingly $lf(z)$ undergoes the change $2i\pi n$, i. e. *in the arbitrarily great circle of convergence there are always n values, for which the rational function*

$$a_0 + a_1 z + a_2 z^2 + \cdots a_n z^n$$

vanishes; some of these vanishing points may coincide, but always the total sum of their nullitudes is n. This theorem, which can also be stated in the words: *Every equation of the nth degree*

$$f(z) = a_0 + a_1 z + a_2 z^2 + \cdots a_n z^n = 0$$

has n complex roots; is known as the **Fundamental Theorem of Algebra.*)** When the coefficients $a_0, a_1 \ldots a_n$ are all real, if the equation have one complex root $z = \alpha + i\beta$, it has also the conjugate complex root $z = \alpha - i\beta$; for then we have in general:

$$f(\alpha - i\beta) = U - iV, \text{ if } f(\alpha + i\beta) = U + iV.$$

But, when $\alpha + i\beta$ is a root, both U and V vanish.

The implicit algebraic function, defined by the equation § 25:

$$A_0 + A_1 y + A_2 y^2 + \cdots A_n y^n = 0,$$

in which $A_0 \ldots A_n$ signify integer polynomials in x, is accordingly, when complex solutions also are taken into account, an n-valued function, i. e. to each value of x, for which the values of the coefficients $A_0 \ldots A_n$ are determinate, belong n equal or different values of y, namely the n roots of this equation of the nth degree.

The calculation of the n values of the roots, i. e. their expression as functions of the coefficients, forms the object of the Theory of Equations. As long as $n \leqq 4$, the roots can be developed in a closed form as functions of the coefficients by help of the explicit algebraical operations of the first six species; if $n > 4$ the solution of the general equation gives rise to new functions whose properties we have to investigate in the next Chapter. But it is in all cases possible when the coefficients of an equation are given in the form of determinate numerical quantities, to express each root numerically with any required degree of approximation, i. e. after the method of inclusion within limits to form two infinite series of rational numbers, whereof one has the real constituent, and the other the factor of the imaginary constituent of a root as its limiting value.

---

*) The theorem was first proved by Gauss in his doctoral dissertation 1799, to this there is a supplement of the year 1849; Gauss published two other proofs in 1815 and 1816 (Werke, Vol. III). The proof detailed in the text, applicable as it is to infinite series of powers, is derived essentially from Cauchy (Journal de l'École polytechnique, Cahier 25, 1837); he had previously, in his Analyse algébrique, chap. X, 1821, given an elementary proof for the existence of the n roots of an equation, that coincides in principle with that developed by Argand (Gergonne Ann., Vol. V, 1815).

Fourth Chapter.

The implicit algebraic function.*)

91. The most general form in which a variable w is defined as an algebraic function of the variable z is by the vanishing of a polynomial consisting of integer powers of z and of w:

1) $f(w^n, z^m) = w^n \varphi_0(z) + w^{n-1}\varphi_1(z) + \cdots w\varphi_{n-1}(z) + \varphi_n(z) = 0,$

the factors $\varphi(z)$ being integer polynomials of arbitrary degree in z with complex coefficients; let the highest power of z in any of them be m. This form is of the nth order in w, assuming that all coefficients in the polynomial $\varphi_0(z)$ do not vanish; let φ_0 be of the degree k in z, $k = 0$ denoting that φ_0 is a constant. It may be assumed that the form 1) is not reducible, i. e. that f cannot be resolved into products of algebraic expressions of lower order; for if it could, each factor equated to zero might be investigated separately.

Integer and fractional rational functions are included in this form, for these $n = 1$; in like manner it includes the explicit irrational function treated above:

$$w = (z - a)^{\frac{p}{q}}, \text{ which in form 1) is: } w^q - (z - a)^p = 0.$$

To each value of z, correspond n determinate values of w, different or equal, the roots of equation 1), as was proved in last Chapter. Let these be denoted by $w_1, w_2, \ldots w_n$; they will vary according to the value of z. The equation 1) presents therefore n functions of z, or in other words: it determines an n-valued function of z. The following investigations have to demonstrate how these n branches of the function may be separated, and how far they are continuous functions with determinate derivates.**)

92. If the function w be considered at a determinate point, and so one of the possible values w calculated for a determinate $z = z_0$,

---

*) In this chapter the theorems of Algebra regarding the resultant and the discriminant are supposed known.

**) Cauchy: Exercices d'analyse et de physique mathématique. Tome II. V. Puiseux (1820—83): Recherches sur les fonctions algébriques. Journal de Mathématiques, T. XV et XVI. 1850—1. (German translation by Fischer, Halle 1861). Briot et Bouquet: Théorie des fonctions elliptiques. Paris 1875.

the question arises, how is this value of w altered as the value z changes. The value z_0, chosen to begin with, can always be assumed finite; for when the change of w from an infinite value z concerns us, let us substitute $z = \frac{1}{z' - a}$, thereby converting 1) into a relation between w and z', and investigate what values w assumes for $z' = a$.

Now we must first establish, what kind of singular points can occur. An equation of the nth degree has always n roots. When its coefficients are variable, as in the present case, these roots can exhibit the peculiarities either of becoming infinite or of some of them becoming equal. These two are the only kinds of singularities. As we pursue the investigation, we shall show, that the number of such points is finite.

1. Let us put $w = \frac{1}{w'}$ and determine the points at which $w' = 0$, then we have instead of equation 1):

$$\varphi_0(z) + w'\varphi_1(z) + \cdots w'^{n-1}\varphi_{n-1}(z) + w'^n\varphi_n(z) = 0.$$

If w' is to converge to zero, $\varphi_0(z)$ must become $= 0$. This is an equation determining k separate or coincident finite points for which one value of w increases beyond any finite amount. These singular points of the function w we shall denote by $\alpha_1, \alpha_2 \ldots \alpha_k$; they are, in the absence of further conditions, *non-essential infinity points (infinities)*; for, it follows from the equation

$$w\varphi_0(z) = -\varphi_1(z) - \frac{1}{w}\varphi_2(z) - \cdots - \frac{1}{w^{n-2}}\varphi_{n-1}(z) - \frac{1}{w^{n-1}}\varphi_n(z),$$

that, though w is infinite for $z = \alpha$, the product $w\varphi_0(z) = -\varphi_1(z)$ still remains finite.

In such a singular point, besides the one infinite value, $n - 1$ further values of w, generally finite, will be found from the equation

$$w^{n-1}\varphi_1(\alpha) + w^{n-2}\varphi_2(\alpha) + \cdots w\varphi_{n-1}(\alpha) + \varphi_n(\alpha) = 0.$$

Only when $\varphi_1(\alpha)$ also vanishes, a second value of w becomes infinite; when $\varphi_2(\alpha)$ also $= 0$, a third; and so on. Such an infinity point α can lose the non-essential character, because it is at the same time a critical point. *Critical points form the second kind of singularity.*

2. To investigate an algebraic expression of the nth degree $f(w)$ at a determinate finite point w_1, let us bring it to the form (§ 87):

$$f(w) = f(w_1) + (w - w_1)f'(w_1) + \frac{(w - w_1)^2}{\underline{|2}} f''(w_1) + \cdots \frac{(w - w_1)^n}{\underline{|n}} f^n(w_1).$$

If f vanish simply for $w = w_1$, we have $f(w_1) = 0$, while the value of its first derivate $f'(w_1)$ at this point is not zero. But if there be λ roots $= w_1$, all the derived functions up to the $(\lambda - 1)^{th}$ inclusive also vanish:

$$f(w_1) = 0, \quad f'(w_1) = 0, \ \ldots \ f^{\lambda-1}(w_1) = 0.$$

If therefore $w = w_1$ be a double or a multiple root of equation 1): $f(w^n, z^m) = 0$, it can be so only for such values of z as make the derived function $\frac{\partial f(w^n, z^m)}{\partial w}$ simultaneously vanish. From this theorem we derive in Algebra the condition which subsists among the coefficients of an equation that has a double root: Forming the resultant of the equation and of its first derivate, either by continued division, or, with Euler, as the determinant of the coefficients of a system of equations, we obtain the discriminant as a rational integer function of the coefficients. But in the equation $f(w^n, z^m) = 0$ the coefficients are integer polynomials in z, accordingly we find on equating the discriminant to zero, a finite number of points, that we shall denote by $\beta_1, \beta_2 \dots \beta_l$, which alone can be critical points of the function w, i.e. points at which two or more values of w coincide.

Besides these points α and β, there remains further only the point infinity, which is to be investigated by means of the substitution $z = \frac{1}{z' - a}$; it can be either a regular point, or a singular point of the first or second kind.

93. It is now possible to show, that each branch of the algebraic function proceeds in general continuously. Let us bound off all critical points in the plane of z by circles of arbitrarily small radius, and likewise the non-essential singular points. Let us consider some one of the possible n values of w, ex. gr. the value w_1, at the point $z = z_0$ that must not be a critical point or a non-essential infinity point for w_1, even though it may be a singular point for some of the other values of w; thus w_1 must be at z_0 a simple finite root of the equation. If now we change z continuously, by making its representative point travel from z_0 to another point Z along an arbitrary curve which does not cross any of the bounding curves just named, we are going to show that w_1 also varies continuously.

Denoting a point upon the curve near z_0 by $z_0 + \Delta z$ and the corresponding value of the function by $w_1 + \Delta w$, if such continuity exist, it must be possible, for any number δ however small, to assign a value h, so that mod Δw shall be less than δ, as long as mod $\Delta z < h$.

Putting $z = z_0 + \Delta z$, $w = w_1 + \Delta w$ in the equation:

$$w^n \varphi_0(z) + w^{n-1} \varphi_1(z) + \cdots w \varphi_{n-1}(z) + \varphi_n(z) = 0,$$

let us suppose it arranged by powers of Δw:

$$f(z_0 + \Delta z, w_1 + \Delta w) = Z_0 + Z_1 \Delta w + Z_2 \Delta w^2 + \cdots Z_n \Delta w^n = 0.$$

The coefficients $Z_0, \dots Z_n$ are functions of Δz and of the constants z_0 and w_1. For $\Delta z = 0$, a single root of this equation is $\Delta w = 0$, therefore Z_0 must vanish, while the other finite roots each increased by w_1, express the other $n-1$ values of the algebraic function at z_0. Since w_1 is not a multiple root at that point, Z_1 certainly does not vanish for $\Delta z = 0$. Now let the modulus of Δz be chosen so small, that, for all values within the circle described with it as radius around z_0, mod Z_0 shall not exceed a determinate arbitrarily small number A; the modulus of Z_1 will then not fall below a certain amount B. This requirement can be satisfied; a superior limit of Δz is determined by the assigned value A, for, in the polynomial Z_0 there is no term free from Δz, whereas Z_1 contains such a term.

Now if we consider the form

$$f(z_0+\Delta z, w_1+\Delta w) = Z_1 \Delta w \left\{\frac{Z_0}{Z_1}\frac{1}{\Delta w} + 1 + \frac{Z_2}{Z_1}\Delta w + \cdots \frac{Z_n}{Z_1}\Delta w^{n-1}\right\}$$
$$= Z_1 \Delta w (1+P)$$

and put mod $\Delta w = \delta$, then, if B denote the smallest amount of Z_1,

$$\text{mod } P = \text{mod}\left[\frac{Z_0}{Z_1}\frac{1}{\Delta w} + \frac{Z_2}{Z_1}\Delta w + \cdots \frac{Z_n}{Z_1}\Delta w^{n-1}\right]$$
$$< \frac{1}{B}\left[\frac{A}{\delta} + \delta \text{ mod } Z_2 + \cdots \delta^{n-1} \text{ mod } Z_n\right],$$

therefore if C denote the greatest of the moduli of $Z_2 \ldots Z_n$ within the limit assumed for Δz,

$$\text{mod } P < \frac{1}{B}\left[\frac{A}{\delta} + C(\delta + \cdots \delta^{n-1})\right] < \frac{1}{B}\left[\frac{A}{\delta} + \frac{C\delta}{1-\delta}\right].$$

We can choose δ so as to make $\frac{1}{B}\frac{C\delta}{1-\delta}$ smaller than $\frac{\varepsilon}{2}$, where ε is arbitrarily small. In like manner we can determine the value of A and thereby the superior limit for Δz, so that however small δ is, we shall have $\frac{A}{B}\frac{1}{\delta} < \frac{\varepsilon}{2}$; for this we must choose $A < \varepsilon\frac{\delta B}{2}$. The corresponding limit for Δz we call h. Now it has to be shown, that within the circle having the radius δ there is one and only one root of the equation:

$$f(z_0 + \Delta z, w_1 + \Delta w) = Z_1 \Delta w (1+P) = 0;$$

for any value of Δz, whose modulus $\leq h$. This we show by taking:

$$\log\{f(z_0 + \Delta z, w_1 + \Delta w)\} = \log Z_1 + \log \Delta w + \log(1+P).$$

When we conduct Δw along the circle round w_1 with radius δ, $\log Z_1$ remains a constant, $\log \Delta w$ increases by $2i\pi$, since the zero point is included, but as mod P remains smaller than the arbitrarily small number ε for all points on the circle, $\log(1+P)$ does not change its imaginary part by the circuit; in the circle consequently there is one root Δw, whose modulus is smaller than δ, as was to be proved.

Accordingly it follows, that on a determinate path, leading from z_0 to Z without crossing the boundary of a singular point, each value of the algebraic function w varies uniquely and continuously.

94. *The algebraic function thus varying is moreover an analytic function,* i. e. at each regular point z it has a determinate differential quotient that is independent of the differential $dz = dx + i\,dy$. Let z and w, $z + \Delta z$ and $w + \Delta w$ denote a pair of corresponding values; if Δz converge to zero, Δw also becomes $= 0$, it is required therefore to determine the ratio $\frac{\Delta w}{\Delta z}$.

Let $\mathrm{mod}\,\Delta z \leqq h$, then $\mathrm{mod}\,\Delta w < \delta$. Now from the equation

$$f(z + \Delta z, w + \Delta w) = 0,$$

that becomes on expanding by powers of Δz and Δw:

$$f(z, w) + \left(\frac{\partial f}{\partial z}\Delta z + \frac{\partial f}{\partial w}\Delta w\right)$$
$$+ \tfrac{1}{2}\left(\frac{\partial^2 f}{\partial z^2}\Delta z^2 + 2\frac{\partial^2 f}{\partial z\partial w}\Delta z\Delta w + \frac{\partial^2 f}{\partial w^2}\Delta w^2\right) + \cdots = 0,$$

in which the first term vanishes and the last are of the dimensions of Δz^m and Δw^n; dividing by Δz we obtain:

$$\left(\frac{\partial f}{\partial z} + \frac{\partial f}{\partial w}\frac{\Delta w}{\Delta z}\right) + \tfrac{1}{2}\Delta z\left(\frac{\partial^2 f}{\partial z^2} + 2\frac{\partial^2 f}{\partial z\partial w}\frac{\Delta w}{\Delta z} + \frac{\partial^2 f}{\partial w^2}\left(\frac{\Delta w}{\Delta z}\right)^2\right)$$
$$+ \frac{1}{\underline{|3}}\Delta z^2(\quad) + \cdots = 0.$$

The quotient $\frac{\Delta w}{\Delta z}$ is by this equation an n-valued algebraic function that for $\Delta z = 0$ has one and only one finite value, namely:

$$\left(\frac{\Delta w}{\Delta z}\right)_{\Delta z = 0} = -\frac{\partial f}{\partial z} : \frac{\partial f}{\partial w},$$

because by hypothesis, z not being a critical point, $\frac{\partial f}{\partial w}$ is not zero. But we have just proved (§ 93), that a simple branch of an algebraic function varies continuously in the neighbourhood of any point; accordingly we have

$$\left(\frac{\Delta w}{\Delta z}\right)_{\Delta z = 0} = \frac{dw}{dz} = -\frac{\partial f}{\partial z} : \frac{\partial f}{\partial w},$$

a value proceeding continuously from the quotient of differences $\frac{\Delta w}{\Delta z}$, and therefore the required differential coefficient (derived function, or, for brevity, derivate). Accordingly we have found for the implicit algebraic function for any complex value the same rule of differentiation as was already established for real arguments. The fact that here the derived function is expressed also in terms of w does not contravene the statement that it is determined exclusively by the value of z, because w depends uniquely upon z.

95. So far we have always had to speak of a determinate path that the argument z has to travel from z_0 to Z, in order to keep in view the variation of one branch w_1. Now the question arises, whether the value of w_1 at the point Z will always be the same, however the path be chosen. For, the algebraic function has n different values at the point Z; it is therefore conceivable, that w_1 may pass over into these n different values at Z, according as z travels by different paths from z_0 to Z.

When two different paths leading from z_0 to Z bound a finite surface and neither enclose nor pass through a critical point, the values acquired by w at the terminal point Z are identical. If on the other hand there be any such points within the surface, the values may be different.

When there is no critical point within the surface, a finite quantity D can be assigned that will be the minimum absolute difference between the various values of w belonging to any single value z. With z_0 as centre and radius h a circle can be drawn, such that for all points within it, one and only one of the values at any such point differs from the value w_1 by a quantity with smaller modulus than $\frac{1}{2}D$, while the other values corresponding to the point must differ from w_1 by more than $\frac{1}{2}D$; for w_1, as was proved, is a unique and continuous function along every continuous curve, and there can be only one value differing from w_1 by less than $\frac{1}{2}D$ at each point in this circle, because if there were two such values they would differ from each other by less than D. Let this circle intersect the curve (I) in the point z' and let us call the corresponding value

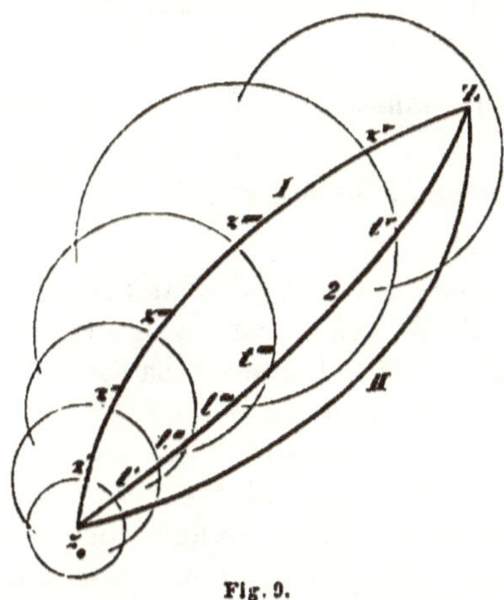

Fig. 9.

$$w_1 + \Delta w = w'.$$

A circle with radius h' can be drawn round z' as centre with a similar property; let it cut the curve (I) in the point z''; we have

$$w'' = w' + \Delta w, \ (\mathrm{mod}\, \Delta w < \tfrac{1}{2}D).$$

Repeating this process a finite number of times we arrive at a point z^{ν} with the value w^{ν} and from this reach the point Z with the value

$$W = w^{\nu} + \Delta w, \quad (\mathrm{mod}\, \Delta w < \tfrac{1}{2}D).$$

Now if we draw between (I) and (II) a curve (2) from z_0 to Z, near enough to (1) to intersect the ν circles in the points $t', t'' \ldots t^{\nu}$, the portion $z_0 t'$ must be within the circle round z_0; $t't''$ within the circle round z'; $\ldots t^{\nu} Z$ within the circle round z^{ν}. To t', then, belongs one and only one value $w = v'$ differing from w_1 by a quantity whose modulus is smaller than $\frac{1}{2}D$. Now we must observe that if z move upon the curve (2) from z_0 to t', w passes continuously from the value w_1

precisely into this value v' and into none of the others. For, at each point through which z passes, there is always only one value that differs from w_1 by less than $\frac{1}{2}D$. If then w were to change continuously to a value differing from w_1 by more than $\frac{1}{2}D$, this difference should somewhere become equal to $\frac{1}{2}D$; but this cannot be for any point in the circle. Thus too since t' lies in the circle round z', the difference between v' and w' is smaller than $\frac{1}{2}D$. Similarly, to t'' belongs one value v'', differing from the value w' by less than $\frac{1}{2}D$, since t'' is on the circumference of the circle drawn round z'. This value is obtained by travelling along the curve (2) from t' to t''; for were w to assume a different final value, it would be one differing from w' by more than $\frac{1}{2}D$. Therefore there should be upon the path $t't''$ a point at which the difference is $\frac{1}{2}D$; and this again is excluded, because t' and t'' are within and upon the circle round z'. Since t'' lies in the circle round the centre z'' we have:

$$\text{mod}\,[v'' - w''] < \tfrac{1}{2}D.$$

In like manner it can be seen that for the values at t^{ν} and z^{ν}:

$$\text{mod}\,[v^{\nu} - w^{\nu}] < \tfrac{1}{2}D,$$

and from these points we arrive at the values W and V. Here the inequalities are:

$$\text{mod}\,[W - w_{\nu}] < \tfrac{1}{2}D, \qquad \text{mod}\,[V - w_{\nu}] < \tfrac{1}{2}D.$$

Hence follows that mod $[W - V] < D$.

Now since by hypothesis the n different values of w at the point Z differ from each other at least by D, W and V cannot signify two different values, so that we must have $V = W$. Therefore the curve (I) and the curve (2) lead to the same final value. From curve (2) we can pass over to a curve (3) closer to (II) and proceeding thus we must ultimately be able to arrive at curve (II). For, all the radii h are of finite assignable magnitude therefore it is not possible that the number of steps can be infinite. Such a *progressus in infinitum* can only occur when the interpolated curves are approaching a critical point of w: for, in the immediate neighbourhood of such a point no circle can be determined within which always one only of the corresponding roots differs from w_1 by less than $\frac{1}{2}D$: the quantity indicated by D, here converges to zero. If in the included domain there be a non-essential singular point that is not also a critical point, the theorem still holds. For, although the algebraic function becomes infinite in this point, yet it retains the character of a rational function and remains unique. In fact if we surround the point by a circle of arbitrarily small radius, then, while z goes round this circle, that root, whose amount increases beyond all limits as the singular point is approached, passes through a continuous series of values, that returns back to its initial value. For in this case in the equation $f(z^m, w^n) = 0$ we have at the point $z = \alpha$, $\varphi_0(\alpha) = 0$, while $\varphi_1(\alpha)$ is different from zero. Therefore if we put $z = \alpha + \Delta\alpha$ and consider the equation:

$$w^n A_0 + w^{n-1} A_1 + \cdots A_n = 0, \text{ writing } A_k = \varphi_k(\alpha + \Delta\alpha),$$

we can choose $\Delta\alpha$ so as to make A_0 smaller than an arbitrarily small number A, while mod $[A_1]$ certainly remains greater than some finite number B. Consequently it is true, in conformity with the proof § 93, of the equation:

$$A_0 + w' A_1 + w'^2 A_2 + \cdots w'^n A_n = 0,$$

that it has for each value of $\Delta\alpha$ one and only one root, for which

$$\text{mod } w' = \text{mod } \frac{1}{w} < \delta, \text{ or, mod } w > \frac{1}{\delta};$$

i. e. when z has gone round the arbitrarily small circle, the final value of w can coincide with no other than the initial value.

From the theorem follows further: *If we make the argument z begin at a point z_0 and go round a closed finite curve not including any critical point, the final value of w_1 at z_0 is the same as that with which it began.*

96. These Theorems do not yet enable us to picture to ourselves a branch of the function. For, having calculated the value w_1 at a point z_0, we can arrive at any other point by very different paths and any two of them may include a critical point; thus it is still possible that we may obtain at each point different values according to the path.

The perfectly unique exposition, already exemplified in the explicit irrational function, is obtained here also by Riemann's method of adopting for the representation of z, instead of a single plane, n planes fastened together along their branching sections. But this requires that we should investigate more closely the properties of the critical points.

Let the point $z = 0$ be a regular point for all values of the function w; i. e. let n simple finite roots of the equation

$$f(z^m, w^n) = 0$$

belong to it. Now conceive n different planes lying one on another; to each of them coordinate in the point $z = 0$ one of the values $w_1^0, \ldots w_n^0$, indexing each plane by a number $1, 2, \ldots n$. Further, mark in each plane all those points β, which are critical points for any values whatever of w, and from these points to the point infinity draw curves, intersecting neither themselves nor one another; in all the n planes the curves proceeding from a point β_k are conceived to be identical.

Let us further coordinate to each point of plane 1 that value of w into which w_1^0 changes continuously when the argument z is made travel along a path not crossing any of the curves starting from the β points. In this manner a perfectly determinate value of w belongs to each point, and this system of values w is in general continuous. Only at the two sides of a β-curve is it possible that the values of w differ by a finite quantity. We shall distinguish these two sides as left and right banks, fixing them by the direction from the point β to the point infinity. We then determine as follows whether there is a finite difference or not. Surround the point β by an arbitrarily small circle and make the argument z go round it, starting towards the left at its point of intersection with the β-curve and returning to that point. If in this circuit w_1 has returned to its initial value and not passed over into some of the $n-1$ other possible values, the values of w also pass over into one another at both sides along the entire β-curve; thus β is not a branching point in leaf 1; the curve starting from β in this leaf can be erased. But if the value of w changes in this circuit, we have on the left bank of the β-curve values that must be denoted by w_1, and on the right other values that we may call w_2. Conceive the leaf perforated along this curve. Let the same thing be done for all points β that are branching points in leaf 1, then after this coordination, w is a continuous function in regard to all continuous paths in leaf 1 that do not cross any section.

Now when on the right side of a β-curve in leaf 1 there are values w_2 different from w_1, among the remaining leaves, for which the values $w_2^0, \ldots w_n^0$ respectively were coordinated to the origin, there must be one and one only with the property, that to the points upon the left side of that β-curve correspond all the values w_2. For, the path along which we arrived at w_1 from w_1^0 in the first leaf, must when reversed lead back w_2 into one of the values that w has at the point $z=0$. In the leaf in question, let it be called 2 and its value at the origin w_2^0, the values w_2 belong therefore to the left side of the β-curve; the same values w_2 cannot however now correspond to the right side, for it was assumed that the path, which leads to a point on the right side, leads w_1^0 over into w_2; therefore it is not possible, reversing the same path, to come from w_2 to w_2^0. Therefore on the right side of the β-curve in the second leaf there are either the values w_1 or new values w_3. I f the former, fasten the left bank in the first leaf to the right in the second and the right bank in the first to the left in the second along the whole β-curve, thus a winding surface of the first order is constructed about the branching point β. The leaves 1 and 2 form a cycle; a single circuit round the branching point leads from the first leaf into the second, a further circuit, from the second leaf into the first. The same point β can determine a winding of the first order also for other leaves.

But in the latter case, when leaf 2 brings new values w_3, we have to determine that third leaf, which presents the values w_3 on the left side of the β-curve. Either, this leaf has then on the right side of the β-curve the values w_1, in which case the three leaves:

1) with the values: w_1 on the left, w_2 on the right,
2) with the values: w_2 on the left, w_3 on the right,
3) with the values: w_3 on the left, w_1 on the right

form a cycle, and constitute a winding surface of the second order: on completing a circle therein round the point β, keeping it on the left, we pass from leaf 1 into 2, in a further circuit from 2 into 3, and in a third from 3 into 1; if we travel in the opposite direction along the circle we come from leaf 1 into 3, from 3 into 2 and from 2 into 1. Or else, the third leaf has new values w_4 on the right; there is then a fourth leaf which either concludes the cycle or leads to new values w_5 and thereby to an enlargement of the cycle.

Accordingly certain cycles of values belong to each branching section β; and an n-leaved Riemann's surface is produced by fastening together the different leaves belonging to each cycle along all sections. Throughout any arbitrary curve drawn upon this surface, whether it cross branching sections or not, the algebraic function w is completely unique and continuous. It can become infinite only in the non-essential singular points α.

97. The only further remark we shall here make is on the values of w at the point infinity. In our distribution of values to each leaf, the point infinity occurs as a many-valued point, i. e. at the two sides of the β-curve that is a branching section in leaf 1, the function w takes for $z = \infty$ both the values which ex. gr. w_1 and w_2 assume for this value of the argument. Within any finite distance however small from the point infinity, taking the illustration (§ 79) from the sphere instead of a plane, or in numerical language, however great may be the value of z, complete uniqueness still prevails.

The character of the point infinity, as a regular point or as a branching point with determinate cycles in regard to the different values of w, reveals itself, when starting from a leaf 1 we construct a circle surrounding only the point infinity and no other branching point, and consider how w changes value along this circle. By a circle surrounding only the point infinity, the Transformation by Geometrical Inversion plainly shows we have to conceive a circle round the origin, whose radius can be arbitrarily great, but at any rate must be so great that it shall include all finite branching points. When we go quite round this circle keeping the finite surface on our right, this signifies a circuit of the point infinity that is included on the left. Accordingly we have the relation:

When the circuit of all finite branching points does not change a value w_1, the point infinity is not a branching point for w_1; and conversely.

When on the other hand the circuit of all finite branching points in a determinate direction brings on a cycle: w_1 into w_2, w_2 into w_3, ... w_{p-1} into w_p, lastly w_p into w_1, the point infinity is a branching point for the p leaves. The circuit of the branching point in the prescribed direction, which becomes reversed in transforming by Inversion, brings on the same cycle.

98. We shall work out the general theory in the following: Examples.

$$1)\quad w^2 - b(z-\beta_1)(z-\beta_2) = 0.$$

Let the quantities b, β_1, β_2 have any arbitrary complex values; and let β_1 be different from β_2. By the substitution $\frac{1}{w}$, it appears that there are no finite non-essential singular points. The critical points can easily be determined by means of the explicit form:

$$w = \{b(z-\beta_1)(z-\beta_2)\}^{\frac{1}{2}}.$$

For, at points at which the two values of a square root are equal, the function under the root must vanish, therefore the critical points are: $z = \beta_1$, $z = \beta_2$.

In going round such a point the values are interchanged. For if we consider a neighbouring point $z = \beta_1 + re^{i\varphi}$, where r is arbitrarily small, the amplitudes of the corresponding pair of values

$$w = \{bre^{i\varphi}(\beta_1 - \beta_2 + re^{i\varphi})\}^{\frac{1}{2}},$$

differ by π. If we choose one value and make φ go through all values from zero to 2π, $e^{2i\pi+i\varphi}$ occurs instead of $e^{i\varphi}$. In the last factor the amplitude will not increase by 2π, since it always differs arbitrarily little from the amplitude of the constant number $\beta_1 - \beta_2$. Accordingly the root undergoes the change by the factor $e^{i\pi}$, i. e. the two values interchange. The same is true for the point β_2. Therefore they are both branching points. The two leaves are connected along sections which start from them.

If we call leaf 1 that which has at the origin O the value:

$$w_1{}^0 = (b\beta_1\beta_2)^{\frac{1}{2}} = (B\mathsf{B}_1\mathsf{B}_2)^{\frac{1}{2}}\left(\cos\frac{\alpha+\psi_1+\psi_2}{2} + i\sin\frac{\alpha+\psi_1+\psi_2}{2}\right),$$

$$b = B(\cos\alpha + i\sin\alpha),\qquad \beta_1 = \mathsf{B}_1(\cos\psi_1 + i\sin\psi_1),$$
$$\beta_2 = \mathsf{B}_2(\cos\psi_2 + i\sin\psi_2),$$

and leaf 2 that for which:

$$w_2{}^0 = (B\mathsf{B}_1\mathsf{B}_2)^{\frac{1}{2}}\left(\cos\frac{\alpha+\psi_1+\psi_2+2\pi}{2} + i\sin\frac{\alpha+\psi_1+\psi_2+2\pi}{2}\right),$$

we can decide what value of w belongs to each point in each leaf, as soon as we have settled about the branching section. If this be drawn from β_1 parallel to the positive axis of abscissæ, we have to determine what values belong to leaf 1 along the right line $O\beta_1$.

Let a point on this right line be $z = r(\cos\psi_1 + i\sin\psi_1)$; for it:

$$w = \left\{B(\cos\alpha + i\sin\alpha)(r - \mathsf{B}_1)(\cos\psi_1 + i\sin\psi_1)\right.$$
$$\left.[r(\cos\psi_1 + i\sin\psi_1) - \mathsf{B}_2(\cos\psi_2 + i\sin\psi_2)]\right\}^{\frac{1}{2}},$$

or putting: $r(\cos\psi_1 + i\sin\psi_1) - \mathsf{B}_2(\cos\psi_2 + i\sin\psi_2)$

$$= \mathsf{P}(\cos\Phi + i\sin\Phi),\quad (\Phi < 2\pi),$$

the values for $r < \mathsf{B}_1$ are:

$$w_2 = (B(\mathsf{B}_1 - r)\mathsf{P})^{\frac{1}{2}}\left\{\cos\frac{\alpha + \psi_1 + \Phi + \pi}{2} + i\sin\frac{\alpha + \psi_1 + \Phi + \pi}{2}\right\},$$
$$w_1 = (B(\mathsf{B}_1 - r)\mathsf{P})^{\frac{1}{2}}\left\{\cos\frac{\alpha + \psi_1 + \Phi + 3\pi}{2} + i\sin\frac{\alpha + \psi_1 + \Phi + 3\pi}{2}\right\}.$$

Now if r converge to zero, P becomes B_2, $\Phi = \psi_2 + \pi$; thus we see that the first value passes over into w_2^0 and the second into w_1^0. The latter lies in the first leaf.

But in order to consider points for which $r > \mathsf{B}_1$, let us surround β_1 with a circle of arbitrarily small radius ϱ and let us travel along it from the point below β_1 to the point above β_1 on $O\beta_1$, in such a way as not to cross the branching section. If we put:

$$r(\cos\varphi + i\sin\varphi) - \mathsf{B}_1(\cos\psi_1 + i\sin\psi_1) = \varrho(\cos\chi + i\sin\chi),$$

a glance at the figure, in which $\varrho\cos\chi$ and $\varrho\sin\chi$ can be constructed geometrically for all values of r and ϱ, shows that when χ begins with the value $\psi_1 + 3\pi$ it changes to the value $\psi_1 + 2\pi$, so that now at a point of the right line $O\beta_1$ the value in the first leaf is: $(\Phi < 2\pi)$

$$w_1 = (B(r - \mathsf{B}_1)\mathsf{P})^{\frac{1}{2}}\left(\cos\frac{\alpha + \psi_1 + \Phi + 2\pi}{2} + i\sin\frac{\alpha + \psi_1 + \Phi + 2\pi}{2}\right).$$

If we surround the two branching points β_1 and β_2 by a curve, and travel along it so as to keep the finite surface on the right, we pass from a point of the first leaf by crossing one branching section into the second, and then by crossing the other, back again into the first leaf, so that therefore w returns into the original value of the function.

Consequently also the point infinity is not a branching point. This appears by substituting $z = \frac{1}{z'}$ in 1); for, the point $z' = 0$ is not a branching point of the function $w = \frac{\{b(1-\beta_1 z')(1-\beta_2 z')\}^{\frac{1}{2}}}{z'}$, but its branching points are the points $z' = \frac{1}{\beta_1}$, $z' = \frac{1}{\beta_2}$.

$$2)\qquad w^3 - 3zw + z^3 = 0.$$

To determine the critical points for this function, we calculate

$$\frac{\partial f}{\partial w} = 3(w^2 - z) = 0$$

and from this we substitute $w^2 = z$ in 2); thus they will be determined by the equation $z^3(z^3 - 4) = 0$.

At the point $z = 0$ three roots of this equation vanish. To decide whether this is a branching point of the three leaves or not, we take an arbitrarily small value of r and put $z = re^{i\varphi}$; then

$$w^3 - 3re^{i\varphi}w + r^3e^{3i\varphi} = 0.$$

Since the algebraic function is continuous, the modulus of each value of w must decrease arbitrarily at the same time as r does; but as the ratio $\frac{w}{r}$ is the root of an equation:

$$\left(\frac{w}{r}\right)^3 - \frac{3}{r}\left(\frac{w}{r}\right)e^{i\varphi} + e^{3i\varphi} = 0,$$

and thus can never become indeterminate, it may tend to a value either finite or zero or ∞.

Now we see by this equation that the limit of the ratio $\frac{w}{r}$ cannot be finite, for, its middle term increases beyond any finite amount when $r = 0$.

If we assume that the limit of $\frac{w}{r}$ is zero, the first term vanishes in comparison with the second and third, so that we have

$$\text{Lim}\left(-\frac{3}{r}\,\frac{w}{r}\,e^{i\varphi}\right) = -e^{3i\varphi}, \text{ therefore: Lim } w = \tfrac{1}{3}r^2e^{2i\varphi} = w_1^0;$$

this shows that close to the origin one root of the equation differs arbitrarily little from $\frac{1}{3}r^2e^{2i\varphi} = \frac{1}{3}z^2$. This root is unique in the neighbourhood of the point zero.

Again, $\frac{w}{r}$ can only tend to an infinitely great value if

$$\text{Lim}\left(\frac{w}{r}\right)^3 - \frac{3}{r}\,\text{Lim}\,\frac{w}{r}\,e^{i\varphi} + e^{3i\varphi} = 0,$$

or, as results from dividing this by $\frac{w}{r}$, if

$$\text{Lim}\left(\frac{w}{r}\right)^2 - \frac{3}{r}\,e^{i\varphi} = 0.$$

It follows hence that either:

$$\text{Lim } w = \sqrt{3}\,.\,r^{\frac{1}{2}}\,e^{\frac{i\varphi}{2}} = w_2^0,$$

or:

$$\text{Lim } w = \sqrt{3}\,.\,r^{\frac{1}{2}}e^{\frac{i\varphi}{2}+i\pi} = w_3^0.$$

Two roots of the equation differ arbitrarily slightly from these values, and for these two the origin is a branching point. Let us choose its branching section along the negative axis of ordinates.

The other branching points are determined by the equation $z^3 - 4 = 0$, which gives three values:

$$\beta_1 = \sqrt[3]{4},\ \beta_2 = \sqrt[3]{4}\left(\cos\frac{2\pi}{3} + i\sin\frac{2\pi}{3}\right) = \sqrt[3]{4}\,e^{\frac{2i\pi}{3}},$$

$$\beta_3 = \sqrt[3]{4}\left(\cos\frac{4\pi}{3} + i\sin\frac{4\pi}{3}\right) = \sqrt[3]{4}\,e^{\frac{4i\pi}{3}}.$$

In order to establish in what manner the roots interchange about the points β, let us start from a point z at the arbitrarily small distance r from the origin on the axis of abscissæ; the corresponding roots are real and *quam proxime:*

$$w_1{}^0 = \tfrac{1}{3}r^2,\quad w_2{}^0 = \sqrt{3}\, r^{\frac{1}{2}},\quad w_3{}^0 = -\sqrt{3}\, r^{\frac{1}{2}}.$$

Two of them are positive and one negative, and all three remain real as z moves along the axis of abscissæ towards the point β_1. For then, since complex roots can occur only as conjugate pairs, and moreover the real constituent and the imaginary vary continuously, a transition to complex values can only occur in points wherein the real constituents are equal and the imaginary constituents vanish, that is to say, in branching points. In the point β_1 the value of the two equal roots is

$$w = +\sqrt{z} = +2^{\frac{1}{3}};$$

the two positive values interchange; it is a branching point for the leaves 1 and 2; let its section be chosen so as not to cut the segments $O\beta_2$, $O\beta_3$.

Let us now consider a point arbitrarily near the origin upon the right line $O\beta_2$; when we proceed to this point along the circle with radius r without crossing the negative axis of ordinates, we have:

$$w_1{}^0 = \tfrac{1}{3}r^2 e^{\frac{4i\pi}{3}},\quad w_2{}^0 = \sqrt{3}r^{\frac{1}{2}}e^{\frac{i\pi}{3}} = -\sqrt{3}r^{\frac{1}{2}}e^{\frac{4i\pi}{3}},\quad w_3{}^0 = \sqrt{3}r^{\frac{1}{2}}e^{\frac{4i\pi}{3}}.$$

A point upon $O\beta_2$ is expressed by $z = \varrho e^{\frac{2i\pi}{3}}$, the value of the double root for the point β_2 is $w = 2^{\frac{1}{3}}e^{\frac{4i\pi}{3}}$. If then we substitute:

$$z = \varrho e^{\frac{2i\pi}{3}},\quad w = w' e^{\frac{4i\pi}{3}}$$

in our equation 2), it takes the form:

$$w'^3 - 3\varrho w' + \varrho^3 = 0,$$

from which it appears that w' behaves along $O\beta_2$ just as w along the radius $O\beta_1$, thus the two positive values with the factor $e^{\frac{4i\pi}{3}}$, viz. w_1 and w_3, interchange; therefore β_2 is a branching point for the first and third leaf; let its section be drawn so as not to cut the segments $O\beta_1$, $O\beta_3$.

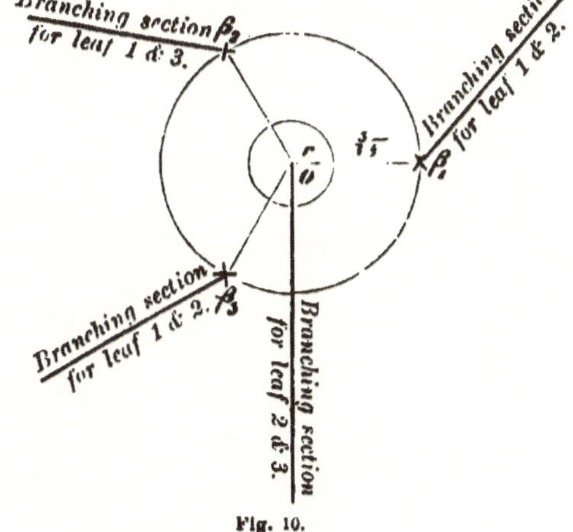

Fig. 10.

Lastly we obtain upon the radius $O\beta_3$:

$$w_1{}^0 = \tfrac{1}{3} r^2 e^{\frac{2i\pi}{3}}, \quad w_2{}^0 = \sqrt{3}\, r^{\frac{1}{2}} e^{\frac{2i\pi}{3}}, \quad w_3{}^0 = \sqrt{3}\, r^{\frac{1}{2}} e^{\frac{5i\pi}{3}} = -\sqrt{3}\, r^{\frac{1}{2}} e^{\frac{2i\pi}{3}},$$

and in the same manner it is established, that β_3 is a branching point for w_1 and w_2.

Accordingly the system of branching sections is established. A circuit of all the branching points leads each value back into the initial value; consequently the point infinity also is only a non-essential singularity.

99. We are not going to enter into the methods of simplifying the system of branching sections — such investigations are important for the theory of algebraic integrals and their periods — but we must raise the question: By what general method can the different values of w, which continuously follow each other as z varies continuously, be calculated? For, the previous investigations have only demonstrated that this problem is determinate, and in our simple examples, rising no higher than the 2nd and the 3rd degree in w, only some methods of treating the values at the branching points have found application. The general problem of calculating the algebraic function therefore still remains to be solved (see Book IV), and will find its accomplishment by means of Taylor's series for complex functions.

We shall have occasion at the same time also for showing how to determine the higher derivates of an algebraic function.

Third Book.

Integrals of functions of real variables.

First Chapter.

The definite and the indefinite integral.

100. Before considering the fundamental problem of the Integral Calculus we must make ourselves acquainted with a theorem which is supplementary to the propositions proved in § 21 and § 22.

From the Theorem of the Mean Value in § 22 we deduced that: A function whose progressive and regressive differential quotients vanish everywhere in an interval, is continuous in this interval and in fact is constant. The example mentioned in § 17 shows that this proposition does not admit of the enunciation: If the progressive differential quotient vanish at each individual point in an interval, the function is constant. For, the discontinuous function $y = G(x)$, where G signifies the greatest integer number contained in x, is discontinuous at the points $1, 2, 3 \ldots$; and yet we must admit that its progressive differential quotient is zero at each individual point. For, however near, ex. gr. we may assume $x = 1 - \varepsilon$ to be to the point 1, still an interval $\Delta x < \varepsilon$ can be assigned such that $x + \Delta x < 1$, therefore

$$\frac{G(x + \Delta x) - G(x)}{\Delta x} = 0.$$

By the help of the Theorem of the Mean Value, however, we can perceive *that a continuous function, whose progressive differential quotient vanishes everywhere in an interval, is constant*, and at the same time prove the Theorem of § 21 in the form:

When in an interval in which $f(x)$ is continuous, its progressive differential quotient is also a continuous function of x, there exists everywhere in this interval a determinate value of its regressive differential quotient that is identical with the progressive.

Whereas therefore the uniform continuity of the quotient of differences in regard to x and Δx formed previously the hypothesis whereon rested both the identity of the two differential quotients and their continuity, here that identity is to be deduced from the continuity of one of them; we shall also find that the uniform continuity of the quotient of differences arises from it.

The propositions we have stated can be proved by the following considerations:

1. When a continuous function has throughout the whole of an interval a positive progressive differential quotient, the values of the function increase in this interval and its initial value is less than its final value.

At each point, at which a continuous function has a progressive differential quotient different from zero, a progressive interval Δx can be assigned, within which the difference $f(x + \Theta \Delta x) - f(x)$ does not change sign, § 20. Hence, if the function were to decrease at a point instead of increasing, $f(x + \Theta \Delta x) - f(x)$, and therefore also the differential quotient, should be negative. Moreover the case is inconceivable, that while x converges to a determinate point x' in the interval, Δx should fall below any assignable limit. For, let us form the difference $f(x' - \varepsilon + \Delta x) - f(x' - \varepsilon)$ and make ε converge to zero, then in case Δx were to converge to zero, this difference should become zero in consequence of the continuity of f, but since at the point x' there is a positive differential quotient, there must at any rate be an assignable interval h, within which $f(x' + h) - f(x')$ remains positive. Therefore $f(x' + h) - f(x' - \varepsilon)$ also is positive, however small ε is chosen.

2. When a continuous function assumes the same value at the extremities of an interval, throughout which it has a determinate progressive differential quotient that is continuous in the entire interval, there must be a point, at which the differential quotient vanishes.

Since the function attains the same value at the extreme points, unless it remain throughout constant, it must undergo alternation in its continuous increase or decrease, i. e. must have points at which its differential quotient is positive and points at which it is negative. But as this latter is continuous, there must be between these a point at which it vanishes.

3. For every continuous function f whose progressive differential quotient f_1 is also continuous, we have in an interval from x_0 to X the equation:

$$\frac{f(X) - f(x_0)}{X - x_0} = f_1\{x_0 + \Theta(X - x_0)\} \qquad (0 < \Theta < 1).$$

For, if we denote the value of

$$\frac{f(X) - f(x_0)}{X - x_0} \text{ by } K,$$

then
$$\varphi(x) = \{f(x) - Kx\} - \{f(x_0) - Kx_0\}$$

is a continuous function of x, which has the same value, zero, at both extremities x_0 and X of the interval, and its differential quotient is continuous as that of f is; therefore there is a point

$$x_0 + \Theta(X - x_0),$$

at which:

$$\varphi_1\{x_0 + \Theta(X - x_0)\} = f_1\{x_0 + \Theta(X - x_0)\} - K = 0.$$

The equation:

$$\frac{f(X) - f(x_0)}{X - x_0} = f_1\{x_0 + \Theta(X - x_0)\}, \text{ or: } \frac{f(x+h) - f(x)}{h} = f_1(x + \Theta h),$$

holds for every value of h, when x and $x + h$ denote any points in the interval. For an arbitrarily small prescribed value of h, we can choose x so that $x + h$ may represent any point x_1 in the interval. Hence we have the result: For values of h however small, the equation

$$\frac{f(x_1) - f(x_1 - h)}{h} = f_1(x_1 - \Theta h)$$

can be fulfilled at every point x_1 in the interval. Making h converge to zero, while maintaining the value x_1, the right side passes over continuously into $f_1(x_1)$, therefore the regressive differential quotient is identical with f_1 at every point, as was to be proved.

Accordingly the Theorem of the Mean Value holds for a continuous function if its progressive differential quotient is likewise continuous; and hence follows: *A continuous function, whose progressive differential quotient vanishes throughout any interval, is constant in this interval.*

We shall further for completeness deduce the uniform continuity of the quotient of differences. It has to be proved, that in consequence of the continuity of f and f_1 for every value of x, a superior limit can be assigned for h and Δx, such that for all smaller values

$$\left[\frac{f(x+h+\Delta x) - f(x+h)}{\Delta x} - \frac{f(x+\Delta x) - f(x)}{\Delta x}\right]$$

remains smaller than an arbitrarily small number δ.

The first quotient can be brought to the form $f_1(x + h + \Theta\Delta x)$, the second is equal to $f_1(x + \Theta'\Delta x)$. Since f_1 is continuous, we are able merely by choice of h and Δx to make the difference

$$f_1(x + h + \Theta\Delta x) - f_1(x + \Theta'\Delta x)$$

smaller than δ.

Thence it follows, that if the function $f(x)$ and its derivate $f_1(x)$ be defined for an entire interval from a to b, a superior limit can be assigned for Δx, sufficient, for a given value of δ, that every smaller interval Δx between a and b shall satisfy the inequality

$$\frac{f(x+\Delta x) - f(x)}{\Delta x} - f_1(x) < \delta.$$

For if, while x converges to a value x', Δx were to fall below any assignable limit, arbitrarily near this point it would become impossible to satisfy the inequality:

$$f_1(x + \Theta \Delta x) - f_1(x) < \delta$$

by any assignable value of Δx, but this would be contrary to the continuity of f_1.

101. The fundamental problem of the Integral Calculus consists in the inversion of the problem of differentiation; it may be expressed: *Any arbitrary unique function $f(x)$ being given in the interval from $x = a$ to $x = b$; it is required to find a continuous function $F(x)$ possessing the property that its derived function is identical with $f(x)$ for all values from $x = a$ to $x = b$.*

Regarding the function $f(x)$ we make here the following restrictive hypotheses: first, $f(x)$ is to be throughout the entire interval finite; second, $f(x)$ is to be throughout the entire interval continuous, or if not, its discontinuities must be finite, and, however numerous, they must occur only at isolated points.

When $f(x)$ is a continuous function, the required function $F(x)$, if it exist, is such that its progressive and regressive differential quotients coincide everywhere in the interval. But when $f(x)$ is discontinuous at separate points, so that at any such point the values $\text{Lim}\, f(x + \delta)$ and $\text{Lim}\, f(x - \delta)$, that by hypothesis are determinate, are different for $\delta = 0$, the function $F(x)$ must be such that its progressive differential quotient at this point is equal to $f(x + 0)$, and its regressive is equal to $f(x - 0)$; these abridged notations being employed for the limiting values above named.

Now the first question to be answered is whether under these conditions and with these data the problem is definite or not; that is, whether there are not different continuous functions whose derived functions coincide in the interval from a to b. Suppose that besides $F(x)$ a second function $\Phi(x)$ were found whose differential quotients in the interval a to b likewise equal $f(x)$; then $\Phi(x) - F(x)$ is a continuous function having its progressive and regressive differential quotients throughout that entire interval zero. Such a function can only be a constant, as was proved in the last Section. Hence:

$$\Phi(x) = F(x) + \text{Const.},$$

i. e. *all continuous functions, that have the same determinate values of the progressive and regressive differential quotients respectively in an interval, differ from each other only by an additive constant whose value is arbitrary.*

This result can also be stated as follows: There is only a single continuous function whose differential quotients coincide with $f(x)$ in the interval a to b, and which has a determinate, arbitrarily chosen value at the point $x = a$. For, the additive constant is uniquely fixed by establishing a value of the function at the point $x = a$.

102. To determine the function $F(x)$ on the hypothesis that $f(x)$ is throughout continuous.

A continuous function $F(x)$ is connected with its progressive differential quotient by the equation:

$$\frac{F(x+\Delta x)-F(x)}{\Delta x}=f(x)+\delta,$$

where δ denotes a continuous function of Δx, that converges to zero with Δx; the value of δ for every finite Δx is unknown as long as the values of F are unknown. We assume the interval from a up to any value $x < b$ to be of finite length, and divide it into n parts by the points $x_1, x_2 \ldots x_{n-1}$; let $d_1, d_2 \ldots d_{n-1}, d_n$ denote the lengths of the parts $x_1-a, x_2-x_1, \ldots x_{n-1}-x_{n-2}, x-x_{n-1}$; at the point a let $F(a) =$ const. be chosen at pleasure, then the required function $F(x)$ must satisfy the equations:

$$\text{I.}\quad \begin{aligned} F(x_1)-F(a) &= d_1 f(a) + d_1\delta_1 \\ F(x_2)-F(x_1) &= d_2 f(x_1) + d_2\delta_2 \\ F(x_3)-F(x_2) &= d_3 f(x_2) + d_3\delta_3 \\ &\ldots\ldots\ldots\ldots \\ &\ldots\ldots\ldots\ldots \\ F(x_{n-1})-F(x_{n-2}) &= d_{n-1} f(x_{n-2}) + d_{n-1}\delta_{n-1} . \\ F(x) - F(x_{n-1}) &= d_n f(x_{n-1}) + d_n\delta_n . \end{aligned}$$

From addition of all these equations we find:

$$\text{II. } F(x)-F(a)=\{d_1 f(a)+d_2 f(x_1)+d_3 f(x_2)\ldots d_{n-1} f(x_{n-2})+d_n f(x_{n-1})\}+\Delta,$$

writing the unknown quantity:

$$d_1\delta_1 + d_2\delta_2 + d_3\delta_3 + \cdots d_{n-1}\delta_{n-1} + d_n\delta_n = \Delta.$$

Now if we denote by δ the greatest in absolute amount of all the values $\delta_1, \delta_2 \ldots \delta_n$, the absolute value of Δ is certainly not greater than the absolute value of the product:

$$\delta(d_1 + d_2 + \cdots d_n) = \delta(x-a);$$

so that for a continuous function $F(x)$ whose derivate is to be $f(x)$, the value of Δ will become smaller than any assignable quantity when the partial intervals d all fall below a certain amount. Therefore should equation II. serve for calculating the value $F(x)$, it is requisite that as the number of partial intervals is arbitrarily increased, the expression within brackets on its right shall converge to a determinate value depending on x and on the constant a, and moreover that this value shall be a continuous function of x with the derived function $f(x)$.

103. In order to show that this first requirement is actually fulfilled, we proceed as follows. Let the sum:

$$S = d_1 f(a) + d_2 f(x_1) + d_3 f(x_2) + \cdots d_{n-1} f(x_{n-2}) + d_n f(x_{n-1})$$

be altered by breaking up each of the intervals $d_1, d_2 \ldots d_n$ anew into subdivisions; let S' denote the corresponding sum of products formed like S by multiplying each new partial interval by the value of f at the beginning of that interval; let n' be the number of the new intervals; after this each of these intervals is to be broken up into an arbitrary number of subdivisions, let the respective value of the sum of products be called S'' and the number of intervals n''; proceeding thus we obtain a series of arbitrarily increasing numbers:

$$n, n', n'' \dots n^{(k)} \dots \text{etc.}$$

and a corresponding series of sums:

$$S, S', S'' \dots S^{(k)} \dots \text{etc.}.$$

This series must represent a determinate limiting value, i. e. for any number δ however small, it must be possible to find a value $n^{(k)}$ such that the difference between $S^{(k)}$ and any following value $S^{(k+\nu)}$ shall be smaller than δ.

We first remark, that a sum of the form S can be represented always by an expression of the form:

$$\text{III.} \qquad S = (x-a)\,f(a+\Theta(x-a)), \text{ where } 0 \leqq \Theta \leqq 1;$$

for if, taking account of the sign, the greatest value among the coefficients $f(a) \dots f(x_{n-1})$ be denoted by G, and the least by K, we have $K(x-a) < S < G(x-a)$, or S is equal to the product of $x-a$ by a value between K and G. Now because $f(x)$ is a continuous function of x, it assumes at least once each value between the least value K and the greatest G, it overleaps none, therefore there must be a point at which f actually has the value that is requisite for equation III.

Now if each of the intervals from a to x_1, from x_1 to x_2, etc. be divided into smaller intervals, new sums come up in place of the products $d_1f(a)$, $d_2f(x_1)\dots$; namely, when the dividing points in the kth interval: from x_{k-1} to x_k, are denoted by $x_1^{(k)}, x_2^{(k)}, \dots x_{\nu-1}^{(k)}$, the product $d_kf(x_{k-1})$ is replaced by the sum:

$$\sum^{(k)} = (x_1^{(k)} - x_{k-1})f(x_{k-1}) + (x_2^{(k)} - x_1^{(k)})f(x_1^{(k)}) + \cdots (x_k - x_{\nu-1}^{(k)})f(x_{\nu-1}^{(k)}).$$

In analogy with equation III. the sum on the right can be brought to the form:

$$\sum_{\nu}^{(k)} = (x_k - x_{k-1})f(x_{k-1} + \Theta_k(x_k - x_{k-1})) = d_kf(x_{k-1} + \Theta_k(x_k - x_{k-1}))\,.$$
$$0 \leqq \Theta_k \leqq 1.$$

Thus, partition of the intervals of S into new subdivisions leads to a value S', that only differs from the former by each term $d_kf(x_{k-1})$ containing in place of

$f(x_{k-1})$ another value of f that belongs to a point within the interval d_k.

In like manner, S'' arises from S' by the occurrence in place of the term $d'_k f(x'_{k-1})$ in S' of another value of f that belongs to a point in the interval d'_k; where d'_k denotes that one of the n' partial intervals that begins at the point x'_{k-1}; and so on.

But now since the function f is continuous, at each point a finite interval can be discovered wherein the various values of f differ by less than an arbitrarily small finite quantity ε. Therefore by continued subdivision the intervals can certainly be made so small, that in each of them the absolute differences of the various values of f shall be smaller than ε; let the number of these intervals be $n^{(k)}$, the respective sum $S^{(k)}$, when we advance to any of the further partitions we have:

$$\text{abs}\,[S^{(k)} - S^{(k+r)}] < \varepsilon\{d_1 + d_2 + \cdots d_{n^{(k)}}\};$$

but as the total interval is always equal to $(x - a)$, this difference is smaller than $\varepsilon(x - a)$.

When therefore the partition has advanced so far, that in each interval the fluctuations of f are smaller than $\frac{\delta}{x-a} = \varepsilon$, any further partitions can alter the amount of $S^{(k)}$ only by less than δ; therefore the series S, S', S'', etc., approximates to a determinate limiting value.

But it must still be investigated, whether this limiting value depends on the original partition into n intervals and the consequent partition of each of these into smaller intervals, or whether it is quite indifferent in what way the total interval from a to x is broken up into subdivisions that ultimately decrease below any finite amount.*) That the latter is the case, appears from the following consideration.

Let the original partition be into m parts, the corresponding sum being S_1. By further dividing these intervals we obtain as before a series of values $S_1^{(1)}, S_1^{(2)}, \ldots S_1^{(k)} \ldots$, the numbers of intervals being $m', m'', \ldots m^{(k)}$ etc.. Let the partition have advanced so far, that each further partition can alter the value of $S_1^{(k)}$ only by less than δ.

Now let us conceive these two partitions: into $m^{(k)}$, and into $n^{(k)}$ intervals, combined into a single one, then to it belongs a sum Σ that differs from $S^{(k)}$ as well as from $S_1^{(k)}$ by a quantity smaller than δ; for, this third partition arising from their combination is to be regarded as a continuation of each of the two former. Abs $[S^{(k)} - S_1^{(k)}]$ is therefore smaller than the arbitrarily small quantity 2δ, i. e. the series S_1 has the same limiting value as the series S.

*) This is investigated with still more detail in § 142.

Thus as a particular case we can conceive the interval from a to x broken up into equal parts Δx, of which the number n increases without limit; the required value is then expressed in the form:

$$\text{Lim}\,\{\Delta x(f(a)+f(a+\Delta x)+f(a+2\Delta x)+\cdots f(a+\overline{n-1}\Delta x))\}$$

$$\left(\frac{x-a}{n}=\Delta x,\quad \text{for } \Delta x=0\right).$$

Employing the sign of the differential $dx = \text{Lim}\,\Delta x$, we follow Leibnitz in denoting the sum by the abridged symbol:

$$\text{IV.}\quad \int_a^x f(x)\,dx = \text{Lim}\,\{\Delta x(f(a)+f(a+\Delta x)+\cdots f(x-\Delta x))\}$$

$$\textit{for } \Delta x = 0,\qquad \left(\frac{x-a}{n}=\Delta x\right);$$

and call it the Definite Integral of the function $f(x)$ taken from the lower limit a to some determinate upper limit x.

The integral sign $\int$ is a sign of a sum; on the left side of the defining equation IV. stands a symbol, on the right an expression that can be calculated. It is to be observed regarding this formula, that x as upper limit represents a definite value, but under the integral sign it signifies a variable, since f is to be formed for the points $f(a)$, $f(a+dx)$, $f(a+2dx)$, etc..

The conception of the integral as a sum gave rise to an erroneous impression. For, if we first put Δx equal to zero on the right side of equation IV., since all terms have the factor Δx, we obtain only summands whose value is zero, and however many of them are added, the resulting value of the sum is necessarily zero. An integral could therefore never have any value but zero, or equation IV. should contain a contradiction. This is not removed but only obscured by the further contradiction: $f(x)dx$ is not zero but an infinitely small quantity. Euler therefore (see foot-note § 105) completely rejected the definition of the integral as a sum, and maintained only the definition that follows from inversion of differentiation. Meanwhile, as the above development shows, this same definition leads unavoidably to the conception of a sum and this contains no contradiction, when we bear in mind, that

$$\int_a^x f(x)\,dx$$

is not the sum of the limiting values of $f(x)\Delta x$, but the limiting value of the sum of the terms $f(x)\Delta x$; in other words: what is required is, first to find the sum for a finite number of terms as a function of Δx, and then to determine its limiting value for $\Delta x = 0$.

We have, ex. gr. for $\Delta x = \frac{b-a}{n}$:

$$\int_a^b x\,dx = \operatorname{Lim} \Delta x (a + (a + \Delta x) + \cdots (a + \overline{n-1}\, \Delta x))$$

$$= \operatorname{Lim} an\Delta x + \operatorname{Lim} \Delta x^2 \frac{n(n-1)}{2}$$

$$= (b-a)a + \operatorname{Lim} \tfrac{1}{2}(b-a)(b-a-\Delta x) = \tfrac{1}{2}b^2 - \tfrac{1}{2}a^2.$$

In equation IV. the proposition holds for any value of Δx however small, that the sum on the right is equal to the product of $x - a$ by some value between the greatest and least values assumed by f at the different points of division; but since by hypothesis f is continuous it must actually assume this mean value at any rate once, i. e. we have also:

$$\text{V.} \qquad \int_a^x f(x)\,dx = (x-a)\, f(a + \Theta(x-a)), \text{ where } 0 \leqq \Theta \leqq 1.$$

104. But the determinate limiting value is moreover (§ 102) a continuous function of its upper limit x; for every number δ however small, a number h can be found such that:

$$\operatorname{abs}\left[\int_a^{x \pm h} f(x)\,dx - \int_a^x f(x)\,dx\right] < \delta,$$

on the hypothesis, that $x \pm h$ also lies within the interval a to b.

For, from the definition as a sum we have:

$$\int_a^{x+h} f(x)\,dx - \int_a^x f(x)\,dx = \operatorname{Lim} \Delta x (f(x) + f(x + \Delta x) + \cdots f(x + h - \Delta x))$$

$$= \int_x^{x+h} f(x)\,dx.$$

But by Equation V.:

$$\int_x^{x+h} f(x)\,dx = hf(x + \Theta h), \qquad 0 \leqq \Theta \leqq 1.$$

Since $f(x)$ is throughout finite, h can be chosen so as to make this expression arbitrarily small. Similarly:

$$\int_a^{x-h} f(x)\,dx - \int_a^x f(x)\,dx = -\operatorname{Lim} \Delta x (f(x-h) + f(x-h+\Delta x) + \cdots f(x - \Delta x))$$

$$= -\int_{x-h}^x f(x)\,dx,$$

and by Equation V.:

$$-\int_{x-h}^{x} f(x)\,dx = -h f(x - \Theta h), \qquad 0 \leqq \Theta \leqq 1.$$

It also follows from these equations, that the integral regarded as a function of its upper limit, has $f(x)$ as its derived. For we have:

$$\operatorname{Lim} \frac{1}{h}\left\{\int_a^{x+h} f(x)\,dx - \int_a^x f(x)\,dx\right\} = \operatorname{Lim} \frac{1}{h}\int_x^{x+h} f(x)\,dx = \operatorname*{Lim}_{\text{for } h=0} f(x + \Theta h),$$

$$\operatorname{Lim} -\frac{1}{h}\left\{\int_a^{x-h} f(x)\,dx - \int_a^x f(x)\,dx\right\} = \operatorname{Lim} \frac{1}{h}\int_{x-h}^{x} f(x)\,dx = \operatorname*{Lim}_{\text{for } h=0} f(x - \Theta h).$$

At points in which $f(x)$ is continuous, both $f(x + \Theta h)$ and $f(x - \Theta h)$ pass over continuously into the value $f(x)$; the progressive and regressive differential quotients are here identical.

Accordingly the conditions are fulfilled which are necessary and sufficient for us to obtain from Equation II. the Theorem:

The continuous function $F(x)$ required to possess the property, that its progressive and regressive derived functions everywhere in the interval from a to b have the same values with the continuous function $f(x)$, is equal to the definite integral $\int_a^x f(x)\,dx$ increased by the addition of an arbitrary constant:

$$F(x) = \int_a^x f(x)\,dx + \text{Const.}$$

The function $F(x)$ is called the Indefinite Integral of $f(x)$. The constant is fixed, once the value is given that F is to have at the point a. For if we put $x = a$ in the equation, we have:

$$\int_a^a f(x)\,dx = 0, \text{ thus } F(a) = \text{Const.}$$

Conversely therefore the definite integral can be described as the difference of the values of the indefinite integral formed for the upper and lower limit:

$$F(x) - F(a) = \int_a^x f(x)\,dx.$$

105. The definite integral $\int_a^x f(x)\,dx$ admits of a simple geometrical interpretation, when the values $f(x)$ are represented as the ordinates of a curve, or more precisely as the ordinates of the corners of a polygon of arbitrarily many sides, for here the condition that f may admit of representation by a curve or of being differentiated is not necessarily fulfilled. If we construct for the points:

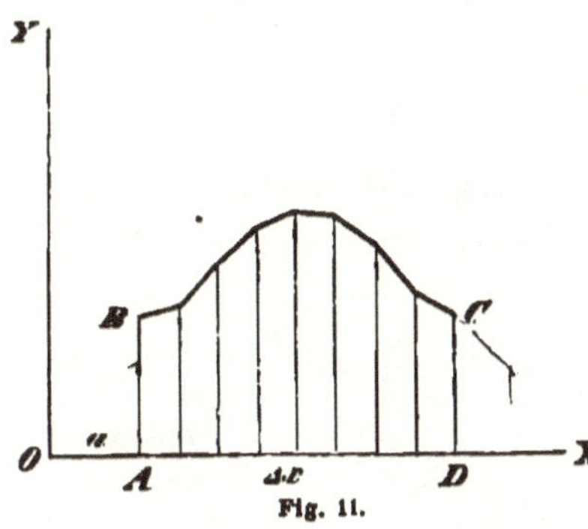

Fig. 11.

$$a,\ a + \Delta x,$$
$$a + 2\Delta x, \dots (a + \overline{n-1}\,\Delta x),\ x$$

the ordinates:

$$f(a),\ f(a + \Delta x),$$
$$f(a + 2\Delta x) \dots f(a + \overline{n-1}\,\Delta x), f(x)$$

and join consecutive extremities of these ordinates by right lines, the area $ABCD$ bounded by the polygon and by the coordinates, being a sum of trapezia, is equal to:

$$\Delta x \left\{ \frac{f(a) + f(a + \Delta x)}{2} + \frac{f(a + \Delta x) + f(a + 2\Delta x)}{2} + \dots \frac{f(a + \overline{n-1}\,\Delta x) + f(x)}{2} \right\}$$

or equal to:

$$S - \frac{\Delta x}{2} \{f(a) - f(x)\},$$

where S stands for a sum of products of the previous form. Now if we make Δx converge to zero, i. e. in geometric terms: if we construct the polygons corresponding to the function f with more and more corners, S passes over into the value of the definite integral:

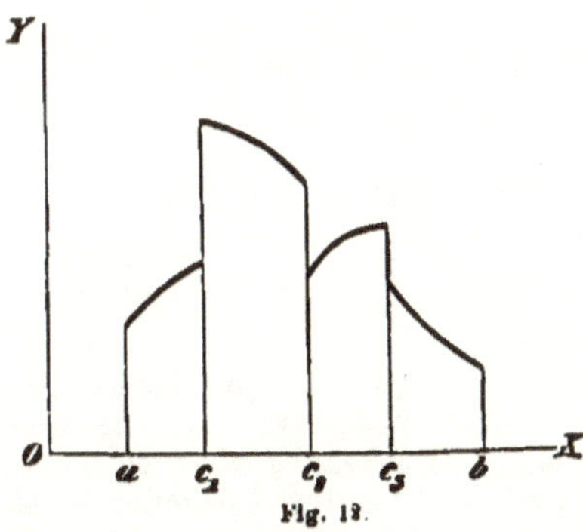

Fig. 12.

$$\int_a^x f(x)\,dx,$$

while the second term of the equation converges to zero.

Thus the definite integral is equal to the number of units of an area when the values of f are interpreted as ordinates of points; of the area bounded by a portion of a curve, by the ordinates of its extremities and by the axis of abscissæ, in the particular case when the polygons of the function f converge to a definite curve.*) If the

*) From the solution of this geometric problem: that of measuring the surface bounded by an arbitrary curve given by a function, the Integral Calculus originated simultaneously with the Differential Calculus from that of the problem

values of f in the interval from a to x differ in sign, the definite integral measures the difference of areas. This geometric intuition brings home to us most simply that the definite integral

$$\int_a^b f(x)\,dx$$

has still a determinate finite value, although the function f is discontinuous at any finite number of points $c_1, c_2, \ldots c_m$ while remaining finite.

In that case:

$$\int_a^b f(x)\,dx = \int_a^{c_1-0} f(x)\,dx + \int_{c_1+0}^{c_2-0} f(x)\,dx + \cdots \int_{c_{(m-1)}+0}^{c_m-0} f(x)\,dx + \int_{c_m+0}^{b} f(x)\,dx$$

is a determinate finite quantity, namely the sum of the areas bounded by the several polygons or portions of curves and the ordinates of their extremities. At such a discontinuity of f, the indefinite integral

$$F(x) = \int_a^x f(x)\,dx + \text{Const.}$$

has as its progressive differential quotient $f(c+0)$, for we have:

$$\frac{F(c+h)-F(c)}{h} = \frac{\int_a^{c+h} f(x)\,dx - \int_a^{c} f(x)\,dx}{h} = \frac{1}{h}\int_c^{c+h} f(x)\,dx = f(c+\Theta h),$$

and as its regressive differential quotient $f(c-0)$, for we have:

$$\frac{F(c-h)-F(c)}{-h} = \frac{\int_a^{c-h} f(x)\,dx - \int_a^{c} f(x)\,dx}{-h} = +\frac{1}{h}\int_{c-h}^{c} f(x)\,dx = f(c-\Theta h),$$

as was originally required when proposing to find the function F.

---

of tangents. Leibnitz and Newton in the writings named in § 23 gave the first propositions of both; before this, Fermat (1608—1665) and Wallis (1616—1703) had evolved the fundamental conception of a summation for measuring areas and applied it to parabolic curves. But the principal merit of the further cultivation of this calculus belongs to the brothers James (1654—1705) and John (1667—1748) Bernoulli of Bâle, who strove to outbid each other in the solution of problems by its means. John Bernoulli compiled in the years 1691 and 1692 at Paris his Lectiones mathematicæ, the first text-book on the integral calculus; in 1742 it appeared in print in the complete collection of his writings. In forwarding them to Euler, John Bernoulli wrote: "Exhibeo enim mathesin sublimem, qualis fuit in infantia, Tu vero eam nobis sistis in virili aetate." Euler's systematic treatment of the integral calculus: Institutiones calculi integralis, appeared at St. Petersburg 1768—70.

106. Hitherto the formula

$$F(x) = \int_a^x f(x)\,dx + \text{Const.}$$

has been derived in the sense of expressing the indefinite integral $F(x)$ from the definite integral; it therefore requires the calculation of the limiting value of a sum with arbitrarily many summands. It can however inversely be employed in calculating this limiting value of a sum, when the indefinite integral $F(x)$ is known. But now since in the differential calculus the derived functions $f(x) = F'(x)$ belonging to whole classes of functions $F(x)$ have been calculated; inversely, the indefinite integral $F(x)$ belonging to each of these derived functions $f(x)$ is also known. On the hypothesis that this integral is unique and continuous in the interval from a to b, we obtain by the difference of the values of this function the definite integral from a to b. This calculation is still valid when in the interval from a to b the function $f(x)$ assumes infinite values, while $F(x)$ remains finite, since at such a point c we have to put:

$$\int_a^c f(x)\,dx = \operatorname*{Lim}_{\text{for } \delta = 0}\left\{\int_a^{c-\delta} f(x)\,dx\right\} = \operatorname*{Lim}_{\text{for } \delta = 0}\{F(c-\delta) - F(a)\} = F(c) - F(a),$$

and it holds even for an infinite interval from a to ∞, or from $-\infty$ to $+\infty$, when the function $F(x)$ retains a determinate finite value actually at these limits, if we introduce the definition:

$$\int_a^{+\infty} f(x)\,dx = \operatorname*{Lim}_{w=\infty}\left\{\int_a^{w} f(x)\,dx\right\} = \operatorname*{Lim}_{w=\infty}\{F(w) - F(a)\} = F(\infty) - F(a).$$

The indefinite integral of $f(x)$ is usually denoted by the symbol $\int f(x)\,dx + \text{const.}$, so that $\frac{d\int f(x)\,dx}{dx} = f(x)$; $f(x)$ is called the function that is to be integrated; $F(x)$ the integral function.

107. Fundamental formulas:

1) $\frac{dx^{m+1}}{m+1} = x^m dx, \ (m \gtrless -1).$

2) $d(lx) = \frac{dx}{x}.$ 3) $\frac{de^{mx}}{m} = e^{mx}dx.$

4) $\frac{d \sin mx}{m} = \cos mx\,dx.$ 5) $-\frac{d \cos mx}{m} = \sin mx\,dx.$

6) $d \tan x = \frac{dx}{(\cos x)^2}.$ 7) $-d \cot x = \frac{dx}{(\sin x)^2}.$

8) $d \sin^{-1} x = \frac{dx}{\sqrt{1-x^2}}.$ 9) $-d \cos^{-1} x = \frac{dx}{\sqrt{1-x^2}}.$

10) $d \tan^{-1} x = \frac{dx}{1+x^2}.$ 11) $-d \cot^{-1} x = \frac{dx}{1+x^2}.$

From these we find by inverting:

$$1)\ \int x^m dx = \frac{x^{m+1}}{m+1} + \text{const.}, \quad (m \gtrless -1).$$

$$2)\ \int \frac{dx}{x} = l(x) + \text{const.} \qquad 3)\ \int e^{mx} dx = \frac{e^{mx}}{m} + \text{const.}$$

$$4)\ \int \cos mx\, dx = \frac{\sin mx}{m} + \text{const.} \qquad 5)\ \int \sin mx\, dx = -\frac{\cos mx}{m} + \text{const.}$$

$$6)\ \int \frac{dx}{\cos^2 x} = \tan x + \text{const.} \qquad 7)\ \int \frac{dx}{\sin^2 x} = -\cot x + \text{const.}$$

$$8)\ \int \frac{dx}{\sqrt{1-x^2}} = \sin^{-1} x + \text{const.} \qquad 9)\ \int \frac{dx}{\sqrt{1-x^2}} = -\cos^{-1} x + \text{const.}$$

$$10)\ \int \frac{dx}{1+x^2} = \tan^{-1} x + \text{const.} \qquad 11)\ \int \frac{dx}{1+x^2} = -\cot^{-1} x + \text{const.}$$

These formulas only state in another form the same thing as is already asserted by the first eleven equations.

But now the definite integrals can be got from these; the region in which they are valid, determined by the requirement that the integral function remain real, continuous and finite, is assigned in the brackets adjoining; the following equations furnish new information since they express the calculation of the limiting values of sums.

$$1)\ \int_a^b x^m dx = \frac{b^{m+1} - a^{m+1}}{m+1}, \ (m \gtrless -1,\ 0 < a < \infty,\ 0 < b < \infty)$$

if $m + 1 > 0$ the formula holds also for a or b equal to 0, if $m + 1 < 0$ the formula holds also for a or b equal to ∞.

$$2)\ \int_a^b \frac{dx}{x} = l\left(\frac{b}{a}\right), \ (0 < a < \infty,\ 0 < b < \infty);\ \int_{-a}^{-b} \frac{dx}{x} = +l\left(\frac{b}{a}\right).$$

$$3)\ \int_a^b e^{mx} dx = \frac{e^{mb} - e^{ma}}{m}, \qquad (\text{if } m > 0,\ -\infty \leqq a < +\infty,\ -\infty \leqq b < +\infty).$$

$$4)\ \int_a^b \cos mx\, dx = \frac{\sin(mb) - \sin(ma)}{m}, \qquad (-\infty < a < +\infty,\ -\infty < b < +\infty).$$

$$5)\ \int_a^b \sin mx\, dx = -\frac{\cos(mb) - \cos(ma)}{m}, \qquad (-\infty < a < +\infty,\ -\infty < b < +\infty).$$

$$6)\ \int_a^b \frac{dx}{\cos^2 x} = \tan b - \tan a,$$ (in every interval from a to b, which does not contain odd multiples of $\frac{1}{2}\pi$).

7) $\int_a^b \frac{dx}{\sin^2 x} = -\cot b + \cot a$, (in every interval from a to b, which does not contain π or multiples of π).

8) and 9) $\int_a^b \frac{dx}{\sqrt{1-x^2}} = \sin^{-1} b - \sin^{-1} a = \quad \cos^{-1} b + \cos^{-1} a,$

$$(-1 \leqq a \leqq +1, \quad -1 \leqq b \leqq +1).$$

10) and 11) $\int_a^b \frac{dx}{1+x^2} = \tan^{-1} b - \tan^{-1} a = -\cot^{-1} b + \cot^{-1} a,$

$$(-\infty \leqq a \leqq +\infty, \quad -\infty \leqq b \leqq +\infty).$$

108. We can also assign the indefinite integral for functions compounded of the simple ones; for this we require the General Rules that can be derived by inverting the Rules of Differentiation (§ 26).

If:

a) $$f(x) = f_1(x) + f_2(x) + f_3(x) + \cdots + f_n(x)$$

we have:

$$\int f(x)\,dx = \int f_1(x)\,dx + \int f_2(x)\,dx + \int f_3(x)\,dx + \cdots \int f_n(x)\,dx,$$

that is, the integral of a sum of functions is equal to the sum of the integrals of the several summands. This is proved by differentiation.

If:

b) $F(x) = \varphi(x)\psi(x)$, then $F'(x) = f(x) = \varphi(x)\psi'(x) + \psi(x)\varphi'(x)$, therefore inverting:

$$\int \varphi(x)\psi'(x)\,dx + \int \psi(x)\varphi'(x)\,dx = \varphi(x)\psi(x).$$

If we write this formula:

$$\int \varphi(x)\psi'(x)\,dx = \varphi(x)\psi(x) - \int \psi(x)\varphi'(x)\,dx$$

or again:

$$\int \varphi(x)\,d(\psi(x)) = \varphi(x)\psi(x) - \int \psi(x)\,d(\varphi(x)),$$

it shows how to reduce the integral of a function, consisting of two factors of which one can be integrated, to another integral. This reduction, called the process of integration by parts, in many cases simplifies the problem.

Special theorem: If:

$$F(x) = a\varphi(x), \text{ then } F'(x) = f(x) = a\varphi'(x),$$

therefore:

$$\int a\varphi'(x)\,dx = a\varphi(x) = a\int \varphi'(x)\,dx.$$

c) Let us introduce a new variable for x into the formula

$$F(x) = \int f(x)\,dx$$

by the equation $x = \varphi(u)$, such that to continuous consecutive values of x from a to b correspond uniquely continuous consecutive values of u not undergoing any alternation of increase or decrease, so that $\frac{du}{dx}$ does not vanish; thus let $F(x)$ become $\Psi(u)$, and let $f(x)$ become $\psi(u)$. We have then the relation:

$$F'(x) = f(x) = \psi(u) = \frac{d\Psi(u)}{du} \cdot \frac{du}{dx},$$

and so:

$$\frac{d\Psi(u)}{du} = \psi(u)\frac{dx}{du} = \psi(u)\varphi'(u),$$

whence:

$$\Psi(u) = \int \psi(u)\varphi'(u)\,du.$$

Substitution of a variable by the equation $x = \varphi(u)$ reduces the determination of the integral of f:

$$\int f(x)\,dx, \text{ to that of: } \int \psi(u)\varphi'(u)\,du;$$

by an apt choice of the formula of substitution this integral may be more simply found than the original one.

When the original is to be taken between the limits $x = a$ and $x = b$, the new integral is to be formed with the limits u_a and u_b for which $a = \varphi(u_a)$, $b = \varphi(u_b)$, so that the values from $x = a$ to $x = b$ are uniquely related to the values from u_a to u_b.

But if the relation between x and u is not uniquely convertible, the total interval must be broken up into partial intervals in which a mutually unique relation can be established.

Thus, when the integral ex. gr. $\int\limits_0^\infty f\left(ax + \frac{b}{x}\right)dx$ is proposed, a and b being positive, if we put $ax + \frac{b}{x} = u$, to each value of x corresponds a unique value of u, but while x passes from 0 to ∞, u which began by decreasing, undergoes a change and subsequently increases; to each value of u correspond two values of x,

$$x = \frac{u}{2a} \pm \frac{1}{2a}\sqrt{u^2 - 4ab}, \qquad dx = \frac{du}{2a}\left(1 \pm \frac{u}{\sqrt{u^2 - 4ab}}\right),$$

as we can realise geometrically by drawing the hyperbola. We can calculate the minimum value of u by means of $\frac{du}{dx} = 0$, that is $a - \frac{b}{x^2} = 0$; we have $x = +\sqrt{\frac{b}{a}}$, $u = 2\sqrt{ab}$.

As x goes from 0 to $\sqrt{\frac{b}{a}}$, u diminishes from ∞ to $2\sqrt{ab}$ and we have:

$$\int_0^{\sqrt{\frac{b}{a}}} f\left(ax + \frac{b}{x}\right) dx = \frac{1}{2a}\int_{\infty}^{+2\sqrt{ab}} f(u)\left(1 - \frac{u}{\sqrt{u^2 - 4ab}}\right) du;$$

on the other hand, in the interval from $x = \sqrt{\frac{b}{a}}$ to ∞, u increases from $2\sqrt{ab}$ to ∞, and therefore:

$$\int_{\sqrt{\frac{b}{a}}}^{\infty} f\left(ax + \frac{b}{x}\right) dx = \frac{1}{2a}\int_{2\sqrt{ab}}^{\infty} f(u)\left(1 + \frac{u}{\sqrt{u^2 - 4ab}}\right) du.$$

Hence adding up we have:

$$\int_0^{\infty} f\left(ax + \frac{b}{x}\right) dx = \frac{1}{a}\int_{2\sqrt{ab}}^{\infty} f(u) \frac{u}{\sqrt{u^2 - 4ab}} du.$$

We proceed to apply the general theorems to various functions $f(x)$ with the view of determining $\int f(x) dx$ by means of the Fundamental Formulas.

Second Chapter.

The integral of rational algebraic functions. Partial fractions.*)

109. The integral of the rational integer function of the nth order:

$$f(x) = a_0 + a_1 x + a_2 x^2 + \cdots + a_n x^n$$

is, by Theorems a) and b) § 108 and by Fundamental Formula 1) § 107:

$$F(x) = \int f(x)dx = a_0 \int dx + a_1 \int x dx + \cdots + a_n \int x^n dx$$

$$= a_0 x + a_1 \frac{x^2}{2} + \cdots a_n \frac{x^{n+1}}{n+1} + \text{Const.}$$

110. A rational fractional function: $f(x) = \frac{\psi(x)}{\varphi(x)}$, ψ being of the mth and φ of the nth order, when $m \geqq n$ can always be resolved into an integer function and a proper fractional function, i. e. a function in which the order of the numerator is at most $n - 1$. This only requires the division by the denominator $\varphi(x)$ to be carried out until the order of the remainder becomes less than n. As the integration of the integer function has been given already, we have only to determine the integral of the form:

$$\int \frac{\psi(x)}{\varphi(x)} dx$$

in which ψ is of lower order than φ and they have no common root.

This proper fractional function can be resolved into a sum of fractions with constant numerators and with denominators that are linear functions or powers of linear functions.

Partial fractions.

111. Let $\alpha_1, \alpha_2, \ldots \alpha_n$ the n roots or vanishing points of

$$\varphi(x) = a_0 + a_1 x + \cdots + x^n$$

be real or complex, but first let them all be different. The coefficient of the highest power of x in φ is supposed unity, it can always be made so by putting the original factor before the entire quotient

*) Leibnitz and John Bernoulli: Acta erud., 1702—1703. Euler: Institutiones calculi integralis. Vol. I. Sect. I. Cap. I.

and therefore also before the integral. The n vanishing points shall be assumed known and $\varphi(x)$ resolved into the product:

$$\varphi(x) = (x - \alpha_1)(x - \alpha_2)\dots(x - \alpha_n).$$

Calling the product of all these factors except the first, $\varphi_1(x)$, so that $\varphi(x) = (x - \alpha_1)\varphi_1(x)$, we are going to show that:

The fraction $\frac{\psi(x)}{\varphi(x)}$ can be resolved in one way only into the form:

1) $$\frac{\psi(x)}{\varphi(x)} = \frac{A_1}{x - \alpha_1} + \frac{\psi_1(x)}{\varphi_1(x)},$$

where A_1 is employed to denote a constant and $\psi_1(x)$ an integer function of the order $n - 2$ at most. For we have identically:

$$\psi(x) = A_1\varphi_1(x) + (x - \alpha_1)\psi_1(x),$$

therefore:

2) $$\psi_1(x) = \frac{\psi(x) - A_1\varphi_1(x)}{x - \alpha_1}.$$

Now for this to be an integer function of the order $n - 2$ at most, the numerator on the right must be divisible by $x - \alpha_1$, i.e. must vanish for $x = \alpha_1$; thence follows:

3) $$0 = \psi(\alpha_1) - A_1\varphi_1(\alpha_1), \quad \text{or } A_1 = \frac{\psi(\alpha_1)}{\varphi_1(\alpha_1)}.$$

The value $\varphi_1(\alpha_1) = (\alpha_1 - \alpha_2)(\alpha_1 - \alpha_3)\dots(\alpha_1 - \alpha_n)$ does not vanish, because all the roots α are different; moreover by means of the derived of $\varphi(x)$ it can be written in the form $\varphi'(\alpha_1)$, since we have for $x = \alpha_1$

$$\operatorname{Lim} \frac{\varphi(x)}{x - \alpha_1} = \varphi'(\alpha_1);$$

accordingly:

4) $$\frac{\psi(x)}{\varphi(x)} = \frac{\psi(\alpha_1)}{\varphi'(\alpha_1)}\,\frac{1}{x - \alpha_1} + \frac{\psi_1(x)}{\varphi_1(x)}.$$

Proceeding in like manner we shall resolve the quotient $\frac{\psi_1(x)}{\varphi_1(x)}$ into the form:

$$\frac{\psi_1(x)}{\varphi_1(x)} = \frac{A_2}{x - \alpha_2} + \frac{\psi_2(x)}{\varphi_2(x)}, \text{ where } \varphi_1(x) = (x - \alpha_2)\varphi_2(x);$$

further:

$$\frac{\psi_2(x)}{\varphi_2(x)} = \frac{A_3}{x - \alpha_3} + \frac{\psi_3(x)}{\varphi_3(x)}, \qquad \varphi_2(x) = (x - \alpha_3)\varphi_3(x); \text{ etc.}.$$

In the second quotient on the right the order of the numerator is at least one lower than that of the denominator, so that ultimately:

$$\frac{\psi_{n-2}(x)}{\varphi_{n-2}(x)} = \frac{A_{n-1}}{x - \alpha_{n-1}} + \frac{A_n}{x - \alpha_n},$$

where A_{n-1} and A_n are constants; thus we have proved:

The proper fractional function can be resolved in only one way into partial fractions of the form:

5) $$\frac{\psi(x)}{\varphi(x)} = \frac{A_1}{x-\alpha_1} + \frac{A_2}{x-\alpha_2} + \cdots \frac{A_n}{x-\alpha_n}.$$

We can also now give a uniform method of determining the coefficients A, independent of the order in which they are calculated. For, multiplying both sides of equation 5) by $\varphi(x)$, we have:

$$\psi(x) = \frac{A_1}{x-\alpha_1}\varphi(x) + \frac{A_2}{x-\alpha_2}\varphi(x) + \cdots + \frac{A_n}{x-\alpha_n}\varphi(x).$$

Substituting for $\varphi(x)$ everywhere its value as a product, each denominator cancels; and if we put in for x any vanishing value α_k of φ, all terms having the factor $x - \alpha_k$ disappear, leaving the term with the coefficient A_k; so that

$$\psi(\alpha_k) = A_k \,.\, \operatorname*{Lim}_{\text{for } x=\alpha_k} \frac{\varphi(x)}{x-\alpha_k} = A_k\varphi'(\alpha_k),$$

thus

6) $$A_k = \frac{\psi(\alpha_k)}{\varphi'(\alpha_k)}.$$

$\varphi'(\alpha_k)$ cannot vanish, since $\varphi(x) = 0$ has only distinct roots. Obviously also, since $\varphi'(x)$ would vanish for a multiple root of $\varphi(x) = 0$, the present method does not apply when there are multiple roots.

Accordingly:

$$\frac{\psi(x)}{\varphi(x)} = \sum_{k=1}^{k=n} \frac{\psi(\alpha_k)}{\varphi'(\alpha_k)} \frac{1}{x-\alpha_k},$$

$$\int\frac{\psi(x)}{\varphi(x)}dx = \frac{\psi(\alpha_1)}{\varphi'(\alpha_1)}\int\frac{dx}{x-\alpha_1} + \frac{\psi(\alpha_2)}{\varphi'(\alpha_2)}\int\frac{dx}{x-\alpha_2} + \cdots + \frac{\psi(\alpha_n)}{\varphi'(\alpha_n)}\int\frac{dx}{x-\alpha_n}.$$

Each integral $\int\frac{dx}{x-\alpha_k}$ is brought to the Fundamental Integral $\int\frac{dz}{z} = l(z)$ by the substitution $x - \alpha_k = z$, $dx = dz$, thus the final formula is:

I. $$\int\frac{\psi(x)}{\varphi(x)}dx = \frac{\psi(\alpha_1)}{\varphi'(\alpha_1)}l(x-\alpha_1) + \frac{\psi(\alpha_2)}{\varphi'(\alpha_2)}l(x-\alpha_2) + \cdots$$
$$\cdots + \frac{\psi(\alpha_n)}{\varphi'(\alpha_n)}l(x-\alpha_n) + \text{Const.}$$

Examples:

1) $$\int\frac{x^3+2}{(x-1)(x+1)x}dx = \int dx + \int\frac{x+2}{(x-1)(x+1)x}dx$$
$$= x + \tfrac{3}{2}l(x-1) + \tfrac{1}{2}l(x+1) - 2l(x) + \text{Const.}$$

The definite integral can be had from this formula for every finite interval to which the vanishing points $-1, 0, +1$, do not belong.

2) $$\int\frac{x+1}{x^2+1}dx = \frac{i+1}{2i}l(x-i) + \frac{i-1}{2i}l(x+i) + C$$
$$= \tfrac{1}{2}l(x^2+1) - \frac{i}{2}l\left(\frac{x-i}{x+i}\right) + C.$$

The logarithm of a complex quantity is many-valued, but as its values differ only by an additive imaginary constant, it is indifferent in the indefinite integral which of its values is attributed to each individual logarithm. The transition to the definite integral by formation of the difference of two values of the function, requires that the value of the function at the one limit shall proceed continuously from its value at the other limit; compare the end of next Section.

112. If the coefficients of $\varphi(x)$ be real, it may have complex roots but they are conjugate in pairs (§ 90). When it has, we can prevent complex values appearing in the final formula, provided the coefficients of ψ are also real, by combining the partial fractions relative to conjugate complex roots.

Let $\alpha + i\beta$ and $\alpha - i\beta$ be two conjugate roots, then

$$\frac{\psi(\alpha + i\beta)}{\varphi'(\alpha + i\beta)} \text{ is a complex quantity: } M + iN,$$

and
$$\frac{\psi(\alpha - i\beta)}{\varphi'(\alpha - i\beta)} \text{ is the conjugate value: } M - iN;$$

therefore:

$$\frac{M + iN}{x - (\alpha + i\beta)} + \frac{M - iN}{x - (\alpha - i\beta)} = \frac{2M(x - \alpha) - 2N\beta}{(x - \alpha)^2 + \beta^2}$$

is a real quotient. The constants of the numerator can also be directly determined, by starting from the identity:

$$\frac{\psi(x)}{\varphi(x)} = \frac{Px + Q}{(x - \alpha)^2 + \beta^2} + \frac{\psi_1(x)}{\varphi_1(x)}, \text{ where } \varphi(x) = (\overline{x - \alpha}^2 + \beta^2)\varphi_1(x),$$

hence:

$$\psi(x) = (Px + Q)\,\varphi_1(x) + (\overline{x - \alpha}^2 + \beta^2)\,\psi_1(x).$$

Substituting for x the two values $\alpha \pm i\beta$ we have the equations:

$$P(\alpha \pm i\beta) + Q = \frac{\psi(\alpha \pm i\beta)}{\varphi_1(\alpha \pm i\beta)},$$

whose real and imaginary parts give two equations to determine P and Q.

The integral:

$$\int^{\cdot} \frac{Px + Q}{(x - \alpha)^2 + \beta^2}\, dx$$

can be resolved into the integrals:

$$P\int^{\cdot} \frac{(x - \alpha)\, dx}{(x - \alpha)^2 + \beta^2} + (P\alpha + Q)\int^{\cdot} \frac{dx}{(x - \alpha)^2 + \beta^2}.$$

Putting $\overline{x - \alpha}^2 + \beta^2 = z$, $(x - \alpha)dx = \frac{1}{2}dz$, the first of these becomes:

$$P\int \frac{(x - \alpha)\, dx}{(x - \alpha)^2 + \beta^2} = \frac{P}{2}\int^{\cdot} \frac{dz}{z} = \frac{P}{2}\, l(z) = \frac{P}{2}\, l(\overline{x - \alpha}^2 + \beta^2).$$

The second can be reduced, by putting $x - \alpha = \beta z$, $dx = \beta dz$, to a Fundamental Integral:

$$\frac{P\alpha+Q}{\beta}\int\frac{dz}{z^2+1}=\frac{P\alpha+Q}{\beta}\tan^{-1}\frac{x-\alpha}{\beta}.$$

Accordingly:

$$\text{II.}\quad\int\frac{Px+Q}{(x-\alpha)^2+\beta^2}\,dx=\frac{P}{2}\,l(\overline{x-\alpha}^2+\beta^2)+\frac{P\alpha+Q}{\beta}\tan^{-1}\frac{x-\alpha}{\beta}+C.$$

Treated thus, the last Example gives the value:

$$\int\frac{x+1}{x^2+1}\,dx=\tfrac{1}{2}\,l(x^2+1)+\tan^{-1}x+C.$$

Moreover the definite integral can now be calculated at once from this for any finite real interval by the formula:

$$\int_a^b\frac{x+1}{x^2+1}\,dx=\tfrac{1}{2}\,l\frac{b^2+1}{a^2+1}+\tan^{-1}b-\tan^{-1}a.$$

From the two different forms found for the same integral it is obvious that we must have for all values of x,

$$\tan^{-1}x=-\frac{i}{2}\,l\left(\frac{x-i}{x+i}\right)+\text{Const.}.$$

But this is in fact involved in § 74 which gives:

$$l\frac{x-i}{x+i}=l\,\frac{x^2-1-2xi}{x^2+1}=-i\tan^{-1}\frac{2x}{x^2-1}+C=2i\tan^{-1}x+C.$$

Example:

$$\int\frac{dx}{a+2bx+cx^2}=c\int\frac{dx}{(cx+b)^2+ac-b^2}=\frac{1}{\sqrt{ac-b^2}}\tan^{-1}\frac{cx+b}{\sqrt{ac-b^2}}+C.$$

This expression is real, provided $ac-b^2>0$, i. e. when the roots are complex. Under the same condition we have the definite integral:

$$\int_{-\infty}^{+\infty}\frac{dx}{a+2bx+cx^2}=\frac{\pi}{\sqrt{ac-b^2}},$$ otherwise this formula does not hold.

113. If the function φ have multiple roots:

$$\varphi(x)=(x-\alpha)^{\lambda_1}(x-\beta)^{\lambda_2}\cdots(x-\varkappa)^{\lambda_m},\quad \lambda_1+\lambda_2+\cdots+\lambda_m=n,$$

the foregoing process of resolution into partial fractions no longer applies; but we have now the theorem: The quotient $\frac{\psi(x)}{\varphi(x)}$ can be always resolved in one way only into the form:

$$1)\quad\frac{\psi(x)}{\varphi(x)}=\frac{A_0}{(x-\alpha)^{\lambda_1}}+\frac{\psi_1(x)}{(x-\alpha)^{\lambda_1-1}\varphi_1(x)},\quad\text{where }\varphi(x)=(x-\alpha)^{\lambda_1}\varphi_1(x),$$

A_0 denoting a constant and $\psi_1(x)$ an integer function of the order $n-2$ at most. This follows from the identity:

$$2)\quad\psi(x)=A_0\varphi_1(x)+(x-\alpha)\psi_1(x)\quad\text{or: }\psi_1(x)=\frac{\psi(x)-A_0\varphi_1(x)}{x-\alpha}.$$

In order that this quotient may be an integer function, its numerator must vanish for α; therefore we have

$$3)\qquad A_0 = \frac{\psi(\alpha)}{\varphi_1(\alpha)} = \frac{\psi(\alpha)\underline{|\lambda_1}}{\varphi^{\lambda_1}(\alpha)},$$

where $\varphi^{\lambda_1}(\alpha)$ means the value of the λ_1^{th} derived function for $x = \alpha$; the constant A_0 is neither zero nor infinite.

Applying the same process to the second quotient on the right of equation 1) we shall obtain:

$$\frac{\psi_1(x)}{(x-\alpha)^{\lambda_1-1}\varphi_1(x)} = \frac{A_1}{(x-\alpha)^{\lambda_1-1}} + \frac{\psi_2(x)}{(x-\alpha)^{\lambda_1-2}\varphi_1(x)}, \quad A_1 = \frac{\psi_1(\alpha)}{\varphi_1(\alpha)}.$$

The value of A_1, however, can vanish, since we may have $\psi_1(\alpha) = 0$. Continuing this method we obtain the equation:

$$4)\quad \frac{\psi(x)}{(x-\alpha)^{\lambda_1}\varphi_1(x)} = \frac{A_0}{(x-\alpha)^{\lambda_1}} + \frac{A_1}{(x-\alpha)^{\lambda_1-1}} + \frac{A_2}{(x-\alpha)^{\lambda_1-2}} + \cdots \frac{A_{\lambda_1-1}}{x-\alpha} + \frac{\chi(x)}{\varphi_1(x)}.$$

The quotient $\frac{\chi(x)}{\varphi_1(x)}$ in which the order of the numerator is less than that of the denominator, can be resolved by the same rule for each multiple factor of φ_1, so that we have in general and in only one way the equation:

$$5)\quad \begin{aligned}\frac{\psi(x)}{\varphi(x)} &= \frac{A_0}{(x-\alpha)^{\lambda_1}} + \frac{A_1}{(x-\alpha)^{\lambda_1-1}} + \cdots \frac{A_{\lambda_1-1}}{x-\alpha}\\ &+ \frac{B_0}{(x-\beta)^{\lambda_2}} + \frac{B_1}{(x-\beta)^{\lambda_2-1}} + \cdots \frac{B_{\lambda_2-1}}{x-\beta}\\ &\quad \cdots\cdots\cdots\cdots\\ &+ \frac{K_0}{(x-\varkappa)^{\lambda_m}} + \frac{K_1}{(x-\varkappa)^{\lambda_m-1}} + \cdots \frac{K_{\lambda_m-1}}{x-\varkappa}.\end{aligned}$$

The constants $A_0, B_0 \ldots K_0$ do not vanish, any of the others may.

The best way of determining the coefficients A in equation 4) is to put $x - \alpha = h$, then since:

$$\frac{\psi(\alpha+h)}{\varphi_1(\alpha+h)} = A_0 + A_1 h + A_2 h^2 + \cdots + A_{\lambda_1-1}h^{\lambda_1-1} + \frac{\chi(\alpha+h)}{\varphi_1(\alpha+h)}\cdot h^{\lambda_1};$$

$A_0, A_1, \ldots A_{\lambda_1-1}$ are the first λ_1 coefficients when the quotient on the left is expanded in ascending powers of h; when this is done by Taylor's series, denoting $\frac{\psi(\alpha)}{\varphi_1(\alpha)}$ by $\Phi_1(\alpha)$ we have:

$$A_0 = \Phi_1(\alpha),\ A_1 = \Phi_1'(\alpha),\ A_2 = \frac{1}{1.2}\Phi_1''(\alpha), \ldots A_{\lambda_1-1} = \frac{1}{\underline{|\lambda_1-1}}\Phi_1^{\lambda_1-1}(\alpha).$$

Hence in the general case:

$$\int\frac{\psi(x)}{\varphi(x)}dx = \sum_{j=0}^{j=\lambda_1-1} A_j\int\frac{dx}{(x-\alpha)^{\lambda_1-j}} + \sum_{j=0}^{j=\lambda_2-1} B_j\int\frac{dx}{(x-\beta)^{\lambda_2-j}} + \cdots$$

$$\cdots + \sum_{j=0}^{j=\lambda_m-1} K_j\int\frac{dx}{(x-\varkappa)^{\lambda_m-j}}.$$

But,

$$\int\frac{dx}{(x-\alpha)^\lambda}=-\frac{1}{\lambda-1}\cdot\frac{1}{(x-\alpha)^{\lambda-1}} \text{ if } \lambda>1, \text{ and } \int\frac{dx}{x-\alpha}=l(x-\alpha),$$

hence:

$$\text{III.}\qquad \int\frac{\psi(x)}{\varphi(x)}\,dx=\sum_{j=0}^{j=\lambda_1-2}-\frac{A_j}{\lambda_1-j-1}\cdot\frac{1}{(x-\alpha)^{\lambda_1-j-1}}+A_{\lambda_1-1}l(x-\alpha)$$
$$+\;\cdot\;\cdot\;\cdot\;\cdot\;\cdot\;\cdot\;\cdot\;\cdot\;\cdot\;\cdot\;\cdot\;\cdot\;\cdot\;\cdot\;\cdot\;\cdot$$
$$+\sum_{j=0}^{j=\lambda_m-2}-\frac{K_j}{\lambda_m-j-1}\cdot\frac{1}{(x-\varkappa)^{\lambda_m-j-1}}+K_{\lambda_m-1}l(x-\varkappa).$$

The integral of any rational algebraic function $\frac{\psi(x)}{\varphi(x)}$ *can be expressed by a combination of rational algebraic functions and logarithms; to integrate it explicitly the vanishing values of* $\varphi(x)$ *must be determined; moreover circular functions can be introduced instead of the logarithms.*

Example:

$$\frac{\psi(x)}{\varphi(x)}=\frac{2x^4-3x^3-x+21}{(x+1)^3(x-2)^2}.$$

If we put $x=-1+h$, then

$$\frac{\psi(-1+h)}{\varphi_1(-1+h)}=\frac{27-18h+21h^2-11h^3+2h^4}{9-6h+h^2}=3+2h^2+\frac{h^3}{9-6h+h^2},$$

$$\frac{\psi(x)}{\varphi(x)}=\frac{3}{(x+1)^3}+\frac{2}{(x+1)}+\frac{1}{(x-2)^2},$$

$$\int\frac{\psi(x)}{\varphi(x)}\,dx=-\tfrac{3}{2}\cdot\frac{1}{(x+1)^2}+2l(x+1)-\frac{1}{x-2}+C.$$

114. It is also possible in the general case just treated to prevent complex quantities appearing in the ultimate result, if, when φ has complex roots, the coefficients of φ and ψ are real; here again the partial fractions belonging to conjugate complex roots are to be combined. We start directly from the theorem:

When $\varphi(x)=(\overline{x-\alpha}^2+\beta^2)^\lambda\varphi_1(x)$, there is only a single way in which to effect the resolution:

$$\frac{\psi(x)}{\varphi(x)}=\frac{P_0x+Q_0}{(\overline{x-\alpha}^2+\beta^2)^\lambda}+\frac{\psi_1(x)}{(\overline{x-\alpha}^2+\beta^2)^{\lambda-1}\varphi_1(x)},$$

where P_0 and Q_0 signify real constants.

We must have

$$\psi(x)=(P_0x+Q_0)\varphi_1(x)+\psi_1(x)\,(\overline{x-\alpha}^2+\beta^2),$$

i. e.

$$P_0(\alpha\pm i\beta)+Q_0=\frac{\psi(\alpha\pm i\beta)}{\varphi(\alpha\pm i\beta)}.$$

Accordingly:

$$\frac{\psi(x)}{\varphi(x)}=\frac{P_0x+Q_0}{(\overline{x-\alpha}^2+\beta^2)^\lambda}+\frac{P_1x+Q_1}{(\overline{x-\alpha}^2+\beta^2)^{\lambda-1}}+\cdots+\frac{P_{\lambda-1}x+Q_{\lambda-1}}{\overline{x-\alpha}^2+\beta^2}+\frac{\chi(x)}{\varphi_1(x)}.$$

An integral of the form:

$$\int \frac{(Px+Q)\,dx}{((x-\alpha)^2+\beta^2)^\lambda}$$

when $\lambda > 1$, consists of the two parts:

$$P\int \frac{(x-\alpha)\,dx}{((x-\alpha)^2+\beta^2)^\lambda} + (P\alpha+Q)\int \frac{dx}{((x-\alpha)^2+\beta^2)^\lambda}.$$

The value of the first of these is at once:

$$P\int \frac{(x-\alpha)\,dx}{((x-\alpha)^2+\beta^2)^\lambda} = -\frac{P}{2\lambda-2}\cdot\frac{1}{((x-\alpha)^2+\beta^2)^{\lambda-1}}.$$

Let us convert the second by substituting $x-\alpha = y$, into the integral:

$$\int \frac{dy}{(y^2+\beta^2)^\lambda}.$$

Then since:

$$\begin{aligned} d\left(\frac{y}{(y^2+\beta^2)^{\lambda-1}}\right) &= \frac{dy}{(y^2+\beta^2)^{\lambda-1}} - 2(\lambda-1)\,\frac{y^2\,dy}{(y^2+\beta^2)^\lambda} \\ &= \frac{dy}{(y^2+\beta^2)^{\lambda-1}} - 2(\lambda-1)\,\frac{(y^2+\beta^2)-\beta^2}{(y^2+\beta^2)^\lambda}\,dy \\ &= -(2\lambda-3)\,\frac{dy}{(y^2+\beta^2)^{\lambda-1}} + (2\lambda-2)\beta^2\,\frac{dy}{(y^2+\beta^2)^\lambda}, \end{aligned}$$

we arrive at the formula:

$$\text{IV.}\quad \int \frac{dy}{(y^2+\beta^2)^\lambda} = \frac{1}{\beta^2(2\lambda-2)}\,\frac{y}{(y^2+\beta^2)^{\lambda-1}} + \frac{2\lambda-3}{\beta^2(2\lambda-2)}\int \frac{dy}{(y^2+\beta^2)^{\lambda-1}}.$$

Repetition of this recurring formula:

$$\int \frac{dy}{(y^2+\beta^2)^{\lambda-1}} = \frac{1}{\beta^2(2\lambda-4)}\,\frac{y}{(y^2+\beta^2)^{\lambda-2}} + \frac{2\lambda-5}{\beta^2(2\lambda-4)}\int \frac{dy}{(y^2+\beta^2)^{\lambda-2}},\ \text{etc.}$$

reduces the proposed integral to algebraic functions and to the integral:

$$\int \frac{dy}{y^2+\beta^2} = \frac{1}{\beta}\tan^{-1}\frac{y}{\beta} + C.$$

Example:

$$\int \frac{x+1}{(x^2+1)^3}\,dx = \int \frac{x\,dx}{(x^2+1)^3} + \int \frac{dx}{(x^2+1)^3}$$

$$= -\tfrac{1}{4}\cdot\frac{1}{(x^2+1)^2} + \tfrac{1}{4}\cdot\frac{x}{(x^2+1)^2} + \tfrac{3}{8}\cdot\frac{x}{x^2+1} + \tfrac{3}{8}\tan^{-1}x + C.$$

115. To integrate:

$$\int \frac{x^{m-1}\,dx}{x^n+e^{\alpha i}}\ \text{*)}$$

$$i = \sqrt{-1},\qquad m-1 < n,\qquad -\pi \leq \alpha \leq +\pi.$$

The n roots of the denominator are complex and all are different:

*) Euler: Inst. calculi integr., Vol. I., Cap. I, § 77—80., Cap. VIII, § 351—355. Dirichlet: Vorlesungen über die Theorie der bestimmten Integrale. Bearbeitet von Meyer. Leipzig 1871. § 27.

$$x = \sqrt[n]{-e^{\alpha i}} = e^{i\frac{\alpha+(2k+1)\pi}{n}} = \cos\frac{\alpha+\overline{2k+1}\pi}{n} + i\sin\frac{\alpha+\overline{2k+1}\pi}{n},$$

$$k = 0, 1, 2, \ldots n-1.$$

Accordingly the type of the constant numerator of its partial fractions is:

$$A_k = \frac{1}{n} e^{i(m-n)\frac{\alpha+\overline{2k+1}\pi}{n}} = -\frac{1}{ne^{\left(1-\frac{m}{n}\right)\alpha i}} e^{(2k+1)\frac{m\pi i}{n}},$$

therefore:

$$\int \frac{x^{m-1}dx}{x^n+e^{\alpha i}} = -\frac{1}{ne^{\left(1-\frac{m}{n}\right)\alpha i}} \sum_{k=0}^{k=n-1} e^{(2k+1)\frac{m\pi i}{n}} l\left(x - e^{i\frac{\alpha+\overline{2k+1}\pi}{n}}\right) + C.$$

If we separate the logarithm of the complex quantity (§ 74) into its real and imaginary parts, supposing x real, and call:

$$P_k = \tfrac{1}{2} l\left(x^2 - 2x\cos\frac{\alpha+\overline{2k+1}\pi}{n} + 1\right),$$

$$Q_k = \tan^{-1}\frac{\sin\frac{\alpha+\overline{2k+1}\pi}{n}}{\cos\frac{\alpha+\overline{2k+1}\pi}{n} - x};$$

incorporating the multiples of πi in the arbitrary constant we may write this indefinite integral:

$$\int \frac{x^{m-1}dx}{x^n+e^{\alpha i}} = -\frac{1}{ne^{\left(1-\frac{m}{n}\right)\alpha i}} \sum_{k=0}^{k=n-1} e^{(2k+1)\frac{m\pi i}{n}} (P_k + iQ_k) + C.$$

Let the definite integral be required between the limits zero and ∞, when all roots of the denominator are complex or negative, thus $-\pi < \alpha < +\pi$. To obtain it, P_k and Q_k have to be determined at the limits 0 and ∞: they must however remain continuous in the entire range from $x = 0$ to $x = \infty$.

The value of each P vanishes for $x = 0$ and continuously increases as the values of x increase. But inasmuch as

$$x^2 > 2x\cos\frac{\alpha+\overline{2k+1}\pi}{n} - 1,$$

if we write:

$$P_k = \tfrac{1}{2} l(x^2) + \tfrac{1}{2} l\left(1 - \frac{2x\cos\frac{\alpha+\overline{2k+1}\pi}{n} - 1}{x^2}\right),$$

and combine all terms multiplied by $\frac{1}{2} l(x^2)$ in the sum, they amount to

$$\tfrac{1}{2} l(x^2) \sum_{k=0}^{k=n-1} e^{(2k+1)\frac{m\pi i}{n}} = \tfrac{1}{2} l(x^2) \frac{1 - e^{2n\frac{m\pi i}{n}}}{1 - e^{\frac{2m\pi i}{n}}} \cdot e^{\frac{m\pi i}{n}} = 0;$$

this formula is no longer valid when $m = n$. In this case the integral between infinite limits is no longer finite; hence we introduce the hypothesis that m is at most equal to $n-1$.

The second part only of each P_k has now to be considered in the summation and all of them converge to zero for $x=\infty$.

In order to determine the value of Q_k let us observe, that since α was assumed $<\pi$, we have

$$\frac{\alpha+\overline{2k+1}\pi}{n}=\mu \quad <2\pi$$

for all values of $k=0, 1, \ldots, n-1$. For $x=0$, $Q_k=\tan^{-1}(\tan\mu)$; let us put it equal to μ in whatever quadrant this be.

If μ lie in the first quadrant ($\cos\mu>0$, $\sin\mu>0$), Q_k increases as x does, it becomes $+\frac{1}{2}\pi$ for $x=\cos\mu$. For greater values of x the argument is negative, for $x=\infty$ we have $Q_k=\pi$.

If μ be in the second quadrant ($\cos\mu<0$, $\sin\mu>0$), the argument is throughout negative, for $x=\infty$, $Q_k=\pi$.

If μ be in the third quadrant ($\cos\mu<0$, $\sin\mu<0$), the argument is throughout positive, for $x=\infty$, $Q_k=\pi$.

If μ be in the fourth quadrant ($\cos\mu>0$, $\sin\mu<0$), Q_k decreases as x increases, it is $\frac{3}{2}\pi$ for $x=\cos\mu$; it becomes less than $\frac{3}{2}\pi$ for greater values of x and for $x=\infty$, $Q_k=\pi$. Accordingly:

$$\int_0^\infty \frac{x^{m-1}dx}{x^n+c^{\alpha i}}=-\frac{i}{nc^{\alpha i}}\sum_{k=0}^{k=n-1}\left(\pi-\frac{\alpha+2k+1\pi}{n}\right)c^{\frac{\alpha+\overline{2k+1}\pi}{n}mi}.$$

The summation is carried out as follows: we have for any value of $\lambda<2\pi$

$$\sum_{k=0}^{k=n-1}c^{k\lambda i}=\frac{c^{n\lambda i}-1}{c^{\lambda i}-1},\qquad \sum_{k=0}^{k=n-1}kc^{k\lambda i}=\frac{nc^{n\lambda i}(c^{\lambda i}-1)-c^{\lambda i}(c^{n\lambda i}-1)}{(c^{\lambda i}-1)^2}.$$

The second formula results by differentiating the first with respect to λ. Putting $n\lambda=2m\pi$ we obtain:

$$\sum_{k=0}^{k=n-1}c^{\frac{2km\pi i}{n}}=\frac{e^{2m\pi i}-1}{e^{\frac{2m\pi i}{n}}-1}=0,\quad \sum_{k=0}^{k=n-1}kc^{\frac{2km\pi i}{n}}=\frac{n}{e^{\frac{2m\pi i}{n}}-1}=\frac{ne^{-\frac{m\pi i}{n}}}{2i\sin\frac{m\pi}{n}}.$$

Accordingly when $m<n$:

$$\int_0^\infty\frac{x^{m-1}dx}{x^n+c^{\alpha i}}=\frac{2\pi ic^{\frac{(\alpha+\pi)mi}{n}}}{n^2c^{\alpha i}}\cdot\sum_{k=0}^{k=n-1}kc^{\frac{2km\pi i}{n}}=\frac{\pi}{n}\cdot\frac{c^{-\left(1-\frac{m}{n}\right)\alpha i}}{\sin\frac{m\pi}{n}}.$$

Substituting z for x^n in this formula and writing a for the proper rational fraction $\frac{m}{n}$, we obtain the fundamental definite integral:

$$\int_0^\infty\frac{z^{a-1}dz}{z+c^{\alpha i}}=\frac{\pi}{\sin a\pi}c^{(a-1)\alpha i},\quad -\pi<\alpha<+\pi;$$ see further § 159.

116. An expression of the form:

$$\int f(x^a, x^b, \ldots x^l)\,dx$$

in which $a, b, \ldots l$ are rational fractions, and f the sign of functionality denotes a rational algebraic combination of the quantities involved, is reduced to the form of a rational integral by a simple substitution.*) If m be the least common multiple of the denominators of $a, b, \ldots l$, and we put $x = z^m$, $dx = mz^{m-1}dz$, then:

$$\int f(x^a, x^b, \ldots x^l)\,dx = m\int f(z^{am}, z^{bm}, \ldots z^{lm})\,z^{m-1}\,dz$$

is the integral of a rational function.

Example:

$$\int \frac{1+\sqrt{x}-\sqrt[3]{x^2}}{1+\sqrt[3]{x}}\,dx = 6\int \frac{1+z^3-z^4}{1+z^2}\,z^5 dz, \qquad z = x^{\frac{1}{6}},$$

$$= 6\int\left(-z^7+z^6+z^5-z^4+z^2-1+\frac{1}{1+z^2}\right)dz$$

$$= -\tfrac{3}{4}z^8+\tfrac{6}{7}z^7+z^6-\tfrac{6}{5}z^5+2z^3-6z+6\tan^{-1}z+C;$$

and replacing the value of z:

$$\int \frac{1+\sqrt{x}-\sqrt[3]{x^2}}{1+\sqrt[3]{x}}\,dx = -\tfrac{3}{4}x\sqrt[3]{x}+\tfrac{6}{7}x\sqrt[6]{x}+x-\tfrac{6}{5}\sqrt[6]{x^5}$$

$$+2\sqrt{x}-6\sqrt[6]{x}+6\tan^{-1}\sqrt[6]{x}+C.$$

*) Euler: Instit. calc. integr., Cap. I, § 27.

Third Chapter.

The integral of explicitly irrational functions.

117. The function to be integrated is said to be explicitly irrational, when, besides integer or fractional powers of the variable x (see last Paragraph), it contains polynomials in x which are affected with fractional exponents. The simplest of these is the linear binomial and it gives rise to:

$$\int f\{x,\ (a+bx)^{\frac{p}{q}}\}\, dx,$$

where p and q denote integers relatively prime. This is converted into the integral of a rational function by substituting:

$$a+bx=z^q, \qquad b\,dx=qz^{q-1}dz,$$

thus:

$$\int f\{x,\ (a+bx)^{\frac{p}{q}}\}\, dx=\frac{q}{b}\int f\left(\frac{z^q-a}{b},\ z^p\right)z^{q-1}dz.$$

We style an integral of the form:

$$\int x^{m-1}(a+bx^n)^p dx,$$

a general binomial integral, where m, n, p stand for any rational fractions.*) Without any restriction of generality, however, we may always suppose m and n integers; for if they be fractions whose denominators have ν as least common multiple, we can put z^ν for x. Further, n may be always assumed positive, for otherwise we may write:

$$x^{m-1}(a+bx^n)^p dx = x^{m+np-1}(ax^{-n}+b)^p dx.$$

If now p also is an integer, the function is at once rational; but when p is a fraction, the function can in certain cases be brought to a rational form. Putting:

$$a+bx^n=z, \quad x=\left\{\frac{z-a}{b}\right\}^{\frac{1}{n}}, \quad dx=\frac{1}{bn}\left\{\frac{z-a}{b}\right\}^{\frac{1}{n}-1}dz,$$

we obtain:

$$\frac{1}{bn}\int\left\{\frac{z-a}{b}\right\}^{\frac{m}{n}-1}z^p dz; \text{ or: } \frac{1}{bn}\int\left\{\frac{1-az^{-1}}{b}\right\}^{\frac{m}{n}-1}z^{p+\frac{m}{n}-1}dz.$$

*) Euler: *loc. cit.*, Cap. II, § 103—124.

When $\frac{m}{n}$ is an integer, the first expression becomes rational by substituting $z = t^q$, q being the denominator of the fraction p. When $p + \frac{m}{n}$ is an integer, the second expression becomes rational by substituting:

$$\frac{1 - az^{-1}}{b} = t^n.$$

118. Now although it is only under these determinate conditions that the binomial integral can be expressed by explicitly irrational algebraic functions and by logarithms*), yet in all cases its integration can be reduced to that of certain simple forms.

Consider the differential $x^{m-1}(a + bx^n)^p dx$ as a product, either of:

$$\frac{d(x^m)}{m} \cdot (a + bx^n)^p, \text{ or of: } \frac{d(a + bx^n)^{p+1}}{nb(p+1)} \cdot x^{m-n},$$

then integrating by parts:

$$\text{I. } \int x^{m-1}(a + bx^n)^p dx = \frac{x^m(a+bx^n)^p}{m} - \frac{bpn}{m}\int x^{m+n-1}(a+bx^n)^{p-1} dx,$$

$$\text{II. } \int x^{m-1}(a+bx^n)^p dx = \frac{x^{m-n}(a+bx^n)^{p+1}}{nb(p+1)} - \frac{m-n}{nb(p+1)}\int x^{m-n-1}(a+bx^n)^{p+1} dx.$$

We can alter these formulas so that in each integral on the right, one exponent only shall be changed. In equation I. let us write:

$$\int x^{m+n-1}(a+bx^n)^{p-1} dx = \frac{1}{b}\int x^{m-1}(a+bx^n)^p dx - \frac{a}{b}\int x^{m-1}(a+bx^n)^{p-1} dx,$$

also in equation II.:

$$\int x^{m-n-1}(a+bx^n)^{p+1} dx = a\int x^{m-n-1}(a+bx^n)^p dx + b\int x^{m-1}(a+bx^n)^p dx,$$

and combine in each case the integral on the right side with the equal integral on the left; we find:

$$\text{III. } \int x^{m-1}(a + bx^n)^p dx = \frac{x^m(a+bx^n)^p}{m+pn} + \frac{apn}{m+pn}\int x^{m-1}(a+bx^n)^{p-1} dx,$$

$$\text{IV. } \int x^{m-1}(a + bx^n)^p dx = \frac{x^{m-n}(a + bx^n)^{p+1}}{(np+m)b} - \frac{(m-n)a}{(np+m)b}\int x^{m-n-1}(a + bx^n)^p dx.$$

Thus the formulas III. and IV. diminish the exponents p and m by the quantities 1 and n. These formulas can also be regarded as reducing the integrals on the right to those on the left. For uniformity let us solve each equation for the integral on the right, having first put p for $p - 1$ in III. and m for $m - n$ in IV. the results are:

*) Tschebychef: Sur l'intégration de différentielles irrationnelles. Liouville Journal, T. XVIII. 1853.

V. $$\int x^{m-1}(a+bx^n)^p\,dx = -\frac{x^m(a+bx^n)^{p+1}}{an(p+1)} + \frac{m+n(p+1)}{an(p+1)}\int x^{m-1}(a+bx^n)^{p+1}dx,$$

VI. $$\int x^{m-1}(a+bx^n)^p\,dx = \frac{x^m(a+bx^n)^{p+1}}{am} - \frac{(np+m+n)b}{am}\int x^{m+n-1}(a+bx^n)^p\,dx.$$

These formulas of reduction become useless when the quantities in the denominators vanish; but in all such cases the integral becomes rational by the substitutions already assigned; the only cases worthy of special notice here are $m=0$ and $np+m=0$; for these:

$$\int\frac{(a+bx^n)^p}{x}dx = \frac{1}{n}\int\frac{z^p}{z-a}dz, \qquad (a+bx^n=z);$$

$$\int x^{m-1}(a+bx^n)^{-\frac{m}{n}}dx = \frac{1}{bn}\int\left(\frac{1-az^{-1}}{b}\right)^{\frac{m}{n}-1}\frac{dz}{z} = \int\frac{t^{m-1}dt}{1-t^n b},$$

$$\left(\frac{x^n}{a+bx^n}=t^n\right).$$

Formulas IV. and VI. show that the binomial integral can be expressed by algebraic functions and by an integral in which m the exponent of x is between zero and n, and therefore the ratio $m:n$ between 0 and 1.

The forms III. and V. show that the binomial integral can be expressed by algebraic functions and by an integral in which the exponent p is likewise a positive fraction between zero and 1.

Finally if neither $\frac{m}{n}$ vanish, nor the two fractions $\frac{m}{n}$ and p supplement to unity, the value of the integral can be expressed only by an infinite series, by expanding in a series of powers the binomial of the function to be integrated, and then forming the integral of this series multiplied by x^{m-1} (Chap. IV).

119. The group of irrational functions next in order consists of those in which the square root of a polynomial of the second degree enters*):

$$\int F(x,\sqrt{R})dx, \qquad R=a+2bx+cx^2,$$

where F means a rational combination of x and $\sqrt{R}$. The sign of $\sqrt{R}$ is to be taken positively unless the opposite sign is prefixed. Arranged by its rational and irrational parts we have:

*) Euler: *loc. cit.*, Cap. II, § 88—93.

$$F(x, \sqrt{R}) = \frac{G(x) + H(x)\sqrt{R}}{G_1(x) + H_1(x)\sqrt{R}},$$

G, H, G_1, H_1 signifying integer rational functions. If we rationalise the denominator of this quotient by multiplying by

$$G_1(x) - H_1(x)\sqrt{R},$$

we have:

$$F(x, \sqrt{R}) = \varphi(x) + \psi(x)\sqrt{R},$$

φ and ψ denoting rational functions. We are concerned henceforth only with the irrational part, which may be presented in the form:

$$\frac{\psi(x) \cdot R}{\sqrt{R}} = \frac{f(x)}{\sqrt{R}} = \sum_{n=0} \frac{A_n x^n}{\sqrt{R}} + \sum \frac{B_n}{(x - \varrho)^n} \cdot \frac{1}{\sqrt{R}},$$

for, the rational function $f(x)$ breaks up into an integer function and a proper fraction that can be resolved into partial fractions.

The integral $\int \frac{x^n dx}{\sqrt{R}}$ is reduced by means of a recurring formula to algebraic functions and to the integral $\int \frac{dx}{\sqrt{R}}$. In fact:

$$d(x^{n-1}\sqrt{a + 2bx + cx^2})$$

$$= \left(\overline{n-1}\, x^{n-2}\sqrt{a + 2bx + cx^2} + \frac{x^{n-1}(b + cx)}{\sqrt{a + 2bx + cx^2}}\right) dx$$

$$= \left(\frac{\overline{n-1}\, x^{n-2}(a + 2bx + cx^2)}{\sqrt{R}} + \frac{x^{n-1}(b + cx)}{\sqrt{R}}\right) dx,$$

hence by integration:

$$\text{I.}\quad x^{n-1}\sqrt{R} = \overline{n-1}\, a \int \frac{x^{n-2} dx}{\sqrt{R}} + \overline{2n-1}\, b \int \frac{x^{n-1} dx}{\sqrt{R}} + nc \int \frac{x^n dx}{\sqrt{R}}.$$

Putting for n the values 1, 2, 3 ... in succession, we have:

$$\int \frac{x\,dx}{\sqrt{R}} = \frac{1}{c}\sqrt{R} - \frac{b}{c}\int \frac{dx}{\sqrt{R}},$$

$$\int \frac{x^2 dx}{\sqrt{R}} = \frac{x}{2c}\sqrt{R} - \frac{3b}{2c}\int \frac{x\,dx}{\sqrt{R}} - \frac{a}{2c}\int \frac{dx}{\sqrt{R}}$$

$$= \left(\frac{x}{2c} - \frac{3b}{2c^2}\right)\sqrt{R} + \left(\frac{3b^2}{2c^2} - \frac{a}{2c}\right)\int \frac{dx}{\sqrt{R}},$$

$$\int \frac{x^3 dx}{\sqrt{R}} = \frac{x^2}{3c}\sqrt{R} - \frac{5b}{3c}\int \frac{x^2 dx}{\sqrt{R}} - \frac{2a}{3c}\int \frac{x\,dx}{\sqrt{R}}$$

$$= \left(\frac{x^2}{3c} - \frac{5bx}{6c^2} + \frac{15b^2 - 4ac}{6c^3}\right)\sqrt{R} - \left(\frac{5b^3}{2c^3} - \frac{3ab}{2c^2}\right)\int \frac{dx}{\sqrt{R}}.$$

In general:

$$\text{II.}\quad \int \frac{x^n dx}{\sqrt{R}} = \sqrt{R} \cdot \sum_{\nu=1}^{\nu=n} \alpha_\nu x^{n-\nu} + \beta \int \frac{dx}{\sqrt{R}}.$$

The coefficients α_ν and β are determined by recurring formulas that can be found directly by differentiating this equation, thus:

$$\frac{x^n}{\sqrt{R}} = \frac{b+cx}{\sqrt{R}}\sum_{\nu=1}^{\nu=n}\alpha_\nu x^{n-\nu} + \sqrt{R}\sum_{\nu=1}^{\nu=n}\alpha_\nu(n-\nu)x^{n-\nu-1} + \frac{\beta}{\sqrt{R}},$$

or, multiplying both sides by $\sqrt{R}$ and arranging by a, b, c:

$$x^n = c\sum_{\nu=1}^{\nu=n}\alpha_\nu(n-\nu+1)x^{n-\nu+1} + b\sum_{\nu=1}^{\nu=n}\alpha_\nu(2n-2\nu+1)x^{n-\nu}$$
$$+ a\sum_{\nu=1}^{\nu=n}\alpha_\nu(n-\nu)x^{n-\nu-1} + \beta.$$

Hence, comparing coefficients we have the relations:

$$1 = c\alpha_1 n, \quad 0 = c\alpha_2(n-1) + b\alpha_1(2n-1),$$
$$0 = c\alpha_3(n-2) + b\alpha_2(2n-3) + a\alpha_1(n-1),$$

in general:

$$\text{III.}\quad \begin{aligned} 0 &= c\alpha_\nu(n-\nu+1) + b\alpha_{\nu-1}(2n-2\nu+3) + a\alpha_{\nu-2}(n-\nu+2), \\ \beta &= -b\alpha_n - a\alpha_{n-1}. \end{aligned}$$

Note: When $b=0$, $R = a + cx^2$, then the middle term in the recurring formula for α_ν disappears; consequently α_2, α_4, and in general every even α vanishes. Further we have:

$$0 = c\alpha_{2m+1}(n-2m) + a\alpha_{2m-1}(n-2m+1)$$

therefore:

$$\alpha_1 = \frac{1}{cn}, \quad \alpha_3 = -\frac{a(n-1)}{c^2 n(n-2)}, \quad \alpha_5 = \frac{a^2}{c^3}\,\frac{(n-1)(n-3)}{n(n-2)(n-4)},$$

$$\alpha_{2m+1} = (-1)^m \cdot \frac{a^m}{c^{m+1}}\,\frac{(n-1)(n-3)\ldots(n-2m+1)}{n(n-2)(n-4)\ldots(n-2m)}.$$

When n is odd, β vanishes; but when n is even $=2p$, we have:

$$\beta = -a\alpha_{2p-1} = (-1)^p\,\frac{a^p(2p-1)(2p-3)\ldots 5.3}{c^p\,2p(2p-2)(2p-4)\ldots 4.2}.$$

In the first of these cases the integral

$$\int\frac{x^{2p+1}dx}{\sqrt{a+cx^2}} = \frac{1}{2}\int\frac{z^p dz}{\sqrt{a+cz}}$$

reduces to algebraic functions only; in the second case to algebraic functions along with

$$\beta\int\frac{dx}{\sqrt{a+cx^2}}.$$

120. The integral

$$\int\frac{dx}{\sqrt{a+2bx+cx^2}}$$

can always be given a rational form. When the coefficients a, b, c

are real, we distinguish between the cases when c is positive and negative, and endeavour whenever possible to express the integral function by means of real quantities only.

1) When $c > 0$, let us substitute:

$$\sqrt{a + 2bx + cx^2} = z - x\sqrt{c}$$

$$x = \frac{z^2 - a}{2(z\sqrt{c} + b)}, \quad \sqrt{R} = \frac{(z^2 + a)\sqrt{c} + 2bz}{2(z\sqrt{c} + b)},$$

thus both x and $\sqrt{R}$ are expressed rationally and uniquely by z. From the equation:

$$a + 2bx = z^2 - 2zx\sqrt{c}$$

it follows that:

$$b\,dx = z\,dz - x\,dz\sqrt{c} - z\,dx\sqrt{c}, \quad (b + z\sqrt{c})\,dx = (z - x\sqrt{c})\,dz,$$

or:

$$\frac{dx}{\sqrt{R}} = \frac{dz}{b + z\sqrt{c}}.$$

Hence we have:

I.
$$\int \frac{dx}{\sqrt{R}} = \int \frac{dz}{b + z\sqrt{c}} = \frac{1}{\sqrt{c}}\, l(b + z\sqrt{c}) + C$$
$$= \frac{1}{\sqrt{c}}\, l(b + \sqrt{Rc} + xc) + C.$$

2) When $c < 0$, if $a > 0$, let us substitute:

$$\sqrt{a + 2bx + cx^2} = \sqrt{a} + zx$$

$$x = \frac{2(z\sqrt{a} - b)}{c - z^2}, \quad \sqrt{R} = \frac{\sqrt{a}(c + z^2) - 2bz}{c - z^2}, \quad \frac{dx}{\sqrt{R}} = \frac{2dz}{c - z^2}.$$

Therefore:

II.
$$\int \frac{dx}{\sqrt{R}} = 2\int \frac{dz}{c - z^2} = \frac{2}{c}\int \frac{dz}{1 + \left(\frac{z}{\sqrt{-c}}\right)^2} = -\frac{2}{\sqrt{-c}} \tan^{-1} \frac{z}{\sqrt{-c}} + C$$
$$= -\frac{2}{\sqrt{-c}} \tan^{-1} \frac{\sqrt{R} - \sqrt{a}}{x\sqrt{-c}} + C.$$

If $a < 0$ and the roots of $a + 2bx + cx^2 = 0$ are complex, so that $ac - b^2 > 0$, R is negative for all real values of x, and $\sqrt{R}$ is imaginary; in this case there is no way of avoiding the introduction of imaginary constants in expressing the integral function.

But if both roots α and β be real, then for values of x between α and β, $\sqrt{R}$ is real; in this case we can give the integral function a real form by substituting:

3) $R = a + 2bx + cx^2 = c(x - \alpha)(x - \beta), \quad \sqrt{R} = (x - \alpha)z,$

consequently:

$$x = \frac{c\beta - \alpha z^2}{c - z^2}, \quad \sqrt{R} = \frac{c(\beta - \alpha)z}{c - z^2}, \quad \frac{dx}{\sqrt{R}} = \frac{2dz}{c - z^2}.$$

Thus we have:

$$\text{III.} \int \frac{dx}{\sqrt{R}} = 2\int \frac{dz}{c - z^2} = \frac{2}{c}\int \frac{dz}{1 + \left(\frac{z}{\sqrt{-c}}\right)^2} = -\frac{2}{\sqrt{-c}} \tan^{-1} \frac{z}{\sqrt{-c}} + C$$

$$= -\frac{2}{\sqrt{-c}} \tan^{-1} \frac{\sqrt{R}}{(x-\alpha)\sqrt{-c}} + C.$$

These substitutions can also be employed from the very beginning to convert

$$\int F(x, \sqrt{R})\, dx$$

into the integral of a rational function.

Note. The case $c = 0$ reduces to the simplest binomial integral, the integral then becomes:

$$\int \frac{dx}{\sqrt{a + 2bx}} = \frac{1}{b}\sqrt{a + 2bx} + C.$$

121. In order to determine the integral:

$$\int \frac{dx}{(x - \rho)^n \sqrt{R}},$$

let us begin by investigating $\int \frac{dx}{x^n \sqrt{R}}$; we find a recurring formula for it from:

$$d\left(\frac{\sqrt{R}}{x^{n-1}}\right) = -\frac{\overline{n-1}\sqrt{R}}{x^n}\, dx + \frac{(b + cx)\, dx}{x^{n-1}\sqrt{R}}$$

$$= -\frac{\overline{n-1}\,(a + 2bx + cx^2)}{x^n \sqrt{R}}\, dx + \frac{(b + cx)\, dx}{x^{n-1}\sqrt{R}},$$

$$= -\frac{(n-1)a\, dx}{x^n \sqrt{R}} - \frac{(2n-3)b\, dx}{x^{n-1}\sqrt{R}} - \frac{(n-2)c\, dx}{x^{n-2}\sqrt{R}};$$

thus if $a \gtrless 0$:

$$\text{Ia.} \int \frac{dx}{x^n \sqrt{R}} = -\frac{1}{(n-1)a}\frac{\sqrt{R}}{x^{n-1}} - \frac{2n-3}{n-1}\frac{b}{a}\int \frac{dx}{x^{n-1}\sqrt{R}} - \frac{n-2}{n-1}\frac{c}{a}\int \frac{dx}{x^{n-2}\sqrt{R}}.$$

When we put for n the values 2, 3, 4, . . . in succession, the integral reduces ultimately to the form $\int \frac{dx}{x\sqrt{R}}$.

The formula is generalised by substituting $x = z - \rho$; then,

$$R = A + 2Bz + Cz^2 = (a - 2b\rho + c\rho^2) + 2(b - c\rho)z + cz^2,$$

thus:

$$c = C, \qquad b = B + C\rho, \qquad a = A + 2B\rho + C\rho^2.$$

Now write on both sides a, b, c, for A, B, C, and for z put x, also for brevity write:

$$a + 2b\rho + c\rho^2 = f(\rho), \quad b + c\rho = \tfrac{1}{2}f'(\rho), \qquad c = \tfrac{1}{2}f''(\rho),$$

then, provided $f(\rho)$ does not vanish, we have:

I. $$\int \frac{dx}{(x-\varrho)^n\sqrt{R}} = -\frac{1}{(n-1)f(\varrho)}\frac{\sqrt{R}}{(x-\varrho)^{n-1}} - \frac{2n-3}{2(n-1)}\frac{f'(\varrho)}{f(\varrho)}\int\frac{dx}{(x-\varrho)^{n-1}\sqrt{R}} - \frac{n-2}{2(n-1)}\frac{f''(\varrho)}{f(\varrho)}\int\frac{dx}{(x-\varrho)^{n-2}\sqrt{R}}.$$

From this we see that:

II. $$\int\frac{dx}{(x-\varrho)^n\sqrt{R}} = \sum_{\nu=1}^{\nu=n-1}\frac{\alpha_\nu\sqrt{R}}{(x-\varrho)^{n-\nu}} + \beta\int\frac{dx}{(x-\varrho)\sqrt{R}}.$$

The quantities α_ν and β can be calculated by recurring formulas that are found by differentiating this equation:

$$\frac{1}{(x-\varrho)^n\sqrt{R}} = \frac{b+cx}{\sqrt{R}}\sum_{\nu=1}^{\nu=n-1}\frac{\alpha_\nu}{(x-\varrho)^{n-\nu}} - \sqrt{R}\sum_{\nu=1}^{\nu=n-1}\frac{(n-\nu)\alpha_\nu}{(x-\varrho)^{n-\nu+1}} + \frac{\beta}{(x-\varrho)\sqrt{R}},$$

which becomes, on multiplying up:

$$1 = (b+cx)\sum_{\nu=1}^{\nu=n-1}\alpha_\nu(x-\varrho)^\nu - (a+2bx+cx^2)\sum_{\nu=1}^{\nu=n-1}(n-\nu)\,\alpha_\nu(x-\varrho)^{\nu-1} + \beta(x-\varrho)^{n-1}$$

or, writing $x-\varrho = z$:

$$1 = \tfrac{1}{2}\{f'(\varrho)+zf''(\varrho)\}\sum_{\nu=1}^{\nu=n-1}\alpha_\nu z^\nu - \{f(\varrho)+zf'(\varrho)+\tfrac{1}{2}z^2f''(\varrho)\}\sum_{\nu=1}^{\nu=n-1}(n-\nu)\,\alpha_\nu z^{\nu-1} + \beta z^{n-1}.$$

Arranged according to the coefficients this equation assumes the form:

$$1 = -f(\varrho)\sum_{\nu=1}^{\nu=n-1}(n-\nu)\,\alpha_\nu z^{\nu-1} - f'(\varrho)\sum_{\nu=1}^{\nu=n-1}(n-\nu-\tfrac{1}{2})\,\alpha_\nu z^\nu - \tfrac{1}{2}f''(\varrho)\sum_{\nu=1}^{\nu=n-1}(n-\nu-1)\,\alpha_\nu z^{\nu+1} + \beta z^{n-1}.$$

Hence:

$$f(\varrho)(n-1)\,\alpha_1 + 1 = 0, \quad f(\varrho)(n-2)\,\alpha_2 + f'(\varrho)(n-\tfrac{3}{2})\,\alpha_1 = 0.$$

In general for $\nu = 3, 4, \ldots n-1$:

III. $$f(\varrho)(n-\nu)\,\alpha_\nu + f'(\varrho)(n-\nu+\tfrac{1}{2})\,\alpha_{\nu-1} + \tfrac{1}{2}f''(\varrho)(n-\nu+1)\,\alpha_{\nu-2} = 0,$$
$$\beta = \tfrac{1}{2}f'(\varrho)\,\alpha_{n-1} + \tfrac{1}{2}f''(\varrho)\,\alpha_{n-2}.$$

Note. The formulas developed in this section undergo a change when $f(\varrho) = 0$, and therefore ϱ is a root of

$$a + 2bx + cx^2 = 0.$$

Multiplying the identity I. by $f(\varrho)$ and then putting $f(\varrho) = 0$, we find:

$$0 = -\frac{1}{n-1}\cdot\frac{\sqrt{R}}{(x-\varrho)^{n-1}} - \frac{2n-3}{2(n-1)}f'(\varrho)\int\frac{dx}{(x-\varrho)^{n-1}\sqrt{R}}$$
$$-\frac{n-2}{2(n-1)}f''(\varrho)\int\frac{dx}{(x-\varrho)^{n-2}\sqrt{R}},$$

or putting $n+1$ for n:

$$\text{I}'.\int\frac{dx}{(x-\varrho)^n\sqrt{R}} = -\frac{2}{(2n-1)f'(\varrho)}\cdot\frac{\sqrt{R}}{(x-\varrho)^n} - \frac{n-1}{2n-1}\cdot\frac{f''(\varrho)}{f'(\varrho)}\int\frac{dx}{(x-\varrho)^{n-1}\sqrt{R}}.$$

Hence:

$$\text{II}'.\qquad \int\frac{dx}{(x-\varrho)^n\sqrt{R}} = \sum_{\nu=0}^{\nu=n-1}\frac{\alpha_\nu\sqrt{R}}{(x-\varrho)^{n-\nu}}.$$

$$\text{III}'.\qquad f'(\varrho)\alpha_0(n-\tfrac{1}{2}) + 1 = 0,$$
$$f'(\varrho)\alpha_\nu(n-\nu-\tfrac{1}{2}) + \tfrac{1}{2}f''(\varrho)(n-\nu)\alpha_{\nu-1} = 0,$$

or:

$$\alpha_\nu = (-1)^{\nu-1}\cdot\frac{(f''(\varrho))^\nu}{(f'(\varrho))^{\nu+1}}\cdot\frac{(n-1)(n-2)\ldots(n-\nu)}{(2n-1)(2n-3)\ldots(2n-2\nu-1)}.$$

In this case therefore the integral can be expressed by algebraic functions only.

122. The substitutions given in § 120 enable us to calculate the integrals:

$$\int\frac{dx}{x\sqrt{R}} \text{ and } \int\frac{dx}{(x-\varrho)\sqrt{R}}.$$

In expressing the former, it makes a difference whether $a > 0$ or $a < 0$.

In case $a > 0$, we employ substitution 2); we find:

$$\text{I}.\int\frac{dx}{x\sqrt{R}} = \int\frac{dz}{z\sqrt{a}-b} = \frac{1}{\sqrt{a}}l(z\sqrt{a}-b) + C = \frac{1}{\sqrt{a}}l\left(\frac{\sqrt{aR}-a-bx}{x}\right) + C.$$

Putting for x the value $z-\varrho$ and working precisely as when we made the same generalisation in Ia. § 121, we get:

$$\text{I}'.\qquad \int\frac{dx}{(x-\varrho)\sqrt{R}} = \frac{1}{\sqrt{f(\varrho)}}l\left(\frac{\sqrt{Rf(\varrho)} - f(\varrho) - \frac{1}{2}f'(\varrho)(x-\varrho)}{x-\varrho}\right) + C,$$

a real formula, if $f(\varrho) > 0$.

In case $a < 0$, we employ substitution 1); we obtain:

$$\text{II}.\int\frac{dx}{x\sqrt{R}} = 2\int\frac{dz}{z^2-a} = \frac{2}{\sqrt{-a}}\tan^{-1}\frac{z}{\sqrt{-a}} + C = \frac{2}{\sqrt{-a}}\tan^{-1}\frac{\sqrt{R}+x\sqrt{c}}{\sqrt{-a}} + C.$$

Hence results by the same process of generalisation:

$$\text{II}'.\qquad \int\frac{dx}{(x-\varrho)\sqrt{R}} = \frac{2}{\sqrt{-f(\varrho)}}\tan^{-1}\frac{\sqrt{R}+(x-\varrho)\sqrt{c}}{\sqrt{-f(\varrho)}} + C.$$

Lastly, in case $a < 0$, $c < 0$, $a + 2bx + cx^2 = 0$ having its roots real, and consequently of like sign, we have further by means of substitution 3):

III. $$\int \frac{dx}{x\sqrt{R}} = 2\int \frac{dz}{c\beta - \alpha z^2} = \frac{2}{c\beta}\int \frac{dz}{1 + \frac{\alpha}{\beta}\left(\frac{z}{\sqrt{-c}}\right)^2}$$

$$= -\frac{2}{\sqrt{\alpha\beta}\sqrt{-c}} \tan^{-1} \frac{z\sqrt{\alpha}}{\sqrt{-c\beta}} + C,$$

or $$\int \frac{dx}{x\sqrt{R}} = -\frac{2}{\sqrt{\alpha\beta}\sqrt{-c}} \tan^{-1} \sqrt{\frac{\alpha}{\beta}} \cdot \frac{\sqrt{R}}{(x-\alpha)\sqrt{-c}} + C.$$

This equation can also be generalised according to the above process: we have

III'. $$\int \frac{dx}{(x-\varrho)\sqrt{R}} = -\frac{2}{\sqrt{(\alpha-\varrho)(\beta-\varrho)}\sqrt{-c}} \tan^{-1} \sqrt{\frac{\alpha-\varrho}{\beta-\varrho}} \cdot \frac{\sqrt{R}}{(x-\alpha)\sqrt{-c}} + C,$$

where α and β are the roots of the quadratic equation $R = 0$.

Note: The integral $\int F(x, \sqrt{R})\,dx$ can be discussed by the help of geometrical considerations. Denoting the value:

$$\sqrt{R} = \sqrt{a + 2bx + cx^2} \text{ by } y,$$

the equation: $y^2 = a + 2bx + cx^2$, referred to Cartesian coordinates represents a curve of the second order; this is a hyperbola if $c > 0$, a parabola if $c = 0$, an ellipse if $c < 0$, but, when at the same time $ac - b^2 > 0$ the curve is completely imaginary. The axis of abscissæ is an axis of the curve, to each value of x correspond two equal and opposite values of y, or $\pm\sqrt{R}$.

The integral: $\int F(x, y)dx$ is uniquely related to the curve, i. e. to each of its points, real or complex, belongs one value of the function to be integrated, for, the sign of the root is determined by the point of the curve. Now our investigations have shown, that such an integral extended along a conic can be transformed into a rational one, that therefore the coordinates x, y, of points on the conic can be expressed as rational functions of a variable z. If conversely we assume this theorem, that is found in the projective geometry of conics to be a fundamental theorem, the methods of treating the irrational integral are simple deductions and lose all appearance of an artificial substitution. Thus ex. gr. for the hyperbola $y^2 = a + 2bx + cx^2$, $c > 0$, the directions of the asymptotes are given by $y^2 - cx^2 = 0$. If now we construct a system of right lines parallel to one asymptote, we have for the equation of this system $y + x\sqrt{c} = z$, where z stands for a variable. Each of these right lines meets the hyperbola only in one finite point, its other intersection being always the same point at infinity, and the coordinates of this single finite intersection are expressed as the following functions of z:

$$x = \frac{z^2 - a}{2(z\sqrt{c} + b)}, \qquad y = \frac{(z^2 + a)\sqrt{c} + 2bz}{2(z\sqrt{c} + b)}.$$

But these are the formulas of the first substitution in § 120. Both the other substitutions of that section are accounted for by analogous considerations. By considering the pencil of rays at any point of the conic, we obtain all possible substitutions by which the integral is made rational. The study of algebraic integrals in general first gains connexion and perspicuity by the geometry of algebraic curves; this was first brought out in the fundamental works of Aronhold and Clebsch (Journal f. Math., Vol. 61. 63. 64).

Regarding the present problem we have still to remark: the two fundamental integrals to which every other was reduced, were:

$$\int \frac{dx}{y} \text{ and } \int \frac{dx}{(x-\varrho)y}, \text{ where } y^2 = a + 2bx + cx^2.$$

Now considering one of the solutions we established for them, ex. gr.:

$$\int \frac{dx}{y} = \frac{1}{\sqrt{c}}\, l(b + cx + \sqrt{cR}) + C,$$

$$\int \frac{dx}{(x-\varrho)y} = \frac{1}{\sqrt{f(\varrho)}}\, l\left\{\frac{\sqrt{Rf(\varrho)} - f(\varrho) - \frac{1}{2}f'(\varrho)(x-\varrho)}{x-\varrho}\right\} + C,$$

we perceive that the integral function does not become infinite at the points $y = 0$. These intersections with the axis of abscissæ, at which the tangents to the conic are parallel to the axis of ordinates, are branching points of the function y but are not infinities for the integral function. On the other hand, in the first formula the argument of the logarithm is infinite when x becomes infinitely great. For the hyperbola there are two real points in which $x = \infty$ and for the ellipse two complex points.

In the second formula when $x = \varrho$ the argument of the logarithm vanishes, and therefore as:

$$\left(\frac{\sqrt{Rf(\varrho)} - f(\varrho) - \frac{1}{2}f'(\varrho)(x-\varrho)}{x-\varrho}\right)_{x=\varrho} = \frac{0}{0} = \left(\frac{(b+cx)\sqrt{f(\varrho)}}{\sqrt{R}} - \frac{1}{2}f'(\varrho)\right)_{x=\varrho} = 0$$

the logarithm itself is infinite. The one fundamental integral is logarithmically infinite at the two points at infinity upon the conic, the other at its two points upon the right line $x = \varrho$. This connexion between the two integrals becomes evident, when we can treat the equation of the line at infinity like that of any other right line. We can do so most simply by using homogeneous coordinates.

Putting $x = \frac{x_1}{x_3}$, $y = \frac{x_2}{x_3}$, the equation of the line at infinity is $x_3 = 0$; that of the conic is $x_2^2 = ax_3^2 + 2bx_1x_3 + cx_1^2$; and

we have:

$$\int \frac{dx}{y} = \int \frac{x_3 dx_1 - x_1 dx_3}{x_3 x_2}, \quad \int \frac{dx}{(x-\varrho) y} = \int \frac{x_3 dx_1 - x_1 dx_3}{(x_1 - \varrho x_3) x_2}.$$

Hence we see, what is lost sight of in the non-homogeneous form, that the first integral is only a special form of the second, since it involves the intersections of the conic with the right line $x_3 = 0$, instead of those with the right line $x_1 - \varrho x_3 = 0$. (Aronhold *loc. cit.*).

When $f(\varrho) = 0$, the right line $x = \varrho$ is a tangent to the curve, for, its two intersections have the same coordinates $x = \varrho$, $y = 0$; but then:

$$\int \frac{dx}{(x-\varrho)\sqrt{R}} = -\frac{2}{f'(\varrho)} \frac{\sqrt{R}}{x-\varrho} + C,$$

and for $x = \varrho$ this algebraic expression becomes:

$$\left(-\frac{2}{f'(\varrho)} \frac{\sqrt{R}}{x-\varrho}\right)_{x=\varrho} = \left(-\frac{2}{f'(\varrho)} \frac{b+cx}{\sqrt{R}}\right)_{x=\varrho} = \left(-\frac{1}{\sqrt{f(\varrho)}}\right) = \infty.$$

The same holds for the parabola where it meets the line at infinity.

123. From the indefinite integral we can form the definite for two limits between which the integral function remains continuous and does not become infinite; thus for instance, if $\varrho^2 > 1$, we have by formula III' § 122:

$$\int_{-1}^{+1} \frac{dx}{(x-\varrho)\sqrt{1-x^2}} = \left(-\frac{2}{\sqrt{\varrho^2-1}} \tan^{-1} \sqrt{\frac{\varrho-1}{\varrho+1}} \cdot \frac{\sqrt{1-x^2}}{x-1}\right)_{x=-1}^{x=+1}.$$

When $x = -1$ the argument of this circular function vanishes; its values are continuous and negative as x increases, when $x = +1$ the value becomes $-\infty$; for, we have

$$\frac{\sqrt{1-x^2}}{x-1} = \frac{\sqrt{1-x} \cdot \sqrt{1+x}}{-\sqrt{1-x} \cdot \sqrt{1-x}} = \left(\frac{\sqrt{1+x}}{-\sqrt{1-x}}\right)_{x=1} = -\infty;$$

the factor $\sqrt{\frac{\varrho-1}{\varrho+1}}$ is positive, therefore we have:

$$\int_{-1}^{+1} \frac{dx}{(x-\varrho)\sqrt{1-x^2}} = \frac{\pi}{\sqrt{\varrho^2-1}}.$$

This reasoning would not be possible if $\varrho^2 < 1$; in this case $\tan^{-1}$ is discontinuous in the interval; the value of its argument is $-i$ when $x = \varrho$.

124. The class of integrals of explicitly irrational functions next in order is given by $\int F(x, \sqrt{R})\, dx$; R being a polynomial of the third or fourth degree in x,

$$R = a + bx + cx^2 + dx^3 + ex^4,$$

in which c may vanish; and F a rational combination of x and $\sqrt{R}$. For, just as it was really indifferent in considering the previous class whether the polynomial under the square root was of the first degree or of the second — the integral could be reduced to a rational form in both cases —, so also here the cases of the third and of the fourth degree are equivalent: the integrals can be transformed into one another; they no longer however in general become rational. This transformation is effected by the substitution:

$$x = k + \frac{l}{z+m} = k + h,$$

by which:

$$R = a + bx + cx^2 + dx^3 + ex^4 = f(x)$$

becomes

$$f(k+h) = f(k) + hf'(k) + \frac{h^2}{1.2}f''(k) + \frac{h^3}{1.2.3}f'''(k) + \frac{h^4}{1.2.3.4}f^{IV}(k),$$

or

$$f(k+h) = \frac{1}{(z+m)^4}\Big\{(z+m)^4 f(k) + (z+m)^3 l f'(k) + \frac{(z+m)^2}{1.2} l^2 f''(k) + \frac{z+m}{1.2.3} l^3 f'''(k) + \frac{l^4}{1.2.3.4} f^{IV}(k)\Big\}.$$

Now if we determine k so that $f(k) = 0$, i. e. that k be a root of the equation $R = 0$, we have:

$$\sqrt{R} = \frac{1}{(z+m)^2}\sqrt{Az^3 + Bz^2 + Cz + D} = \frac{1}{(z+m)^2}\sqrt{Z},$$

$$A = lf'(k), \quad B = 3mlf'(k) + \tfrac{1}{2}l^2 f''(k),$$

$$C = 3m^2 lf'(k) + ml^2 f''(k) + \frac{l^3}{1.2.3}f'''(k),$$

$$D = m^3 lf'(k) + \tfrac{1}{2}m^2 l^2 f''(k) + \frac{ml^3}{1.2.3}f'''(k) + \frac{l^4}{1.2.3.4}f^{IV}(k).$$

The quantities l and m remain arbitrary.

Under the square root there is now only a cubic polynomial, and since x is rationally expressed by z, $\int F(x, \sqrt{R})\,dx$ is transformed into $\int \Phi(z, \sqrt{Z})\,dz$, as was stated. Denoting the roots of $R = 0$ by $\alpha, \beta, \gamma, \delta$, and making k coincide with α, the corresponding values of z are: ∞, $\frac{l}{\beta-\alpha} - m$, $\frac{l}{\gamma-\alpha} - m$, $\frac{l}{\delta-\alpha} - m$. Therefore to one vanishing value of R corresponds a value of z that becomes infinite; to the others correspond the roots of $Z = 0$.

*The evaluation of the elliptic integral:**)

*) The geometric problem, to determine the length of the arc of an ellipse or of a hyperbola between arbitrarily given terminal points, led to integrals of this form. The Italian mathematician Fagnano (1682—1766) (Produzioni matematiche, t. II, 1750) first found geometric relations among arcs of one of these curves by

$$\int F(x, \sqrt{R})\,dx, \text{ where } R = a + bx + cx^2 + dx^3, \quad (c = 0),$$

can be reduced to algebraic and logarithmic functions and to three fundamental integrals:

$$\int \frac{dx}{\sqrt{R}}, \quad \int \frac{x\,dx}{\sqrt{R}}, \quad \int \frac{dx}{(x-\varrho)\sqrt{R}}.$$

When arranged according to its rational and irrational parts, we have (compare § 119):

$$F(x, \sqrt{R}) = \frac{G(x) + H(x)\sqrt{R}}{G_1(x) + H_1(x)\sqrt{R}} = \varphi(x) + \psi(x)\sqrt{R},$$

φ and ψ denoting rational functions; the integration of φ leads to algebraic and logarithmic expressions; the second part gives us:

$$\frac{\psi(x) R}{\sqrt{R}} = \frac{f(x)}{\sqrt{R}} = \sum_{n=0} \frac{A_n x^n}{\sqrt{R}} + \sum \frac{B_n}{(x-\varrho)^n} \frac{1}{\sqrt{R}}.$$

The integral $\int \frac{x^n\,dx}{\sqrt{R}}$ can be reduced by a recurring formula: From

$$d(x^{n-2}\sqrt{R}) = \frac{(n-2)x^{n-3}(a+bx+cx^2+dx^3)}{\sqrt{R}}\,dx + \frac{x^{n-2}(b+2cx+3dx^2)}{2\sqrt{R}}\,dx,$$

we find by integration:

$$\text{I.} \qquad x^{n-2}\sqrt{R} = a(n-2)\int \frac{x^{n-3}\,dx}{\sqrt{R}} + b(n-\tfrac{3}{2})\int \frac{x^{n-2}\,dx}{\sqrt{R}} + c(n-1)\int \frac{x^{n-1}\,dx}{\sqrt{R}} + d(n-\tfrac{1}{2})\int \frac{x^n\,dx}{\sqrt{R}}.$$

Now putting in succession for n the values 2, 3, 4 . . ., we get:

$$\int \frac{x^2\,dx}{\sqrt{R}} = \frac{2}{3d}\sqrt{R} - \frac{b}{3d}\int \frac{dx}{\sqrt{R}} - \frac{2c}{3d}\int \frac{x\,dx}{\sqrt{R}},$$

$$\int \frac{x^3\,dx}{\sqrt{R}} = \frac{2}{5d}x\sqrt{R} - \frac{2a}{5d}\int \frac{dx}{\sqrt{R}} - \frac{3b}{5d}\int \frac{x\,dx}{\sqrt{R}} - \frac{4c}{5d}\int \frac{x^2\,dx}{\sqrt{R}}.$$

In general for $n \geqq 2$:

$$\int \frac{x^n\,dx}{\sqrt{R}} = \sqrt{R}\sum_{\nu=1}^{\nu=n-1} \alpha_\nu x^{n-1-\nu} + \beta \int \frac{dx}{\sqrt{R}} + \gamma \int \frac{x\,dx}{\sqrt{R}},$$

means of the integral; for these relations Euler (in the Nov. Com. Petrop., 1761, Vols. VI, VII; see also his Inst. calc. integral., Vol. I, Sect. 2, Cap. VI, §§ 606—649) discovered the general ground in the theorem of addition: "par une combinaison qu'on peut regarder comme fort heureuse, quoique ces hasards n'arrivent qu'à ceux qui savent les faire naître", as Legendre says in the Introduction to his great work. Euler perceived that by these integrals new functions are introduced into analysis, so that the group of transcendental functions (logarithmic and circular, and their inverses: the exponential and trigonometric) becomes legitimately enlarged. Legendre (1752—1833) established a theory of these new transcendents by his: Traité des fonctions elliptiques et des intégrales Eulériennes, 1825—26.

so that the reduction to the first two fundamental integrals is effected.

Putting the values 1, 0, — 1, — 2, . . . for n in the same recurring formula, it enables us to express integrals of the form $\int\frac{dx}{x^n\sqrt{R}}$ by means of the first two fundamental integrals and by the integral $\int\frac{dx}{x\sqrt{R}}$. To show this, let us put $-n+3$ for n; we find:

$$\text{II.}\qquad \frac{\sqrt{R}}{x^{n-1}} = -a(n-1)\int\frac{dx}{x^n\sqrt{R}} - b(n-\tfrac{3}{2})\int\frac{dx}{x^{n-1}\sqrt{R}}$$
$$-c(n-2)\int\frac{dx}{x^{n-2}\sqrt{R}} - d(n-\tfrac{5}{2})\int\frac{dx}{x^{n-3}\sqrt{R}}.$$

Now provided a does not vanish, this gives for $n = 2, 3$, etc.:

$$\int\frac{dx}{x^2\sqrt{R}} = -\frac{1}{a}\frac{\sqrt{R}}{x} - \frac{b}{2a}\int\frac{dx}{x\sqrt{R}} + \frac{d}{2a}\int\frac{x\,dx}{\sqrt{R}},$$
$$\int\frac{dx}{x^3\sqrt{R}} = -\frac{1}{2a}\frac{\sqrt{R}}{x^2} - \frac{3b}{4a}\int\frac{dx}{x^2\sqrt{R}} - \frac{c}{2a}\int\frac{dx}{x\sqrt{R}} - \frac{d}{4a}\int\frac{dx}{\sqrt{R}}.$$

In general for $n > 2$:

$$\int\frac{dx}{x^n\sqrt{R}} = \sqrt{R}\sum_{\nu=1}^{\nu=n-1}\frac{\alpha_\nu}{x^{n-\nu}} + \beta\int\frac{dx}{x\sqrt{R}} + \gamma\int\frac{dx}{\sqrt{R}} + \delta\int\frac{x\,dx}{\sqrt{R}}.$$

In case a vanishes, $\int\frac{dx}{x\sqrt{R}}$ is expressed by the first two fundamental integrals, as appears from the recurring formula II..

Formula II. admits of generalisation according to the method already applied (§ 121), by writing for x, $x-\varrho$ and for $a+bx+cx^2+dx^3$, $A+Bx+Cx^2+Dx^3$, where:

$$a = A + \varrho B + \varrho^2 C + \varrho^3 D = f(\varrho),$$
$$b = B + 2\varrho C + 3\varrho^2 D = f'(\varrho),$$
$$c = C + 3\varrho D = \tfrac{1}{2}f''(\varrho),$$
$$d = D = \tfrac{1}{6}f'''(\varrho).$$

Then if we imagine the letters a, b, c, d, put for A, B, C, D, we have:

$$\text{III.}\qquad \frac{\sqrt{R}}{(x-\varrho)^{n-1}} = -f(\varrho)(n-1)\int\frac{dx}{(x-\varrho)^n\sqrt{R}} - f'(\varrho)(n-\tfrac{3}{2})\int\frac{dx}{(x-\varrho)^{n-1}\sqrt{R}}$$
$$-\tfrac{1}{2}f''(\varrho)(n-2)\int\frac{dx}{(x-\varrho)^{n-2}\sqrt{R}} - \tfrac{1}{6}f'''(\varrho)(n-\tfrac{5}{2})\int\frac{dx}{(x-\varrho)^{n-3}\sqrt{R}}.$$

In accordance with this formula we have for $n \geq 2$:

$$\int\frac{dx}{(x-\varrho)^n\sqrt{R}} = \sqrt{R}\sum_{\nu=1}^{\nu=n-1}\frac{\alpha_\nu}{(x-\varrho)^{n-\nu}} + \beta\int\frac{dx}{(x-\varrho)\sqrt{R}} + \gamma\int\frac{dx}{\sqrt{R}}$$
$$+\delta\int\frac{(x-\varrho)\,dx}{\sqrt{R}},$$

i. e. it is expressed, as already stated, by the three fundamental integrals:

$$\int \frac{dx}{\sqrt{R}}, \quad \int \frac{x\,dx}{\sqrt{R}}, \quad \int \frac{dx}{(x-\varrho)\sqrt{R}};$$

these are called the elliptic integrals of the first, second and third kind.

Employing the substitution which served to convert a polynomial of the fourth degree under the square root into one of the third degree, by transforming these integrals back again, we can also state our result as follows:

Every elliptic integral $\int F(x, \sqrt{R})\,dx$, in which:

$$R = a + bx + cx^2 + dx^3 + ex^4,$$

can be expressed by three fundamental integrals:

$$\int \frac{dx}{\sqrt{R}}, \quad \int \frac{dx}{(x-\alpha)\sqrt{R}}, \quad \int \frac{dx}{(x-\varrho)\sqrt{R}}$$

and by algebraic and logarithmic functions.

In the integral of the second kind, α denotes a root of $R = 0$. It can also be shown, by developing a formula of reduction analogous to III., that:

$$\int \frac{x\,dx}{\sqrt{R}} \text{ and } \int \frac{x^2\,dx}{\sqrt{R}}$$

can be introduced instead of the last two integrals.

125. We now proceed still on the basis of these reductions to establish the three normal integrals to the calculation of which Legendre reduced the general elliptic integral.

Let the coefficients in $R = a + bx + cx^2 + dx^3 + ex^4$ be real, then whether the four roots of $R = 0$ be real or complex, we can always by a real linear substitution cause the odd powers in the polynomial to vanish.

When the four roots are all real, let us name them so that $\alpha > \beta > \gamma > \delta$; when complex let α be conjugate to β, and γ to δ. Putting

$$R = c(x-\alpha)(x-\beta)(x-\gamma)(x-\delta) = c(x^2 - 2\lambda x + \mu)(x^2 - 2\varrho x + \sigma),$$

then:

$$\alpha + \beta = 2\lambda, \; \alpha\beta = \mu, \; \gamma + \delta = 2\varrho, \; \gamma\delta = \sigma.$$

From the substitution:

$$x = \frac{p + qz}{1 + z}$$

we have:

$$x^2 - 2\lambda x + \mu = \frac{Az^2 + Bz + C}{(1+z)^2}, \quad x^2 - 2\varrho x + \sigma = \frac{A'z^2 + B'z + C'}{(1+z)^2}$$

where:

$$A = q^2 - 2\lambda q + \mu, \qquad A' = q^2 - 2\varrho q + \sigma,$$
$$B = 2(pq - \lambda(p+q) + \mu), \qquad B' = 2(pq - \varrho(p+q) + \sigma),$$
$$C = p^2 - 2\lambda p + \mu, \qquad C' = p^2 - 2\varrho p + \sigma;$$

p and q can always be determined so that B and B' may vanish. For from the equations:

$$pq - \lambda(p+q) + \mu = 0, \qquad pq - \varrho(p+q) + \sigma = 0,$$

unless $\lambda - \varrho = 0$, we have:

$$p + q = \frac{\mu - \sigma}{\lambda - \varrho}, \qquad pq = \frac{\varrho\mu - \lambda\sigma}{\lambda - \varrho},$$

and hence:

$$(p-q)^2 = \frac{(\mu-\sigma)^2}{(\lambda-\varrho)^2} - \frac{4(\varrho\mu - \lambda\sigma)}{\lambda - \varrho} = \frac{(\mu+\sigma-2\lambda\varrho)^2 - 4(\sigma-\varrho^2)(\mu-\lambda^2)}{(\lambda-\varrho)^2}.$$

Since the values $p+q$ and pq are real, the values of p and q will separately be real unless $(p-q)^2$ be negative.

When two of the four roots are real and two complex, $\lambda^2 - \mu$ and $\varrho^2 - \sigma$ are different in sign, hence the numerator is positive.

When the four roots are real, developing the numerator we find:

$$\left(\alpha\beta + \gamma\delta - \frac{(\alpha+\beta)(\gamma+\delta)}{2}\right)^2 - 4\left(\gamma\delta - \left(\frac{\gamma+\delta}{2}\right)^2\right)\left(\alpha\beta - \left(\frac{\alpha+\beta}{2}\right)^2\right)$$
$$= \left(\alpha\beta + \gamma\delta - \frac{(\alpha+\beta)(\gamma+\delta)}{2}\right)^2 - \tfrac{1}{4}(\gamma-\delta)^2(\alpha-\beta)^2.$$

This expression is symmetrical in α and β, γ and δ, it vanishes for $\alpha = \gamma$, thus has $(\alpha - \gamma)$ as a factor, it has therefore also the factors $(\alpha-\delta)$, $(\beta-\gamma)$, $(\beta-\delta)$. Being of the fourth degree, it consists of their product multiplied only by a numerical factor, which is found to be 1 as the term $\alpha^2\beta^2$ occurs in both with that factor, thus:

$$\left(\alpha\beta+\gamma\delta-\frac{(\alpha+\beta)(\gamma+\delta)}{2}\right)^2 - \tfrac{1}{4}(\gamma-\delta)^2(\alpha-\beta)^2 = (\alpha-\gamma)(\alpha-\delta)(\beta-\gamma)(\beta-\delta).$$

This product is positive since $\alpha > \beta > \gamma > \delta$.

When all the four roots are complex, let them be:

$$\alpha = \lambda + i\alpha', \quad \beta = \lambda - i\alpha', \quad \gamma = \varrho + i\gamma', \quad \delta = \varrho - i\gamma',$$

then:

$$\alpha - \gamma = (\lambda - \varrho) + i(\alpha' - \gamma'), \quad \alpha - \delta = (\lambda - \varrho) + i(\alpha' + \gamma'),$$
$$\beta - \gamma = (\lambda - \varrho) - i(\alpha' + \gamma'), \quad \beta - \delta = (\lambda - \varrho) - i(\alpha' - \gamma').$$

Thus the product is $((\lambda-\varrho)^2 + (\alpha'-\gamma')^2)\,((\lambda-\varrho)^2 + (\alpha'+\gamma')^2)$, i. e. positive as before.

When $\lambda = \varrho$:

$$R = c(x^2 - 2\lambda x + \mu)(x^2 - 2\lambda x + \sigma),$$

if we put $x = z + \lambda$:

$$R = c(z^2 + \mu - \lambda^2)(z^2 + \sigma - \lambda^2).$$

By the assumed substitution the reduced elliptic integral becomes:

$$\int\frac{F(x)}{\sqrt{R}}\,dx = \frac{q-p}{e}\int F\left(\frac{p+qz}{1+z}\right)\frac{dz}{\sqrt{Z}},\ Z=(Az^2+C)(A'z^2+C'),$$

where:

$$A=(q-\alpha)(q-\beta),\ C=(p-\alpha)(p-\beta),\ A'=(q-\gamma)(q-\delta),$$

$$C'=(p-\gamma)(p-\delta),\ p+q=\frac{2(\alpha\beta-\gamma\delta)}{(\alpha+\beta)-(\gamma+\delta)},\ pq=\frac{\alpha\beta(\gamma+\delta)-\gamma\delta(\alpha+\beta)}{2((\alpha+\beta)-(\gamma+\delta))}.$$

All these quantities are real, when the coefficients in R are real.

Developing the rational function $F\left(\frac{p+qz}{1+z}\right)$ we can collect the odd and the even terms in its numerator as well as in its denominator:

$$F=\frac{G(z^2)+zH(z^2)}{G_1(z^2)+zH_1(z^2)}.$$

If we multiply this above and below by $G_1 - zH_1$, we get in the denominator only even powers, so that:

$$\frac{F}{\sqrt{Z}}=\frac{\varphi(z^2)}{\sqrt{Z}}+\frac{z\psi(z^2)}{\sqrt{Z}}.$$

The integral of the second term is converted by the substitution $z^2=t$ into an integral of the form:

$$\tfrac{1}{2}\int\frac{\psi(t)\,dt}{\sqrt{(At+C)(A't+C')}},$$

and is therefore evaluated by logarithmic and algebraic functions.

In the integral of the first term the polynomial Z can be brought to the form:

$$(1-y^2)(1-k^2y^2),$$

where $0<k^2<1$. In order to examine this, let us write:

$$Z=CC'\left(1+\frac{A}{C}z^2\right)\left(1+\frac{A'}{C'}z^2\right)=\pm\gamma^2(1\pm\alpha^2z^2)(1\pm\beta^2z^2),$$

taking $\alpha^2>\beta^2$. Of course α, β, γ here are not to be confounded with the former notation for the roots of R. Now according to the signs eight cases arise; in each we consider those values of z, with values corresponding for y, for which Z remains positive, so that both the function to be integrated and the integral function are real for real values of the variable.

1) $Z=+\gamma^2(1+\alpha^2z^2)(1+\beta^2z^2)$ remains positive for all real values of z. Let us put:

$$\alpha z=\frac{y}{\sqrt{1-y^2}},\quad \alpha\,dz=\frac{dy}{(1-y^2)^{\frac{3}{2}}}.$$

If z increase from $-\infty$ to 0, y increases from -1 to 0; and if z increase from 0 to $+\infty$, y increases from 0 to $+1$; thus we have:

$$\frac{dz}{\sqrt{Z}}=\frac{dy}{\alpha\gamma\sqrt{(1-y^2)(1-k^2y^2)}},\quad k^2=\frac{\alpha^2-\beta^2}{\alpha^2}.$$

The radicals in this equation as well as in all the analogous results from 2) to 8) have the same sign on both sides.

2) $Z = +\gamma^2(1+\alpha^2 z^2)(1-\beta^2 z^2)$ remains positive as long as z lies between $-\frac{1}{\beta}$ and $+\frac{1}{\beta}$. Let us put:

$$\beta z = \sqrt{1-y^2}, \quad \beta dz = -\frac{y\,dy}{\sqrt{1-y^2}}.$$

If z increase from $-\frac{1}{\beta}$ to 0, let us make y increase from 0 to $+1$, taking the square root in the formula of substitution negative; but if z increase from 0 to $+\frac{1}{\beta}$, let us make y increase from -1 to 0, the square root having the positive sign; then:

$$\frac{dz}{\sqrt{Z}} = \frac{dy}{\gamma\sqrt{\alpha^2+\beta^2}\sqrt{(1-y^2)(1-k^2y^2)}}, \quad k^2 = \frac{\alpha^2}{\alpha^2+\beta^2}.$$

3) $Z = +\gamma^2(1-\alpha^2 z^2)(1+\beta^2 z^2)$ remains positive as long as z lies between $-\frac{1}{\alpha}$ and $+\frac{1}{\alpha}$. Let us put:

$$\alpha z = \sqrt{1-y^2}, \quad \alpha dz = -\frac{y\,dy}{\sqrt{1-y^2}}.$$

If z increase from $-\frac{1}{\alpha}$ to 0, y increases from 0 to $+1$, the square root is negative; if z go from 0 to $+\frac{1}{\alpha}$, y increases from -1 to 0, the square root is positive; then:

$$\frac{dz}{\sqrt{Z}} = \frac{dy}{\gamma\sqrt{\alpha^2+\beta^2}\sqrt{(1-y^2)(1-k^2y^2)}}, \quad k^2 = \frac{\beta^2}{\alpha^2+\beta^2}.$$

4) $Z = +\gamma^2(1-\alpha^2 z^2)(1-\beta^2 z^2)$ remains positive as long as z lies either between $-\frac{1}{\alpha}$ and $+\frac{1}{\alpha}$, between $+\infty$ and $+\frac{1}{\beta}$, or between $-\infty$ and $-\frac{1}{\beta}$. Let us put: $\alpha z = y$, $\alpha dz = dy$ in the first case, $-\frac{1}{\beta z} = y$, $\frac{dz}{\beta z^2} = dy$ in the others, then:

$$\frac{dz}{\sqrt{Z}} = \frac{dy}{\gamma\alpha\sqrt{(1-y^2)(1-k^2y^2)}}, \quad k^2 = \frac{\beta^2}{\alpha^2}.$$

5) $Z = -\gamma^2(1+\alpha^2 z^2)(1+\beta^2 z^2)$; in this case Z is negative for all real values of z; the integral function cannot be expressed by means of real quantities. If we put the factor $\sqrt{-1}$ before the integral, this case is in other respects reduced to the first.

6) $Z = -\gamma^2(1+\alpha^2 z^2)(1-\beta^2 z^2)$ remains positive, for z from $-\infty$ to $-\frac{1}{\beta}$, and from $+\frac{1}{\beta}$ to $+\infty$. Let us put:

$$\beta z = \frac{1}{\sqrt{1-y^2}}, \quad \beta dz = \frac{y\,dy}{(1-y^2)^{\frac{3}{2}}}.$$

If z increase from $-\infty$ to $-\frac{1}{\beta}$, y increases from -1 to 0, the square root is to be taken negatively; if z increase from $+\frac{1}{\beta}$ to $+\infty$, y increases from 0 to $+1$, the square root is positive. We have:

$$\frac{dz}{\sqrt{Z}} = \frac{1}{\gamma\sqrt{\alpha^2+\beta^2}} \cdot \frac{dy}{\sqrt{(1-y^2)(1-k^2y^2)}}, \quad k^2 = \frac{\beta^2}{\alpha^2+\beta^2},$$

7) $Z = -\gamma^2(1-\alpha^2 z^2)(1+\beta^2 z^2)$ remains positive, for z from $-\infty$ to $-\frac{1}{\alpha}$, and from $+\frac{1}{\alpha}$ to $+\infty$. Let us put:

$$\alpha z = \frac{1}{\sqrt{1-y^2}}, \quad \alpha dz = \frac{y\,dy}{(1-y^2)^{\frac{3}{2}}}, \text{ thus we have:}$$

$$\frac{dz}{\sqrt{Z}} = \frac{1}{\gamma\sqrt{\alpha^2+\beta^2}} \cdot \frac{dy}{\sqrt{(1-y^2)(1-k^2y^2)}}, \quad k^2 = \frac{\alpha^2}{\alpha^2+\beta^2}.$$

8) $Z = -\gamma^2(1-\alpha^2 z^2)(1-\beta^2 z^2)$ is positive, for z from $-\frac{1}{\beta}$ to $-\frac{1}{\alpha}$, and from $+\frac{1}{\alpha}$ to $+\frac{1}{\beta}$. Let us put:

$$\beta z = \sqrt{1-\frac{\alpha^2-\beta^2}{\alpha^2}y^2}, \quad \beta dz = \frac{\alpha^2-\beta^2}{\alpha^2} \cdot \frac{-y\,dy}{\sqrt{1-\frac{\alpha^2-\beta^2}{\alpha^2}y^2}}.$$

If z increase from $-\frac{1}{\beta}$ to $-\frac{1}{\alpha}$, y increases from 0 to $+1$, the square root is negative; if z increase from $+\frac{1}{\beta}$ to $+\frac{1}{\alpha}$, y increases from -1 to 0, the square root is positive. We have:

$$\frac{dz}{\sqrt{Z}} = \frac{dy}{\beta\gamma\sqrt{(1-y^2)(1-k^2y^2)}}, \quad k^2 = \frac{\alpha^2-\beta^2}{\alpha^2}.$$

This concludes the proof of our assertion for all cases. To all values of z for which $\sqrt{Z}$ remains real, correspond values $y^2 \leq 1$.

But if we observe that all the substitutions we have employed are included in the form:

$$z^2 = \frac{\delta+\delta_1 y^2}{\varepsilon+\varepsilon_1 y^2},$$

we obtain the result: The elliptic integral $\int \frac{\varphi(z^2)}{\sqrt{Z}}\,dz$, disregarding constant factors, can always be brought to the form:

$$\int \varphi\left(\frac{\delta+\delta_1 y^2}{\varepsilon+\varepsilon_1 y^2}\right)\frac{dy}{\sqrt{Y}}, \quad \text{where } Y = (1-y^2)(1-k^2y^2),$$

in which, provided Z has real coefficients, k — the modulus of the elliptic integral — signifies a real positive proper fraction.

Putting $y^2 = t$ in order to complete the reduction of this integral to the normal forms of Legendre, we have:

$$\int \varphi\left(\frac{\delta+\delta_1 y^2}{\varepsilon+\varepsilon_1 y^2}\right)\frac{dy}{\sqrt{Y}} = \tfrac{1}{2}\int \frac{F(t)\,dt}{\sqrt{t(1-t)(1-k^2t)}}.$$

It was proved (§ 124) of this integral, in which the polynomial is only of the third degree, that it can be reduced to the three fundamental forms:

$$\int \frac{dt}{\sqrt{T}}, \quad \int \frac{t\,dt}{\sqrt{T}}, \quad \int \frac{dt}{(t-\varrho)\sqrt{T}};$$

accordingly Legendre's normal integrals expressed in y are:

$$\int \frac{dy}{\sqrt{(1-y^2)(1-k^2y^2)}}, \int \frac{y^2\,dy}{\sqrt{(1-y^2)(1-k^2y^2)}}, \int \frac{dy}{(y^2-\varrho)\sqrt{(1-y^2)(1-k^2y^2)}},$$

or as Legendre, restricting himself primarily to the investigation of real values of the integral, by substituting

$$y = \sin\varphi, \quad dy = \cos\varphi\, d\varphi,$$

wrote them:

$$F(\varphi) = \int_0^\varphi \frac{d\varphi}{\Delta\varphi}, \quad Z(\varphi) = \int_0^\varphi \frac{\sin^2\varphi\, d\varphi}{\Delta\varphi}, \quad \Pi(\varphi) = \int_0^\varphi \frac{d\varphi}{(1+n\sin^2\varphi)\Delta\varphi},$$

$$\Delta\varphi = \sqrt{1-k^2\sin^2\varphi}.$$

The coefficient $n = -\frac{1}{\varrho}$ in the third integral is called the Parameter of the third normal integral.*)

Note. From the central equation of the ellipse

$$\frac{x^2}{a^2} + \frac{y^2}{b^2} = 1, \quad a^2 > b^2,$$

introducing φ the eccentric anomaly reckoned from the axis minor, we have:

$$x = a\sin\varphi, \quad y = b\cos\varphi,$$

$$\int \sqrt{dx^2+dy^2} = \int d\varphi \sqrt{a^2\cos^2\varphi + b^2\sin^2\varphi} = a\int d\varphi\sqrt{1-k^2\sin^2\varphi},$$

$$k^2 = \frac{a^2-b^2}{a^2} < 1.$$

Thus the length of the elliptic arc that belongs to the values 0 and φ depends on the calculation of the integral:

$$aE(\varphi) = a\int_0^\varphi d\varphi\,\Delta\varphi = a\left\{\int_0^\varphi \frac{d\varphi}{\Delta\varphi} - k^2\int_0^\varphi \frac{\sin^2\varphi\, d\varphi}{\sqrt{1-k^2\sin^2\varphi}}\right\} = a\{F(\varphi) - k^2Z(\varphi)\}.$$

From the central equation of the hyperbola $\frac{x^2}{a^2} - \frac{y^2}{b^2} = 1$, in which a means the length of the real semiaxis, putting $x = a\sec\varphi$, we find $y = b\tan\varphi$,

$$\sqrt{dx^2+dy^2} = d\varphi\sec^2\varphi\sqrt{b^2+a^2\sin^2\varphi}.$$

---

*) A more compendious account of the transformation to the normal form, due to Weierstrass, in which the coefficients of a fractional linear substitution are determined so that the values $y = \pm 1, \pm \frac{1}{k}$ shall correspond to the four roots of $R = 0$, is communicated by Schellbach: Die Lehre von den elliptischen Integralen und den Thetafunctionen; and Königsberger: Vorlesungen über die Theorie der elliptischen Integrale.

To obtain integrals of the previous form, however, we introduce the distance from the centre to the focus, $c^2 = a^2 + b^2$, and put $cy = b^2 \tan \varphi$, then:

$$x = a \sec \varphi \sqrt{1 - k^2 \sin^2 \varphi}, \quad c^2 k^2 = a^2,$$

thus:

$$\int \sqrt{dx^2 + dy^2} = \int dy \sqrt{1 + \frac{y^2}{x^2} \cdot \frac{a^4}{b^4}} = \frac{a}{b^2} \int \frac{dy \sqrt{b^4 + y^2 (a^2 + b^2)}}{x}$$

$$= \frac{b^2}{c} \int_0^{\varphi} \frac{d\varphi}{\cos^2 \varphi \Delta \varphi}, \quad \Delta \varphi = \sqrt{1 - k^2 \sin^2 \varphi}.$$

Therefore the integral $\Upsilon(\varphi)$ by which the length of the hyperbolic arc from $\varphi = 0$ is measured, is directly equal to that third normal integral $\Pi(\varphi)$ in which the parameter $n = -1$. But when $n = -1$, this third integral can be reduced to the second and first, it is the case of the vanishing value ϱ of the linear function coinciding with a root of $Y = 0$. From the identity:

$$d(\Delta \varphi \,.\, \tan \varphi) = \frac{\Delta \varphi \, d\varphi}{\cos^2 \varphi} - \frac{k^2 \sin^2 \varphi \, d\varphi}{\Delta \varphi}$$

$$= (1 - k^2) \frac{d\varphi}{\cos^2 \varphi \Delta \varphi} + k^2 \frac{d\varphi}{\Delta \varphi} - k^2 \frac{\sin^2 \varphi \, d\varphi}{\Delta \varphi},$$

we have:

$$\Upsilon(\varphi) = c \Delta \varphi \tan \varphi - \frac{a^2}{c} F(\varphi) + \frac{a^2}{c} Z(\varphi),$$

or:

$$\Upsilon(\varphi) = c \Delta \varphi \tan \varphi + \frac{b^2}{c} F(\varphi) - c E(\varphi).$$

The first term has a simple geometric meaning. It is equal to the length along the tangent to the curve at the point belonging to φ, measured from that point to the foot of the perpendicular let fall from the centre.*)

126. Elliptic integrals of the three kinds, or the three normal integrals of Legendre are calculated by means of expansions obtained by converting the function to be integrated into a series of powers and forming the integral of this infinite series. But this method requires some preliminary general investigations.

---

*) Legendre: Traité des fonctions elliptiques, p. 16.

Fourth Chapter.

Uniform convergence, Differentiation and Integration of an infinite series.

127. In the General Theorems concerning series of powers, (§ 44. IV), it was indicated that the proof of the continuity of a function expressed by a series of powers, as well as the rule for its differentiation, is based on a definite property of these series, namely on their uniform or equable convergence.

We are now going to discuss this conception more closely for any arbitrary infinite series which converges for a real interval.

Let the infinite series:

$$f_1(x) + f_2(x) + f_3(x) + \cdots + f_n(x) + f_{n+1}(x) + \cdots$$

be convergent in the interval from a to b; let its sum be denoted by $F(x)$. The functions $f_n(x)$ can be continued unrestrictedly according to some law, and we assume that however many of them are formed they are all continuous in the assigned interval. This hypothesis is to be maintained in all the following theorems.

The convergence of the infinite series requires, that, for any number δ however small, it shall be possible to find a place n in the series, such that every remainder:

$$R_n = f_n(x) + f_{n+1}(x) + \cdots, \quad R_{n+1} = f_{n+1}(x) + f_{n+2}(x) + \cdots,$$
$$R_{n+k} = f_{n+k}(x) + \cdots$$

from R_n onwards shall be smaller in amount than δ (§ 39).

*The infinite series is said to be uniformly or equably convergent in the entire interval without exception, when this criterion of convergence is satisfied by the same n for any given δ, while x passes through all values from a to b; we need not therefore take a different value for n until another value is assigned to δ.**)

A sufficient, although not a necessary, criterion of uniform convergence is presented in the theorem: When the series of the numerically greatest values assumed by the terms of the infinite

*) Heine: Ueber trigonometrische Reihen. Crelle's Journal, Vol. 71.

series in the interval from a to b converges, the series is uniformly convergent for all values of x.

For then the number n of the place from which onwards the remainders P_n, P_{n+1} ... of the newly formed series are constantly less than δ, assigns also a place such as is required for every x.

Example:

The series: $\frac{\sin x}{1^2} - \frac{\sin 3x}{3^2} + \frac{\sin 5x}{5^2} - \frac{\sin 7x}{7^2} + \cdots$

is uniformly convergent for all values of x, because:

$$\frac{1}{1^2} + \frac{1}{3^2} + \frac{1}{5^2} + \frac{1}{7^2} + \cdots$$

is convergent (§ 47, foot-note p. 82).

For a series whose terms alternate in sign within an interval, the following also is a sufficient criterion: The series is uniformly convergent when for each number δ a place n can be discovered in the series, such that for every value of x the numerical values of the terms from $f_n(x)$ onwards decrease and are constantly less than δ. For, putting:

$$R_n(x) = f_n(x) - f_{n+1}(x) + f_{n+2}(x) - f_{n+3}(x) + \cdots$$

where the quantities f are all positive, $R_n(x)$ is greater than 0 but less than $f_n(x)$, because:

$$R_n(x) = [f_n(x) - f_{n+1}(x)] + [f_{n+2}(x) - f_{n+3}(x)] + \cdots$$
$$R_n(x) = f_n(x) - [f_{n+1}(x) - f_{n+2}(x)] - [f_{n+3}(x) - f_{n+4}(x)] - \cdots$$

128. *When an infinite series is uniformly convergent in the neighbourhood of a point of its convergency, the infinite series expresses a continuous function at this point.*

Denoting the sum of the first $n - 1$ terms of the series by $\Sigma(x)$, we have:

$$F(x) = \Sigma(x) + R_n(x).$$

Since the series is uniformly convergent in the neighbourhood of x, a value n can be found such that abs $R_n(x)$ shall be less than $\frac{1}{3}\delta$ for any value from x to $x \pm h$, δ being an arbitrarily small given number. Hence, as:

$$\text{abs}[F(x \pm h) - F(x)] = \text{abs}[\Sigma(x \pm h) - \Sigma(x) + R_n(x \pm h) - R_n(x)],$$

we have:

$$\text{abs}[F(x \pm h) - F(x)] \leq \text{abs}[\Sigma(x \pm h) - \Sigma(x)] + \tfrac{2}{3}\delta.$$

But since the functions f are continuous and Σ is a sum of terms finite in number, we can always choose a finite h so small that we may have:

$$\text{abs}[\Sigma(x \pm h) - \Sigma(x)] < \tfrac{1}{3}\delta;$$

hence for any δ however small, a value h can be found for which:

$$\text{abs}\,[F(x \pm h) - F(x)] < \delta.$$

Accordingly the criterion of continuity is satisfied.

The theorem we have proved can also be stated thus: If the function expressed by the series be discontinuous at a point of its convergency, the series must converge unequably in the neighbourhood of this point.

From the theorem it follows: If the series converge uniformly without exception in its convergency, it expresses a function everywhere continuous in the same interval.

These theorems admit of conversion only on a certain hypothesis.

When the infinite series expresses a continuous function at a point, to any point x belongs a finite value n such that the remainder R_n and all that follow it are less than $\frac{1}{3}\delta$; moreover a value of h can be assigned for which:

$$\text{abs}\,[F(x \pm h) - F(x)] < \frac{\delta}{3}$$

and likewise:

$$\text{abs}\,[\Sigma(x \pm h) - \Sigma(x)] < \frac{\delta}{3}\cdot$$

Accordingly from the equation:

$$F(x \pm h) - F(x) = \Sigma(x \pm h) - \Sigma(x) + R_n(x \pm h) - R_n(x)$$

it follows that:

$$\text{abs}\,[R_n(x \pm h)] \text{ also } < \delta.$$

But it does not follow from this, that all the following remainders in the interval $\pm h$ continue less than δ. This will only be certainly the case when all terms have the same sign in the interval x to $x \pm h$, because then the amounts of the remainders form a decreasing series. Accordingly the statement of the theorem is:

If in the neighbourhood of a point a finite value n can be assigned such that the nth and all following terms in the series retain the same sign in the interval, then the uniform convergence of the series is a consequence of its continuity at this point; or again: If an infinite series converge absolutely in the neighbourhood of a point, its uniform convergence is a consequence of its continuity at this point.*)

*) It can be shown as a fact that the continuity of a convergent series alone is not a sufficient condition for its uniform convergence, by examples of continuous but unequably convergent series recently formed by Du Bois-Reymond, Darboux and Cantor. An example given by Cantor is: The infinite series having its general term:

$$f_n(x) = \frac{nx}{n^2x^2+1} - \frac{(n+1)x}{(n+1)^2x^2+1}$$

129. We come now in the second place to investigate, under what condition the differential quotient of an infinite series is expressed by the series formed of the differential quotients of its several terms. Here we assume that all the functions $f(x)$ can be differentiated and that their derived functions are continuous; but moreover since the infinite series:

$$f_1'(x) + f_2'(x) + f_3'(x) + \cdots f_n'(x) + \cdots$$

cannot possibly be convergent unless $\operatorname{Lim} f_n'(x)$ vanish for $n = \infty$,*) our investigation must be based on the hypothesis that it does vanish.

Example. It is shown in the Theory of Trigonometric Series, that for $-\pi < x < +\pi$ the infinite series:

$$\sin x - \frac{\sin 2x}{2} + \frac{\sin 3x}{3} - \frac{\sin 4x}{4} + \cdots$$

expresses the value $\frac{1}{2}x$; that is to say, a function whose derived is $\frac{1}{2}$. But this value is not presented by the series got by differentiating the several terms:

$$\cos x - \cos 2x + \cos 3x - \cos 4x + \cdots$$

which does not even converge, but is completely indeterminate, because:

$$\operatorname{Lim} f_n'(x) = \pm \operatorname{Lim} \cos nx$$

assumes for $n = \infty$ all possible values between -1 and $+1$.

In order to determine the differential quotient of the function $F(x)$ at a point in which $F(x)$ is continuous, let us first form the quotient of differences:

---

converges, for, the remainder:

$$R_n(x) = \frac{nx}{n^2x^2 + 1}$$

is zero for $n = \infty$, and the sum of the series is the continuous function:

$$F(x) = \frac{x}{x^2 + 1}.$$

But in the neighbourhood of the point $x = 0$ this series converges unequably; for, the function $R_n(x)$ has maximum and minimum values $\pm\frac{1}{2}$ for $x = \pm\frac{1}{n}$. Therefore near zero no interval can be assigned however small, within which the amounts of all remainders after a certain one continue smaller than an arbitrarily small number.

*) $\operatorname{Lim} f_n'(x)$ denotes that $f_n'(x)$ is first formed and then n put $= \infty$. This is of course to be distinguished from $\frac{d}{dx}\operatorname{Lim} f_n(x)$ in which we first put $n = \infty$ and then differentiate.

If $f_n(x) = \frac{\sin nx}{n}$, $\operatorname{Lim} f_n'(x) = \operatorname{Lim} \cos nx$ is completely indeterminate, while on the other hand:

$$\frac{d}{dx}\operatorname{Lim}\frac{\sin nx}{n} = 0.$$

$$1)\quad \frac{F(x+\Delta x)-F(x)}{\Delta x}=\frac{\Sigma(x+\Delta x)-\Sigma(x)}{\Delta x}+\frac{R_n(x+\Delta x)-R_n(x)}{\Delta x}.$$

For any finite value Δx however small, this continuous expression in Δx has a determinate finite value; we can write it:

$$2)\quad \frac{F(x+\Delta x)-F(x)}{\Delta x}=f_1'(x+\Theta\Delta x)+f_2'(x+\Theta\Delta x)+\cdots f_{n-1}'(x+\Theta\Delta x) + \frac{R_n(x+\Delta x)-R_n(x)}{\Delta x}.$$

Now making Δx converge to zero and denoting the differential quotient of the remainder function $R_n(x)$ by $R_n'(x)$ we get:

$$3)\quad F'(x)=f_1'(x)+f_2'(x)+\cdots f_{n-1}'(x)+R_n'(x).$$

If then the remainder of the original series be so constituted, that for any number δ however small, a place n can be assigned from which onwards not only $R_n(x)$ but also $R_n'(x)$ remains smaller than δ, however large n be, this equation passes over into the infinite series:

$$4)\quad F'(x)=f_1'(x)+f_2'(x)+\cdots f_{n-1}'(x)+f_n'(x)+\cdots$$

The statement of the result is: *If the remainder of an infinite series possess the property, that for a given value of x by choice of a lower limit for n, $R_n'(x)$ becomes arbitrarily small, the series formed of the differential quotients of the several terms is convergent and expresses at this point the value of the derived function.*

Since this property of the remainder in any arbitrary series cannot be recognised at once, we may usefully establish another criterion, not indeed necessary, but still sufficient, by which in many cases the question is decided:

If the series of the derived terms converge uniformly in an interval, it expresses everywhere in this interval the derived of the given series.

In order to examine this, let us denote the value of the supposed uniformly convergent series

$$f_1'(x)+f_2'(x)+\cdots f_n'(x)+\cdots$$

by $\Phi(x)$ and its remainder from the nth place by $\mathrm{P}_n(x)$, then equation 2) passes over into the form:

$$\frac{F(x+\Delta x)-F(x)}{\Delta x}=\{\Phi(x+\Theta\Delta x)-\mathrm{P}_n(x+\Theta\Delta x)\}+\frac{R_n(x+\Delta x)-R_n(x)}{\Delta x}.$$

Now here let n first become infinite and then put $\Delta x=0$. When n becomes infinite, the value of Θ also changes. But whatever be its value, since the derived series converges uniformly, a value n can be assigned, from which onwards we shall have $\mathrm{P}_n(x+\Theta\Delta x)<\delta$ for all values of Θ. In like manner n can be chosen so great that the last expression also shall amount to less than δ. Therefore a point can be found in the interval from x to $x+\Delta x$, at which the continuous function Φ shall differ inappreciably from the quotient of differences; there must therefore also be a point at which:

$$\frac{F(x+\Delta x)-F(x)}{\Delta x}=\Phi(x+\Theta\Delta x);$$

and since Φ as a uniformly convergent series signifies a continuous function, we shall have for $\Delta x=0$:

$$F'(x)=\Phi(x).$$

Series, to which these criteria do not apply, cannot be differentiated except by attempting to sum directly the infinite series for the quotient of differences $\frac{F(x+\Delta x)-F(x)}{\Delta x}$ and then passing to the limit for Δx.

Examples.

1) It was seen in § 47 that the infinite series:

$$x-\frac{x^2}{2}+\frac{x^3}{3}-\frac{x^4}{4}+\cdots$$

has for $-1<x\leqq+1$ the value:

$$l(1+x).$$

This series converges uniformly. The series formed of its derived terms:

$$1-x+x^2-x^3+\cdots$$

converges uniformly for $-1<x<+1$, and is thus a continuous function. It expresses the derived function $\frac{1}{1+x}$; but this connexion does not hold for $x=1$, although the differential quotient of the logarithm has the determinate value $\frac{1}{2}$.

2) It is shown in the Theory of Trigonometric Series, that the signification of the infinite series:

$$F(x)=\frac{4}{\pi}\left\{\frac{\sin x}{1^2}-\frac{\sin 3x}{3^2}+\frac{\sin 5x}{5^2}-\cdots\right\}$$

when $0\leqq x\leqq\frac{\pi}{2}$, is $F(x)=x$; and when $\frac{\pi}{2}<x\leqq\pi$, is $F(x)=\pi-x$; it is uniformly convergent. Further, we have:

$$F'(x)=\frac{4}{\pi}\left\{\frac{\cos x}{1}-\frac{\cos 3x}{3}+\frac{\cos 5x}{5}-\cdots\right\},$$

except for $x=\frac{1}{2}\pi$ for which the derived series is discontinuous and expresses the value zero, while the progressive differential quotient of $F(x)$ is -1 and the regressive $+1$.

3) The infinite series whose general term is:

$$f_n(x)=\frac{1}{2n}\,l(n^2x^2+1)-\frac{1}{2(n+1)}\,l(\overline{n+1}^2x^2+1)$$

is uniformly convergent for all finite values of x, since:

$$R_n(x)=\frac{1}{2n}\,l(n^2x^2+1).$$

For, evidently $R_n(x)$ is a function, which for a given x decreases when the values of n increase, and for a fixed value of n increases when the values of x increase. $R_n(x)$ vanishes for $x = 0$ for all values of n. Moreover at that point, for an interval from $x = 0$ to $x = h$, we can determine n so that $R_n(x)$ shall be less than an arbitrary number δ; we have only to choose n, so that:

$$\frac{1}{2n} l(n^2h^2 + 1) < \delta.$$

The series expresses the value:

$$F(x) = \tfrac{1}{2} l(x^2 + 1).$$

We have also:

$$F'(x) = \frac{x}{x^2+1} = \sum_{n=1}^{n=\infty}\left\{\frac{nx}{n^2x^2+1} - \frac{\overline{n+1}\,x}{\overline{n+1}^2x^2+1}\right\},$$

although this series converges unequably in the neigbourhood of $x = 0$; but the remainder of the original series has here the property, that:

$$\text{Lim}\; R_n'(x) = \text{Lim}\; \frac{nx}{n^2x^2+1} = 0.$$

4) The infinite series:

$$F(x) = \sum_{n=0}^{n=\infty} b^n \cos(a^n x \pi)$$

converges uniformly if $0 < b < 1$; its differential quotient, however, cannot be calculated, when the product $ab > 1$, from the derived series:

$$-\pi \sum_{n=0}^{n=\infty} (ab)^n \sin(a^n x \pi),$$

for this series does not converge, because $\text{Lim}\,(ab)^n \sin(a^n x \pi)$ is not zero for $n = \infty$. The differential quotient of $F(x)$ has no determinate value. (Communication of Weierstrass in Du Bois-Reymond's memoir, Journal f. Math., Vol. 79. Darboux, Annal. de l'école normale, T. VIII, p. 195, gives a further class of examples of this kind.)

130. The rules for the integration of an infinite series result without further investigation from the Theorems established for its differentiation. In accordance with our investigations in § 127, we assume that the infinite series expresses a function everywhere continuous in the interval from a to b. Let us denote the definite integral taken between two values x_0 and x_1 in the interval, by:

$$\int_{x_0}^{x_1} F(x)dx = \Phi(x_1) = (x_1 - x_0)\, F(x_0 + \Theta(x_1 - x_0)),$$

and likewise each continuous function $f_n(x)$ when integrated, by

$$\int_{x_0}^{x_1} f_n(x)\,dx = \varphi_n(x_1).$$

Now first of all, the series of these integral functions cannot possibly converge unless:

$$\text{Lim}\;\varphi_n(x_1) = \text{Lim}\int_{x_0}^{x_1} f_n(x)\,dx$$

vanish for $n = \infty$. We cannot infer that this condition is fulfilled of itself because

$$\int_{x_0}^{x_1} f_n(x)\,dx = (x_1 - x_0) f_n(x_0 + \Theta(x_1 - x_0))$$

and $\text{Lim}\, f_n = 0$. For instance:

$$\text{Lim}\int_{0}^{x_1} nxe^{-nx^2}dx = \tfrac{1}{2}\,\text{Lim}(1 - e^{-nx_1^2}) = \tfrac{1}{2}$$

and is not equal to

$$x_1\,\text{Lim}(n\Theta x_1 e^{-n\Theta^2 x_1^2}) = 0.$$

When the series formed of the integral functions of the several terms is a continuous function of x, and its derived series for every value in the interval from x_0 to x_1 is equal to the series $F(x)$, it expresses the integral of the original series.

According to the first Theorem of last § *this requires the series of the integral functions to have the property that for any number δ however small, a place n can be found such that for it and all higher values the derived of the remainder term $R_n'(x)$ shall be less than δ.*

But by the second Theorem it is a sufficient condition, *that the given series converges uniformly.*

This second theorem can be seen directly as follows: If

$$F(x) = f_1(x) + f_2(x) + f_3(x) + \cdots f_{n-1}(x) + \mathsf{P}_n(x),$$

and for the entire interval from x_0 to x_1 a single n can be found for which and for all greater values the continuous function $\mathsf{P}_n(x)$ shall remain less in amount than δ, we shall have:

$$\int_{x_0}^{x_1} \mathsf{P}_n(x)\,dx = (x_1 - x_0)\,\mathsf{P}_n(x_0 + \Theta\overline{x_1 - x_0}),\quad\text{therefore}$$

$$\int_{x_0}^{x_1} F(x)\,dx = \varphi_1(x_1) + \varphi_2(x_1) + \cdots \varphi_{n-1}(x_1) + (x_1 - x_0)\,\mathsf{P}_n(x_0 + \Theta\overline{x_1 - x_0}).$$

If now n be arbitrarily increased, we have

$$\int_{x_0}^{x_1} F(x)\,dx = \varphi_1(x_1) + \varphi_2(x_1) \cdots + \varphi_{n-1}(x_1) \cdots,$$

and this series likewise converges uniformly in the entire interval from x_0 to x_1.

The Theorems proved concerning the definite integral show that these investigations can be extended to series representing functions that are discontinuous or infinite in separate points, or again, to the definite integral with an infinite limit, always on the hypothesis that the series of the integral functions remains convergent.

The examples adduced in last § can be regarded inversely also as examples for the integration of infinite series.

We cite the following, due to Darboux, as an example in which the integration is not effected, although the series of the integral functions converges:

$$F(x) = xe^{-x^2} - \sum_{n=1}^{n=\infty}(nxe^{-nx^2} - \overline{n+1}xe^{-\overline{n+1}x^2})$$

is a convergent series for all values of x and a continuous function, although the series converges unequably in the neighbourhood of the point $x = 0$. In fact $R_n(x) = nxe^{-nx^2}$ and for $x = \frac{1}{\sqrt{2n}}$ it becomes

$$R_n = \frac{\sqrt{n}}{\sqrt{2}}\, e^{-\frac{1}{2}}.$$

By integrating the individual terms between 0 and x we obtain:

$$\int_0^x nxe^{-nx^2}dx - \int_0^x (n+1)xe^{-\overline{n+1}x^2}dx = -\tfrac{1}{2}e^{-nx^2} + \tfrac{1}{2}e^{-\overline{n+1}x^2}.$$

The infinite series formed of these integral functions:

$$\sum_{n=1}^{n=\infty}(-\tfrac{1}{2}e^{-nx^2} + \tfrac{1}{2}e^{-\overline{n+1}x^2})$$

converges, it expresses the function $-\frac{1}{2}e^{-x^2}$ for every finite value of x, but for $x = 0$ its value is 0, it is therefore a discontinuous function at this point and not in general equal to:

$$\int_0^x F(x)\,dx = \int_0^x xe^{-x^2}dx = \tfrac{1}{2}(1 - e^{-x^2}).$$

131. Applying these Theorems to a series ascending by powers of any continuous function $f(x)$:

$$a_0 + a_1 f(x) + a_2\{f(x)\}^2 + \cdots + a_n\{f(x)\}^n + \cdots,$$

we see:

First: Such a series is a continuous function of x within its convergency; for, if we put $f(x) = z$, the absolutely convergent series:

$$a_0 + a_1 z + a_2 z^2 + \cdots + a_n z^n + \cdots$$

is a uniformly convergent series. Moreover even at the limits of the convergency this series of powers is a continuous function, although only semiconvergent (§ 44. IV); thus it always converges uniformly.

Second: The series derived by differentiating with respect to x:

$$f'(x)\{a_1 + 2a_2 f(x) + 3a_3(f(x))^2 + \cdots n a_n (f(x))^{n-1} + \cdots\},$$

as long as it converges, is continuous and expresses the derived of the given series; but it certainly converges within the interval of the original series.

Third: The integral of the given series, taken between two values x_0 and x_1 in the convergency, is formed by the uniformly convergent series:

$$a_0 + a_1\int_{x_0}^{x_1} f(x)\,dx + a_2\int_{x_0}^{x_1}\{f(x)\}^2 dx + \cdots + a_n\int_{x_0}^{x_1}\{f(x)\}^n dx + \text{etc.}.$$

If this series converge also at the limits of the convergency and remain continuous there, it expresses the integral up to and including the limits.

132. Expression of the function $\sin^{-1} x = \int_0^x \frac{dx}{\sqrt{1-x^2}}$ by a series.

If $x^2 < 1$ we have the expansion:

$$(1-x^2)^{-\frac{1}{2}} = 1 + \frac{1}{2}x^2 + \frac{1.3}{2.4}x^4 + \frac{1.3.5}{2.4.6}x^6 + \cdots + \frac{1.3.5\ldots(2n-1)}{2.4.6\ldots 2n}x^{2n} + \cdots,$$

hence:

$$\sin^{-1}x = x + \frac{1}{2}\frac{x^3}{3} + \frac{1.3}{2.4}\frac{x^5}{5} + \frac{1.3.5}{2.4.6}\frac{x^7}{7} + \cdots + \frac{1.3\ldots(2n-1)}{2.4\ldots 2n}\frac{x^{2n+1}}{2n+1} + \text{etc.}.$$

This series continues to converge even for $x^2 = 1$, although the above binomial series is no longer convergent. For, the terms of the series:

$$\frac{1}{1} + \frac{1}{2}\frac{1}{3} + \frac{1.3}{2.4}\frac{1}{5} + \frac{1.3.5}{2.4.6}\frac{1}{7} + \frac{1.3.5.7}{2.4.6.8}\frac{1}{9} + \cdots$$

are smaller than the corresponding terms of the series:

$$1 + \frac{1}{2} + \frac{1}{2.4} + \frac{1.3}{2.4.6} + \frac{1.3.5}{2.4.6.8} + \cdots + \text{etc.}.$$

But this series converges and its value is 2. For:

$$\sqrt{1-x} = 1 - \frac{x}{2} - \frac{1}{2}\frac{x^2}{4} - \frac{1.3}{2.4}\frac{x^3}{6} - \cdots$$

converges even for the value $x = 1$. Therefore:

$$\sin^{-1}(1) = \frac{\pi}{2} = \frac{1}{1} + \frac{1}{2}\frac{1}{3} + \frac{1.3}{2.4}\frac{1}{5} + \frac{1.3.5}{2.4.6}\frac{1}{7} + \cdots + \text{etc.}.$$

The definite integral $\int_0^1 \frac{dx}{\sqrt{1-x^2}}$, as already indicated in § 107, has the finite value $\frac{1}{2}\pi$, although the function to be integrated becomes infinite at the upper limit.

133. The elliptic integral:

$$\int_0^x \frac{dx}{\sqrt{(1-x^2)(1-k^2x^2)}} = \int_0^\varphi \frac{d\varphi}{\sqrt{1-k^2\sin^2\varphi}} \qquad (k^2 < 1)$$

can be developed by powers of k. Expanding:

$$(1-k^2\sin^2\varphi)^{-\frac{1}{2}} = 1 + \frac{1}{2}k^2\sin^2\varphi + \frac{1.3}{2.4}k^4\sin^4\varphi + \frac{1.3.5}{2.4.6}k^6\sin^6\varphi + \cdots,$$

we have therefore:

$$F(\varphi) = \int_0^\varphi \frac{d\varphi}{\Delta\varphi} = \varphi + \frac{1}{2}k^2\int_0^\varphi \sin^2\varphi\, d\varphi + \frac{1.3}{2.4}k^4\int_0^\varphi \sin^4\varphi\, d\varphi$$

$$+ \frac{1.3.5}{2.4.6}k^6\int_0^\varphi \sin^6\varphi\, d\varphi + \ldots, \qquad (k^2\sin^2\varphi < 1).$$

The integrals in this series belong to the binomial integrals investigated in § 118 as is seen by substituting $\sin\varphi = x$, they are determined by integrating by parts; we have:

$$\int \sin^{2m}\varphi\, d\varphi = -\int \sin^{2m-1}\varphi\, d(\cos\varphi)$$

$$= -\sin^{2m-1}\varphi\cos\varphi + (2m-1)\int \sin^{2m-2}\varphi\cos^2\varphi\, d\varphi,$$

hence replacing in the last term $\cos^2\varphi$ by $1 - \sin^2\varphi$, transposing and inserting the limits we find:

$$\int_0^\varphi \sin^{2m}\varphi\, d\varphi = -\frac{\sin^{2m-1}\varphi\cos\varphi}{2m} + \frac{2m-1}{2m}\int_0^\varphi \sin^{2m-2}\varphi\, d\varphi.$$

For the limits zero and $\frac{1}{2}\pi$ we have:

$$\int_0^{\frac{1}{2}\pi} \sin^{2m}\varphi\, d\varphi = \frac{2m-1}{2m}\int_0^{\frac{1}{2}\pi} \sin^{2m-2}\varphi\, d\varphi = \frac{2m-1}{2m}\,\frac{2m-3}{2m-2}\,\frac{2m-5}{2m-4}\cdots\frac{3}{4}\,\frac{1}{2}\cdot\frac{\pi}{2}\,.$$

Hence the "complete integral" is:

$$F\left(\frac{\pi}{2}\right) = \int_0^{\frac{1}{2}\pi} \frac{d\varphi}{\Delta\varphi} = \frac{\pi}{2}\left\{1 + (\tfrac{1}{2})^2k^2 + (\tfrac{1}{2}\cdot\tfrac{3}{4})^2k^4 + (\tfrac{1}{2}\cdot\tfrac{3}{4}\cdot\tfrac{5}{6})^2k^6 + \cdots\right\}.$$

Similarly we obtain for the integral:

$$E(\varphi) = \int_0^\varphi \Delta\varphi\, d\varphi = F(\varphi) - k^2 Z(\varphi)$$

by expanding:

$$(1-k^2\sin^2\varphi)^{\frac{1}{2}} = 1 - \frac{1}{2}k^2\sin^2\varphi - \frac{1}{2.4}k^4\sin^4\varphi - \frac{1.3}{2.4.6}k^6\sin^6\varphi - \cdots,$$

the series:

$$E(\varphi)=\varphi-\frac{1}{2}k^2\int_0^\varphi\sin^2\varphi\,d\varphi-\frac{1}{2.4}k^4\int_0^\varphi\sin^4\varphi\,d\varphi-\frac{1.3}{2.4.6}k^6\int_0^\varphi\sin^6\varphi\,d\varphi-\cdots$$

$$(k^2\sin^2\varphi\leqq 1),$$

$$E\left(\frac{\pi}{2}\right)=\int_0^{\frac{1}{2}\pi}\Delta\varphi\,d\varphi=\frac{\pi}{2}\left\{1-\left(\frac{1}{2}\right)^2k^2-\left(\frac{1}{2.4}\right)^2 3k^4-\left(\frac{1.3}{2.4.6}\right)^2 5k^6-\cdots\right\}$$

$$(k^2\leqq 1).$$

These series converge slowly when the value of k^2 is nearly 1; for this case Legendre (Traité, p. 65) established more rapidly convergent series ascending by powers of the complementary modulus

$$k'=\sqrt{1-k^2}.$$

134. To discover the law of the explicit expression of $F(\varphi)$ and $E(\varphi)$, it is convenient to introduce the forms by which the powers of $\sin\varphi$ are expressed by sines or cosines of multiples of φ.*) In consequence of the equations (§ 67):

$$e^{i\varphi}=\cos\varphi+i\sin\varphi,\quad e^{-i\varphi}=\cos\varphi-i\sin\varphi,$$

let us put:

1) $$\Delta^2=1-k^2\sin^2\varphi=\frac{(1+ce^{2i\varphi})(1+ce^{-2i\varphi})}{(1+c)^2}=1-\frac{4c}{(1+c)^2}\sin^2\varphi,$$

$$k=\frac{2\sqrt{c}}{1+c},\qquad c=\frac{1-\sqrt{1-k^2}}{1+\sqrt{1-k^2}}<1.$$

Now, for $c<1$:

2) $$P=(1+ce^{2i\varphi})^{-\frac{1}{2}}=1-\frac{1}{2}ce^{2i\varphi}+\frac{1.3}{2.4}c^2e^{4i\varphi}-\frac{1.3.5}{2.4.6}c^3e^{6i\varphi}+\cdots$$
$$Q=(1+ce^{-2i\varphi})^{-\frac{1}{2}}=1-\frac{1}{2}ce^{-2i\varphi}+\frac{1.3}{2.4}c^2e^{-4i\varphi}-\frac{1.3.5}{2.4.6}c^3e^{-6i\varphi}+\cdots$$

are absolutely convergent series, therefore the value of

$$\frac{1}{\Delta}=(1+c)PQ$$

calculated by the rule of multiplication (§ 78), will be likewise an absolutely convergent series, which can be arranged by cosines of multiples of φ when we replace

$$e^{im\varphi}+e^{-im\varphi}=2\cos(m\varphi).$$

The result is:

3) $$\frac{1}{\Delta}=A-2A_1\cos 2\varphi+4A_2\cos 4\varphi-6A_3\cos 6\varphi+\cdots$$

$$A=(1+c)\left\{1+\left(\frac{1}{2}\right)^2c^2+\left(\frac{1.3}{2.4}\right)^2c^4+\left(\frac{1.3.5}{2.4.6}\right)^2c^6+\left(\frac{1.3.5.7}{2.4.6.8}\right)^2c^8+\cdots\right\}$$

$$A_1=\frac{1+c}{1}\left\{\frac{1}{2}c+\frac{1}{2}\cdot\frac{1.3}{2.4}c^3+\frac{1.3}{2.4}\cdot\frac{1.3.5}{2.4.6}c^5+\frac{1.3.5}{2.4.6}\cdot\frac{1.3.5.7}{2.4.6.8}c^7+\cdots\right\}$$

*) Legendre, *loc. cit.*, p. 273.

$$A_2=\frac{1+c}{2}\left\{\frac{1.3}{2.4}c^2+\frac{1}{2}\cdot\frac{1.3.5}{2.4.6}c^4+\frac{1.3}{2.4}\cdot\frac{1.3.5.7}{2.4.6.8}c^6+\frac{1.3.5}{2.4.6}\cdot\frac{1.3.5.7.9}{2.4.6.8.10}c^8+\cdots\right\}$$

$$A_3=\frac{1+c}{3}\left\{\frac{1.3.5}{2.4.6}c^3+\frac{1}{2}\cdot\frac{1.3.5.7}{2.4.6.8}c^5+\frac{1.3}{2.4}\cdot\frac{1.3.5.7.9}{2.4.6.8.10}c^7+\cdots\right\}$$

. . . .

These numbers $A, A_1, 2A_2, 3A_3, 4A_4$, etc., decrease continually and have zero as limit, for:

$$A_1<cA,\quad 2A_2<cA_1,\quad 3A_3<2cA_2,\quad 4A_4<3cA_3,\ldots$$

consequently:

$$A_1<cA,\quad 2A_2<c^2A,\quad 3A_3<c^3A,\quad 4A_4<c^4A,\ldots \text{ etc..}$$

Hence, as $c<1$, series 3) converges even when we give each of its terms the absolutely greatest value it admits of; therefore 3) converges uniformly (§ 127) and expresses a continuous function in the interval from zero to $\frac{1}{2}\pi$. Accordingly we obtain by integration:

4) $$F(\varphi)=\int_0^{\varphi}\frac{d\varphi}{\Delta\varphi}=A\varphi-A_1\sin 2\varphi+A_2\sin 4\varphi-A_3\sin 6\varphi+\cdots$$

In particular for $\varphi=\frac{1}{2}\pi$:

5) $$A=\frac{2}{\pi}\int_0^{\frac{1}{2}\pi}\frac{d\varphi}{\Delta\varphi}=\frac{2}{\pi}F\left(\frac{\pi}{2}\right).$$

But the other coefficients of this series can also be expressed as definite integrals. For, if we multiply series 3) in turn by:

$$\cos 2\varphi,\quad \cos 4\varphi,\quad \cos 6\varphi,\ldots$$

and integrate these products between zero and $\frac{1}{2}\pi$, since for $m\gtrless n$:

$$\int_0^{\frac{1}{2}\pi}\cos 2m\varphi\cos 2n\varphi\,d\varphi=\frac{1}{2}\int_0^{\frac{1}{2}\pi}\{\cos 2\overline{m+n}\varphi+\cos 2\overline{m-n}\varphi\}\,d\varphi=0,$$

and for $m=n$: $\int_0^{\frac{1}{2}\pi}(\cos 2m\varphi)^2d\varphi=\frac{1}{2}\frac{\pi}{2}$, we thus obtain:

6) $$\int_0^{\frac{1}{2}\pi}\frac{\cos 2\varphi\,d\varphi}{\Delta\varphi}=-A_1\frac{\pi}{2},\ \int_0^{\frac{1}{2}\pi}\frac{\cos 4\varphi\,d\varphi}{\Delta\varphi}=2A_2\frac{\pi}{2},\ \int_0^{\frac{1}{2}\pi}\frac{\cos 6\varphi\,d\varphi}{\Delta\varphi}=-3A_3\frac{\pi}{2},\cdots$$

The calculation of A_1 introduces the values $F\left(\frac{\pi}{2}\right)$ and $E\left(\frac{\pi}{2}\right)$, for:

7) $$\frac{\pi}{2}A_1=-\int_0^{\frac{1}{2}\pi}\frac{(1-2\sin^2\varphi)\,d\varphi}{\Delta\varphi}=\frac{2}{k^2}\left\{F\left(\frac{\pi}{2}\right)-E\left(\frac{\pi}{2}\right)\right\}-F\left(\frac{\pi}{2}\right).$$

Recurring formulas can be found for the other coefficients by differentiating series 3). Since the series thus obtained:

$$\frac{k^2\sin\varphi\cos\varphi}{\Delta^3}=4A_1\sin 2\varphi-16A_2\sin 4\varphi+36A_3\sin 6\varphi-\cdots$$

is uniformly convergent, the differentiation is admissible (§ 129). If we multiply the left side of this equation by $\frac{2\Delta^2}{k^2}$, and the right by the equal value:

$$\frac{2-k^2}{k^2} + \cos 2\varphi = \lambda + \cos 2\varphi$$

and arrange by sines of multiples of φ, since:

$$2\sin 2m\varphi\cos 2\varphi = \sin 2\overline{m+1}\varphi + \sin 2\overline{m-1}\varphi,$$

we shall have:

$$\frac{\sin 2\varphi}{\Delta\varphi} = (4A_1\lambda - 8A_2)\sin 2\varphi - (16A_2\lambda - 2A_1 - 18A_3)\sin 4\varphi$$
$$+ (36A_3\lambda - 8A_2 - 32A_4)\sin 6\varphi - (64A_4\lambda - 18A_3 - 50A_5)\sin 8\varphi$$
$$\cdots\cdots\cdots\cdots\cdots$$
$$(-1)^m\left((2m-2)^2A_{m-1}\lambda - \frac{(2m-4)^2}{2}A_{m-2} - \frac{(2m)^2}{2}A_m\right)\sin 2\overline{m-1}\varphi$$
$$\cdots\cdots\cdots\cdots\cdots$$

On the other hand we obtain by multiplying equation 3) by $\sin 2\varphi$:

$$\frac{\sin 2\varphi}{\Delta\varphi} = (A - 2A_2)\sin 2\varphi - (A_1 - 3A_3)\sin 4\varphi + (2A_2 - 4A_4)\sin 6\varphi$$
$$\cdots(-1)^m((m-2)A_{m-2} - mA_m)\sin 2\overline{m-1}\varphi\cdots$$

This series must be identical with the previous one; and since both series converge uniformly, the only way in which this identity can subsist, is, that the coefficients of corresponding sines coincide in both. This becomes evident when, as in the deduction of equations 6), we express each coefficient by a definite integral. Hence:

$$A - 2A_2 = 4A_1\lambda - 8A_2, \quad A_1 - 3A_3 = 16A_2\lambda - 2A_1 - 18A_3,$$
$$(m-2)A_{m-2} - mA_m = (2m-2)^2A_{m-1}\lambda - \frac{(2m-4)^2}{2}A_{m-2} - \frac{(2m)^2}{2}A_m,$$

or:

8) $2m(2m-1)A_m = 2(2m-2)^2A_{m-1}\lambda - (2m-3)(2m-4)A_{m-2}$.

Accordingly the coefficients of series 4) are determined by the equations:

$$A = \frac{2}{\pi}F\left(\frac{\pi}{2}\right), \quad A_1 = \frac{2}{\pi}\left\{\lambda F\left(\frac{\pi}{2}\right) - \frac{2}{k^2}E\left(\frac{\pi}{2}\right)\right\}, \quad A_2 = \frac{4A_1\lambda - A}{6},$$
$$A_3 = \frac{16A_2\lambda - 3A_1}{15}, \text{ etc.}.$$

Similarly an explicit expression is got for the integral $E(\varphi)$.

The third normal integral $\Pi(\varphi)$ requires special investigations, upon which we do not here enter since these series can in general be replaced by more rapidly convergent developments, investigations that demand a detailed theory of elliptic integrals.

Fifth Chapter.

Integrals of transcendental functions.*)

135. If $f(e^x)$ denote a rational function of e^x, $\int f(e^x)dx$ is transformed into a rational integral by substituting $z=e^x$, $\frac{dz}{z}=dx$.

The integral of a rational function of $\sin x$ and $\cos x$ can be converted into one of the above form by substituting:

$$\cos x=\frac{e^{ix}+e^{-ix}}{2},\quad \sin x=\frac{e^{ix}-e^{-ix}}{2i},$$

and therefore also into the integral of a rational function.

However as this introduces imaginary quantities, we ordinarily prefer to substitute:

$$\tan\tfrac{1}{2}x=z,\quad \sin x=\frac{2z}{1+z^2},\quad \cos x=\frac{1-z^2}{1+z^2},\quad dx=\frac{2dz}{1+z^2}.$$

Since $\int x\,df=xf-\int f\,dx$, we can also calculate $\int x\,df$ by the rule of rational functions, when f is any rational function of $\sin x$ and $\cos x$.

136. By partial integration:

$$\int e^x x^m\,dx=x^m e^x-m\int x^{m-1}e^x\,dx,$$

$$\int\frac{e^x}{x^m}\,dx=-\frac{e^x}{m-1}\frac{1}{x^{m-1}}+\frac{1}{m-1}\int\frac{e^x}{x^{m-1}}\,dx.$$

If m be a positive integer, we obtain by the first formula:

$$\int e^x x^m\,dx=\underline{|m}\,e^x\sum_{\nu=0}^{\nu=m}\frac{(-1)^\nu x^{m-\nu}}{\underline{|m-\nu}}.$$

The second formula when m is an integer leads to the equation:

$$\int\frac{e^x}{x^m}\,dx=-e^x\sum_{\nu=1}^{\nu=m-1}\frac{1}{\underline{|\nu}\,(m-1)_\nu x^{m-\nu}}+\frac{1}{\underline{|m-1}}\int\frac{e^x}{x}\,dx.$$

*) Without entering on a general investigation, under what conditions the integrals of transcendental functions can be evaluated in finite terms, we only collect in this chapter those formulas to which the simplest applications of analysis lead us. Euler: *ibid.*, Ch. IV and V. General investigations of these integrals were given by Hermite: Cours d'Analyse, p. 320.

New forms of the same integral are found by substitutions:

$$\int x^m e^{kx}\,dx = \frac{1}{k^{m+1}}\int y^m e^y\,dy, \qquad \left(x = \frac{y}{k}\right),$$

$$\int x^m e^{kx}\,dx = \int (ly)^m y^{k-1}\,dy, \qquad (x = l(y)).$$

137. The logarithmic integral: $\int \frac{e^x\,dx}{x} = \int \frac{dy}{l(y)}$, when $x = l(y)$. Since:

$$\frac{e^x}{x} = \frac{1}{x} + 1 + \frac{1}{\lfloor 2}\,x + \frac{1}{\lfloor 3}\,x^2 + \cdots + \frac{1}{\lfloor n}\,x^{n-1} + \cdots,$$

we have:

$$1) \qquad \int \frac{e^x\,dx}{x} = l(x) + x + \frac{1}{\lfloor 2}\,\frac{x^2}{2} + \frac{1}{\lfloor 3}\,\frac{x^3}{3} + \cdots + \frac{1}{\lfloor n}\,\frac{x^n}{n} + \cdots$$

The definite integral can be taken between two limits that do not include zero, that are thus either both positive or both negative:

$$\int_a^b \frac{e^x}{x}\,dx = l\left(\frac{b}{a}\right) + (b - a) + \frac{1}{2\lfloor 2}\,(b^2 - a^2) + \frac{1}{3\lfloor 3}\,(b^3 - a^3) + \cdots$$

Likewise when $x = -l(y)$:

$$2) \int \frac{e^{-x}\,dx}{x} = \int \frac{dy}{l(y)} = l(x) - x + \frac{x^2}{2\lfloor 2} - \frac{x^3}{3\lfloor 3} + \cdots + (-1)^n \frac{x^n}{n\lfloor n} + \cdots$$

But here a special case presents itself. Since the function $\frac{1}{l(y)}$ is finite and continuous in a finite interval that does not include unity, ex. gr. from $y = 0$ up to any proper fraction y', the value of the integral $\int_0^{y'} \frac{dy}{l(y)}$ which is the same as $\int_\infty^{x'} \frac{e^{-x}\,dx}{x}$, where $x' = -l(y')$ may be any positive number, must be determinate (§ 106). Series 2) however is no longer convergent at the limit $x = +\infty$. Hence we must have:

$$3) \quad \int_{+\infty}^{x} \frac{e^{-x}\,dx}{x} = l(x) - x + \frac{x^2}{2\lfloor 2} - \frac{x^3}{3\lfloor 3} + \cdots + (-1)^n \frac{x^n}{n\lfloor n} + \cdots + C,$$

where the value of the constant C is still to be found. If we denote the convergent series:

$$l(x) - x + \frac{x^2}{2\lfloor 2} - \frac{x^3}{3\lfloor 3} + \cdots$$

by $F(x)$, since the integral vanishes for $x = \infty$, we must have:

$$F(\infty) + C = 0.$$

If then we put a large number a for x, we obtain an approximate value for $C = -F(a)$, of which we can estimate the error as follows:

Forming the function:

$$\varphi(x) = \frac{1}{a}(e^{-a} - e^{-x}),$$

its derived is $\varphi'(x) = \frac{e^{-x}}{a}$. Now since $F'(x) = \frac{e^{-x}}{x}$, $\varphi'(x)$ is $> F'(x)$ if $x > a$. Hence the functions $F(x) - F(a)$ and $\varphi(x)$ both increase continuously from zero for $x \geqq a$, so that the second function is always greater than the first; hence

$$\varphi(\infty) = \frac{e^{-a}}{a} > F(\infty) - F(a),$$

therefore

$$F(\infty) < F(a) + \frac{e^{-a}}{a}.$$

Thus the error incurred in putting $C = -F(a)$ is less than $\frac{e^{-a}}{a}$.*) In this manner, assuming ex. gr. $a = 10$, we determine the value of C from that of $F(10)$ with a defect less than

$$\frac{e^{-10}}{10} < 0{,}00001.$$

The value of C, the Eulerian constant, for which we shall give a more rapidly convergent series § 165, is: 0.5772156649 . . .

138. Integrating by parts:

$$\int x^m \cos x\,dx = x^m \cos(x - \tfrac{1}{2}\pi) - m\int x^{m-1}\cos(x - \tfrac{1}{2}\pi)\,dx,$$

$$\int x^m \sin x\,dx = x^m \sin(x - \tfrac{1}{2}\pi) - m\int x^{m-1}\sin(x - \tfrac{1}{2}\pi)\,dx.$$

Hence for a positive integer m:

$$\int x^m \cos x\,dx = \underline{|m}\sum_{\nu=0}^{\nu=m}\frac{x^\nu}{\underline{|\nu}}(-1)^{m-\nu}\cos\left(x - \frac{m-\nu+1}{2}\pi\right),$$

$$\int x^m \sin x\,dx = \underline{|m}\sum_{\nu=0}^{\nu=m}\frac{x^\nu}{\underline{|\nu}}(-1)^{m-\nu}\sin\left(x - \frac{m-\nu+1}{2}\pi\right).$$

Likewise:

$$\int\frac{\cos x\,dx}{x^m} = -\frac{\cos x}{(m-1)x^{m-1}} - \frac{1}{m-1}\int\frac{\sin x\,dx}{x^{m-1}},$$

$$\int\frac{\sin x\,dx}{x^m} = -\frac{\sin x}{(m-1)x^{m-1}} + \frac{1}{m-1}\int\frac{\cos x\,dx}{x^{m-1}}.$$

In this case a positive integer value of m leads to the integrals:

$$\int\frac{\cos x\,dx}{x}, \quad \int\frac{\sin x\,dx}{x},$$

that can only be found by expansion. We get:

*) Minding: Handbuch der Differential- und Integralrechnung, p. 191.

$$\int \frac{\cos x}{x}\,dx = l(x) - \frac{x^2}{2\underline{|2}} + \frac{x^4}{4\underline{|4}} - \frac{x^6}{6\underline{|6}} + \cdots + (-1)^n \cdot \frac{x^{2n}}{2n\underline{|2n}} \cdots$$

$$\int \frac{\sin x}{x}\,dx = x - \frac{x^3}{3\underline{|3}} + \frac{x^5}{5\underline{|5}} - \frac{x^7}{7\underline{|7}} + \cdots + (-1)^n \cdot \frac{x^{2n+1}}{2n+1\underline{|2n+1}} \text{ etc.}.$$

The first holds for every interval that does not include the number zero, the second without restriction. Moreover, determinate values must result even for infinite limits (§ 155).

139. The integral $\int \sin^m x \cos^n x\,dx$ is converted by the substitution:

$$\sin x = z^{\frac{1}{2}}, \quad \cos x = (1-z)^{\frac{1}{2}}, \quad dx = \frac{\frac{1}{2}dz}{z^{\frac{1}{2}}(1-z)^{\frac{1}{2}}}$$

into the binomial integral:

$$\frac{1}{2}\int z^{\frac{m-1}{2}}(1-z)^{\frac{n-1}{2}}\,dz.$$

The Theorems in § 117 show that this integral can be brought to a rational form when one of the numbers:

$$\tfrac{1}{2}(m-1), \quad \tfrac{1}{2}(n-1), \quad \tfrac{1}{2}(m+n)$$

is integer; one of these equations must be fulfilled if m and n are both integers. In other cases the recurring formulas of § 118 are applicable to the present integral. We can write down these six recurring formulas retaining the trigonometric shape, directly thus:

$$\text{I.} \int \sin^m x \cos^n x\,dx = \frac{\sin^{m+1}x\cos^{n-1}x}{m+1} + \frac{n-1}{m+1}\int \sin^{m+2}x\cos^{n-2}x\,dx,$$

$$\text{II.} \int \sin^m x \cos^n x\,dx = -\frac{\sin^{m-1}x\cos^{n+1}x}{n+1} + \frac{m-1}{n+1}\int \sin^{m-2}x\cos^{n+2}x\,dx.$$

Putting on the right in the first equation $\sin^{m+2}x = \sin^m x(1-\cos^2 x)$, and in the second $\cos^{n+2}x = \cos^n x(1-\sin^2 x)$, we find:

$$\text{III.} \int \sin^m x \cos^n x\,dx = \frac{\sin^{m+1}x\cos^{n-1}x}{m+n} + \frac{n-1}{m+n}\int \sin^m x\cos^{n-2}x\,dx,$$

$$\text{IV.} \int \sin^m x \cos^n x\,dx = -\frac{\sin^{m-1}x\cos^{n+1}x}{m+n} + \frac{m-1}{m+n}\int \sin^{m-2}x\cos^n x\,dx.$$

If we solve these equations for the integrals on the right, and replace in III. $n-2$ by n, and in IV. $m-2$ by m, we obtain:

$$\text{V.} \int \sin^m x \cos^n x\,dx = -\frac{\sin^{m+1}x\cos^{n+1}x}{n+1} + \frac{m+n+2}{n+1}\int \sin^m x\cos^{n+2}x\,dx,$$

$$\text{VI.} \int \sin^m x \cos^n x\,dx = \frac{\sin^{m+1}x\cos^{n+1}x}{m+1} + \frac{m+n+2}{m+1}\int \sin^{m+2}x\cos^n x\,dx.$$

Equations III. and IV. cannot be employed when $m+n=0$. Here we have for $n = -m = -\frac{\mu}{\nu}$:

$$\int \left(\frac{\sin x}{\cos x}\right)^m dx = \int (\tan x)^{\frac{\mu}{\nu}}\,dx = \nu\int \frac{z^{\mu+\nu-1}\,dz}{1+z^{2\nu}}, \text{ putting } \tan x = z^\nu.$$

Similarly the other equations are inapplicable in the cases of m or n being -1. But here too the condition of being integrable rationally holds.

The recurring formulas show that in all other cases the exponents m and n can be brought down to numbers between -1 and $+1$, or 0 and 2. If m and n are integers, we are in all cases led by repeated application of the recurring formulas to one of the eight integrals:

$$\int \sin x\,dx = -\cos x + C. \qquad \int \cos x\,dx = \sin x + C.$$

$$\int \frac{dx}{\sin x} = \int \frac{dz}{z} = l(z) + C = l(\operatorname{tang} \tfrac{1}{2}x) + C. \qquad (\S\ 135.)$$

$$\int \frac{dx}{\cos x} = \int \frac{dy}{\sin y} = l(\tan(\tfrac{1}{2}x + \tfrac{1}{4}\pi)) + C, \qquad (y = x + \tfrac{1}{2}\pi).$$

$$\int \frac{\sin x\,dx}{\cos x} = -l(\cos x) + C. \qquad \int \frac{\cos x\,dx}{\sin x} = l(\sin x) + C.$$

$$\int \sin x \cos x\,dx = \tfrac{1}{2}\int \sin 2x\,dx = -\tfrac{1}{4}\cos 2x + C.$$

$$\int \frac{dx}{\sin x \cos x} = 2\int \frac{dx}{\sin 2x} = l(\operatorname{tang} x) + C.$$

140. Putting $e^{kx}dx = d\left(\frac{e^{kx}}{k}\right)$ and integrating by parts we find:

$$1.\quad \int e^{kx}\sin^n x\,dx = \frac{e^{kx}\sin^n x}{k} - \frac{n}{k}\int e^{kx}\sin^{n-1}x\cos x\,dx.$$

Likewise we find for this new integral:

$$\int e^{kx}\sin^{n-1}x\cos x\,dx = \frac{e^{kx}\sin^{n-1}x\cos x}{k} - \frac{(n-1)}{k}\int e^{kx}\sin^{n-2}x\cos^2 x\,dx + \frac{1}{k}\int e^{kx}\sin^n x\,dx,$$

or, as $\cos^2 x = 1 - \sin^2 x$:

$$2.\quad \int e^{kx}\sin^{n-1}x\cos x\,dx = \frac{e^{kx}\sin^{n-1}x\cos x}{k} - \frac{(n-1)}{k}\int e^{kx}\sin^{n-2}x\,dx + \frac{n}{k}\int e^{kx}\sin^n x\,dx.$$

If we combine equations 1. and 2. we have:

$$3.\quad \int e^{kx}\sin^n x\,dx = \frac{e^{kx}\sin^{n-1}x(k\sin x - n\cos x)}{k^2+n^2} + \frac{n(n-1)}{k^2+n^2}\int e^{kx}\sin^{n-2}x\,dx.$$

For $n = 1$ this is:

$$\int e^{kx}\sin x\,dx = \frac{e^{kx}(k\sin x - \cos x)}{k^2+1}.$$

For $n = 0$ we had already:

$$\int e^{kx}\,dx = \frac{e^{kx}}{k}.$$

Every integral of this kind, in which n is a positive integer, is reduced by formula 3. to one of these two integrals.

When k is a negative number, the integral can be taken up to a positive infinite limit, for although $\sin x$ and $\cos x$ become quite indeterminate between -1 and $+1$, yet e^{kx} in the integral function will pass over continuously into zero. Thus we have for $k < 0$:

$$\int_0^\infty e^{kx}\sin x\,dx = \frac{1}{k^2+1}.$$

In like manner we find:

$$4. \quad \int e^{kx}\cos^n x\,dx = \frac{e^{kx}\cos^{n-1}x\,(k\cos x + n\sin x)}{k^2+n^2} + \frac{n(n-1)}{n^2+k^2}\int e^{kx}\cos^{n-2}x\,dx.$$

Here

$$\int e^{kx}\cos x\,dx = \frac{e^{kx}(k\cos x + \sin x)}{k^2+1},$$

and for $k < 0$:

$$\int_0^\infty e^{kx}\cos x\,dx = -\frac{k}{k^2+1}.$$

141. If circular functions occur in the function to be integrated, the process of integration by parts leads likewise in many cases to a solution or simplification of the problem. If in the integral

$$\int X\sin^{-1}x\cdot dx$$

the function X be integrable we have:

$$\int X\sin^{-1}x\,dx = \sin^{-1}x\int X\,dx - \int\left\{\frac{dx}{\sqrt{1-x^2}}\int X\,dx\right\}.$$

Ex. gr.: $$\int x^n\sin^{-1}x\,dx = \frac{x^{n+1}}{n+1}\sin^{-1}x - \frac{1}{n+1}\int\frac{x^{n+1}\,dx}{\sqrt{1-x^2}}.$$

This new binomial integral can be expressed in finite terms when n is an integer. We can also get rid of the circular functions by introducing algebraic and trigonometric functions, putting

$$\sin^{-1}x = z, \quad x = \sin z, \quad dx = \cos z\,dz.$$

Thus ex. gr.: $$\int(\sin^{-1}x)^n\,dx = \int z^n\cos z\,dz.$$

Sixth Chapter.

General theorems concerning the definite integral as the limiting value of a sum.

142. The fundamental problem of the Integral Calculus in its simplest statement (§ 101) leads to the evaluation of the limiting value of a sum with arbitrarily many summands. Independently therefore of the differential conception, the problem of the integral calculus opens up the question: *What must be the nature of a function $f(x)$ in the interval from $x = a$ to $x = b$, in order that the sum:*

$$S = d_1 f(a + \Theta_1 d_1) + d_2 f(x_1 + \Theta_2 d_2) + d_3 f(x_2 + \Theta_3 d_3) + \cdots + d_n f(x_{n-1} + \Theta_n d_n)$$

may have a determinate finite limiting value, when the subdivision of the interval from a to b by the points:

$$x_1 = a + d_1, \quad x_2 = x_1 + d_2, \quad x_3 = x_2 + d_3 \ldots, \quad b = x_{n-1} + d_n$$

is continued arbitrarily, while the lengths d converge to zero? Such is the most general form in which this question can be proposed. The quantities Θ denote proper fractions, they may also be zero or unity, so that the values of the function are always chosen anywhere within or at the limits of an interval. The limiting value must be quite independent of the arbitrary quantities Θ. Still more generally we may denote by $f(x + \Theta_p d_p)$ any value whatever from the greatest to the least of the values assumed by the function in the interval d_p. If it be discontinuous in the interval, this selected value may not occur among those of the function. It is a secondary question that must be answered by itself, whether, when there is a limiting value, this limiting value regarded as a function of the upper limit has $f(x)$ as its derived function or not.

Now while in § 102 the investigation admitted of a simple form, because $f(x)$ was assumed continuous, it will have to be conducted differently here, since we have first to ascertain the hypotheses necessary regarding the function $f(x)$. Riemann*) who was the first

*) Riemann: Ueber die Darstellbarkeit einer Function durch eine trigonometrische Reihe (Werke, pp. 213—253). Some details in the following proof have been rendered more precise by Du Bois-Reymond (J. f. M., Vol. 79).

to formulate the problem precisely, has also supplied its solution. We restrict ourselves at first to functions which do not become infinite anywhere in the finite interval from a to b, so that all the values of the function are included between a superior limit that may be denoted by G and an inferior that may be denoted by g, these being positive or negative.

The function must be one that is defined without exception for this interval; i. e. its value belonging to each point is actually given.

The question proposed above may now be abridged into the words: *Under what hypotheses is such a function integrable?* The answer is:

If we denote the greatest fluctuation of the function, i. e. the positive difference of its greatest and least values, in the interval from a to x_1 including those limits by D_1, likewise between x_1 and x_2 by $D_2, \ldots$ between x_{n-1} and b by D_n, then the limit of the sum:

$$d_1 D_1 + d_2 D_2 + \cdots + d_n D_n \qquad (d_1 + d_2 + \cdots + d_n = b - a)$$

must be zero as the values of n increase, when simultaneously all the quantities d converge to zero.

This is the necessary and sufficient condition. What takes place is:

When the above sum converges to zero for any law by which the number of intervals increases arbitrarily, it always converges to zero in whatever manner the quantities d are chosen and arbitrarily diminished.

We prove this last statement in the first place as follows:

Suppose the number n already chosen so large, that the absolute amount of $\sum_{p=1}^{p=n} d_p D_p$, since its limiting value vanishes, is less than σ. Choosing another completely different subdivision into m parts, where $m > n$, and the quantities $d_1', d_2', \ldots d_m'$ are arbitrarily smaller than the least of the quantities d, we shall show that $\sum_{p=1}^{p=m} d_p' D_p'$ also converges to zero. Let us consider the length ab simultaneously divided into n intervals d and into m intervals d', then there will be in each part d a certain number of the intervals d'; but, in general, extremities of the intervals d' will not coincide with extremities of the intervals d. Suppose:

$$a + d_1' + d_2' + \cdots d_\lambda' < a + d_1 = x_1 < a + d_1' + \cdots d'_{\lambda+1},$$
$$a + d_1' + d_2' + \cdots d_\mu' < x_1 + d_2 = x_2 < a + d_1' + \cdots d'_{\mu+1},$$
$$a + d_1' + d_2' + \cdots d_\nu' < x_2 + d_3 = x_3 < a + d_1' + \cdots d'_{\nu+1},$$
$$\cdots\cdots\cdots\cdots\cdots\cdots;$$

then we can separate the sum $\sum_{p=1}^{p=m} d_p' D_p'$ into parts as follows:

$$\sum_{p=1}^{p=\lambda} d_p' D_p' + d'_{\lambda+1} D'_{\lambda+1} + \sum_{p=\lambda+2}^{p=\mu} d_p' D_p' + d'_{\mu+1} D'_{\mu+1} + \cdots,$$

isolating the terms that refer to intervals that contain the dividing points of the first partition; they are $n-1$ in number. But since D_1 denotes the greatest fluctuation in the whole interval d_1, we have:

$$\sum_{p=1}^{p=\lambda} d_p' D_p' \leqq d_1 D_1, \quad \sum_{p=\lambda+2}^{p=\mu} d_p' D_p' \leqq d_2 D_2, \text{ etc..}$$

Further, the sum of all the isolated terms is certainly smaller than $n-1$ times the product of the greatest d' of their intervals by the greatest fluctuation D' occurring among them. Therefore:

$$\sum_{p=1}^{p=m} d_p' D_p' \leqq \sum_{p=1}^{p=n} d_p D_p + (n-1) d' D'.$$

Now since we can arbitrarily diminish the values d', we can always choose them so small that the product $(n-1)d'D' = \varepsilon$ shall become arbitrarily small; therefore we have:

$$\sum_{p=1}^{p=m} d_p' D_p' \leqq \sigma + \varepsilon,$$

i. e. this sum also becomes arbitrarily small by suitable choice of m.

We now proceed with the proof of the above theorem as follows:

Let the entire interval from a to b be divided, in succession, first into n_1, then into $n_2, n_3, \ldots n_\nu, \ldots$ parts; and let:

each segment of the first partition: $d_1^{(1)}, d_2^{(1)}, \ldots d_{n_1}^{(1)}$ be smaller than δ_1,
each segment of the second partition: $d_1^{(2)}, d_2^{(2)}, \ldots d_{n_2}^{(2)}$ be smaller than δ_2,
each segment of the third partition: $d_1^{(3)}, d_2^{(3)}, \ldots d_{n_3}^{(3)}$ be smaller than δ_3,
. . . .
each segment of the νth partition: $d_1^{(\nu)}, d_2^{(\nu)}, \ldots d_{n_\nu}^{(\nu)}$ be smaller than δ_ν,
. . . . ;

let $\delta_1, \delta_2, \delta_3, \ldots \delta_\nu \ldots$ form a series of positive numbers converging to zero; and let the dividing points of the second partition include all the dividing points of the first, and likewise let those of each further partition include all the dividing points of the preceding one, so that each interval $d_1, d_2 \ldots$ is divided into new subdivisions.

Let $G_\mu^{(\nu)}$ denote the superior limit of the function $f(x)$ within the interval $d_\mu^{(\nu)}$, taking the sign into account, and similarly $g_\mu^{(\nu)}$ its inferior limit. Then let us form the sums:

$$\sum_{\mu=1}^{\mu=n_\nu} d_\mu^{(\nu)} G_\mu^{(\nu)}, \qquad \sum_{\mu=1}^{\mu=n_\nu} d_\mu^{(\nu)} g_\mu^{(\nu)},$$

and denote their values by A_ν and B_ν. In the series of numbers: $A_1, A_2, A_3, \ldots A_\nu \ldots$, each number is less than, or at most equal to, the preceding, for, whereas one interval ex. gr. $d_2^{(1)}$ contributes $d_2^{(1)} G_2^{(1)}$ in A_1, the same interval in the sum A_2 consists of several parts. But these partial intervals certainly do not contribute more to the sum A_2 than the product $d_2^{(1)} G_2^{(1)}$, because $G_2^{(1)}$ denotes the greatest number that occurs in the entire interval $d_2^{(1)}$, and therefore also in any of its subdivisions.

The quantities $B_1, B_2, B_3, \ldots B_\nu \ldots$ form a series of increasing numbers; and since each A is greater than each B, the series of numbers A as well as the series B has each a definite limiting value. These limiting values become identical when a place ν can be assigned for which and for all greater values of ν the difference $A_\nu - B_\nu$ is less than an arbitrarily small number σ, i. e. such that:

$$A_\nu - B_\nu = \sum_{\mu=1}^{\mu=n_\nu} d_\mu^{(\nu)} (G_\mu^{(\nu)} - g_\mu^{(\nu)}) = \sum_{\mu=1}^{\mu=n_\nu} d_\mu^{(\nu)} D_\mu^{(\nu)} < \sigma.$$

In whatever way the quantities Θ are assumed, the sum:

$$S = d_1^{(\nu)} f(a + \Theta_1 d_1^{(\nu)}) + d_2^{(\nu)} f(a + d_1^{(\nu)} + \Theta_2^{(\nu)} d_2^{(\nu)}) + \cdots$$
$$+ d_{n_\nu}^{(\nu)} f(b - \Theta_n d_{n_\nu}^{(\nu)}),$$

whose limiting value defines the integral, always lies between the limiting values of A_ν and B_ν and when these are equal, this sum has also the same finite and determinate value; as we undertook to prove.

The condition enunciated is sufficient; but it is also necessary. For if both series of numbers A and B had not the same limiting value, the limiting value of the integral sum could, by varying the quantities Θ, be brought to coincide with the limiting value either of A or of B; thus it would not be independent of the quantities Θ.

But the proof is not yet complete; for it was assumed, that the successive partitions are always carried out so that the extremities of a partial interval occur also as extremities in the subsequent partitions. The questions therefore arise: Is the value of S quite independent of the choice of dividing points? and is it also independent of the manner of continuing the partition? Let:

$$a, \; x_1, \; x_2, \ldots x_{n-1}, \; b$$

and:

$$a, \; x_1', \; x_2', \ldots x'_{m-1}, \; b$$

be the extremities in two partitions. Suppose these in whatever way commenced to have been carried on quite independently according to our process so far, that the value S_n of the sum belonging to the former differs from its limiting value S only by the arbitrarily small quantity ε, whereas S_m' likewise differs from its limiting value S' only by ε'; for, as was proved at first, every partition must lead to a definite limiting value provided there be one for any such partition. If now we imagine the two partitions combined into a single one and form the corresponding sum $S_{m,n}$ relative to this single partition resulting from their combination, this may be regarded as a step forward as well in the series S_n as also in the series S_m'; hence $S_{m,n}$ only differs from S by a quantity $\eta < \varepsilon$, and from S' by a quantity $\eta' < \varepsilon'$:

$$S = S_{m,n} \pm \eta, \quad S' = S_{m,n} \pm \eta'.$$

The absolute difference $S - S'$ will therefore not be more than $\eta + \eta'$, it is less than the arbitrarily small quantity $\varepsilon + \varepsilon'$; the limiting values S and S' consequently are identical. The second question also is answered by the same process. If we consider a succession of different independent partitions: in the first let each of the intervals be less than $\delta^{(1)}$, and the sum of their products by the fluctuations be less than $\varepsilon^{(1)}$; in the νth let each of the intervals be less than $\delta^{(\nu)}$, and the sum of their products by the fluctuations be less than $\varepsilon^{(\nu)}$; further let:

$$S^{(1)}, \quad S^{(2)}, \ldots S^{(\nu)} \ldots$$

be the respective values of the sum; we can again combine the νth partition with the first and regard this combination as a continuation of the first as well as of the νth partition. Denoting the value of the sum relative to the combination by S', we have:

$$S' = S^{(1)} \pm (< \varepsilon^{(1)}), \quad S' = S^{(\nu)} \pm (< \varepsilon^{(\nu)}).$$

Therefore $S^{(1)}$ and $S^{(\nu)}$ differ by less than the arbitrarily small quantity $\varepsilon^{(1)} + \varepsilon^{(\nu)}$; i. e. the series of the sums S has a determinate limiting value.

The limiting value of the sums S is called the definite integral and is denoted by the symbol:

$$\int_a^b f(x)\,dx.$$

143. The condition thus established is fulfilled:

First: when the function $f(x)$ is throughout; continuous; this was proved directly in §§ 102 and 103. For, in this case a quantity δ can be found such that at all points, in intervals that are equal to or less than δ, the fluctuations of the function:

$$\text{abs}\,[f(x) - f(x + \Theta\delta)]$$

are smaller than an arbitrarily small number σ. Hence we have:

$$\sum dD < (b - a)\sigma.$$

Second: when the function $f(x)$ has finite discontinuities at a finite number of separate points (§ 105). For, these separate points, suppose n in number, can be included within arbitrarily small intervals δ, such that when D denotes the largest value among the sudden changes, these intervals of discontinuity do not contribute more to the sum S than $n\delta D$. Since n and D are finite, δ can be chosen so as to make this product arbitrarily small.

We can likewise see that, at each of an arbitrary finite number of points, the function can have, within an interval however small, infinitely many maxima and minima with finite fluctuations, (as ex. gr. $\sin\frac{1}{x-a}$ at the point $x = a$), or even that it can be left altogether indeterminate, i. e. that we are at liberty to attribute any finite value whatever to the function at such a number of points, without the value of the definite integral being thereby altered.

Third: when at an infinite number of points the function is discontinuous or indeterminate between finite limits, or else, within an interval however small, has infinitely many maxima and minima with arbitrary finite fluctuations; provided this infinite system of points answer to a certain definite description. Into this we shall enter in the next Paragraph because the investigation presents an occasion for us to extend considerably our conception of a function.

144. To grasp the conception of an infinite number of points, we must first of all dwell upon the difference: a finite length contains infinitely many points, but infinitely many points do not necessarily fill up a length, or in purely arithmetical language: the continuous series of numbers between any two limits, contains infinitely many numbers between these limits, but yet infinitely many numbers between two limits do not fill up the series of numbers. In order to characterise this difference we introduce the following definitions:*)

Naming the interval from $x-\varepsilon$ to $x+\varepsilon$, whose length is any arbitrarily small finite quantity 2ε, the neighbourhood of a point x, we shall call an infinite multiplicity of points a discrete set or mass of points, when it is possible to include all of these points within neighbourhoods whose sum can be made smaller than an arbitrarily small length, while the number of the neighbourhoods can increase arbitrarily.

*) The investigation of infinite sets of points first given concisely (1871) by G. Cantor, Math. Annal., Vol. V, is developed also in Dini: Fondamenti per la teorica delle funzioni di variabili reali. Pisa 1878. The above distinction of discrete and linear sets of points differs however from Cantor's definition of sets of the first and second species (Math. Annal., Vol. XV, p. 2). In strictness, by the phrase "discrete set" of points or values, we imply that for the problems of the integral calculus such a set has the same property as a finite number of separate points or values, often using it briefly whether the number is finite or infinite. See Ex. 1).

But on the other hand we shall describe the infinite system as a **linear set or mass of points**, when the sum of the neighbourhoods cannot be arbitrarily diminished.

From these definitions it follows: (1) In case of a discrete set of points it is always possible to assign intervals, whose sum differs arbitrarily little from $b - a$, such that there is no point of the given set in any of these intervals; for we need only exclude the points along with their neighbourhoods, of which the sum is arbitrarily small. In case of a linear set of points the sum of such intervals always differs by a finite quantity from $b - a$.

(2) In case of a discrete set of points it is always possible, arbitrarily near any position whatever, to assign a finite interval within which there is no point of the given set. For, if α be an arbitrary point of the interval, it could only be impossible to find arbitrarily near α an interval devoid of any point of the set, if in the neighbourhood of each point within the finite distance δ of α there were infinitely many points of the set. But then it must also have been impossible originally to have included all the points of the set within neighbourhoods having their sum less than δ; i. e. the set could not have been discrete.

In case of a linear set of points this is not possible everywhere.

These differences will become clear by the following Examples:

1) Every finite number of points in an interval of finite length is discrete; considering it as a set, we denote its order by zero.

2) The infinite set of points in the interval from 0 to 1, which is determined by the numbers:

$$1,\ \tfrac{1}{2},\ (\tfrac{1}{2})^2,\ (\tfrac{1}{2})^3,\ (\tfrac{1}{2})^4 \ldots (\tfrac{1}{2})^n \ldots$$

is discrete, for, the points of this set concentrate only at zero. If we separate off an arbitrarily small interval beginning from the point zero, we leave only a finite number of points of the set in the remaining length, so that the total sum of all the neighbourhoods can be made arbitrarily small.

The positions at which points of any set concentrate or condense themselves infinitely are called its **limiting points**; the set of the limiting points is called the **first derived set of points**. In the present case the first derived set is of the order zero; the order of the original set is therefore denoted by unity.

3) A discrete set of points can have more derived sets than one, or be of higher order. The points:

$$1,\ \tfrac{1}{2},\ (\tfrac{1}{2})^2,\ \tfrac{1}{2} + (\tfrac{1}{2})^2,\ (\tfrac{1}{2})^3,\ \tfrac{1}{2} + (\tfrac{1}{2})^3,\ (\tfrac{1}{2})^2 + (\tfrac{1}{2})^3,\ (\tfrac{1}{2})^4,\ (\tfrac{1}{2}) + (\tfrac{1}{2})^4, \cdots$$

concentrate at infinitely many positions, namely at the points:

$$0,\ \tfrac{1}{2},\ (\tfrac{1}{2})^2,\ (\tfrac{1}{2})^3, \cdots \text{ etc.}.$$

Nevertheless they form a discrete set. For if we lay off an arbitrarily small interval from the zero, there remains only a finite number of points at which concentrations of the given set of points occur; and if we include these within arbitrarily small intervals, there remains further only a finite number of points of the given set. The first derived set is here of the order one; the order of the original set is two.

4) Considering all rational numbers in the interval from a to b as a set of points, we have a set of the second kind, a linear mass of points. For, at no position can a finite interval be assigned, within which there are not infinitely many points of this set; such a linear set answers to the description "everywhere dense" within a finite interval. Similarly all irrational numbers form a linear set, and so do also all numbers that when reduced to their lowest terms have as denominator a power of any number n.

Here moreover no derived set of lower order can be found, because each point of a segment however small is a limiting point.

It can be shown on the other hand quite generally: Whenever a set of points has a finite number of derived sets, it is discrete.

For, starting from the last derived set of order zero, i. e. from a finite number of points $a_1, a_2, \ldots a_m$, the set of points of which this is the derived set, contains only at these positions concentrations of infinitely many points, and further a finite number of points $b_1, b_2, \ldots b_n$. The sum total of their neighbourhoods can be made arbitrarily small. The set of the first order is discrete. From this let us proceed to the set of the next higher order. Having included the positions $a_1, a_2, \ldots a_m$, and $b_1, b_2, \ldots b_n$, within arbitrarily small neighbourhoods, the new set contains further only a finite number of points $c_1, c_2, \ldots c_p$. It accordingly is likewise discrete; therefore the character of the discrete mass of points is preserved through any finite number of such ascending processes.*)

145. A function that is generally continuous yet in infinitely many points is either discontinuous or completely indeterminate between finite limits, or else, within an interval however small has infinitely many finite maxima and minima, we call discretely discontinuous, whenever the points at which the fluctuations of the function exceed a determinate finite number σ, form only a discrete set of points.**)

*) The conception of derived sets of points was introduced by Cantor (Math. Annal., Vol. V, p. 129). His example (*ib.* Vol. XVII, p. 358) shows that a discrete set of points can also have infinitely many derived sets, being thus of his second species.

**) These definitions are essentially connected with those given by H. Hankel (1839—73) in his: Untersuchungen über die unendlich oft oscillirenden und unstetigen Functionen, Tübingen 1870. Reprinted Math. Annal., Vol. XX, p. 63.

By "fluctuation" of the function is meant the magnitude of the breaches of continuity, or, the difference of the limits between which the indeterminate values lie, or lastly the difference between the maximum and minimum values.

On the other hand we call a function linearly discontinuous, in which such points form a linear set of points. Now it is easy to see from Riemann's Theorem that within its interval:

A discretely discontinuous function is integrable.

For, σ being a prescribed arbitrarily small finite number, we can carry the partition of the length ab so far, that in the partial intervals generally the fluctuations become less than σ, while s, the sum of the neighbourhoods of all the points at which the fluctuations exceed σ, can be arbitrarily diminished. Let m be the greatest value among these fluctuations, then:

$$\sum_{\mu=1}^{\mu=n} dD < \sigma(b-a) + sm.$$

But the sum on the right can be diminished arbitrarily, since σ may be assumed arbitrarily small, and likewise s in consequence of the property of a discrete set of points.

Linearly discontinuous functions are not integrable.

Examples of functions infinitely often discontinuous that are integrable.

1) Let the value of the function $f(x)$ be zero everywhere in the interval from 0 to 1, except in the infinite series of points:

$$\tfrac{1}{2},\ (\tfrac{1}{2})^2,\ (\tfrac{1}{2})^3,\ (\tfrac{1}{2})^4, \ldots (\tfrac{1}{2})^n \ldots$$

in which its value is to be $\frac{1}{2}$. This function is infinitely often discontinuous within an arbitrarily small interval from zero; but the sum of the intervals in which the fluctuations are $\frac{1}{2}$ can be made arbitrarily small. The value of the integral is determinate, it vanishes.

2) We can likewise construct a function that is integrable, although it is discontinuous in every interval however small, and though the number of points, at which it has discontinuities greater than some finite number, is not finite. Determining, ex. gr., that the function $f(x)$ is to vanish generally in the interval from zero to unity, but yet that at all points of a discrete set of which the point $\frac{1}{2}$ is the derived set its value is to be $\frac{1}{2}$; at the set of points whose derived set are the points $\frac{1}{3}$, $\frac{2}{3}$, its value is to be $\frac{1}{3}$; at the set of points with the derived set $\frac{1}{5}$, $\frac{2}{5}$, $\frac{3}{5}$, $\frac{4}{5}$, its value is to be $\frac{1}{5}$; and generally: if p be a prime number and q denote each number smaller than p, at the set of points with the derived set $\frac{q}{p}$, its value is to be $\frac{1}{p}$,

the integral of the function so defined is zero. The sets of points in question we may ex. gr. conceive formed by the series:

$$\frac{q}{p} + (\tfrac{1}{2})^m, \quad \frac{q}{p} + (\tfrac{1}{2})^{m+1}, \quad \frac{q}{p} + (\tfrac{1}{2})^{m+2}, \cdots \text{ etc..}$$

For, the points at which the discontinuities exceed some given finite number, form always only a finite number of discrete sets of points. The sum of their neighbourhoods becomes arbitrarily small.

3) The first example of a discretely discontinuous integrable function was given by Riemann.*) Let (x) denote the positive or negative excess of x over the nearest integer less or greater; when x is midway between two integers, let $(x) = 0$. The series:

$$f(x) = \frac{(x)}{1} + \frac{(2x)}{2^2} + \frac{(3x)}{3^2} + \cdots = \sum_{m=1}^{m=\infty} \frac{(mx)}{m^2}$$

converges; for, $\frac{1}{1} + \frac{1}{2^2} + \frac{1}{3^2} + \cdots$ is convergent (§ 47) and its sum found from the expansions of $\tan x$ and $\cot x$ is $\frac{1}{6}\pi^2$; also the value $\frac{1}{2}$ is a superior limit of (mx). Each term of the series $f(x)$ is generally continuous, only when $2mx =$ an odd integer p, neighbouring values in the function (mx) differ from each other *quam proxime* by 1. When $x = \frac{p}{2m}$ this takes place not only with the term (mx) but also with the terms $(3mx)$, $(5mx)$, etc.. Hence follows: When x is of the form $\frac{p}{2m}$, where p is prime to m:

$$f(x+0) = f(x) - \frac{\pi^2}{16m^2}, \quad f(x-0) = f(x) + \frac{\pi^2}{16m^2}.$$

For, when $x = \frac{p}{2m}$ the terms named contribute nothing, they vanish; while when x begins to increase, they increase each *quam proxime* by $-\frac{1}{2}$, and when x decreases, each increases by $+\frac{1}{2}$. But:

$$\frac{1}{2m^2}\left\{\frac{1}{1} + \frac{1}{3^2} + \frac{1}{5^2} + \cdots\right\} = \frac{1}{2m^2}\left\{\frac{1}{1} + \frac{1}{2^2} + \frac{1}{3^2} + \cdots - \frac{1}{2^2}\left(\frac{1}{1} + \frac{1}{2^2} + \frac{1}{3^2} + \cdots\right)\right\}$$

$$= \frac{1}{2m^2}\left\{\frac{\pi^2}{6} - \frac{\pi^2}{24}\right\} = \frac{\pi^2}{16m^2}.$$

For each rational value of x, that in its lowest terms is a fraction with an even denominator $2m$, there is therefore a discontinuity of $f(x)$; and thus, infinitely many between any two limits however close. But the number of such discontinuities whose value exceeds a given limit is finite; for, if $\frac{\pi^2}{8m^2}$ must be $> \sigma$, m must be $< \frac{\pi}{2\sqrt{2\sigma}}$. But in

*) Gesammelte Werke, p. 228.

a finite interval there are only a finite number of fractions having denominators below a given finite limit.

The series converges uniformly; it is not continuous, because its separate terms are not continuous functions; the integral is obtained by integration of the separate terms.

146. The fundamental theorems concerning the definite integral follow immediately from the equation of definition, whose shortest form, independent moreover of the quantities Θ, is:

$$\int_a^b f(x)dx = \mathrm{Lim}\{(x_1 - a)f(a) + (x_2 - x_1)f(x_1)\cdots + (b - x_{n-1})f(x_{n-1})\}.$$

We have:

I. $$\int_a^b cf(x)dx = c\int_a^b f(x)dx.$$

II. Interchanging a with b, and keeping the same partition of the interval, the integral:

$$\int_b^a f(x)\,dx$$

is obviously equal to:

$$\mathrm{Lim}\{(x_{n-1} - b)f(b) + (x_{n-2} - x_{n-1})f(x_{n-1}) + \cdots + (x_1 - x_2)f(x_2) + (a - x_1)f(x_1)\}$$

and also to:

$$\mathrm{Lim}\{(x_{n-1} - b)f(x_{n-1}) + (x_{n-2} - x_{n-1})f(x_{n-2}) + \cdots + (x_1 - x_2)f(x_1) + (a - x_1)f(a)\}.$$

For, as was shown, it is indifferent at what points within an interval the values of the function are chosen.

It becomes evident by this second equality that:

$$\int_a^b f(x)dx = -\int_b^a f(x)dx,$$

i. e. the integral changes only in sign by interchanging its upper and lower limits.

III. $$\int_a^c f(x)dx + \int_c^b f(x)\,dx = \int_a^b f(x)dx.$$

This equation holds even when c lies outside the interval a to b, provided only the function remains integrable. For, when:

$$a < b < c$$

we have:

$$\int_a^c f(x)\,dx = \int_a^b f(x)\,dx + \int_b^c f(x)\,dx,$$

therefore:

$$\int_a^b f(x)\,dx = \int_a^c f(x)\,dx - \int_b^c f(x)\,dx = \int_a^c f(x)\,dx + \int_c^b f(x)\,dx.$$

IV. The sum of integrable functions is itself integrable, we have:

$$\int_a^b \{f_1(x) \pm f_2(x) \pm \cdots \pm f_n(x)\}\,dx = \int_a^b f_1(x)\,dx \pm \int_a^b f_2(x)\,dx \pm \cdots \pm \int_a^b f_n(x)\,dx.$$

V. The product of two or more integrable functions is itself integrable.*)

We must remember that the foregoing discussions only deal with functions that do not become infinite. An extension of theorem V. will be found in § 149.

In the interval d_p let the value of the greatest fluctuation of the function $\varphi(x)$ be D_p and of $\psi(x)$ be D_p'. We have by hypothesis:

$$\sum_{p=1}^{p=n} d_p D_p = 0, \quad \sum_{p=1}^{p=n} d_p D_p' = 0, \qquad \text{for } n = \infty.$$

The product $\varphi(x) \,.\, \psi(x)$ is subject in the same interval to fluctuations, which, if $x + \Theta d_p$ and $x + \Theta' d_p$ denote the places of its greatest and least values, are measured by the difference:

$$\begin{aligned} \varphi(x + \Theta d_p)\psi(x + \Theta d_p) &- \varphi(x + \Theta' d_p)\psi(x + \Theta' d_p) \\ &= \varphi(x + \Theta d_p)\{\psi(x + \Theta d_p) - \psi(x + \Theta' d_p)\} \\ &+ \psi(x + \Theta' d_p)\{\varphi(x + \Theta d_p) - \varphi(x + \Theta' d_p)\}. \end{aligned}$$

This form shows that the fluctuation of the product is certain not to exceed $G_p D_p' + G_p' D_p$, if G_p and G_p' denote the greatest absolute amounts which the functions φ and ψ assume in the interval d_p. When G and G' are the greatest of all the absolute amounts which these functions assume in the entire interval of integration, we have:

$$\sum_{p=1}^{p=n} \{\varphi(x + \Theta d_p)\,\psi(x + \Theta d_p) - \varphi(x + \Theta' d_p)\,\psi(x + \Theta' d_p)\}\,d_p$$
$$< G \sum_{p=1}^{p=n} d_p D_p' + G' \sum_{p=1}^{p=n} d_p D_p$$

therefore in consequence of our hypothesis, it vanishes. Q. E. D.

---

*) Du Bois-Reymond, Journal f. Math., Vol. 79, p. 21.

VI. Integration by parts. (Partial integration.)

Let the functions $\varphi(x)$, $f(x)$, as also their product be integrable; further let $\varphi(x)$ be a function everywhere continuous and have the integrable derived function $\varphi'(x)$, so that therefore (see § 147):

$$\int_c^x \varphi'(y)dy = \varphi(x) - \varphi(c),$$

where c and x mean arbitrary points in the interval of integration; we have then:

$$\int_a^b f(x)\varphi(x)dx = \int_a^b f(x)\left\{\int_c^x \varphi'(y)dy + \varphi(c)\right\}dx$$

$$= \varphi(c)\int_a^b f(x)dx + \int_a^b dx\left\{f(x)\int_c^x \varphi'(y)dy\right\}.$$

Now putting $c = a$, we find:

I) $$\int_a^b f(x)\varphi(x)dx = \varphi(a)\int_a^b f(x)dx + \int_a^b dx\left\{f(x)\int_a^x \varphi'(y)dy\right\}.$$

Putting $c = b$, and $\int_b^x \varphi'(y)dy = -\int_x^b \varphi'(y)dy$, we find:

II) $$\int_a^b f(x)\varphi(x)dx = \varphi(b)\int_a^b f(x)dx - \int_a^b dx\left\{f(x)\int_x^b \varphi'(y)dy\right\}.$$

It will be proved in § 168 that the order of these integrations on the right can be interchanged, at least, if $f(x)$ and $\varphi'(y)$ remain finite within the domain of integration; that in fact we have:

$$\int_a^b dx\left\{f(x)\int_a^x \varphi'(y)dy\right\} = \int_a^b dy\left\{\varphi'(y)\int_y^b f(x)dx\right\},$$

$$\int_a^b dx\left\{f(x)\int_x^b \varphi'(y)dy\right\} = \int_a^b dy\left\{\varphi'(y)\int_a^y f(x)dx\right\}.$$

Accordingly we obtain from I) and II) the formulas:

$$\int_a^b f(x)\varphi(x)dx = \varphi(a)\int_a^b f(x)dx + \int_a^b dy\left\{\varphi'(y)\int_y^b f(x)dx\right\},$$

$$\int_a^b f(x)\varphi(x)dx = \varphi(b)\int_a^b f(x)dx - \int_a^b dy\left\{\varphi'(y)\int_a^y f(x)dx\right\}.$$

These are known as the formulas of integration by parts; they contain the application to a definite integral of the method of § 108 b.*)

VII. The First Theorem of the Mean Value. (Cf. § 103.) From the definition by the sum immediately follows:

$$\int_a^b f(x)\,dx = \{g + \Theta(G - g)\}\,(b - a) \qquad (0 \leqq \Theta \leqq 1),$$

g denoting the least and G the greatest value of $f(x)$ in the interval. When $f(x)$ is continuous, it cannot overleap any intermediate value; therefore the equation can be written:

$$\int_a^b f(x)\,dx = (b - a)\,f(a + \Theta\,(b - a)).$$

The following generalisation of the Theorem of the Mean Value is often useful: If $f(x)$ and $\varphi(x)$ are integrable, further, if within the interval of integration the function $\varphi(x)$ has always the same sign, suppose positive, then g denoting the inferior and G the superior limit of $f(x)$, we have:

$$g\int_a^b \varphi(x)\,dx \leqq \int_a^b f(x)\,\varphi(x)\,dx \leqq G\int_a^b \varphi(x)\,dx$$

or:

$$\int_a^b f(x)\,\varphi(x)\,dx = \{g + \Theta(G - g)\}\int_a^b \varphi(x)\,dx \qquad (0 \leqq \Theta \leqq 1).$$

When the function $f(x)$ is continuous, this can be written:

$$\int_a^b f(x)\,\varphi(x)\,dx = f(a + \Theta(b - a))\int_a^b \varphi(x)\,dx \qquad (0 \leqq \Theta \leqq 1).$$

VIII. From the Theorem of the Mean Value in connexion with theorem III. (as in § 104) follows that: The definite integral is a continuous function of its upper limit.

For, x and $x \pm h$ denoting any values within the interval ab, if we call the integral as a function of its upper limit x briefly $F(x)$, we have:

---

*) These formulas also hold when $f(x)$ and $\varphi'(x)$ become infinite within the domain of integration, provided only their integrals as well as that of the product $f(x)\varphi(x)$ are finite and therefore also $\varphi(x)$ remains a continuous function. The theorem is not liable to any further restrictions than those just mentioned. Cf. Du Bois-Reymond, Abhandl. der k. bayer. Akad. d. Wissensch., II. Classe, XII. Vol., I. Abthl., p. 133.

$$F(x\pm h)-F(x)=\int_a^{x\pm h} f(x)dx-\int_a^x f(x)dx=\int_x^{x\pm h} f(x)dx=\pm h.(g+\Theta(G-g)),$$

G denoting the greatest and g the least value of $f(x)$ within the interval $+h$ or $-h$. Since the function f is finite throughout, an interval $\pm h$ can be assigned at each point x, within which this difference becomes less than an arbitrarily small number.

IX. When $f(x)$ and $\varphi(x)$ are integrable, and further the function $f(x)$ within the interval of integration is a positive quantity decreasing everywhere as x passes through all values from a to b, then if M and m denote respectively the greatest and the least value the integral:

$$\int_a^x \varphi(x)\,dx$$

assumes as x thus varies, we have the relation:

$$f(a)\,.\,m \leq \int_a^b f(x)\varphi(x)\,dx \leq f(a)\,.\,M.$$

By the definition we have:

$$\int_a^b f(x)\varphi(x)\,dx = \text{Lim}\,((x_1-a)f(a)\varphi(a)+(x_2-x_1)f(x_1)\varphi(x_1)\ldots$$
$$+(b-x_{n-1})f(x_{n-1})\varphi(x_{n-1})).$$

Now let the dividing points $x_1 \ldots x_{n-1}$ be chosen so close together that the sum of the products of fluctuations by partial intervals: $\Sigma d_p D_p$ shall be $< \sigma$. Then inasmuch as $f(a)$, $f(x_1)\ldots f(x_{n-1})$ are positive decreasing quantities, we can apply to the finite sum:

$$d_1 f(a)\varphi(a)+d_2 f(x_1)\varphi(x_1)+\cdots d_n f(x_{n-1})\varphi(x_{n-1}),$$

the Lemma (§ 44 IV.) proved by Abel, and it shows that this sum is less than the product $f(a)G$, but greater than the product $f(a)g$, where G and g are the greatest and least values algebraically in the series:

$$d_1\varphi(a),\ d_1\varphi(a)+d_2\varphi(x_1),\ d_1\varphi(a)+d_2\varphi(x_1)+d_3\varphi(x_2),\ \ldots \text{ etc.}.$$

But in consequence of our hypothesis as to the partial intervals, these values differ from the definite integrals:

$$\int_a^{a+d_1}\varphi(x)dx,\ \int_a^{a+d_1+d_2}\varphi(x)dx,\ldots\ \int_a^{a+\Sigma d_p}\varphi(x)dx,\ldots\ \int_a^b\varphi(x)dx,$$

by quantities that are certainly less than the arbitrarily small quantity σ, because the total sum of all the products of the fluctuations by their intervals is $<\rho$. If then we denote by g' the least and by G' the greatest value in this series, we have the inequality:

$$f(a)(g' \pm \sigma) < d_1 f(a)\varphi(a) + d_2 f(x_1)\varphi(x_1) \ldots d_n f(x_{n-1})\varphi(x_{n-1}) < f(a)(G' \pm \sigma).$$

If n be arbitrarily increased in this inequality, σ becomes zero, while g' and G' do not pass beyond the least and greatest of the values which the integral

$$\int_a^x \varphi(x)\,dx$$

assumes while x is given all values between a and b. Hence we have:

$$f(a)\,.\,m \leqq \int_a^b f(x)\varphi(x)\,dx \leqq f(a)\,.\,M.$$

Since the definite integral is a continuous function of its upper limit, it actually assumes each value lying between its maximum and its minimum, it overleaps none; there is therefore certainly a value between a and b such that:

$$\int_a^b f(x)\varphi(x)\,dx = f(a)\int_a^{a+\Theta(b-a)} \varphi(x)\,dx \qquad (0 \leqq \Theta \leqq 1)^{*}).$$

X. Waiving the hypothesis that $f(x)$ retains the same sign, we obtain from this last equation a more general theorem usually called the Second Theorem of the Mean Value. If $f(x)$ and $\varphi(x)$ are everywhere finite integrable functions and $f(x)$ in the interval from a to b a decreasing quantity that can also become negative, $f(x) - f(b)$ is a positive quantity in the interval from a to b, therefore also by theorem IX.:

$$\int_a^b (f(x) - f(b))\varphi(x)\,dx = (f(a) - f(b))\int_a^{a+\Theta(b-a)} \varphi(x)\,dx,$$

or:

$$\int_a^b f(x)\varphi(x)\,dx = f(a)\int_a^{a+\Theta(b-a)} \varphi(x)\,dx + f(b)\int_{a+\Theta(b-a)}^b \varphi(x)\,dx.$$

If $f(x)$ be an increasing function, therefore $-(f(x) - f(b))$ positive and decreasing, it follows that:

$$-\int_a^b (f(x) - f(b))\varphi(x)\,dx = -(f(a) - f(b))\int_a^{a+\Theta(b-a)} \varphi(x)\,dx,$$

*) O. Bonnet: Remarques sur quelques intégrales définies. Liouv. Journ., XIV.

or likewise:

$$\int_a^b f(x)\varphi(x)\,dx = f(a)\int_a^{a+\Theta(b-a)}\varphi(x)\,dx + f(b)\int_{a+\Theta(b-a)}^b \varphi(x)\,dx.^{*)}$$

Note. The values of f at the extremities a and b can also be indeterminate. From our deduction it follows, that then in case of a decreasing function, $f(a)$ is to be replaced by the greatest value and $f(b)$ by the least value to which this function approximates in the neighbourhood of these points; the reverse holds in the case of an increasing function.

147. Having learned that the definite integral is a continuous function of its upper limit; we naturally ask, whether its progressive and regressive differential quotients have a single determinate value? From the Equation VIII.:

$$F(x \pm h) - F(x) = \int_a^{x\pm h} f(x)dx - \int_a^x f(x)dx = \int_x^{x\pm h} f(x)dx = \pm h(g + \Theta(G - g))$$

we find:

$$\frac{F(x \pm h) - F(x)}{\pm h} = g + \Theta\,(G - g).$$

Now making h converge to zero, it is evident that wherever the progressive value of $f(x)$ is continuous, where we can therefore put

$$g + \Theta(G - g) = f(x + \Theta h),$$

the function $F(x)$ has the progressive differential quotient

$$F'(x) = f(x + 0);$$

and wherever the regressive value of $f(x)$ is continuous, the function $F(x)$ has the regressive differential quotient $f(x - 0)$.

The same holds also at every point at which $f(x)$ differs from the values $f(x + 0)$ and $f(x - 0)$ by an arbitrary finite quantity; and there may be a discrete set of such points, provided, arbitrarily near each such point, the values $f(x + 0)$ and $f(x - 0)$ follow from $f(x + h)$ and $f(x - h)$ by continuous transition. As a particular case we have under these hypotheses:

$$\left\{\frac{d}{dx}\int_a^b f(x)\,dx\right\}_{x=b} = \text{Lim} - \frac{1}{h}\int_{b-h}^b f(x)\,dx = f(b).$$

Regarding the integral as a function of its lower limit, the differential quotient can be determined either by means of the inversion of limits:

$$\int_a^b f(x)\,dx = -\int_b^a f(x)\,dx,$$

*) Du Bois-Reymond, Journal f. Mathem., Vol 69.

leading to:

$$\frac{d}{da}\int_a^b f(x)\,dx = -f(a);$$

or directly from the formula:

$$\operatorname{Lim} -\frac{1}{h}\int_a^{a+h} f(x)\,dx = -f(a).$$

At all points in whose neighbourhood $f(x)$ has infinitely many maxima and minima with finite fluctuations, moreover such points also can only form a discrete set, the function $F(x)$ has no differential quotient.

Summing up it may therefore be said: *Every definite integral expresses within its interval of integration a continuous function of its upper limit, whose progressive differential quotient has generally a determinate value, and its regressive also one identical with this. It is only in discrete sets of points that the progressive and regressive differential quotients can differ or can be altogether indeterminate.*

Since the definite integral is quite independent of the nature of the function in discrete points, all integrable functions that coincide within their interval except in such points give rise to the same value of the definite integral.

From this appears, further, what is the form of the connexion between the definite and the indefinite integral. For, if $F(x)$ be any continuous function, whose derived function $F'(x)$ is integrable, (so that if this derivate be discontinuous or indeterminate or else have infinitely many maxima and minima with finite fluctuations it is so only at infinitely many discrete points,) then, the difference:

$$\int_a^x F'(x)\,dx - F(x)$$

is a continuous function of x, whose differential quotient, generally zero, is *primâ facie* indeterminate only in discrete points. But such a function is a constant. For in every interval however small, after separating out the singular points, finite intervals will still remain within which not only is the function continuous but also its derived function vanishes. By the Theorem § 100 therefore the function is constant within such an interval; and because the limits of the interval can be brought arbitrarily close to the singular points, since the function is continuous it will have the same value also at the singular points. Therefore it does not undergo any change of value on passing from one interval into the next; i.e. it remains constant throughout the entire interval of integration:

$$\int_a^x F'(x)dx - F(x) = C.$$

The value of this constant is determined by putting $x = a$; thence follows:

$$-F(a) = C \quad \text{or:} \int_a^x F'(x)dx = F(x) - F(a).$$

Therefore when the derivate $F'(x)$ of a continuous function $F(x)$ is known and is finite and integrable, the value of its integral is always $F(x) - F(a)$; even though we may alter arbitrarily the value of $F'(x)$ in infinitely many discrete points.

148. When the function $f(x)$ becomes infinitely great, either determinately, or, oscillating between arbitrary limits, as the variable x approaches a certain value c in the interval from a to b, the sum, whose limiting value is the definite integral, can assume any value whatever for any finite partition of the interval, it has therefore no limiting value, and $\int_a^b f(x)dx$ as hitherto defined would have no meaning. But if under these circumstances the sum:

$$\int_a^{c-\alpha_1} f(x)dx + \int_{c+\alpha_2}^b f(x)dx$$

assume a fixed limiting value while α_1 and α_2 independently converge to zero: $\int_a^b f(x)dx$ is understood to mean this limiting value. This was indicated in § 106. Examples in which the function to be integrated becomes infinite occur in § 122 Note and § 132.

Now the necessary and sufficient condition that each of the two integrals may have a determinate value, is that

$$\int_{c-\alpha_1}^{c-\varepsilon_1} f(x)dx \quad \text{and} \quad \int_{c+\alpha_2}^{c+\varepsilon_2} f(x)dx$$

shall vanish, when ε is always smaller than α, and α converges to zero.

In case the function becomes determinately infinite at a point, this condition is certainly fulfilled, when in the neighbourhood of this point it becomes algebraically infinite in lower than the first order; taking as unity the order of $\frac{1}{x}$ for $x = 0$.

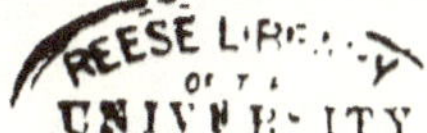

For, if $f(x)$ be a function that increases as c is approached, but in such a way, that in the neighbourhood of the point c, for $x < c$:

$$\text{abs } f(x) \text{ is } < \frac{A}{(c-x)^\nu},$$

A being any finite quantity, and ν a positive proper fraction; then:

$$\int_{c-\alpha}^{c-\varepsilon} f(x)\,dx \text{ will be smaller than } A\int_{c-\alpha}^{c-\varepsilon} \frac{dx}{(c-x)^\nu} = -\frac{A}{1-\nu}\,(\varepsilon^{1-\nu} - \alpha^{1-\nu}),$$

and as long as $1 - \nu > 0$ this expression converges to zero along with α and ε. Even when the order of becoming infinite (infinitude) differs from unity by no assignable number, when ex. gr.*)

$$\text{abs } f(x) < \frac{A}{c-x} \cdot \frac{1}{(\log(c-x))^{1+\nu}},$$

the quantity:

$$\int_{c-\alpha}^{c-\varepsilon} f(x)\,dx < A\int_{c-\alpha}^{c-\varepsilon} \frac{dx}{(c-x)\,(\log(c-x))^{1+\nu}} = \frac{A}{\nu}\left\{(\log(\varepsilon))^{-\nu} - (\log(\alpha))^{-\nu}\right\}$$

converges to zero along with α, provided ν is positive.

But the above condition for a determinate limiting value of the definite integral is certainly not fulfilled, when the function that is to be integrated becomes determinately infinite in the first or in a higher order; i. e. when its infinitude $\overline{>}$ 1. For, if:

$$\text{abs } f(x) > \frac{A}{c-x},$$

$$\int_{c-\alpha}^{c-\varepsilon} f(x)\,dx \text{ is } > A\int_{c-\alpha}^{c-\varepsilon} \frac{dx}{c-x} = A\,(\log\alpha - \log\varepsilon),$$

then as α converges to zero, these logarithms become infinite and their difference is completely indeterminate.**)

In case of a function that becomes indeterminately infinite at a point, the condition is satisfied without requiring any restriction as to the order of becoming infinite; thus ex. gr. the function:

---

*) Riemann: ges. Werke, p. 229. See a general remark on the universality of logarithmic criteria by Du Bois-Reymond, Journal f. Math., Vol. 76, p. 88.

**) $\log\alpha - \log\varepsilon = \log(\alpha : \varepsilon)$ assumes arbitrary values, according to the way in which the ratio of the vanishing quantities $\alpha : \varepsilon$ is determined. Making $\alpha = \varepsilon$, gives a value for which the logarithm vanishes. We could therefore in this sense speak of a finite value of the integral $\int \frac{dx}{c-x}$, that results from a determinate way of approaching the infinity point. Such special determinations, which were frequently employed by Cauchy, are styled singular integrals, but, as Riemann noticed, they have not been adopted in framing the general conception, because they require special arbitrary investigations in each calculation.

$$\cos\left(e^{\frac{1}{x}}\right) + \frac{1}{x}\, e^{\frac{1}{x}} \sin\left(e^{\frac{1}{x}}\right),$$

which for every finite value of x is identical with the derivate:

$$\frac{d}{dx}\left(x \cos\left(e^{\frac{1}{x}}\right)\right),$$

becomes completely indeterminate for $x = 0$, oscillating between arbitrarily great positive and negative values as x approaches zero. Its infinitude is infinite. Nevertheless we have:

$$\int^{x} \left\{\cos\left(e^{\frac{1}{x}}\right) + \frac{1}{x}\, e^{\frac{1}{x}} \sin\left(e^{\frac{1}{x}}\right)\right\} dx = x \cos\left(e^{\frac{1}{x}}\right) + C,$$

even for the value $x = 0$; because:

$$\int_{\alpha}^{\varepsilon} \frac{d}{dx} \left\{x \cos\left(e^{\frac{1}{x}}\right)\right\} dx = \left\{\varepsilon \cos\left(e^{\frac{1}{\varepsilon}}\right) - \alpha \cos\left(e^{\frac{1}{\alpha}}\right)\right\}$$

converges to zero along with α and ε.

Corollary. When the function becomes infinite in infinitely many discrete points of its interval of integration, we can resolve this interval into a finite number of finite intervals that contain none of the infinity points, while these latter are included within intervals whose sum becomes arbitrarily small. The integral has a meaning, provided the partial integrals, formed for the intervals containing no infinity points, converge to fixed values when the limits of these intervals are brought arbitrarily close to the infinity points.

When the infinity points form a linear set, such a definition is no longer possible, since there are then finite intervals containing everywhere infinitely many points of the kind.

149. The Theorems given in § 146 are somewhat modified by the occurrence of infinity points. We confine our investigation to the assumption of a finite number of such points.

Instead of Theorem V., whose proof essentially required that the function to be integrated should be finite, we obtain the theorem:

When $\varphi(x)$ and $\psi(x)$ are two functions integrable from a to b, each becoming infinite at certain points c, but without any infinity point of φ coinciding with one of ψ, the product $\varphi(x) \,.\, \psi(x)$ can be integrated in the same interval whenever the functions $\varphi_1(x)$ and $\psi_1(x)$ formed of the absolute values of φ and ψ remain integrable.

In proving this theorem we need only consider an interval from a to c, within which neither function becomes infinite, while the limit c is an infinity point for one, ex. gr. for $\varphi(x)$. Evidently then: the amount of

$$\int_a^{c-\delta} \varphi(x)\psi(x)\,dx$$

is smaller than:

$$M\int_a^{c-\delta} \varphi_1(x)\,dx\,,$$

where M signifies the greatest amount $\psi(x)$ assumes in this interval. But by hypothesis the second factor remains finite even for $\delta = 0$.

Scholium: When the integrable functions become infinite always in a determinate manner, their product is integrable; for then all the hypotheses of the theorem are fulfilled.

But when infinity points coincide, we cannot immediately conclude anything concerning integrability. Thus, from the integrability of a function that becomes infinite, we cannot conclude that of the square of this function: Ex. gr. we have the value of

$$\int_0^1 \frac{dx}{\sqrt{1-x^2}} = \frac{\pi}{2}, \quad \text{but that of} \int_0^1 \frac{dx}{(\sqrt{1-x^2})^2}$$

becomes infinite.

The extension of the First Theorem of the Mean Value to the product of two functions, VII., holds although $\varphi(x)$ becomes determinately infinite, provided $f(x)$ remains finite. Here moreover the definite integral is a continuous function of its upper limit, Theorem VIII.. For, c being an infinity point, we have:

$$\int_a^c f(x)\,dx = \operatorname*{Lim}_{\delta=0}\left(\int_a^{c-\delta} f(x)\,dx\right) = \operatorname*{Lim}_{\delta=0}\left(F(c-\delta)\right).$$

Now since F is a continuous function of δ, we can choose δ so that

$$\text{abs}\,[F(c) - F(c-\delta)]$$

shall become smaller than an arbitrarily small prescribed number ε, provided there be any limiting value $F(c)$.

The Second Theorem of the Mean Value, X., continues valid even when the function $\varphi(x)$ becomes infinite; provided only $\varphi(x)$ and $f(x)\varphi(x)$ are integrable. The differential quotient of the integral with respect to its upper limit (§ 147) becomes determinately infinite at each point at which in continuous increase the function that is to be integrated becomes determinately infinite; when this is not the case, the differential quotient of the integral needs not coincide with the value of the function that is to be integrated; it also can become indeterminate. We have ex. gr. in conformity with the equation:

$$\frac{d}{dx}\left(x^2\cos\left(e^{\frac{1}{x}}\right)\right) = 2x\cos\left(e^{\frac{1}{x}}\right) + \sin\left(e^{\frac{1}{x}}\right)\cdot e^{\frac{1}{x}},$$

$$\int_a^x \left(2x\cos\left(c^{\frac{1}{x}}\right)+\sin\left(c^{\frac{1}{x}}\right)c^{\frac{1}{x}}\right)dx = x^2\cos\left(c^{\frac{1}{x}}\right)-a^2\cos\left(c^{\frac{1}{a}}\right)=F(x)-F(a).$$

The differential quotient of the function $F(x)$ for $x=0$ has the value:

$$\operatorname{Lim}\left(\frac{F(0+h)-F(0)}{h}\right)=\operatorname{Lim}_{h=0}\left(h\cos\left(c^{\frac{1}{h}}\right)\right)=0,$$

while the function to be integrated becomes indeterminately infinite at this point.

150. As already shown in § 106, the definite integral can still have a finite value, even when its limits become infinite, provided we understand by $\int_a^\infty f(x)dx$ the limiting value assumed by $\int_a^b f(x)dx$ when $b=\infty$. Similarly we define:

$$\int_{-\infty}^b f(x)dx = \operatorname{Lim}\int_a^b f(x)dx, \text{ when } a=-\infty^{*)}.$$

By the substitution $x=-z$ we can always reduce the investigation of the negatively infinite limit to that of the positive. Examples of the existence of such limiting values have also been already given, see specially § 107 and § 137.

But now the necessary and sufficient condition that there may be a determinate limiting value, is: that

$$\int_u^w f(x)dx$$

shall be smaller than an arbitrarily small quantity σ, when u is assumed sufficiently great and w greater than u.

When, for arbitrarily increasing values of x, the function does not oscillate infinitely often, this condition is certainly fulfilled if, for $x=\infty$, $f(x)$ vanish algebraically in an order higher than the first, taking as unity the order of $\frac{1}{x}$ for $x=\infty$. In other words, $f(x)$ can be integrated up to $x=\infty$ when this limit is a zero or nullity of $f(x)$ whose nullitude >1). For, if the absolute values of $f(x)$ form a decreasing series, such that:

$$\text{abs } f(x) < \frac{A}{x^\nu},$$

A being an arbitrary finite quantity, and ν a number greater than 1:

*) How a definite integral with infinite limits can be considered directly as the limiting value of a sum $\sum_{p=1}^{p=n} d_p f_p$, is shown by Dini: Fondamenti, p. 338 etc..

$\int_u^w f(x)\,dx$ will be smaller than $A\int_u^w \frac{dx}{x^\nu} = \frac{A}{1-\nu}(w^{1-\nu} - u^{1-\nu})$,

and while $1-\nu < 0$ this expression converges to zero as the values of u and w increase.

But the condition stated is not fulfilled when the function vanishes in the first or a lower order, or when it remains finite. For, if

$$\text{abs}\, f(x) > \frac{A}{x^\nu}, \quad (\nu \leq 1)$$

we shall have

$$\int_u^w f(x)\,dx > A\int_u^w \frac{dx}{x^\nu} = \frac{A}{1-\nu}(w^{1-\nu} - u^{1-\nu}),$$

and here the exponents of the arbitrarily increasing values u and w are positive.

On the whole the investigation is evidently quite analogous to that in § 148, because by the substitution $x = \frac{1}{z}$, $\int f(x)dx$ passes over into

$$-\int f\left(\frac{1}{z}\right)\frac{dz}{z^2};$$

accordingly the behaviour of the new integral at the point $z = 0$ must be examined.

Our criterion shows for instance, without any substitution, that the integral (§ 137):

$$\int_\infty^{+x'} \frac{e^{-x}\,dx}{x}$$

must have a finite value; for, the function:

$$\frac{e^{-x}}{x} \text{ becomes smaller than } \frac{A}{x^\nu}.$$

for every $\nu > 1$, because:

$$\lim_{x=\infty}(x^{\nu-1}\cdot e^{-x}) = 0.$$

But when the function makes infinitely many oscillations as the values of x become infinite, its nullitude needs no restriction in order that the condition may be fulfilled. Thus ex. gr. the value of:

$$\int_0^\infty \sin(x^2)\,dx$$

is finite and determinate, although for $x = \infty$ the function to be integrated becomes quite indeterminate between the limits -1 and $+1$.*) For:

*) Dirichlet, Journal f. Math., Vol. 17.

$$\int_u^w \sin(x^2)\,dx = -\int_u^w d\{\cos(x^2)\}\cdot\frac{1}{2x} = -\left(\frac{\cos(x^2)}{2x}\right)_u^w - \frac{1}{2}\int_u^w \frac{\cos(x^2)}{x^2}\,dx$$

$$= -\frac{\cos(w^2)}{2w} + \frac{\cos(u^2)}{2u} - \frac{1}{2}\int_u^w \frac{\cos(x^2)}{x^2}\,dx.$$

The absolute difference of these two first terms does not exceed $\frac{1}{u}$. Further, since the function $\frac{1}{x^2}$ does not change sign, we see by the First Theorem of the Mean Value that the amount of:

$$\frac{1}{2}\int_u^w \frac{\cos(x^2)}{x^2}\,dx = \frac{M}{2}\int_u^w \frac{dx}{x^2} = \frac{M}{2}\left[\frac{1}{u} - \frac{1}{w}\right] \text{ is smaller than } \frac{1}{2u},$$

since M signifies a mean value of $\cos(x^2)$, and is therefore a proper fraction. Accordingly:

$$\text{abs}\int_u^w \sin(x^2)\,dx < \frac{3}{2u},$$

and tends to zero as u increases.

We have:

$$\int_0^\infty \sin(x^2)\,dx = \frac{1}{2}\int_0^\infty \frac{\sin z}{\sqrt{z}}\,dz = \frac{1}{2}\sqrt{\frac{1}{2}\pi}\,;\quad \text{(see § 158.).}$$

Another example is:

$$\int_0^\infty \frac{\sin x}{x}\,dx,$$

which can similarly be proved to be finite. It is still more instructive to consider the following process: Taking k an integer and $\alpha < \pi$, let $w = k\pi + \alpha$ be an arbitrarily great number, then:

$$\int_0^w \frac{\sin x}{x}\,dx = \int_0^\pi + \int_\pi^{2\pi} + \cdots \int_{(k-1)\pi}^{k\pi} + \int_{k\pi}^{k\pi+\alpha} \frac{\sin x}{x}\,dx.$$

The terms of this infinite series (for $k = \infty$) alternate in sign and decrease in amount; for, comparing:

$$\int_{(k-1)\pi}^{k\pi} \frac{\sin x}{x}\,dx \text{ with } \int_{k\pi}^{(k+1)\pi} \frac{\sin x}{x}\,dx,$$

by substituting $x = y + \pi$ and so making the limits of the second integral the same as those of the first, we find:

$$\int_{k\pi}^{(k+1)\pi} \frac{\sin x}{x}\,dx = \int_{(k-1)\pi}^{k\pi} \frac{\sin(y+\pi)}{y+\pi}\,dy = -\int_{(k-1)\pi}^{k\pi} \frac{\sin y}{y+\pi}\,dy.$$

As k increases, these integrals converge in amount to zero; hence the infinite series has for $x = \infty$ a finite value. It is worthy of notice further that:

$$\int_0^\infty \frac{\sin x}{x} dx = \int_0^\infty \frac{\sin \alpha z}{z} dz, \qquad (x = \alpha z, \ \alpha > 0)$$

so that the value of the integral is independent of α; it is $\frac{1}{2}\pi$ (§ 155).

The convergence of these integrals is only conditional, i.e. arises only by the changes in sign of the functions to be integrated; the integrals formed of the absolute amounts are divergent. It has however been proved by Du Bois-Reymond in a general class of examples (Math. Ann., Vol. XIII) that integrals can be convergent even in case of oscillations between zero and positive limits.

151. Differentiation of a definite integral with respect to a parameter.*)

A theorem stating how the definite integral in certain cases can be differentiated with respect to a quantity that is contained in the function to be integrated, is of importance in calculation with definite integrals.

First of all we remark: When $f(x, \alpha)$ is a continuous function of both variables x and α (§ 52) within the domain that is determined by the values $x = a$ to $x = b$, and $\alpha = \beta$ to $\alpha = \gamma$, the definite integral:

$$\int_a^b f(x, \alpha) dx$$

is also a continuous function of α. For if, whatever be the value of x, a value h can be assigned such that in the interval α to $\alpha \pm h$:

$$\text{abs}\,[f(x, \alpha \pm h) - f(x, \alpha)] < \delta,$$

then will also the absolute value of:

$$\int_a^b f(x, \alpha \pm h) dx - \int_a^b f(x, \alpha) dx = \int_a^b [f(x, \alpha \pm h) - f(x, \alpha)]\, dx$$

be smaller than $\delta(b - a)$; therefore it can be arbitrarily diminished by choice of h. This is a sufficient condition; but the theorem cannot be converted.

The differential quotient of the definite integral for a determinate value of α is to be calculated as a limiting value:

$$\text{Lim}\, \frac{\int_a^b f(x, \alpha + h) dx - \int_a^b f(x, \alpha) dx}{h} = \text{Lim} \int_a^b \frac{f(x, \alpha + h) - f(x, \alpha)}{h} dx;$$

*) Thomae: Einleitung in die Theorie der bestimmten Integrale. p. 20 etc..

i. e. this definite integral must first be evaluated for a finite value of h and then the limit found for $h = 0$. The question arises, whether these processes may also be applied in reversed order; when they can, we have the differential quotient of the definite integral expressed by a new integral, namely:

$$\int_a^b \frac{\partial f(x, \alpha)}{\partial \alpha} dx;$$

and this will present an important method of calculating definite integrals.

In order to examine the condition of this theorem, let us put:

$$\frac{f(x, \alpha + h) - f(x, \alpha)}{h} = \frac{\partial f(x, \alpha)}{\partial \alpha} + \varphi(x, \alpha, h),$$

thus φ is a quantity variable with x and h, that vanishes for $h = 0$. Now we shall assume that both f and its differential quotient $\frac{\partial f}{\partial \alpha}$ are continuous functions of x within the interval $x = a$ to $x = b$; then too, for every value of h, φ is a continuous function of x, so that both functions are also integrable.*) Hence:

$$\int_a^b \frac{f(x, \alpha + h) - f(x, \alpha)}{h} dx = \int_a^b \frac{\partial f(x, \alpha)}{\partial \alpha} dx + \int_a^b \varphi(x, \alpha, h) dx.$$

Making the value h converge to zero in this equation, it will only pass over into the desired equation:

$$\lim_{h=0} \int_a^b \frac{f(x, \alpha + h) - f(x, \alpha)}{h} dx = \int_a^b \frac{\partial f(x, \alpha)}{\partial \alpha} dx,$$

provided:

$$\lim_{h=0} \int_a^b \varphi(x, \alpha, h) dx$$

continuously converges to the limit zero. Therefore the necessary and sufficient condition is, that for each value of δ however small, it shall be possible to find a value h such that:

$$\text{abs}\left[\int_a^b \varphi(x, \alpha, h) dx\right] < \delta.$$

This condition shows, that the theorem of the interchange of integration and differentiation is by no means always valid. It will

---

*) If $\frac{\partial f}{\partial \alpha}$ be not an integrable function, the possibility of the theorem is completely excluded.

not hold, ex. gr. for any functions f for which, while φ has always the same sign, there is a linear set of points x at which the inequality:

$$\text{abs}\,[\varphi(x, \alpha, h)] = \text{abs}\left[\frac{f(x, \alpha + h) - f(x, \alpha)}{h} - \frac{\partial f(x, \alpha)}{\partial \alpha}\right] < \delta$$

cannot be satisfied by an assignable value h.*)

But the course of our investigation reveals to us a condition that is **sufficient**. In fact if $\frac{\partial f(x, \alpha)}{\partial \alpha}$ for every value of x is also a continuous function of α, the above inequality can be written by the help of the Theorem of the Mean Value in the form:

$$\text{abs}\left[\frac{\partial f(x, \alpha + \Theta h)}{\partial \alpha} - \frac{\partial f(x, \alpha)}{\partial \alpha}\right] < \delta.$$

Now when this condition is to be fulfilled generally for all values of x within the interval of integration (with exception, possibly, of only a discrete set of points), $\frac{\partial f}{\partial \alpha}$ must be generally a **uniformly** continuous function of α, and therefore also generally a **continuous function of both variables.** We have therefore the theorem: *When for*

---

*) Suppose the integral $\int\limits_0^x f(x, \alpha)\,dx = x \sin\left(4 \tan^{-1} \frac{\alpha}{x}\right)$ and therefore is a continuous function of both variables, we have:

$$f(x, \alpha) = \sin\left(4 \tan^{-1} \frac{\alpha}{x}\right) - \frac{4\alpha x}{x^2 + \alpha^2} \cos\left(4 \tan^{-1} \frac{\alpha}{x}\right).$$

At $x = 0$, $\alpha = 0$, $f(0, 0)$ is to be $= 0$; this convention has no influence upon the integral. Moreover the equation holds for every value of x; for $\alpha = 0$ we have $f(x, 0) = 0$; for $x = 0$, $f(0, \alpha) = 0$. If we differentiate the integral with respect to α we obtain:

$$\frac{\partial}{\partial \alpha} \int\limits_0^x f(x, \alpha)\,dx = \frac{4}{1 + \frac{\alpha^2}{x^2}} \cos\left(4 \tan^{-1} \frac{\alpha}{x}\right),$$

and for $\alpha = 0$ this value is equal to 4. But the value of:

$$\int\limits_0^x \left(\frac{\partial f(x, \alpha)}{\partial \alpha}\right)_{\alpha = 0} dx = 0,$$

because:

$$\left(\frac{\partial f(x, \alpha)}{\partial \alpha}\right)_{\alpha = 0} = \lim_{\Delta\alpha = 0} \left\{\frac{\sin\left(4 \tan^{-1} \frac{\Delta\alpha}{x}\right)}{\Delta\alpha} - \frac{4x}{x^2 + \Delta\alpha^2} \cos\left(4 \tan^{-1} \frac{\Delta\alpha}{x}\right)\right\} = 0.$$

The theorem of the interchange therefore does not hold, although for every value of x both $f(x, \alpha)$ and $\frac{\partial f(x, \alpha)}{\partial \alpha}$ are continuous functions of α.

a determinate value of α an interval can be assigned such that $\frac{\partial f(x,\alpha)}{\partial \alpha}$ within the domain from α to $\alpha + h$ and from $x = a$ to $x = b$ is generally a continuous function of both variables, we have:

$$\frac{\partial}{\partial \alpha}\int_a^b f(x, \alpha)\,dx = \int_a^b \frac{\partial f(x, \alpha)}{\partial \alpha}\,dx.$$

Moreover also it is sufficient for the validity of the theorem, that the definite integral be such a function of its upper limit x and of the parameter α as to have its first derived functions with respect to x and α continuous and the order of differentiations with respect to x and α interchangeable (§ 54). For, writing:

$$\int_a^x f(x, \alpha)\,dx = F(x, \alpha),$$

we have then:

$$\frac{\partial F}{\partial x} = f(x, \alpha), \quad \frac{\partial^2 F}{\partial \alpha \partial x} = \frac{\partial f(x, \alpha)}{\partial \alpha}, \quad \frac{\partial^2 F}{\partial x \partial \alpha} = \frac{\partial^2}{\partial x \partial \alpha}\int_a^x f(x, \alpha)\,dx,$$

and from the equation:

$$\frac{\partial f(x, \alpha)}{\partial \alpha} = \frac{\partial^2}{\partial x \partial \alpha}\int_a^x f(x, \alpha)\,dx$$

follows by integration:

$$\int_a^x \frac{\partial f(x, \alpha)}{\partial \alpha}\,dx = \frac{\partial}{\partial \alpha}\int_a^x f(x, \alpha)\,dx.$$

When the limits of the integral likewise depend on the parameter α, we obtain its derived function with respect to α, on the hypothesis that differentiation is admissible under the integral sign, by the formula:

$$\frac{\partial}{\partial \alpha}\int_a^b f(x, \alpha)\,dx = f(b, \alpha)\,\frac{db}{d\alpha} - f(a, \alpha)\,\frac{da}{d\alpha} + \int_a^b \frac{\partial f(x, \alpha)}{\partial \alpha}\,dx.$$

152. The preceding theorems require to be supplemented, when the limits of the integral or the functions to be integrated become infinite. Without exhausting all possibilities, we consider the following cases.

a) Although $f(x, \alpha)$ be a continuous function of both variables in the domain from $x = a$ to $x = \infty$ and from $\alpha = \beta$ to $\alpha = \gamma$, still it is only on a certain hypothesis that the integral:

$$\int_a^\infty f(x, \alpha)\,dx$$

is a continuous function of α. We have:

$$\int_a^\infty [f(x, \alpha \pm h) - f(x, \alpha)]\,dx = \int_a^w [f(x, \alpha \pm h) - f(x, \alpha)]\,dx$$

$$+ \int_w^\infty [f(x, \alpha \pm h) - f(x, \alpha)]\,dx.$$

In order that by choice of h this expression may become smaller than δ, the function must be so constituted that, for all values in the interval from $\alpha - h$ to $\alpha + h$, one and the same w shall be sufficient to make:

$$\int_w^\infty [f(x, \alpha \pm h) - f(x, \alpha)]\,dx < \delta.$$

Thus ex. gr. (§ 155) for every finite value of α:

$$\int_0^\infty \frac{\sin \alpha x}{x}\,dx = \pm \tfrac{1}{2}\pi,$$

but for $\alpha = 0$ the value of the integral vanishes. The definite integral is therefore not a continuous function of α, although the function that is to be integrated is continuous in both variables.

But the hypothesis is fulfilled, if, whatever value α may have, the function f vanish determinately for $x = \infty$ in an order higher than the first. For then we can first assume w so great as to make:

$$\int_w^\infty [f(x, \alpha \pm h) - f(x, \alpha)]\,dx < \frac{\delta}{2},$$

and afterwards determine the value of h so that we may have:

$$\int_a^w [f(x, \alpha \pm h) - f(x, \alpha)]\,dx < \frac{\delta}{2}.$$

In this case differentiation under the integral sign is certainly possible, if $\frac{\partial f(x, \alpha)}{\partial \alpha}$ also is a continuous function of both variables that vanishes in higher than the first order for $x = \infty$. For, putting:

$$\frac{f(x, \alpha + h) - f(x, \alpha)}{h} = \frac{\partial f(x, \alpha + \Theta h)}{\partial \alpha},$$

we have:

$$\int_a^\infty \left[\frac{f(x, \alpha + h) - f(x, \alpha)}{h} - \frac{\partial f(x, \alpha)}{\partial \alpha}\right] dx = \int_a^w \left[\frac{\partial f(x, \alpha + \Theta h)}{\partial \alpha} - \frac{\partial f(x, \alpha)}{\partial \alpha}\right] dx$$

$$+ \int_w^\infty \left[\frac{\partial f(x, \alpha + \Theta h)}{\partial \alpha} - \frac{\partial f(x, \alpha)}{\partial \alpha}\right] dx.$$

The quantity on the right side of this equation, as was just shown, converges with h to zero and therefore we have:

$$\lim_{h=0}\int_a^\infty \frac{f(x,\alpha+h)-f(x,\alpha)}{h}\,dx = \int_a^\infty \frac{\partial f(x,\alpha)}{\partial\alpha}\,dx.$$

b) If either or both of the functions $f(x, \alpha)$ and $\frac{\partial f}{\partial\alpha}$ become infinite at the point $x = c$ within the interval, but in such a manner that integration is permitted for each, while α varies within an interval $\alpha - h$ to $\alpha + h$, then it is once more a sufficient condition for the validity of the Theorem of differentiation under the integral sign, that $\frac{\partial f}{\partial\alpha}$ shall become determinately infinite in an order lower than the first but otherwise shall remain continuous. For we have:

$$\frac{\partial}{\partial\alpha}\int_a^c f(x,\alpha)\,dx = \lim_{h=0}\int_a^{c-\delta}\frac{f(x,\alpha+h)-f(x,\alpha)}{h}\,dx + \lim_{h=0}\int_{c-\delta}^c \frac{f(x,\alpha+h)-f(x,\alpha)}{h}\,dx.$$

The first of these limiting values passes over, however small δ may be, into the value of the integral:

$$\int_a^{c-\delta}\frac{\partial f(x,\alpha)}{\partial\alpha}\,dx,$$

the second takes the form:

$$\int_{c-\delta}^c \frac{\partial f(x,\alpha+\Theta h)}{\partial\alpha}\,dx$$

and can by hypothesis, however small h is chosen, exclusively by choice of δ be made smaller than an arbitrarily small number. Therefore:

$$\frac{\partial}{\partial\alpha}\int_a^c f(x,\alpha)\,dx$$

differs arbitrarily little from:

$$\int_a^{c-\delta}\frac{\partial f(x,\alpha)}{\partial\alpha}\,dx,$$

as δ converges to zero; establishing the equation we desired to prove:

$$\frac{\partial}{\partial\alpha}\int_a^c f(x,\alpha)\,dx = \int_a^c \frac{\partial f(x,\alpha)}{\partial\alpha}\,dx.$$

153. Integration of a definite integral with respect to a parameter.

If the definite integral is a continuous function of the parameter

α within certain limits β and γ, then it is also between these limits undoubtedly integrable with respect to α.

Calling: $\int_a^b f(x, \alpha)\,dx = F(\alpha)$, we have:

$$\int_\beta^\gamma F(\alpha)\,d\alpha = \int_\beta^\gamma d\alpha \int_a^b f(x, \alpha)\,dx.$$

To such an expression the name definite double integral has been given; it implies, that first the integration regarding x has to be effected and then that regarding α. The general theory of double integrals will be treated in Chapter VIII; one question only is to be solved here. Assuming the limits a and b, β and γ, to be independent determinate constants, and $f(x, \alpha)$ to be a continuous function of both variables, does the following equality hold between the integrals:

$$\int_\beta^\gamma d\alpha \int_a^b f(x, \alpha)\,dx = \int_a^b dx \int_\beta^\gamma f(x, \alpha)\,d\alpha\,;$$

or, does the order in which we integrate influence the result? For $b = a$ both expressions vanish. They are therefore equal functions of b, if their derivates with respect to b coincide. Differentiating each with respect to the upper limit b, we find on equating:

$$\frac{\partial}{\partial b}\int_\beta^\gamma d\alpha \int_a^b f(x, \alpha)\,dx = \left\{\int_\beta^\gamma f(x, \alpha)\,d\alpha\right\}_{\text{for } x=b} = \int_\beta^\gamma f(b, \alpha)\,d\alpha.$$

Now this requires the integral with respect to α on the left side to admit of differentiation under the integral sign. But it does admit of it, because the derivate of

$$\int_a^b f(x, \alpha)\,dx$$

with respect to b, whose value is $f(b, \alpha)$, is by hypothesis a continuous function of both α and b. The order of the integrations can therefore be inverted for a continuous function of two variables.

When the limits b and γ are infinite, we have:

$$\int_\beta^\infty d\alpha \int_a^\infty f(x, \alpha)\,dx = \operatorname*{Lim}_{w=\infty} \int_\beta^w d\alpha \left\{ \operatorname*{Lim}_{u=\infty} \int_a^u f(x, \alpha)\,dx \right\}$$

$$\int_a^\infty dx \int_\beta^\infty f(x, \alpha)\,d\alpha = \operatorname*{Lim}_{u=\infty} \int_a^u dx \left\{ \operatorname*{Lim}_{w=\infty} \int_\beta^w f(x, \alpha)\,d\alpha \right\}.$$

Now when $f(x, \alpha)$, as the values of x and α increase arbitrarily, is a continuous function of both variables, we have the equation:

$$\int_\beta^w d\alpha \int_a^u f(x, \alpha)\, dx = \int_a^u dx \int_\beta^w f(x, \alpha)\, d\alpha.$$

First let u increase arbitrarily, while the arbitrarily great value of w is kept fixed, then if the functions $f(x, \alpha)$ and $\int_\beta^w f(x, \alpha)\, d\alpha$ are integrable with respect to x up to infinite limits, we have the relation:

$$\int_\beta^w d\alpha \int_a^\infty f(x, \alpha)\, dx = \int_a^\infty dx \int_\beta^w f(x, \alpha)\, d\alpha.$$

If now when w also increases arbitrarily, the integral on the left side pass over into a determinate value, the quantity w on the right side may also be replaced by the value ∞, provided the further condition is fulfilled, that an upper limit w can be found such that:

$$\int_a^\infty dx \int_w^\infty f(x, \alpha)\, d\alpha$$

shall remain smaller than an arbitrarily small number δ. For then both sides are continuous functions of w. As a special case this will occur, when it is a property of the function $f(x, \alpha)$, that, whatever value x may have, an inferior limit can be assigned for α, such that the function $f(x, \alpha)$ shall remain absolutely smaller than $\frac{\varphi(x)}{\alpha^\nu}$, where $\nu > 1$, and $\varphi(x)$ signifies a function of x integrable between the limits 0 and ∞.

When the function $f(x, \alpha)$ becomes infinite at a point $x = c$, but in such a way that, for all values of α between β and γ integration is admissible up to the point $x = c$, we have:

$$\int_\beta^\gamma d\alpha \int_a^{c-\delta} f(x, \alpha)\, dx = \int_a^{c-\delta} dx \int_\beta^\gamma f(x, \alpha)\, d\alpha$$

however small we choose δ. Now when we make δ converge to zero, the integral on the left side will pass over into the value:

$$\int_\beta^\gamma d\alpha \int_a^c f(x, \alpha)\, dx,$$

provided a quantity δ can be assigned that is sufficient generally for all values of α between β and γ, to make:

$$\int_{c-\delta}^{c} f(x, \alpha)\, dx$$

smaller than an arbitrarily small quantity ε. The integral on the right side then, as a continuous function of δ, also passes over into the same value, provided it has any determinate limiting value.

These sufficient conditions are fulfilled, for instance, when:

$$\text{abs}\, f(x, \alpha) < \frac{A}{(x-c)^\nu}, \quad (\nu < 1)$$

while A as a function of α remains finite.

Example. $\int_0^\infty \frac{\sin \alpha x}{x}\, dx = \frac{1}{2}\pi$ (§§ 150, 152, 155) for every finite value of α; except for $\alpha = 0$, where the value of the integral vanishes. Nevertheless an integration with respect to α is possible ex. gr. between the limits 0 and 1:

$$\int_0^1 d\alpha \int_0^\infty \frac{\sin \alpha x}{x}\, dx = \frac{1}{2}\pi \int_0^1 d\alpha = \frac{1}{2}\pi.$$

Interchanging the order of integrations we obtain:

$$\int_0^\infty \frac{dx}{x} \int_0^1 \sin \alpha x\, d\alpha = \int_0^\infty \frac{1-\cos x}{x^2}\, dx = \int_0^\infty \frac{(\sin x)^2}{x^2}\, dx = \frac{1}{2}\pi.$$

The following Example shows that an interchange of the order of integrations in discontinuous functions gives rise to a different value*):

$$\int_0^1 d\alpha \int_0^1 \frac{(\alpha^2 - x^2)\, dx}{(\alpha^2 + x^2)^2} \text{ is not equal to } \int_0^1 dx \int_0^1 \frac{(\alpha^2 - x^2)\, d\alpha}{(\alpha^2 + x^2)^2}.$$

Here the function to be integrated is discontinuous for $x = 0$, $\alpha = 0$.

We have:

$$\int_0^1 \frac{(\alpha^2 - x^2)\, dx}{(\alpha^2 + x^2)^2} = \left(\frac{x}{\alpha^2 + x^2}\right)_0^1 = \frac{1}{1+\alpha^2}.$$

Hence the first double integral is equal to:

$$\int_0^1 \frac{d\alpha}{1+\alpha^2} = \left(\tan^{-1}\alpha\right)_0^1 = \frac{\pi}{4}.$$

On the other hand:

$$\int_0^1 \frac{(\alpha^2 - x^2)\, d\alpha}{(\alpha^2 + x^2)^2} = -\left(\frac{\alpha}{\alpha^2 + x^2}\right)_0^1 = -\frac{1}{1+x^2},$$

therefore the second double integral is equal to $-\frac{\pi}{4}$.

---

*) Cauchy, Leçons de calcul différentiel et intégral, rédigées par Moigno. p. 85.

It is to be noticed, that this difference arises from the value of

$$\int_0^1 \frac{(\alpha^2 - x^2)\,dx}{(\alpha^2 + x^2)^2},$$

which is generally $\frac{1}{1+\alpha^2}$ for finite values of α, becoming $-\int_0^1 \frac{dx}{x^2}$, that is to say, infinite, for $\alpha = 0$; the value of:

$$\lim_{\alpha=0} \left(\int_0^1 \frac{(\alpha^2 - x^2)\,dx}{(\alpha^2 + x^2)^2} \right)$$

is different from that of:

$$\int_0^1 \lim_{\alpha=0} \left(\frac{(\alpha^2 - x^2)}{(\alpha^2 + x^2)^2} \right) dx.$$

Consequently we have in the neighbourhood of the point $x = 0$, $\alpha = 0$:

$$\int_0^{\delta'} d\alpha \int_0^{\delta} \frac{(\alpha^2 - x^2)\,dx}{(\alpha^2 + x^2)^2} = \int_0^{\delta'} \frac{\delta}{\alpha^2 + \delta^2}\,d\alpha = \left(\tan^{-1}\frac{\alpha}{\delta}\right)_0^{\delta'} = \tan^{-1}\frac{\delta'}{\delta};$$

on the other hand:

$$\int_0^{\delta} dx \int_0^{\delta'} \frac{(\alpha^2 - x^2)\,d\alpha}{(\alpha^2 + x^2)^2} = -\int_0^{\delta} \frac{\delta'}{\delta'^2 + x^2}\,dx = \left(-\tan^{-1}\frac{x}{\delta'}\right)_0^{\delta} = -\tan^{-1}\frac{\delta}{\delta'}.$$

The values of the double integrals are quite indeterminate, depending on the way in which δ' and δ converge to zero. (Cf. § 168.)

Seventh Chapter.

Examples on the calculation of definite integrals. The fundamental formulas of Eulerian integrals.

154. First group.

1) $$\int_0^1 x^{a-1}\,dx = \left(\frac{x^a}{a}\right)_0^1 = \frac{1}{a}, \quad (a > 0).$$

By successive differentiations with respect to the parameter a we obtain the following integrals:

2) $$\int_0^1 x^{a-1}l(x)\,dx = -\frac{1}{a^2},$$

$$\int_0^1 x^{a-1}(lx)^2\,dx = \frac{1.2}{a^3}, \dots, \int_0^1 x^{a-1}(lx)^{n-1}\,dx = \frac{(-1)^{n-1}\cdot \lfloor n-1}{a^n}.$$

By the Theorems § 152b and § 148, all these integrals hold for $a > 0$. By integration with respect to the parameter a between the limits $a = \alpha$ and β, it is found that:

$$\int_\alpha^\beta da \int_0^1 x^{a-1}dx = \int_0^1 dx \int_\alpha^\beta x^{a-1}da = \int_0^1 dx\, \frac{x^{\beta-1} - x^{\alpha-1}}{l(x)} = l\left(\frac{\beta}{\alpha}\right).$$

This formula holds good, provided α and $\beta > 0$.

For, so long as a is positive, we have not only x^{a-1} but also $\frac{x^{a-1}}{l(x)}$ integrable at the points $x = 0$ and $x = 1$, moreover the condition stated at the end of § 153 is fulfilled.

Putting $\alpha = 1$, we find:

3) $$\int_0^1 dx\,\frac{x^{\beta-1} - 1}{l(x)} = l(\beta), \quad (\beta > 0).$$

Substituting in these integrals the function e^{-x} for x, the limits become ∞ and 0, instead of 0 and 1, and we obtain for $a > 0$:

1a) $$\int_0^\infty e^{-ax}\,dx = \frac{1}{a}.$$

2a) $\int_0^\infty e^{-ax} x\,dx = \frac{1}{a^2}, \ldots, \int_0^\infty e^{-ax} x^{n-1}\,dx = \frac{\lfloor n-1}{a^n}$; n a positive integer.

3a) $$\int_0^\infty \frac{e^{-x} - e^{-ax}}{x}\,dx = l(a).$$

155. Second group. (§ 140.)

1) $$\int_0^\infty e^{-ax} \sin bx\,dx = \frac{b}{a^2 + b^2}$$ $(a > 0)$
$$\int_0^\infty e^{-ax} \cos bx\,dx = \frac{a}{a^2 + b^2}.$$

Hence follows by differentiation with respect to a:

2) $$\int_0^\infty e^{-ax} x \sin bx\,dx = \frac{2ab}{(a^2 + b^2)^2}, \quad \int_0^\infty e^{-ax} x \cos bx\,dx = \frac{a^2 - b^2}{(a^2 + b^2)^2}.$$

Again by integration with respect to a between the limits α and β:

3) $$\int_0^\infty \frac{e^{-\alpha x} - e^{-\beta x}}{x} \sin bx\,dx = \tan^{-1}\frac{\beta}{b} - \tan^{-1}\frac{\alpha}{b}.$$ $(\alpha$ and $\beta > 0)$.
$$\int_0^\infty \frac{e^{-\alpha x} - e^{-\beta x}}{x} \cos bx\,dx = \frac{1}{2} l \frac{\beta^2 + b^2}{\alpha^2 + b^2}.$$

Making α converge to zero in the first of these two integrals, we find:

4) $$\int_0^\infty \frac{1 - e^{-\beta x}}{x} \sin bx\,dx = \tan^{-1}\frac{\beta}{b}.$$

For, the integral 3) is a continuous function of α inclusive of the value $\alpha = 0$. In fact, separating the integral from 0 to w and from w to ∞, as § 152 requires, we can evidently by choice of w alone, independently of the value h, make:

$$\int_w^\infty \frac{e^{-hx} - 1}{x} \sin bx\,dx$$

less than an arbitrarily small number, because (§ 150):

$$\int_w^\infty \frac{\sin bx}{x}\,dx$$

can be arbitrarily diminished by choice of w, and moreover by the Second Theorem of the Mean Value we have:

$$\int_w^\infty \frac{e^{-hx}\sin bx}{x}\,dx = e^{-hw}\int_w^u \frac{\sin bx}{x}\,dx,$$

the upper limit u denoting a number between w and ∞. In like manner, integral 4) is a continuous function of β. Therefore making β become infinite, we obtain:

$$5)\qquad \int_0^\infty \frac{\sin bx}{x}\,dx = \pm\frac{\pi}{2}$$

according as $b > 0$ or < 0. But this definite integral, as already remarked, is not a continuous function of b when $b = 0$; for, at that point its value is zero.

The formula 5) can be expressed generally. Replacing b in it successively by $b + a$ and by $b - a$ and assuming that $b > a > 0$, we have:

$$\int_0^\infty \frac{\sin (b+a)x}{x}\,dx = \frac{\pi}{2} \qquad \int_0^\infty \frac{\sin (b-a)x}{x}\,dx = \frac{\pi}{2}\cdot$$

Hence also, as we find by addition and subtraction:

$$\int_0^\infty \frac{\sin bx \cos ax}{x}\,dx = \frac{\pi}{2} \qquad \int_0^\infty \frac{\sin ax \cos bx}{x}\,dx = 0 \qquad (b > a),$$

or:

$$6)\qquad \frac{2}{\pi}\int_0^\infty \frac{\sin bx \cos ax}{x}\,dx = 1 \text{ or } = 0, \text{ according as } b > \text{ or } < a.$$

For $a = b$: $\displaystyle\frac{1}{\pi}\int_0^\infty \frac{\sin 2bx}{x}\,dx = \frac{1}{2}\cdot$

156. Third group. (Laplace's integrals.)

Writing $u = \displaystyle\int_0^\infty \frac{\sin bx \cos ax}{x}\,dx$ (a and $b > 0$), we have $u = 0$ when $b < a$, and $u = \frac{1}{2}\pi$ when $b > a$. Multiplying both sides by e^{-bc}, where c is positive, and integrating from $b = 0$ to $b = \infty$, we find:

$$\int_0^\infty u e^{-bc}\,db = \frac{\pi}{2}\int_0^\infty e^{-bc}\,db = \frac{\pi}{2}\,\frac{e^{-ac}}{c} = \int_0^\infty e^{-bc}\,db\int_0^\infty \frac{\sin bx \cos ax}{x}\,dx.$$

Interchanging the order of integrations in this integral, it becomes:

$$\int_0^\infty \frac{dx}{x} \cos ax \int_0^\infty e^{-bc} \sin bx\,db = \int_0^\infty \frac{\cos ax}{c^2+x^2}\,dx.$$

Hence:

$$\int_0^\infty \frac{\cos ax}{c^2+x^2}\,dx = \frac{\pi}{2} \cdot \frac{e^{-ac}}{c}.$$

The integral is a continuous function of a. If a become negative, the function under the integral sign remains unaltered, therefore:

1) $$\int_0^\infty \frac{\cos ax}{c^2+x^2}\,dx = \frac{\pi}{2} \cdot \frac{e^{\mp ac}}{c}$$ (according as $a > 0$ or < 0) this is true also for $a = 0$.

If we differentiate with respect to a, the result is:

2) $$\int_0^\infty \frac{x \sin ax\,dx}{c^2 + x^2} = \pm \frac{\pi}{2} e^{\mp ac}$$ (according as $a > 0$ or < 0) this does not hold for $a = 0$.

This differentiation with respect to a is permitted (§ 152); but a further differentiation under the integral sign with respect to a is not possible, because the derived function is no longer integrable between the limits 0 and ∞. Integrating equation 1) with respect to a between the limits 0 and a positive, we have:

$$\int_0^\infty \frac{\sin ax}{x(c^2+x^2)}\,dx = \frac{\pi}{2}\left(-\frac{e^{-ac}}{c^2}\right)_0^a = \frac{\pi}{2c^2}(1 - e^{-ac}), \qquad (a > 0)$$

on the other hand for a negative:

$$\int_0^\infty \frac{\sin ax}{x(c^2+x^2)}\,dx = \frac{\pi}{2}\left(\frac{e^{ac}}{c^2}\right)_0^a = \frac{\pi}{2c^2}(e^{ac} - 1). \qquad (a < 0)$$

Successive differentiations with respect to the parameter c can also be effected on integral 1).

157. Fourth group. (§ 139.)

$$\int \sin^n x\,dx = -\frac{\sin^{n-1} x \cos x}{n} + \frac{n-1}{n}\int \sin^{n-2} x\,dx,$$

$$\int_0^{\frac{1}{2}\pi} \sin^n x\,dx = \frac{n-1}{n}\int_0^{\frac{1}{2}\pi} \sin^{n-2} x\,dx.$$

When n is an even integer $= 2m$, we have as in § 133:

$$\int_0^{\frac{1}{2}\pi} \sin^{2m} x\,dx = \frac{2m-1}{2m} \cdot \frac{2m-3}{2m-2} \cdots \frac{3}{4} \cdot \frac{1}{2} \cdot \frac{\pi}{2}.$$

When n is an odd integer $= 2m + 1$:

$$\int_0^{\frac{1}{2}\pi} \sin^{2m+1} x\, dx = \frac{2m}{2m+1} \cdot \frac{2m-2}{2m-1} \cdots \frac{4}{5} \cdot \frac{2}{3}.$$

Now $\sin x$ between the limits 0 and $\frac{1}{2}\pi$ is a positive proper fraction, hence:

$$\int_0^{\frac{1}{2}\pi} \sin^{2m-1} x\, dx > \int_0^{\frac{1}{2}\pi} \sin^{2m} x\, dx > \int_0^{\frac{1}{2}\pi} \sin^{2m+1} x\, dx,$$

$$\frac{2}{3} \cdot \frac{4}{5} \cdots \frac{2m-2}{2m-1} > \frac{\pi}{2} \cdot \frac{1}{2} \cdot \frac{3}{4} \cdots \frac{2m-1}{2m} > \frac{2}{3} \cdot \frac{4}{5} \cdots \frac{2m-2}{2m-1} \cdot \frac{2m}{2m+1}.$$

Dividing across by the coefficient of $\frac{1}{2}\pi$, we find:

$$\frac{2}{1} \cdot \frac{2}{3} \cdot \frac{4}{3} \cdot \frac{4}{5} \cdots \frac{2m-2}{2m-3} \cdot \frac{2m-2}{2m-1} \cdot \frac{2m}{2m-1} > \frac{\pi}{2}$$

$$> \frac{2}{1} \cdot \frac{2}{3} \cdot \frac{4}{3} \cdot \frac{4}{5} \cdots \frac{2m-2}{2m-3} \cdot \frac{2m-2}{2m-1} \cdot \frac{2m}{2m-1} \cdot \frac{2m}{2m+1}.$$

$$A_m > \frac{\pi}{2} > A_m \cdot \frac{2m}{2m+1}.$$

As m increases, the quantities A_m form a series of decreasing numbers greater than $\frac{1}{2}\pi$, they must therefore have a definite limit. But on the other hand we can choose m so large that A_m and

$$A_m \cdot \frac{2m}{2m+1}$$

may differ inappreciably, therefore the limiting value of A_m differs inappreciably from $\frac{1}{2}\pi$, or we have:

$$\frac{\pi}{2} = \frac{2}{1} \cdot \frac{2}{3} \cdot \frac{4}{3} \cdot \frac{4}{5} \cdot \frac{6}{5} \cdot \frac{6}{7} \cdots \text{ in infinitum *)}.$$

We have also:

$$\int_0^{\frac{1}{2}\pi} \cos^n x\, dx = \int_0^{\frac{1}{2}\pi} \sin^n x\, dx.$$

158. Fifth group.

The following process serves for the evaluation of the integral

$$\int_0^{\infty} e^{-x^2} dx.$$

Let us denote its value by A and introduce a new variable z by the equation $x = \alpha z$, then:

---

*) Wallis: Arithmetica infinitorum. The earliest expression of a number in the form of an infinite product. (Cf. § 39 and § 163.)

$$A = \int_0^\infty e^{-\alpha^2 z^2}\alpha\, dz.$$

Multiplying this equation by $e^{-\alpha^2} d\alpha$ let us then integrate from $\alpha = 0$ to $\alpha = \infty$, we find:

$$A\int_0^\infty e^{-\alpha^2} d\alpha = A^2 = \int_0^\infty dz \int_0^\infty e^{-\alpha^2(1+z^2)}\alpha\, d\alpha = \frac{1}{2}\int_0^\infty \frac{dz}{1+z^2} = \frac{\pi}{4},$$

therefore we have:

1) $$A = \int_0^\infty e^{-x^2} dx = \frac{\sqrt{\pi}}{2}, \qquad \int_{-\infty}^{+\infty} e^{-x^2} dx = \sqrt{\pi}.$$

If in the second integral we put for x the value $x\sqrt{\alpha}$, we have:

2) $$\int_{-\infty}^{+\infty} e^{-\alpha x^2} dx = \frac{\sqrt{\pi}}{\sqrt{\alpha}}, \quad (\alpha > 0).$$

We obtain by differentiating n times with respect to α:

3) $$\int_{-\infty}^{+\infty} e^{-\alpha x^2} x^{2n} dx = \sqrt{\pi}\cdot\frac{1.3.5\ldots(2n-1)}{2^n}\,\alpha^{-(n+\frac{1}{2})} \quad (\alpha > 0).$$

Substituting $x \pm \alpha$ for x in integral 1), we have:

4) $$\int_{-\infty}^{+\infty} e^{-x^2 \mp 2\alpha x} dx = \sqrt{\pi}\cdot e^{\alpha^2}.$$

Putting in this for x, $x\sqrt{a}$ and for $2\alpha\sqrt{a}$ the value b, we find:

5) $$\int_{-\infty}^{+\infty} e^{-a x^2 \mp b x} dx = \frac{\sqrt{\pi}}{\sqrt{a}}\cdot e^{\frac{b^2}{4a}}, \qquad (a > 0).$$

From equation 2) in the form:

$$\frac{1}{\sqrt{\alpha}} = \frac{2}{\sqrt{\pi}}\int_0^\infty e^{-\alpha x^2} dx, \qquad (\alpha > 0)$$

follows:

$$\int_0^\infty \frac{\sin\alpha}{\sqrt{\alpha}}\, d\alpha = \frac{2}{\sqrt{\pi}}\int_0^\infty \sin\alpha\, d\alpha \int_0^\infty e^{-\alpha x^2} dx.$$

These integrations can be interchanged; for, $\sin\alpha\,.\,e^{-\alpha x^2}$ is everywhere a continuous function, that for $\alpha = \infty$ becomes infinitely small in a higher order than any algebraic function; hence:

$$\int_0^\infty \frac{\sin\alpha}{\sqrt{\alpha}}\, d\alpha = \frac{2}{\sqrt{\pi}}\int_0^\infty dx \int_0^\infty e^{-\alpha x^2}\sin\alpha\, d\alpha = \frac{2}{\sqrt{\pi}}\int_0^\infty \frac{dx}{x^4+1} = \frac{2}{\sqrt{\pi}}\cdot\frac{\pi}{4\sin\frac{\pi}{4}},$$

therefore:

6) $\int_0^\infty \frac{\sin\alpha}{\sqrt{\alpha}}\,d\alpha = \sqrt{\frac{\pi}{2}}$. We likewise find: $\int_0^\infty \frac{\cos\alpha}{\sqrt{\alpha}}\,d\alpha = \sqrt{\frac{\pi}{2}}$.*) .

159. Sixth group.

We proved in § 115 the formula:

1) $$\int_0^\infty \frac{x^{m-1}\,dx}{x^n + e^{\alpha i}} = \frac{\pi}{n}\cdot\frac{e^{-\left(1-\frac{m}{n}\right)\alpha i}}{\sin\frac{m}{n}\pi},$$ (m and n positive integers, $m < n$) $-\pi < \alpha < +\pi$.

Putting $x^n = z$ and denoting the rational fraction $\frac{m}{n}$ by a, we find:

2) $$\int_0^\infty \frac{z^{a-1}\,dz}{z + e^{\alpha i}} = \frac{\pi}{\sin a\pi}\cdot e^{(a-1)\alpha i}, \qquad (0 < a < 1,\ -\pi < \alpha < +\pi).$$

Now this equation was proved only for rational proper fractions. But since the definite integral, as well as the function on the right side, is a continuous function of a (proof as in § 152), the equation is still true for every irrational number less than 1. For $\alpha = 0$ we find:

3) $$\int_0^\infty \frac{x^{a-1}}{x+1}\,dx = \frac{\pi}{\sin a\pi} \qquad (0 < a < 1),$$

or, separating the integrals between 0 and 1, and between 1 and ∞, and introducing into the second $\frac{1}{x}$ instead of x:

$$\int_0^1 \frac{x^{a-1}\,dx}{x+1} + \int_1^\infty \frac{x^{a-1}\,dx}{x+1} = \int_0^1 \frac{x^{a-1}\,dx}{x+1} + \int_0^1 \frac{x^{-a}\,dx}{x+1}.$$

4) $$\int_0^1 \frac{x^{a-1} + x^{-a}}{x+1}\,dx = \frac{\pi}{\sin a\pi}, \qquad (0 < a < 1).$$

Although integral 1) ceases to be finite for $\alpha = \pm\pi$, because the function to be integrated becomes infinite in the first order at the point $x = 1$, still the integral

$$\int_0^\infty \frac{x^{m-1} - x^{m'-1}}{x^n + e^{\alpha i}}\,dx$$

must exist even for $\alpha = \pi$, because the factor $x - 1$ cancels in

*) The problem, of determining curves having the radius of curvature in inverse proportion to the length of the arc, conducted Euler to these integrals. (Salomon's Uebersetzung der Euler'schen Integralrechnung. Vol. IV. Suppl., p. 321.)

the numerator and denominator. We can express the value of this integral as the difference of two integrals of the form 1), thus:

$$\int_0^\infty \frac{x^{m-1} - x^{m'-1}}{x^n + e^{\alpha i}}\, dx = \frac{\pi}{n}\left\{\frac{e^{-\left(1-\frac{m}{n}\right)\alpha i}}{\sin\frac{m}{n}\pi} - \frac{e^{-\left(1-\frac{m'}{n}\right)\alpha i}}{\sin\frac{m'}{n}\pi}\right\}.$$

Both sides are continuous functions of α inclusive of the value $\alpha = \pi$, consequently we find:

5) $$\int_0^\infty \frac{x^{m-1} - x^{m'-1}}{x^n - 1}\, dx = \frac{\pi}{n}\left\{\frac{e^{\frac{m}{n}\pi i}}{\sin\frac{m}{n}\pi} - \frac{e^{\frac{m'}{n}\pi i}}{\sin\frac{m'}{n}\pi}\right\} e^{-\pi i}$$

$$= \frac{\pi}{n}\left(\cot\frac{m'}{n}\pi - \cot\frac{m}{n}\pi\right).$$

Putting $x^n = z$, $\frac{m}{n} = a$, $\frac{m'}{n} = b$ this becomes:

6) $$\int_0^\infty \frac{z^{a-1} - z^{b-1}}{1 - z}\, dz = \pi(\cot a\pi - \cot b\pi), \qquad (0 < a \text{ and } b < 1).$$

Let $b = 1 - a$:

7) $$\int_0^\infty \frac{z^{a-1} - z^{-a}}{1 - z}\, dz = 2\pi \cot a\pi.$$

160. Eulerian integrals (Gamma functions).

As found in § 154, $\int_0^\infty x^n e^{-x} dx$ has the value $\underline{|n} = 1 \cdot 2 \ldots n$ for every positive integer value of n. Withdrawing this restriction as to the exponents of x, the problem arises, what value has the integral

$$\int_0^\infty x^{a-1} e^{-x} dx$$

for an arbitrary value of a? With any finite value for the upper limit its value can be expressed by a series, the exponential being replaced by its expansion; but for an infinite limit this procedure affords no direct solution.

The value of the integral is finite only for $a > 0$; for, when this condition is not satisfied, the function to be integrated becomes infinite for $x = 0$ in an order higher than the first (§ 148). After Legendre, the integral is called the Eulerian integral of the second kind, and the required value as a function of the exponent a briefly denoted by $\Gamma(a)$ its function Gamma*). Therefore:

*) Euler: Inst. calc. integr., P. I, Cap. 4. 8. 9. Also: Nov. Comment. Acad. Petrop., T. XVI; in Salomon *ibidem* Vol. IV, Supplement III. Legendre: Traité des fonctions elliptiques et des intégrales Eulériennes. Tome II.

I) $$\int_0^\infty x^{a-1}e^{-x}dx = \Gamma(a)$$

is an equation of definition.

Replacing x successively by $x^{\frac{1}{a}}$, by $l\frac{1}{x}$ and by kx where $k > 0$:

dx is replaced by $\frac{1}{a}x^{\frac{1}{a}-1}dx$, therefore: 1) $\int_0^\infty e^{-\left(x^{\frac{1}{a}}\right)}dx = a\Gamma(a)$, $(a>0)$;

dx is replaced by $-\frac{dx}{x}$, therefore: 2) $\int_0^1 \left(l\frac{1}{x}\right)^{a-1}dx = \Gamma(a)$, $(a>0)$;

dx is replaced by $k\,dx$, therefore: 3) $\int_0^\infty e^{-kx}x^{a-1}dx = \frac{\Gamma(a)}{k^a}$, $(a>0, k>0)$;

accordingly these integrals also are calculated, as soon as the value of $\Gamma(a)$ is known.

Moreover, for complex values of a the value of the function $\Gamma(a)$ is finite, provided only the real part of a is positive.

For if we put $a = \alpha + i\beta$, bearing in mind that:

$$x^{i\beta} = e^{i\beta l(x)} = \cos(\beta\,.\,l(x)) + i\sin(\beta\,.\,l(x)),$$

we have:

$$\Gamma(\alpha+i\beta) = \int_0^\infty x^{\alpha-1}\cos(\beta\,.\,l(x))\,e^{-x}dx + i\int_0^\infty x^{\alpha-1}\sin(\beta\,.\,l(x))\,e^{-x}dx.$$

But:

$$\operatorname{Lim}\left(\int_\delta^\varepsilon x^{\alpha-1}\cos(\beta\,.\,l(x))\,e^{-x}dx\right) = \operatorname{Lim}\left(M\int_\delta^\varepsilon x^{\alpha-1}e^{-x}dx\right) = 0$$

for $\delta = 0$, $\varepsilon = 0$,

$$\operatorname{Lim}\left(\int_u^w x^{\alpha-1}\cos(\beta\,.\,l(x))\,e^{-x}dx\right) = \operatorname{Lim}\left(M'\int_u^w x^{\alpha-1}e^{-x}dx\right) = 0,$$

for $u = \infty$, $w = \infty$,

where M and M' respectively denote mean values of the universally finite function $\cos(\beta\,.\,l(x))$; the like holds good for the second integral.

We shall however in what follows only aim at solving the problem of calculating the function gamma for real arguments; although some of our theorems hold good also for a complex argument.

161. First property. From the formula:

$$d(e^{-x}x^a) = ae^{-x}x^{a-1}dx - e^{-x}x^a dx$$

follows by integration between the limits 0 and ∞:

$$0 = a\int_0^\infty e^{-x}x^{a-1}\,dx - \int_0^\infty e^{-x}x^a dx, \text{ or: II)}\quad \Gamma(1+a) = a\Gamma(a).$$

Accordingly, for the series of integers we have:

$$\Gamma(1) = \int_0^\infty e^{-x}dx = 1,\quad \Gamma(2) = 1\Gamma(1) = 1,\quad \Gamma(3) = 2\Gamma(2) = 2\,.\,1,$$

$$\Gamma(4) = 3\Gamma(3) = \underline{|3},\ \cdots\ \Gamma(n) = \underline{|n-1}.$$

Similarly, substituting for a the value $a+n-1$, n being an integer:

II') $\Gamma(a+n) = (a+n-1)(a+n-2)\ldots a\Gamma(a)$*).

This equation shows, that once the function gamma is known for all proper fractions, it can be calculated without difficulty for every other value of a; thus, for instance:

$$\Gamma\left(\tfrac{7}{2}\right) = \Gamma\left(\tfrac{1}{2}+3\right) = \tfrac{5}{2}\cdot\tfrac{3}{2}\cdot\tfrac{1}{2}\,\Gamma\left(\tfrac{1}{2}\right).$$

Second property. From the formula:

$$\int_0^\infty e^{-kx}x^{a-1}\,dx = \frac{\Gamma(a)}{k^a}$$

we find on putting $k = c+y$, c and $y > 0$:

$$\int_0^\infty e^{-(c+y)x}x^{a-1}\,dx = \frac{\Gamma(a)}{(c+y)^a}.$$

Multiplying both sides by $e^{-ky}y^{b-1}dy$, where b and k are positive, there follows by integration:

$$\int_0^\infty e^{-ky}y^{b-1}dy\int_0^\infty e^{-(c+y)x}x^{a-1}\,dx = \Gamma(a)\int_0^\infty \frac{e^{-ky}y^{b-1}}{(c+y)^a}\,dy.$$

It is allowable to interchange the order of integrations on the left side (§ 153), thus we have:

$$\int_0^\infty e^{-cx}x^{a-1}dx\int_0^\infty e^{-(k+x)y}y^{b-1}\,dy = \Gamma(b)\int_0^\infty \frac{e^{-cx}x^{a-1}}{(k+x)^b}\,dx,$$

therefore:

III) $$\Gamma(b)\int_0^\infty \frac{e^{-cx}x^{a-1}dx}{(k+x)^b} = \Gamma(a)\int_0^\infty \frac{e^{-ky}y^{b-1}}{(c+y)^a}\,dy.\text{**)}$$

---

*) Euler: *loc. cit.*, Supplement III, § 10.

**) Dirichlet, Journal f. Math., Vol. 15.

Provided $b > a$, both sides are continuous functions of c, the value $c = 0$ included (§ 152a). For $c = 0$ and $k = 1$ the equation assumes the form:

$$\Gamma(b)\int_0^\infty \frac{x^{a-1}dx}{(1+x)^b} = \Gamma(a)\Gamma(b-a) \qquad (b > a),$$

that can also be written:

IV)
$$\frac{\Gamma(a)\Gamma(b)}{\Gamma(a+b)} = \int_0^\infty \frac{x^{a-1}dx}{(1+x)^{a+b}} \qquad (b > 0,\ a > 0).$$

This formula shows how to calculate a new integral, of the binomial class, by means of the functions gamma. Legendre styles it the Eulerian integral of the first kind. Putting:

$$x = \frac{y}{1-y}, \quad dx = \frac{dy}{(1-y)^2}, \quad y = \frac{x}{1+x}, \quad 1+x = \frac{1}{1-y},$$

we have $y = 0$ for $x = 0$, $y = 1$ for $x = \infty$; and when y is replaced again by x or by $1 - x$, IV) changes into:

$$\frac{\Gamma(a)\Gamma(b)}{\Gamma(a+b)} = \int_0^1 x^{a-1}(1-x)^{b-1}dx = \int_0^1 (1-x)^{a-1}x^{b-1}dx.\text{*)}$$

When we put $a + b = 1$, taking a therefore as a proper fraction, since $\Gamma(1) = 1$, formula IV) becomes:

V)
$$\int_0^\infty \frac{x^{a-1}}{1+x}dx = \Gamma(a)\Gamma(1-a) = \frac{\pi}{\sin a\pi}, \qquad \text{§ 159 Formula 3)}.$$

This formula reduces the calculation of all values of $\Gamma(a)$ for arguments greater than $\frac{1}{2}$ to that of values for arguments between zero and $\frac{1}{2}$.

We have for $a = \frac{1}{2}$ the special value:

$$\Gamma(\tfrac{1}{2}) = \int_0^\infty e^{-x}x^{-\frac{1}{2}}dx = \sqrt{\pi}.$$

Replacing x by y^2, we obtain the integral treated in § 158.

162. Expression of the function gamma by an infinite product.

Integral I) can be differentiated with respect to the parameter a (§ 152) and we get:

$$\frac{d\Gamma(a)}{da} = \Gamma'(a) = \int_0^\infty e^{-x}x^{a-1}l(x)\,dx.$$

If we replace $l(x)$ by its value as an integral, § 154 Formula 3a):

*) Euler, *loc. cit.*, § 25.

$$l(x) = \int_0^\infty \frac{e^{-y} - e^{-xy}}{y}\, dy,$$

we find:

$$\Gamma'(a) = \int_0^\infty e^{-x} x^{a-1} dx \int_0^\infty \frac{e^{-y} - e^{-xy}}{y}\, dy.$$

Here the order of integrations can be interchanged, for the function:

$$f(x, y) = \frac{e^{-x} x^{a-1}(e^{-y} - e^{-xy})}{y}$$

for $x = \infty$, $y = \infty$, becomes infinitely small in a higher order than any algebraic expression, for $x = 0$ it becomes infinitely great in an order lower than the first; for $y = 0$ we have:

$$\operatorname*{Lim}_{y=0} \left(\frac{e^{-y} - e^{-xy}}{y}\right) = \operatorname*{Lim}_{y=0} (-e^{-y} + e^{-xy} x) = -1 + x.$$

Accordingly let us begin by putting the above integral:

$$\Gamma'(a) = \int_0^\infty \frac{dy}{y} \int_0^\infty e^{-x} x^{a-1} (e^{-y} - e^{-xy})\, dx$$

$$= \operatorname*{Lim}_{\delta=0} \int_\delta^\infty \frac{dy}{y} \int_0^\infty e^{-x} x^{a-1} (e^{-y} - e^{-xy})\, dx.$$

But we have:

$$\int_\delta^\infty \frac{dy}{y} \int_0^\infty e^{-x} x^{a-1} e^{-y} dx = \Gamma(a) \int_\delta^\infty \frac{e^{-y}}{y}\, dy,$$

$$\int_\delta^\infty \frac{dy}{y} \int_0^\infty e^{-x(1+y)} x^{a-1}\, dx = \Gamma(a) \int_\delta^\infty \frac{dy}{y(1+y)^a}.$$

Therefore:

$$\Gamma'(a) = \Gamma(a) \operatorname*{Lim}_{\delta=0} \int_\delta^\infty \left(\frac{e^{-y}}{y} - \frac{1}{y(1+y)^a}\right) dy = \Gamma(a) \int_0^\infty \left(\frac{e^{-y}}{y} - \frac{1}{y(1+y)^a}\right) dy.$$

For, it is evident from the fact that

$$\operatorname{Lim} \left(\frac{e^{-y}}{y} - \frac{1}{y(1+y)^a}\right)_{\text{for } y=0} = \operatorname{Lim} \left(\frac{e^{-y}(1+y)^a - 1}{y(1+y)^a}\right)$$

$$= \operatorname*{Lim}_{y=0} \left(\frac{e^{-y}(1+y)^{a-1}}{(1+y)^{a-1}} \cdot \frac{-(1+y)+a}{(1+y)+ay}\right) = -1 + a,$$

and is therefore finite, that while $a > 0$ this integral remains finite and determinate even at the limit $y = 0$.

This investigation was necessary, because the integrals

$$\int_0^\infty \frac{e^{-y}}{y}\,dy, \qquad \int_0^\infty \frac{dy}{y(1+y)^a}$$

are not separately finite.

Accordingly we have:

VI) $$\frac{d\log\Gamma(a)}{da} = \int_0^\infty \left(\frac{e^{-y}}{y} - \frac{1}{y(1+y)^a}\right) dy.$$

Differentiating with respect to a and then putting $1 + y = e^x$ we find:

VII) $$\frac{d^2 l\Gamma(a)}{da^2} = \int_0^\infty \frac{l(1+y)}{y(1+y)^a}\,dy = \int_0^\infty \frac{xe^{-ax}}{1-e^{-x}}\,dx \qquad (a > 0).$$

This is an integral adapted to integration by expanding. We have:

$$\frac{1}{1-e^{-x}} = 1 + e^{-x} + e^{-2x} + e^{-3x} + \cdots e^{-nx} + R_n, \quad R_n = \frac{e^{-(n+1)x}}{1-e^{-x}},$$

therefore:

$$\int_0^\infty \frac{xe^{-ax}}{1-e^{-x}}\,dx$$

$$= \int_0^\infty xe^{-ax}dx + \int_0^\infty xe^{-(a+1)x}dx + \cdots + \int_0^\infty xe^{-(a+n)x}dx + \int_0^\infty R_n xe^{-ax}\,dx.$$

Now:

$$\int_0^\infty xe^{-(a+n)x}dx = \left(-\frac{xe^{-(a+n)x}}{(a+n)} - \frac{e^{-(a+n)x}}{(a+n)^2}\right)_0^\infty = \frac{1}{(a+n)^2},$$

and we can choose n (cf. also § 130), so that

$$\int_0^\infty R_n xe^{-ax}dx = \int_0^\infty \frac{xe^{-(a+n+1)x}}{1-e^{-x}}\,dx$$

shall become arbitrarily small, because:

$$\int_0^1 \frac{xe^{-(a+n+1)x}}{1-e^{-x}}\,dx = M\int_0^1 e^{-(a+n+1)x}dx = M\left(-\frac{e^{-(a+n+1)x}}{a+n+1}\right)_0^1,$$

$$\int_1^\infty \frac{xe^{-(a+n+1)x}}{1-e^{-x}}\,dx = M'\int_1^\infty xe^{-(a+n+1)x}dx = M'\left(-\frac{xe^{-(a+n+1)x}}{a+n+1} - \frac{e^{-(a+n+1)x}}{(a+n+1)^2}\right)_1^\infty,$$

where M denotes a mean value of $\frac{x}{1-e^{-x}}$ in the interval from 0 to 1, and M' a mean value of $\frac{1}{1-e^{-x}}$ in the interval from 1 to ∞; thus:

$$\text{VIII)}\quad \frac{d^2 l\Gamma(a)}{da^2} = \frac{1}{a^2} + \frac{1}{(a+1)^2} + \frac{1}{(a+2)^2} + \cdots \frac{1}{(a+n)^2} + \cdots = \sum_{n=0}^{n=\infty} \frac{1}{(a+n)^2}.$$

$$(a > 0).$$

This is a uniformly convergent series for every positive value of a; integrating it between the limits 1 and a, the result is:

$$\text{IX)}\quad \frac{dl\Gamma(a)}{da} - \left(\frac{dl\Gamma(a)}{da}\right)_{\text{for } a=1} = \sum_{n=0}^{n=\infty}\left(\frac{1}{1+n} - \frac{1}{a+n}\right) = (a-1)\sum_{n=0}^{n=\infty}\frac{1}{(1+n)(a+n)}$$

$$(a > 0),$$

or:

$$\frac{dl\Gamma(a)}{da} = (a-1)\sum_{n=0}^{n=\infty}\frac{1}{(1+n)(a+n)} + C,$$

where C, denotes the value of $\left(\frac{dl\Gamma(a)}{da}\right)_{\text{for } a=1} = \frac{\Gamma'(1)}{\Gamma(1)}$, therefore by VI) the value of the integral:

$$\text{X)}\qquad C = \int_0^\infty \left(\frac{e^{-y}}{y} - \frac{1}{y(1+y)}\right) dy = \Gamma'(1).$$

Integrating equation IX) again between the limits a and 1, we find:

$$l\Gamma(a) = \sum_{n=0}^{n=\infty}\int_1^a \frac{(a-1)\,da}{(1+n)(a+n)} + C(a-1) \qquad (l\Gamma(1) = 0),$$

or:

$$\text{XI)}\qquad l\Gamma(a) = \sum_0^\infty\left(\frac{a-1}{1+n} - l\,\frac{a+n}{1+n}\right) + C(a-1).$$

The constant C can be eliminated: Putting $a = 2$ this becomes:

$$l\Gamma(2) = 0 = \sum_0^\infty\left(\frac{1}{1+n} - l\,\frac{2+n}{1+n}\right) + C^{*});$$

multiplying this equation by $a - 1$ and subtracting from XI) we obtain:

---

*) Writing: $\quad -C = \sum_0^\infty \frac{1}{1+n} - \sum_0^\infty l\,\frac{2+n}{1+n}$, where however each of these series apart is divergent, $-C$ is evidently the same number that occurs in the Logarithmic integral (§ 137), as can also be directly inferred from the connexion of this integral with formula X) above.

$$l\Gamma(a) = \sum_{n=0}^{n=\infty}\left((a-1)l\frac{2+n}{1+n} - l\frac{a+n}{1+n}\right)$$

$$= \sum_{m=1}^{m=\infty}\left\{(a-1)l\left(1+\frac{1}{m}\right) - l\left(1+\frac{a-1}{m}\right)\right\},$$

or:

$$\text{XII)} \qquad l\Gamma(a) = \sum_{m=1}^{m=\infty} l\frac{m\left(\frac{m+1}{m}\right)^{a-1}}{a+m-1}.$$

Accordingly, when we pass over from the logarithm to the number, we have $\Gamma(a)$ expressed for calculation by an infinite product:

$$\Gamma(a) = \prod_{m=1}^{m=\infty}\frac{m\left(\frac{m+1}{m}\right)^{a-1}}{(a+m-1)} = \frac{1.2.3\ldots m(m+1)^{a-1}}{a(a+1)(a+2)\ldots(a+m-1)}$$

$$= \frac{(m+1)^{a-1}m}{a(1+a)\left(1+\frac{a}{2}\right)\cdots\left(1+\frac{a}{m-1}\right)}$$

$$(\text{for } m = \infty),$$

or, as this expression can be written, when m is replaced in the numerator by the value $(m+1)\left(1-\frac{1}{m+1}\right)$, and it is remembered that as m increases arbitrarily, the factor $1-\frac{1}{m+1}$ and likewise $\left(1+\frac{1}{m}\right)^a$ differ inappreciably from unity:

$$\text{XIII)} \quad \Gamma(a) = \frac{(m+1)^a}{a(1+a)\left(1+\frac{a}{2}\right)\cdots\left(1+\frac{a}{m-1}\right)} = \frac{m^a}{a\prod_{m=2}\left(1+\frac{a}{m-1}\right)}.$$

$$(m = \infty).$$

Gauss*) employed this formula as the definition of a function and derived all its properties from this infinite product. This can be shown to converge for every finite value of a, which does not make a factor of the denominator vanish, so that this definition is more comprehensive, than the Eulerian integral.

163. We are going to prove this, by answering in general the question: Under what condition does an infinite product converge? This is an important question; for as was indicated in § 38 and is here worked out for a definite function, the formation of an infinite product is a second instrument for the expression of a function, not symbolically, but suitably for numerical computation. The following

*) Gauss: Disquisitiones generales circa seriem infinitam. Werke Vol. III.

investigations apply also in case there are complex factors. If, in an infinite product, we give the form:

$$(1 + u_1)(1 + u_2) \ldots (1 + u_n)(1 + u_{n+1}) \ldots$$

to the factors that can be unrestrictedly continued in accordance with some law, the successive values obtained by multiplying, first n factors, then $n + 1, \ldots n + k, \ldots$:

$$P_n = (1 + u_1)(1 + u_2) \ldots (1 + u_n),$$
$$P_{n+1} = (1 + u_1)(1 + u_2) \ldots (1 + u_n)(1 + u_{n+1}),$$
$$\cdots\cdots\cdots\cdots\cdots\cdots\cdots\cdots\cdots$$
$$P_{n+k} = (1 + u_1)(1 + u_2) \ldots (1 + u_n)(1 + u_{n+1}) \ldots (1 + u_{n+k})$$

must form a sequence of numbers with a determinate finite limiting value. For this it is requisite: first that none of the products P, and therefore also none of the terms u shall become infinite; second that for any number δ however small, there shall be a place n such that:

$$\text{abs}\,[P_{n+k} - P_n] < \delta$$

for every value of k. From this inequality follows, provided P_n does not fall below any finite assignable limit, that:

$$\text{abs}\left[\frac{P_{n+k}}{P_n} - 1\right] < \frac{\delta}{P_n}, \text{ or: } \text{abs}\left[\frac{P_{n+k}}{P_n}\right] < 1 + \text{abs}\,\frac{\delta}{P_n};$$

in other words: There must be a place n from which onwards the ratio of the values P differs inappreciably from unity; this must also be the case in particular for:

$$P_{n+1} : P_n = 1 + u_{n+1},$$

i. e. the terms u must necessarily converge to zero.

The case, that the quantities P sink below any finite amount, or that separate factors vanish and the limiting value of the product is therefore zero, must be excluded both here and in the following investigations.

If a product still converge even when we give all the terms u their absolute values, it is called absolutely convergent.

This definition requires a preliminary proof, that the convergence of the product $\Pi(1 + u_n)$ is always a necessary consequence of the convergence of the product formed of the absolute values v of the terms u; the case that even a single quantity u is $= -1$ is excluded.

Denoting by v_r the absolute amount of u_r, and by Q_n the product:

$$(1 + v_1)(1 + v_2) \ldots (1 + v_n),$$

we have:

$$\left(\frac{Q_{n+k}}{Q_n} - 1\right) = (1 + v_{n+1})(1 + v_{n+2}) \ldots (1 + v_{n+k}) - 1,$$
$$\left(\frac{P_{n+k}}{P_n} - 1\right) = (1 + u_{n+1})(1 + u_{n+2}) \ldots (1 + u_{n+k}) - 1.$$

Multiplying out these products, the absolute amount resulting from the first equation is easily seen to be not less than the absolute amount resulting from the second; therefore we have the relation:

$$\text{abs}\left[\frac{Q_{n+k}}{Q_n}-1\right] \geqq \text{abs}\left[\frac{P_{n+k}}{P_n}-1\right].$$

By hypothesis we can choose n so as to make the amount on the left side arbitrarily small; therefore also a value n can be found, such that

$$\text{abs}\left[\frac{P_{n+k}}{P_n}-1\right] \text{ shall be } < \delta.$$

Hence follows that P_n tends to a determinate finite limiting value that is not zero. For were zero the limiting value of P, to each value of n however great could be found a value k, for which the ratio $P_{n+k} : P_n$ would be arbitrarily small.

A necessary and sufficient criterion for the absolute convergence of the product:

$$P = (1+u_1)(1+u_2)(1+u_3)\ldots$$

is the convergence of the infinite series formed of the absolute amounts:

$$v_1+v_2+v_3+\cdots+v_n+\cdots \text{ etc..}$$

For, because the product:

$$Q_n = (1+v_1)(1+v_2)\ldots(1+v_n) \text{ is } > v_1+v_2+\cdots v_n,$$

the convergence of the infinite series is necessary, and because:

$$1+v_1 < e^{v_1},\quad 1+v_2 < e^{v_2}, \ldots 1+v_n < e^{v_n},$$

so that:

$$Q_n \text{ is } < e^{v_1+v_2+\ldots v_n},$$

therefore the convergence of the infinite series is sufficient.

The value of an absolutely convergent product is independent of the arrangement of its factors.

For, writing:

$$P_n = (1+u_1)(1+u_2)\ldots(1+u_n),$$
$$P_m' = (1+u_1')(1+u_2')\ldots(1+u_m'),$$

this second product consisting of factors occurring in P only arranged differently, we can choose m so large, that all the factors contained in P_n shall also occur in P_m'. Then:

$$P_m' = P_n(1+u_k)(1+u_l)\ldots(1+u_\nu),$$

where $k, l \ldots \nu$ denote indices that are greater than n; or:

$$\frac{P_m'}{P_n} = (1+u_k)(1+u_l)\ldots(1+u_\nu).$$

But the amount of the right side is not greater than the amount of $(1+v_k)(1+v_l)\ldots(1+v_\nu)$ and this is less than

$$e^{v_k+v_l+\ldots v_\nu}.$$

Since this expression approximates arbitrarily to unity, merely by choice of n, we have:

$$\operatorname{Lim} P_m' = \operatorname{Lim} P_n.\text{*)}$$

164. The convergence of the function gamma for all values of a, for which no factor vanishes, is now demonstrated as follows. Let us form:

$$\frac{1}{\Gamma(a)} = \frac{a\Pi\left(1+\frac{a}{m-1}\right)}{m^a} = a\,\frac{1+a}{(1+1)^a}\cdot\frac{1+\frac{a}{2}}{\left(1+\frac{1}{2}\right)^a}\cdot\frac{\left(1+\frac{a}{3}\right)}{\left(1+\frac{1}{3}\right)^a}\cdot\frac{\left(1+\frac{a}{4}\right)}{\left(1+\frac{1}{4}\right)^a}\cdots$$

Writing the factor $\dfrac{1+\frac{a}{n}}{\left(1+\frac{1}{n}\right)^a}$ in the form $1-u_n$, and applying the Binomial series (§ 46), we have:

$$u_n = \frac{\left(1+\frac{1}{n}\right)^a-\left(1+\frac{a}{n}\right)}{\left(1+\frac{1}{n}\right)^a} = \frac{1+\frac{a}{n}+\frac{a(a-1)}{1.2}\left(\frac{1}{n}\right)^2\left(1+\frac{\Theta}{n}\right)^{a-2}-\left(1+\frac{a}{n}\right)}{\left(1+\frac{1}{n}\right)^a},$$

or:

$$u_n = \frac{a(a-1)}{1.2}\left(\frac{1}{n}\right)^2\left(\frac{1+\frac{\Theta}{n}}{1+\frac{1}{n}}\right)^a\cdot\frac{1}{\left(1+\frac{\Theta}{n}\right)^2} \qquad (0<\Theta<1).$$

If a be positive, $\left(\dfrac{1+\frac{\Theta}{n}}{1+\frac{1}{n}}\right)^a\cdot\dfrac{1}{\left(1+\frac{\Theta}{n}\right)^2}$ is certainly a proper fraction, and the series:

$$u_n+u_{n+1}+u_{n+2}+\cdots<\frac{a(a-1)}{1.2}\left(\frac{1}{n^2}+\frac{1}{(n+1)^2}+\frac{1}{(n+2)^2}+\cdots\right)$$

converges absolutely.

If a be negative, we can choose n so large, that for any possible value of Θ, $\left(\dfrac{1+\frac{\Theta}{n}}{1+\frac{1}{n}}\right)^a\cdot\dfrac{1}{\left(1+\frac{\Theta}{n}\right)^2}$ shall be less than some determinate number, ex. gr. less than 2, so that the series:

$$u_n+u_{n+1}+u_{n+2}\cdots<\frac{a(a-1)}{1.2}\,2\cdot\left(\frac{1}{n^2}+\frac{1}{(n+1)^2}+\frac{1}{(n+2)^2}\cdots\right)$$

likewise converges absolutely.

It is accordingly proved, that the infinite product $\frac{1}{\Gamma(a)}$, therefore

*) Although infinite products were introduced almost simultaneously with infinite series, the fundamental theorems regarding them were first proved by Weierstrass: Ueber die Theorie der analytischen Facultäten, Journ. f. Math., Vol. 51; reprinted in his: Abhandlungen aus der Functionenlehre, p. 183, 1886.

also the product $\Gamma(a)$, converges absolutely for all finite values of a except negative integers.

165. Legendre's series for calculating $l\Gamma(a)$.

Writing Equation VIII) in the form:

$$\frac{d^2 l\Gamma(1+a)}{da^2} = \frac{1}{(a+1)^2} + \frac{1}{(a+2)^2} + \frac{1}{(a+3)^2} + \cdots \quad (a > -1)$$

and differentiating it $n-2$ times, which is allowable since the derived series likewise converge absolutely, the result is:

$$\frac{1}{\lfloor n} \frac{d^n l\Gamma(1+a)}{da^n} = \frac{(-1)^n}{n} \left\{ \frac{1}{(a+1)^n} + \frac{1}{(a+2)^n} + \frac{1}{(a+3)^n} + \cdots \right\}.$$

Let the sum $\frac{1}{1^n} + \frac{1}{2^n} + \frac{1}{3^n} + \frac{1}{4^n} + \cdots$ (§ 47 foot-note p. 82), be denoted by S_n, we have:

$$\frac{1}{\lfloor n} \left\{ \frac{d^n l\Gamma(1+a)}{da^n} \right\}_{a=0} = (-1)^n \frac{S_n}{n},$$

further:

$$\left\{ \frac{dl\Gamma(1+a)}{da} \right\}_{a=0} = \Gamma'(1) = C \qquad \text{and: } l\Gamma(1) = 0.$$

Therefore by Mac Laurin's theorem:

$$l\Gamma(1+a) = aC + \frac{a^2}{2} S_2 - \frac{a^3}{3} S_3 + \frac{a^4}{4} S_4 - \cdots + \frac{a^n}{\lfloor n} \left\{ \frac{d^n l\Gamma(1+a)}{da^n} \right\}_{\Theta a}.$$

The remainder converges to zero when the absolute amount of a is less than 1, accordingly, omitting the remainder, the infinite series is absolutely convergent for all values of a between -1 and $+1$.

But this series is unsuitable for numerical calculation, because the coefficients S do not decrease rapidly enough, and moreover the value of C is still unknown. A more rapidly convergent series is found by expanding the value of $l(1+a)$ and adding:

$$0 = -l(1+a) + a - \frac{a^2}{2} + \frac{a^3}{3} - \frac{a^4}{4} + \cdots$$

thus:

$$l\Gamma(1+a) = -l(1+a) + a(1+C) + \frac{1}{2}(S_2 - 1)a^2 - \frac{1}{3}(S_3 - 1)a^3 + \frac{1}{4}(S_4 - 1)a^4 - \cdots;$$

likewise, taking a with the opposite sign:

$$l\Gamma(1-a) = -l(1-a) - a(1+C) + \frac{1}{2}(S_2 - 1)a^2 + \frac{1}{3}(S_3 - 1)a^3 + \frac{1}{4}(S_4 - 1)a^4 + \cdots$$

Now because by Equation V):

$$l\Gamma(1+a) + l\Gamma(1-a) = l\frac{\pi a}{\sin \pi a},$$

we have finally:

$$\text{XIV)}\qquad l\Gamma(1+a) = \frac{1}{2}\, l\frac{\pi a}{\sin \pi a} - \frac{1}{2}\, l\frac{1+a}{1-a}$$
$$+ a(1+C) - \frac{1}{3}(S_3-1)a^3 - \frac{1}{5}(S_5-1)a^5 - \cdots$$
$$(-1 < a < +1).$$

From this formula the values of $l\Gamma(a)$ can be calculated for the arguments 0 to 2, when the quantities $S_3, S_5 \ldots$ have been previously determined. Legendre has given these values from S_3 to S_{35} to 16 places of decimals. Also the value of C must be expressed, this is done most rapidly by putting $a = \frac{1}{2}$, because:

$$l\Gamma\left(\tfrac{3}{2}\right) = l\left(\tfrac{1}{2}\Gamma\left(\tfrac{1}{2}\right)\right) = l\left(\tfrac{1}{2}\sqrt{\pi}\right)$$

is known. By the series:

$$1 + C = l\,\frac{3}{2} + \frac{1}{3 \cdot 2^2}(S_3 - 1) + \frac{1}{5 \cdot 2^4}(S_5 - 1) + \frac{1}{7 \cdot 2^6}(S_7 - 1) + \cdots$$

we obtain $\qquad C = -0{,}577\,215\,664\,901\,532\ldots$*).

Accordingly the initial numerical terms of series XIV) are:

$$l\Gamma(1+a) = \frac{1}{2}\, l\frac{\pi a}{\sin \pi a} - \frac{1}{2}\, l\frac{1+a}{1-a}$$
$$+ 0{,}422\,784\,3\,a - 0{,}067\,353\,0\,a^3 - 0{,}007\,385\,5\,a^5 - \cdots$$

*) Euler obtained this number correctly to 15 places of decimals; annotating Euler's Calc. integ., Mascheroni calculated it further, Legendre also gives 26 places but both differ in the 20th place from Gauss who had it recalculated by Nicolai and gives the result to 40 places of decimals. Werke, Vol. III, p. 154.

Eighth Chapter.

General theorems concerning the double integral.

166. Definition of the double integral.

Let $f(x, y)$ be a function of the two variables, that is uniquely defined in any way for any domain T, but for the present in such a way as to be everywhere finite. Let the domain be conceived to be in the plane xy surrounded by some continuous and closed succession of points, or, stated analytically, bounded by a curve whose equation is $\varphi(x, y) = 0$. The domain can also be bounded by more closed lines than one, as ex. gr. a circular ring by two circles. The simplest case of boundary of a domain is a rectangle with its sides parallel to the axes of rectangular coordinates; then x takes all values from a to b, y all values from α to β. Should the function be defined for the entire infinite plane, we can always express this: it is defined for a surface, whose boundary can be arbitrarily extended. Let us resolve the domain T, at first on the hypothesis that it is finite, into n small parts or superficial elements, and call them τ_1, $\tau_2, \ldots \tau_n$. All these elements are conceived as positive quantities. Such a resolution is effected, ex. gr., when we cover over the domain with a net having its lines parallel to the coordinate axes at the distances Δx and Δy. In this case all superficial elements are equal, being rectangles whose magnitude is $\Delta x \,.\, \Delta y$. Only at the bounds of the domain are these rectangles cut by the boundary line. Let us select any arbitrary value among those assumed by the function within or at the limits of such a superficial element. For simplicity, let such a value in each be denoted by $f(\tau_1), f(\tau_2), \ldots f(\tau_n)$; thus the question arises:

Under what hypotheses does the value of the sum:

$$S_n = f(\tau_1) \cdot \tau_1 + f(\tau_2) \cdot \tau_2 + \cdots + f(\tau_n) \cdot \tau_n$$

approximate to a determinate limit, altogether independent of the choice both of the superficial elements and of the value of the function in any such element, when the number of the elements is arbitrarily increased according to any law in such a way that each element tends to the limit zero?

The enunciation of the answer is as it was for the simple integral:

If we denote the greatest fluctuation of the function, i. e. the positive difference of its greatest and least values in the element τ_μ *or at its limits, by* D_μ; *the sum:* $\tau_1 D_1 + \tau_2 D_2 + \cdots + \tau_n D_n$ *must converge to zero along with the quantities* τ.

In the first place it can be seen that when for any one continued process of partition this sum converges to zero, it converges likewise to zero for every other. The proof is the same as for the simple integral, only that the conception of the superficial element everywhere replaces that of the linear interval.*)

In the second place we prove that the condition enunciated is necessary, by starting from a determinate partition and continuing it by resolving each element into further elements. The sums formed with the greatest and least values of the function: G_μ and g_μ in the interval τ_μ,

$$\sum G_\mu \tau_\mu \quad \text{and} \quad \sum g_\mu \tau_\mu$$

approximate in this process, the first by continued decrease, the second by continued increase, each to a determinate limiting value, and these two limiting values become equal only when we have:

$$\sum (G_\mu - g_\mu)\tau_\mu = \sum D_\mu \tau_\mu = 0.$$

In the third place it can be seen that the same limiting value is obtained in another partition of the same kind, when two different partitions, each already pushed so far as to yield a value differing arbitrarily little from its limiting value, are considered simultaneously, and this partition resulting from their combination is regarded as a continuation as well of the one as of the other.

Finally we perceive that, provided the above condition is fulfilled, we may also complete the process of partition without retaining the limits of a former partition, because the series of values formed in this way takes also a determinate limiting value, and each term of this series ultimately differs arbitrarily little from the limiting value reached by the previous process.

167. The necessary condition is fulfilled:

First: when $f(x, y)$ is everywhere a continuous function (§ 52).

---

*) Here as in the case of linear intervals it is quite indifferent according to what law the succession of the summands is formed. In that case, as in this, since we are dealing with finite sums and want to demonstrate that it is a property of such a sum to have a limiting value, there is no necessity that we should take the intervals only in the exact order in which they are arranged in the interval a to b. It is otherwise in the transition to the limit for integrals in which the function to be integrated becomes infinite.

Second: when $f(x, y)$ in separate points or in separate lines (∞^1 places) is discontinuous or indeterminate between finite limits, or again, when it has infinitely many maxima and minima with finite fluctuations.

Third: when $f(x, y)$ in infinitely many lines (∞^2 places) is finitely discontinuous or indeterminate or fluctuates infinitely often, but when the sum of the superficial elements in which the fluctuations D_μ exceed an arbitrarily small number σ, can be made arbitrarily small.

This third requirement, which embraces the first two, suggests an extension of the distinction established by Cantor of sets of points in a domain of two dimensions, rising from linear sets to plane sets. Infinitely many lines do not give rise to a plane set of points when their initial elements form a discrete set of points; on the other hand we do obtain a plane set of points, when the initial elements of the lines belong to a linear set or mass. The following investigation, however, is restricted to functions that satisfy the first or second requirement and the third possibility will only be cited incidentally.

The limiting value of the sum:

$$\operatorname{Lim}\{f(\tau_1)\cdot\tau_1+f(\tau_2)\cdot\tau_2+\cdots+f(\tau_n)\cdot\tau_n\} \quad \textit{for } n=\infty$$

is usually denoted by:

$$\iint f(x, y)\,dx\,dy, \quad \text{or: } \int^{(2)} f(x, y)\,dT,$$

and is called the definite double integral in the domain T.

The definite double integral, as well as the simple integral, admits of a geometrical interpretation. If we lay off the value of the function $z=f(x, y)$ perpendicular to the plane xy, the integral:

$$\int^{(2)} f(x, y)\,dT$$

expresses the volume of the cylinder whose base is the area enclosed by the curve $\varphi(x, y)$ and that is bounded above the plane xy by the surface $z=f(x, y)$.*)

Among the theorems resulting from this definition of the double integral we only notice specially the following.

1. When $\varphi(x, y)$ and $\psi(x, y)$ are any finite integrable functions within a certain domain, their product also is integrable within the same domain. (Proof as in § 146.)

2. The First Theorem of the Mean Value: When the

---

*) By the geometrical problems: the determination of volume and the measurement of curved surfaces, the analytical conception of the double integral was introduced. Riemann's investigations on the definite integral established the fundamental principle both for the double and also for multiple integrals.

function φ has the same sign everywhere in the domain of integration, if G denote the greatest and g the least value of f in that domain, we have:

$$\int^{(2)} f(x, y) \cdot \varphi(x, y)\, dT = \{g + \Theta(G - g)\} \int^{(2)} \varphi(x, y)\, dT \qquad (0 \leqq \Theta < 1).$$

In particular, if f be a continuous function throughout, the value $g + \Theta(G - g)$ actually occurs among the values that f assumes in the domain.

168. In order to ascertain the value of the double integral, we endeavour to reduce it to two simple integrations.

Since the manner of partition into superficial elements is quite indifferent, let us conceive it as a network parallel to the axes of coordinates. Each line parallel to the abscissæ must cross the boundary curve in an even finite number of points, likewise each line parallel to the ordinates. The domain can then be resolved by means of a finite number of parallels into finite areas, each having its boundary crossed by the lines of the net only in two places. To a part having such a simple form, an elementary surface, the following investigation has reference.

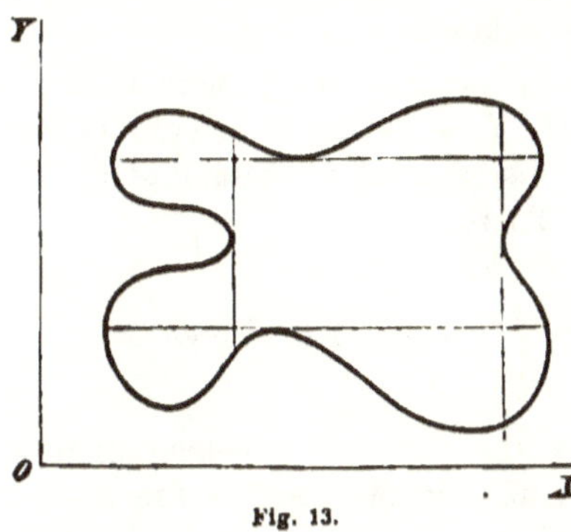

Fig. 13.

Let a and b be the extreme values of x, α and β those of y, that belong to points of the boundary curve. If then x_μ, y_ν denote any arbitrary point in the region and $x_{\mu+1} - x_\mu$, $y_{\nu+1} - y_\nu$ be the lengths of the sides of the rectangle having this point as a vertex, we have to form the double sum:

$$\sum\sum (x_{\mu+1} - x_\mu)(y_{\nu+1} - y_\nu) f(x_\mu, y_\nu).$$

Now we can carry out this summation in two different ways: either by first summing all the values that have the same factor $x_{\mu+1} - x_\mu$, and then adding these quantities; or in reverse order by combining the terms with the same factor $y_{\nu+1} - y_\nu$, and then summing these values. These two different processes are symbolically indicated by:

$$\sum_\mu (x_{\mu+1} - x_\mu) \sum_\nu (y_{\nu+1} - y_\nu) f(x_\mu, y_\nu)$$

and:

$$\sum_\nu (y_{\nu+1} - y_\nu) \sum_\mu (x_{\mu+1} - x_\mu) f(x_\mu, y_\nu).$$

Proceeding successively to the limits, as the differences $(y_{\nu+1} - y_\nu)$ and $(x_{\mu+1} - x_\mu)$ converge to zero, let us in the first case conceive the differences $(y_{\nu+1} - y_\nu)$ chosen so small, that the value of:

$$\sum_\nu (y_{\nu+1} - y_\nu) f(x_\mu, y_\nu)$$

shall differ from the value of the definite integral:

$$\int_{y_0^{(\mu)}}^{y_1^{(\mu)}} f(x_\mu, y)\, dy$$

only by a quantity $\Delta(x_\mu)$ whose absolute amount can be rendered arbitrarily small by diminishing the distances $y_{\nu+1} - y_\nu$.

For inasmuch as the function f, assumed everywhere finite, is subject to the conditions of § 167, f is generally integrable with respect to y, and it is only for a discrete set of values of x_μ that the integral can lose its meaning. Therefore:

$$\sum_\nu (y_{\nu+1} - y_\nu) f(x_\mu, y_\nu) = \int_{y_0^{(\mu)}}^{y_1^{(\mu)}} f(x_\mu, y)\, dy + \Delta(x_\mu).$$

The quantities $y_0^{(\mu)}$ and $y_1^{(\mu)}$ signify the values of y belonging to x_μ at the limits of the domain, we can also denote them as functions of x_μ by the equations:

$$y_0^{(\mu)} = \varphi(x_\mu), \quad y_1^{(\mu)} = \psi(x_\mu).$$

From the above equation we have further:

$$\sum_\mu (x_{\mu+1} - x_\mu) \sum_\nu (y_{\nu+1} - y_\nu) f(x_\mu, y_\nu) = \sum_\mu (x_{\mu+1} - x_\mu) \int_{y_0^{(\mu)}}^{y_1^{(\mu)}} f(x_\mu, y)\, dy$$
$$+ \sum_\mu (x_{\mu+1} - x_\mu)\, \Delta(x_\mu).$$

Now letting each interval $x_{\mu+1} - x_\mu$ converge to zero, the limiting values of the quantities upon the right side become:

$$\int_a^b dx \int_{\varphi(x)}^{\psi(x)} f(x, y)\, dy + \int_a^b \Delta(x)\, dx.$$

But since the absolute amount of $\Delta(x)$ for all values of x, with possibly a discrete set of exceptions that do not influence the value of the integral, can be made arbitrarily small, the value of this second integral is also arbitrarily small, i. e.:

$$\operatorname{Lim} \sum_{\mu}' (x_{\mu+1} - x_\mu) \sum_{\nu} (y_{\nu+1} - y_\nu) f(x_\mu, y_\nu) = \iint f(x, y)\, dx\, dy$$

$$= \int_a^b dx \int_{y=\varphi(x)}^{y=\psi(x)} f(x, y)\, dy.\text{*)}$$

By the second process likewise we find the double integral equal to:

$$\int_\alpha^\beta dy \int_{x=\varphi_1(y)}^{x=\psi_1(y)} f(x, y)\, dx,$$

when $\varphi_1(y)$ and $\psi_1(y)$ denote the values that x assumes at the limits of the domain for the different values of y.

When the boundary of the domain ex. gr. is the ellipse:

$$\frac{x^2}{a^2} + \frac{y^2}{b^2} = 1,$$

we have:

$$\int_{-a}^{+a} dx \int_{y=-\frac{b}{a}\sqrt{a^2-x^2}}^{y=+\frac{b}{a}\sqrt{a^2-x^2}} f(x, y)\, dy = \int_{-b}^{+b} dy \int_{x=-\frac{a}{b}\sqrt{b^2-y^2}}^{x=+\frac{a}{b}\sqrt{b^2-y^2}} f(x, y)\, dx = \iint f(x, y)\, dx\, dy.$$

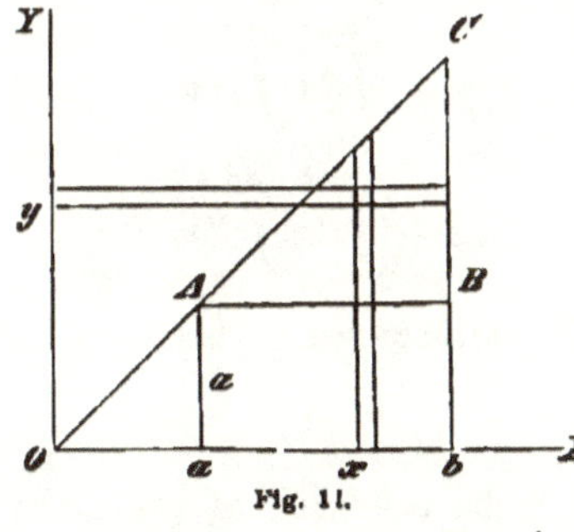

Fig. 11.

When the boundary of the domain is an isosceles right angled triangle ABC with x ranging from a to b; then while for any value of x, y passes from a to x, so on the other hand for any value of y, x varies between the limits y and b, thus we have:

$$\iint f(x, y)\, dx\, dy = \int_a^b dx \int_a^x f(x, y)\, dy = \int_a^b dy \int_y^b f(x, y)\, dx;$$

a rule already applied in § 146, VI.

When the boundary of the domain is a rectangle; for each value of x, y has the same limits α and β; for each value of y, x has the same limits a and b; and we obtain from the theorem under discussion the special theorem:

---

*) It is to be noticed, that this equation assumes the existence of the definite double integral; and on this hypothesis proves its identity with the value resulting from the successive integrations of the right side. It is not allowable to argue conversely from the right side to the left. (§ 171.)

When $f(x, y)$ is an integrable function, we have:

$$\int_a^b dx \int_\alpha^\beta f(x, y)\,dy = \int_\alpha^\beta dy \int_a^b f(x, y)\,dx.$$

This theorem contains an extension of the condition under which the order of integrations can be interchanged in the integration of a function with respect to a parameter, as it was given in § 153. For this only requires as a sufficient condition, that the function $f(x, y)$ should be finite and doubly integrable, and no longer its complete continuity in a domain.

169. A double integral having independent limits:

$$\int_a^b dx \int_\alpha^\beta f(x, y)\,dy$$

when regarded as a function of its upper limits $\Phi(b, \beta)$ is a continuous function of both these quantities. For, on the hypothesis that b and β are within the domain of integration of f, we have:

$$\begin{aligned}\Phi(b + h, \beta + k) - \Phi(b, \beta) &= \int_a^{b+h} dx \int_\alpha^{\beta+k} f(x, y)\,dy - \int_a^b dx \int_\alpha^\beta f(x, y)\,dy \\ &= \int_a^b dx \int_\beta^{\beta+k} f(x, y)\,dy + \int_b^{b+h} dx \int_\alpha^{\beta+k} f(x, y)\,dy \\ &= \int_\beta^{\beta+k} dy \int_a^b f(x, y)\,dx + \int_b^{b+h} dx \int_\alpha^{\beta+k} f(x, y)\,dy,\end{aligned}$$

since the order of integrations may be interchanged. This equation leads to the form:

$$\Phi(b + h, \beta + k) - \Phi(b, \beta) = kM + hN,$$

in which M and N are finite quantities; the condition of continuity therefore is fulfilled.

Moreover the partial derived functions with respect to b and β are found from this equation. Putting $k = 0$ we find:

$$\Phi(b + h, \beta) - \Phi(b, \beta) = \int_b^{b+h} dx \int_\alpha^\beta f(x, y)\,dy.$$

If then $\int_\alpha^\beta f(x, y)\,dy$ is a continuous function of x in the interval from $x = b$ to $x = b + h$ and we denote a mean value of it by:

$$\left\{\int_\alpha^\beta f(x, y)\,dy\right\}_{x=b+\vartheta h},$$

we have:

$$\lim_{h=0} \frac{\Phi(b+h, \beta) - \Phi(b, \beta)}{h} = \frac{\partial \Phi}{\partial b} = \left\{ \int_a^\beta f(x, y)\, dy \right\}_{x=b}.$$

When at $x = b$ the function f is generally for all values of y from α to β a uniformly continuous function of x, so that, with exception possibly of a discrete set of values of y, a value h can be assigned for which we have independently of the value y:

$$\operatorname{abs}[f(b + \Theta h, y) - f(b, y)] < \delta$$

(cf. § 52), this equation may also be written:

$$\frac{\partial \Phi}{\partial b} = \int_a^\beta f(b, y)\, dy.$$

Similarly, under the analogous condition:

$$\lim_{k=0} \frac{\Phi(b, \beta + k) - \Phi(b, \beta)}{k} = \frac{\partial \Phi}{\partial \beta} = \left\{ \int_a^b f(x, y)\, dx \right\}_{y=\beta},$$

or, when f is a uniformly continuous function of y at $y = \beta$ in the entire interval from $x = a$ to $x = b$:

$$\frac{\partial \Phi}{\partial \beta} = \int_a^b f(x, \beta)\, dx.$$

In this case, as the above equation shows, the Theorem of the Total Differential:

$$d\Phi = d\beta \frac{\partial \Phi}{\partial \beta} + db \frac{\partial \Phi}{\partial b},$$

also holds. We have moreover:

$$\frac{\partial^2 \Phi}{\partial \beta \partial b} = f(b, \beta) = \frac{\partial^2 \Phi}{\partial b \partial \beta}.$$

It is accordingly proved: *The definite double integral with constant limits is a continuous function of its upper limits, for which, provided the function that is to be integrated is in the neighbourhood of $x = b$ a uniformly continuous function of x for all values of y and in the neighbourhood of $y = \beta$ a uniformly continuous function of y for all values of x, the Theorems of the total differential and of the interchange of the order of differentiations hold.*

The assigned condition is fulfilled in particular for each point in the domain, when f is a continuous function of the two variables in the entire domain without exception.

The condition that a function may admit of differentiation with respect to a parameter under the integral sign (§ 151) may be stated in a new form by means of these theorems: In order that the result of differentiating the integral $\int_a^b f(x, \alpha)\,dx$ with respect to the parameter α may for a definite value $\alpha = \alpha'$ be equal to:

$$\left\{\int_a^b \frac{\partial f(x, \alpha)}{\partial \alpha}\,dx\right\}_{\alpha=\alpha'}$$

for all values of α' between β and γ, it is sufficient, that $\frac{\partial f(x, \alpha)}{\partial \alpha}$ be an integrable function within the domain from $x = a$ to $x = b$, $\alpha = \beta$ to $\alpha = \gamma$ and that its integral:

$$\int_a^b \frac{\partial f(x, \alpha)}{\partial \alpha}\,dx$$

be a continuous function of α. For then:

$$\iint \frac{\partial f(x, \alpha)}{\partial \alpha}\,dx\,d\alpha = \int_a^b dx \int_\beta^\gamma \frac{\partial f(x, \alpha)}{\partial \alpha}\,d\alpha = \int_\beta^\gamma d\alpha \int_a^b \frac{\partial f(x, \alpha)}{\partial \alpha}\,dx,$$

or, since:

$$\int_\beta^\gamma \frac{\partial f(x, \alpha)}{\partial \alpha}\,d\alpha = f(x, \gamma) - f(x, \beta)$$

with exception possibly of a discrete set of values of x, we have:

$$\int_a^b f(x, \gamma)\,dx - \int_a^b f(x, \beta)\,dx = \int_\beta^\gamma d\alpha \int_a^b \frac{\partial f(x, \alpha)}{\partial \alpha}\,dx.$$

Differentiating this equation with respect to γ, we have under the assigned conditions:

$$\frac{\partial}{\partial \gamma}\int_a^b f(x, \gamma)\,dx = \left\{\int_a^b \frac{\partial f(x, \alpha)}{\partial \alpha}\,dx\right\}_{\alpha=\gamma}.$$

The integral on the right side is not as a matter of course equal to

$$\int_a^b \left\{\frac{\partial f(x, \alpha)}{\partial \alpha}\right\}_\gamma dx;$$

this is its value, however, when $\frac{\partial f(x, \alpha)}{\partial \alpha}$ is in general for the values from $x = a$ to $x = b$ a uniformly continuous function of α in the neighbourhood of the point $\alpha = \gamma$.

170. Substitution of new variables in the double integral. The partition into superficial elements can be accomplished by any arbitrary net of curves. Putting:

$$x = \varphi(p, q), \quad y = \psi(p, q),$$

there is a curve belonging to each constant value of p, and likewise to each value of q. Let us suppose each p-curve to cut each q-curve in one point within the domain of validity of the double integral, and let us consider the rectilinear quadrilateral determined by the points of intersection having the coordinates:

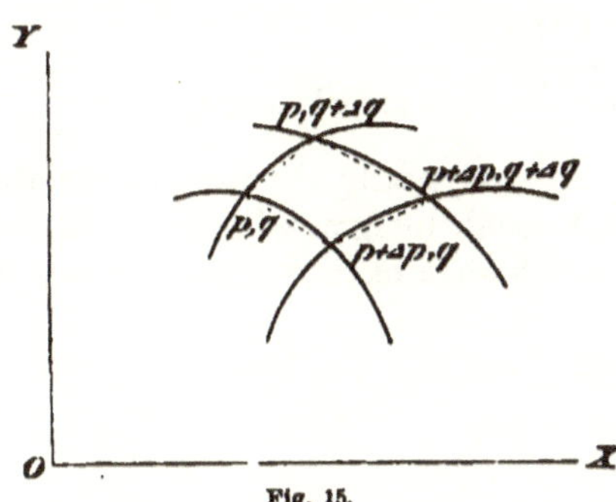

Fig. 15.

$$x_1 = \varphi(p,q),\ x_2 = \varphi(p+\Delta p, q),\ x_3 = \varphi(p, q+\Delta q),\ x_4 = \varphi(p+\Delta p, q+\Delta q)$$
$$y_1 = \psi(p,q),\ y_2 = \psi(p+\Delta p, q),\ y_3 = \psi(p, q+\Delta q),\ y_4 = \psi(p+\Delta p, q+\Delta q).$$

The area of this quadrilateral is:

$$\tau = \tfrac{1}{2} \text{ abs } [(x_4 - x_1)(y_3 - y_2) + (x_2 - x_3)(y_4 - y_1)].$$

The vanishing limit to which this expression tends when the quantities Δp and Δq converge to zero is at the same time the limit of the quadrilateral bounded by the curves, and since:

$$x_4 - x_1 = \varphi(p + \Delta p, q + \Delta q) - \varphi(p, q) = \frac{\partial \varphi}{\partial p} dp + \frac{\partial \varphi}{\partial q} dq$$

$$x_2 - x_3 = \varphi(p + \Delta p, q) - \varphi(p, q + \Delta q)$$
$$= \left\{\varphi(p,q) + \frac{\partial \varphi}{\partial p} dp\right\} - \left\{\varphi(p,q) + \frac{\partial \varphi}{\partial q} dq\right\} = \frac{\partial \varphi}{\partial p} dp - \frac{\partial \varphi}{\partial q} dq$$

$$y_4 - y_1 = \psi(p + \Delta p, q + \Delta q) - \psi(p, q) = \frac{\partial \psi}{\partial p} dp + \frac{\partial \psi}{\partial q} dq$$

$$y_3 - y_2 = \psi(p, q + \Delta q) - \psi(p + \Delta p, q) = \frac{\partial \psi}{\partial q} dq - \frac{\partial \psi}{\partial p} dp,{}^{*)}$$

it is expressed by the differential:

$$dT = \text{abs}\left[dp\,dq\left(\frac{\partial \varphi}{\partial p}\frac{\partial \psi}{\partial q} - \frac{\partial \varphi}{\partial q}\frac{\partial \psi}{\partial p}\right)\right].$$

Accordingly:

$$\iint f(x, y)\,dx\,dy = \iint f(\varphi, \psi) \text{ abs}\left[\left(\frac{\partial \varphi}{\partial p}\frac{\partial \psi}{\partial q} - \frac{\partial \varphi}{\partial q}\frac{\partial \psi}{\partial p}\right)\right] dp\,dq.$$

Employing polar coordinates:

---

*) The theorem of the total differential is applicable to the functions φ and ψ, when in each of the two systems the continuously variable direction of the tangent depends also continuously on the points of the plane.

$$x = r\cos\varphi, \quad y = r\sin\varphi,$$

as:

$$\frac{\partial x}{\partial r} = \cos\varphi, \quad \frac{\partial x}{\partial\varphi} = -r\sin\varphi, \quad \frac{\partial y}{\partial r} = \sin\varphi, \quad \frac{\partial y}{\partial\varphi} = r\cos\varphi;$$

$$dT = r\,dr\,d\varphi.$$

$$\iint f(x, y)\,dx\,dy = \iint f(r\cos\varphi, r\sin\varphi)\,r\,dr\,d\varphi.$$

It is important to prove purely analytically the formula for the substitution of new variables in the double integral without employing geometrical conceptions, in order to obtain a method applicable also to multiple integrals.

Having brought the double integral into the form of successive integrations:

$$\iint f(x, y)\,dx\,dy = \int_a^b dx \int_{y_0}^{y_1} f(x, y)\,dy,$$

we can proceed as follows, employing the theorems for the simple integral.

First: If $x = \varphi(p)$, $y = \psi(q)$, and to the values y_0 and y_1 correspond the values q_0 and q_1 so that while y increases from y_0 to y_1, q likewise increases from q_0 to q_1, then:

$$\int_a^b dx \int_{y_0}^{y_1} f(x, y)\,dy = \int_a^b dx \int_{q_0}^{q_1} f(x, \psi)\frac{\partial\psi}{\partial q}\,dq = \int_{p_0}^{p_1} \frac{\partial\varphi}{\partial p}\,dp \int_{q_0}^{q_1} f(\varphi, \psi)\frac{\partial\psi}{\partial q}\,dq$$

or:

$$\iint f(x, y)\,dx\,dy = \iint f(\varphi, \psi)\frac{\partial\varphi}{\partial p}\,\frac{\partial\psi}{\partial q}\,dp\,dq;$$

both integrals being extended over the same domain.

Second: If $x = \varphi(p)$, $y = \psi(p, q)$, we can first suppose p to be eliminated and y calculated as a function of x and q:

$$y = \chi(x, q)$$

and for each value of x let increasing values of q belong to increasing values of y; we have then:

$$\int_a^b dx \int_{y_0}^{y_1} f(x, y)\,dy = \int_a^b dx \int_{q_0}^{q_1} f(x, \chi)\frac{\partial\chi}{\partial q}\,dq = \int_{p_0}^{p_1} \frac{\partial\varphi}{\partial p}\,dp \int_{q_0}^{q_1} f(\varphi, \psi)\frac{\partial\psi}{\partial q}\,dq.$$

or:

$$\iint f(x, y)\,dx\,dy = \iint f(\varphi, \psi)\,\mathrm{abs}\left[\frac{\partial\varphi}{\partial p}\,\frac{\partial\psi}{\partial q}\right]dp\,dq,$$

since χ passes over into ψ by the substitution $x = \varphi(p)$, and thus $\frac{\partial\chi}{\partial q} = \frac{\partial\psi}{\partial q}$. Hence follows:

Third: the general case, when, retaining q, we introduce a function $p = \chi(p', q)$ instead of p, and then employ the equation of transformation obtained in the second case. Let us denote:

$$x = \varphi(\chi) = \Phi(p', q), \quad y = \psi(\chi, q) = \Psi(p', q),$$

thus corresponding to the second case we have:

$$\iint f(\varphi,\psi)\left[\frac{\partial\varphi(p)}{\partial p}\,\frac{\partial\psi(p,q)}{\partial q}\right]dp\,dq = \iint f(\Phi,\Psi)\left[\frac{\partial\varphi(p)}{\partial p}\,\frac{\partial\psi(p,q)}{\partial q}\right]_{p=\chi(p',q)}\frac{\partial\chi}{\partial p'}dp'dq.$$

Now since when p is regarded as a function of p' and q:

$$\varphi(p) = \Phi(p', q), \quad \psi(p, q) = \Psi(p', q)$$

we have:

$$\frac{\partial\Phi}{\partial p'} = \frac{\partial\varphi}{\partial p}\cdot\frac{\partial p}{\partial p'}, \quad \frac{\partial\Psi}{\partial p'} = \frac{\partial\psi}{\partial p}\cdot\frac{\partial p}{\partial p'},$$

$$\frac{\partial\Phi}{\partial q} = \frac{\partial\varphi}{\partial p}\cdot\frac{\partial p}{\partial q}, \quad \frac{\partial\Psi}{\partial q} = \frac{\partial\psi}{\partial p}\cdot\frac{\partial p}{\partial q} + \frac{\partial\psi}{\partial q},$$

therefore as a function of p' and q:

$$\left[\frac{\partial\Phi}{\partial p'}\,\frac{\partial\Psi}{\partial q} - \frac{\partial\Psi}{\partial p'}\,\frac{\partial\Phi}{\partial q}\right] = \left[\frac{\partial\varphi}{\partial p}\,\frac{\partial p}{\partial p'}\,\frac{\partial\psi}{\partial q}\right] = \left[\frac{\partial\varphi}{\partial p}\cdot\frac{\partial\psi}{\partial q}\cdot\frac{\partial\chi}{\partial p'}\right]_{p=\chi(p',q)},$$

accordingly the above double integral for $x = \Phi(p', q)$, $y = \Psi(p', q)$:

$$\iint f(x, y)\,dx\,dy = \iint f(\Phi, \Psi)\ \text{abs}\left[\frac{\partial\Phi}{\partial p'}\,\frac{\partial\Psi}{\partial q} - \frac{\partial\Psi}{\partial p'}\,\frac{\partial\Phi}{\partial q}\right]dp'\,dq.$$

171. When the function to be integrated becomes determinately infinitely great or indeterminate between infinite limits at isolated points in the domain, let us suppose each of these points surrounded by a closed boundary curve. The question then arises: under what condition is the function $f(x, y)$ integrable in such a domain containing a single infinity point? Let the coordinates of the point be $x = a$, $y = b$ and suppose it the centre of an arbitrarily small circle with radius r_1, the coordinates of any point upon or within this circle are:

$$x = a + r\cos\varphi, \quad y = b + r\sin\varphi,$$

where:

$$0 < r < r_1, \quad 0 < \varphi \leqq 2\pi.$$

Now the necessary and sufficient condition that must be satisfied in order that the double integral may present a determinate finite value in this domain from its external boundary up to the circumference of this circle, however small its radius r_1 be taken, is: that the double integral

$$\iint f(a + r\cos\varphi,\ b + r\sin\varphi)\,r\,dr\,d\varphi$$

extended over the interior of the circle with radius r_1 must converge to zero simultaneously with r_1, or in other words, a superior limit must be ascertainable for r_1, such that the double integral extended throughout a ring bounded by the concentric circles r_1 and $r_2 < r_1$ shall remain smaller than any assignable quantity δ, however small r_2 be taken.

When, for $r = 0$, the function $f(a + r\cos\varphi, b + r\sin\varphi)$ becomes determinately infinite for all values of φ, a limit can be assigned that its infinitude must not exceed if the integral is to continue finite. *The order in which f becomes infinite must be lower than the second,* i. e. in the neighbourhood of this point we must have for all values of φ:

$$f(x, y) < \frac{C}{r^\alpha},$$

where C denotes a constant and α a number less than 2. For, were

$$f(x, y) > \frac{C}{r^2},$$

we should have within the above circular ring:

$$\int\int f(x, y)\,dT \geq C\int_0^{2\pi} d\varphi \int_{r_2}^{r_1} \frac{dr}{r} = C\,2\pi\{l(r_1) - l(r_2)\},$$

and this expression becomes infinite for $r_1 = 0$.

In such a case, whenever the double integral has a determinate value, this value is independent of the succession of the integrations.

On the other hand it is to be noticed, that when by the function becoming infinite the value of the double integral is no longer finite, although each succession of integrations can give rise to a finite value, these need not both be equal. In this case the double integral defined by a determinate succession of integrations is said to be **singular**.

Thus ex. gr. in the last example given in § 153 p. 272, we found:

$$\int_0^1 dy \int_0^1 \frac{y^2 - x^2}{(y^2 + x^2)^2}\,dx = \int_0^1 dy \left(\frac{x}{y^2 + x^2}\right)_0^1 = \int_0^1 \frac{dy}{y^2 + 1} = \frac{\pi}{4},$$

$$\int_0^1 dx \int_0^1 \frac{y^2 - x^2}{(y^2 + x^2)^2}\,dy = \int_0^1 dx \left(\frac{-y}{x^2 + y^2}\right)_0^1 = \int_0^1 \frac{-dx}{1 + x^2} = -\frac{\pi}{4};$$

while the double integral:

$$\int\int \frac{y^2 - x^2}{(y^2 + x^2)^2}\,dx\,dy$$

in the rectangle from $x = 0$ to $x = 1$, $y = 0$ to $y = 1$ is unmeaning; for at the point $x = 0$, $y = 0$ the function to be integrated becomes quite indeterminate and assumes in the neighbourhood of this point values infinitely great in the second order. In fact, putting:

$$x = r\cos\varphi,\quad y = r\sin\varphi,$$

we have:

$$\frac{y^2 - x^2}{(y^2 + x^2)^2} = \frac{-\cos 2\varphi}{r^2},$$

therefore in a quadrant round the vertex $x = 0$, $y = 0$, the integral:

$$\int_0^{\frac{1}{2}\pi} -\cos 2\varphi\, d\varphi \int_{r_0}^{r_1} \frac{dr}{r}$$

increases logarithmically beyond any limit. As another example:

$$\int_0^1\int_0^1 \frac{y^2 - x^2}{y^2 + x^2}\, dx\, dy = \int\int - \cos 2\varphi\, r\, dr\, d\varphi$$

has a finite value, because although the function to be integrated is discontinuous and indeterminate in the point $x = 0$, $y = 0$, it still remains finite.

Again, the double integral:

$$\int\int \frac{dx\,dy}{x^2 + y}$$

has a finite value for $x > 0$, $y > 0$, although the function becomes infinite in the second order at the point $x = 0$ in the direction of the axis of abscissæ, but in this direction only. To demonstrate this, let us calculate the value of the double integral for a rectangle from $x = 0$ to $x = a$, $y = b$ to $y = c$, we have:

$$\int_0^a dx \int_b^c \frac{dy}{x^2+y} = \int_0^a dx\, l\frac{x^2+c}{x^2+b} = a\, l\frac{a^2+c}{a^2+b} + 2\sqrt{c}\tan^{-1}\frac{a}{\sqrt{c}} - 2\sqrt{b}\tan^{-1}\frac{a}{\sqrt{b}}.$$

Now making c and $b < c$ converge to zero in any way whatever, this expression on the right side converges to zero.

The values resulting from the two successions of integrations are not necessarily different, even when the double integral is unmeaning. We have in the first example:

$$\int_0^\infty dy \int_0^\infty \frac{y^2 - x^2}{(y^2 + x^2)^2}\, dx = \int_0^\infty dx \int_0^\infty \frac{y^2 - x^2}{(y^2 + x^2)^2}\, dy = 0,$$

while the double integral for the domain, that is here infinite, has no existence.

When the function f becomes infinite along an entire curve in the domain, let this be taken as the line $p =$ const. Then the product:

$$f(p, q)\left(\frac{\partial\varphi}{\partial p}\frac{\partial\psi}{\partial q} - \frac{\partial\varphi}{\partial q}\frac{\partial\psi}{\partial p}\right)(p - c)$$

in the neighbourhood of the line $p = c$ must not be equal to nor greater than a finite number A, for otherwise the integral:

$$\iint f dT \text{ would be } \geqq A \int dq \int \frac{dp}{p-c}.$$

Thus ex. gr. the double integral:

$$\iint \frac{dx\,dy}{x^2 + y}$$

has no finite value in a domain in which the parabola $x^2 + y = 0$ is situated.

Conversely, when f is such that in the neighbourhood of the line $p = c$ we have the product:

$$f(p,q)\left(\frac{\partial\varphi}{\partial p}\frac{\partial\psi}{\partial q} - \frac{\partial\varphi}{\partial q}\frac{\partial\psi}{\partial p}\right)(p-c)^{\alpha} < A \text{ for } 0 < \alpha < 1,$$

the value of the integral is finite, for it is less than

$$A\int dq \int \frac{dp}{(p-c)^{\alpha}}.$$

In other words: *When the function becomes determinately infinite along an entire curve, provided its infinitude is lower than unity, the double integral will remain finite.* The theorem of the interchangeability of successive integrations is then maintained.

172. When the domain of integration is infinite, the double integral extended over the infinite surface is understood to mean the finite limiting value that the double integral assumes when it is first evaluated for a finite surface and then this finite domain so extended as to pass over into the infinite region. Thus we have to investigate, under what conditions there is a finite limiting value.

The transition from the finite to the infinite region is effected in various ways according to the definition of the latter. When it is to embrace the entire plane, let the double integral be formed for any rectangle from $x = a$ to $x = b$, $y = \alpha$ to $y = \beta$, and let a and α increase negatively and b and β positively beyond all limits.

When the infinite domain is only an infinite section of the plane that is bounded by right lines or by an unclosed curve, we must guide ourselves by this boundary curve in the transition to the limit.

When ex. gr. the double integral is to be extended throughout the parabola whose equation is $y^2 = 2px$, we can put:

$$\int^{(2)} f(x,y)\,dT = \lim_{(a=\infty)} \int_0^a dx \int_{y=-\sqrt{2px}}^{y=+\sqrt{2px}} f(x,y)\,dy = \lim_{(b=\alpha)} \int_{-b}^{+b} dy \int_{x=\frac{y^2}{2p}}^{x=\frac{b^2}{2p}} f(x,y)\,dx.$$

When it is to be formed for the inside of one branch of a hyperbola whose equation is: $x > 0$, $xy = k$, the double integral can be defined as:

$$\int^{(2)} f(x, y)\,dT = \operatorname*{Lim}_{(a=\infty,\, b=\infty)} \int_{x=\frac{k}{b}}^{a} dx \int_{y=\frac{k}{x}}^{b} f(x, y)\,dy = \operatorname*{Lim}_{(a=\infty,\, b=\infty)} \int_{y=\frac{k}{a}}^{b} dy \int_{x=\frac{k}{y}}^{a} f(x, y)\,dx.$$

The value of the double integral becomes singular, when it depends on the way in which a and b become infinite.

When for arbitrarily increasing value of x and y, the function to be integrated $f(x, y) = f(r\cos\varphi, r\sin\varphi)$ has ultimately neither maxima nor minima, so that, within its domain of integration for every value of φ as r increases infinitely, f converges to zero; then, if its order of vanishing (nullitude) be higher than the second, since from a certain value r_1, $f(r\cos\varphi, r\sin\varphi)$ is constantly $< \frac{A}{r^\alpha}$ where $\alpha > 2$, the double integral is finite and completely independent of the method of the transition to the infinite region. For, the part of the integral:

$$\int^{(2)} f(x, y)\,dT$$

relative to a domain for which $r > r_1$, is less than:

$$A\int d\varphi \int \frac{dr}{r^{\alpha-1}},$$

and this expression vanishes for arbitrarily increasing values of r.

But the double integral has no finite value whatever, when the function $f(x, y)$, having neither maxima nor minima as the values of x and y increase, becomes infinitely small in an order lower than the second or continues finite.

In case the function undergoes incessant oscillations as the values of x and y increase infinitely, the double integral exists, as did the simple integral, without involving any limit as to the order in which the function becomes infinitely small.

When the double integral has no existence, there may yet be a singular value for a determinate succession of integrations. One example of this was given in last §.; another important example is the following: The function $f(x, y) = \cos(xy)$ is not integrable in the infinite strip from $x = 0$ to $x = b$, $y = 0$ to $y = \infty$, for we have:

$$\int_0^h dy \int_0^b \cos(xy)\,dx = \int_0^b dx \int_0^h \cos(xy)\,dy = \int_0^b \left(\frac{\sin xy}{x}\right)_0^h dx = \int_0^b \frac{\sin hx}{x}\,dx,$$

and if we first make h increase arbitrarily in the function to be integrated, this becomes quite indeterminate; but the integral:

$$\int_0^h dy \int_0^b \cos(xy)\, dx = \int_0^h \frac{\sin by}{y}\, dy$$

has for $h = \infty$ the determinate values $\pm \frac{1}{2}\pi$, according as $b >$ or < 0 (§ 155). Inasmuch therefore as the double integral does not exist,

$$\int_0^b dx \lim_{h=\infty} \left\{ \int_0^h \cos(xy)\, dy \right\}$$

also becomes indeterminate, while the value of:

$$\lim_{h=\infty} \int_0^h dy \int_0^b \cos(xy)\, dx = \pm \tfrac{1}{2}\pi, \text{ according as } b \gtrless 0.$$

It is to be noticed therefore that although the equation:

$$\int_0^b dx \int_0^h \cos(xy)\, dy = \int_0^h dy \int_0^b \cos(xy)\, dx,$$

holds for every finite value of h, yet we cannot conclude from it that:

$$\int_0^b dx \lim_{h=\infty} \int_0^h \cos(xy)\, dy = \lim_{h=\infty} \int_0^h dy \int_0^b \cos(xy)\, dx\text{*)},$$

because the formula on the left side has no definite meaning.

173. It is further of importance, finally, to recognise that the product of two simple integrals can always be considered as a double integral. When $f(x)$ and $\psi(x)$ are two integrable functions, we have:

$$\int_a^b f(x)\, dx \int_\alpha^\beta \psi(y)\, dy = \iint f(x)\psi(y)\, dx\, dy,$$

this double integral being extended over the rectangle between the limits from $x = a$ to $x = b$, $y = \alpha$ to $y = \beta$.

For:

$$\int_a^b f(x)dx = (x_1 - a)f(a) + (x_2 - x_1)f(x_1)\cdots(b - x_{n-1})f(x_{n-1}) + \Delta = S + \Delta,$$

$$\int_\alpha^\beta \varphi(y)dy = (y_1 - \alpha)\varphi(\alpha) + (y_2 - y_1)\varphi(y_1)\cdots(\beta - y_{m-1})\varphi(y_{m-1}) + \Delta' = S' + \Delta',$$

therefore:

---

*) In the theory of Fourier's integrals as of the more general class to which they belong (Du Bois Reymond, Journal f. Math., Vol. 69), the conventions concerning the succession of integrations and the corresponding transitions to the limit are essential.

$$\int_a^b f(x)\,dx \int_a^\beta \varphi(y)\,dy = \sum\sum f(x_\mu)\,\varphi(y_\nu)(x_{\mu+1} - x_\mu)(y_{\nu+1} - y_\nu) + S\Delta' + S'\Delta + \Delta\Delta'.$$

Now since as the values of m and n increase, the quantities Δ and Δ' converge to zero, it is evident that the left side actually represents the limiting value of the double sum. This theorem is also true, when the functions $f(x)$ and $\varphi(y)$ become infinite within the interval of integration, provided each alone remains integrable. For if $f(x)$ become infinite for $x = c$, and $\varphi(y)$ for $y = c'$, then by the theorem just proved we have:

$$\int_a^{c-\delta} f(x)\,dx \int_a^{c'-\delta'} \varphi(y)\,dy = \iint f(x)\,\varphi(y)\,dx\,dy.$$

As δ and δ' converge to zero, the product on the left side passes over, by hypothesis, into a determinate finite value that at the same time represents the value of the double integral in the rectangle up to the limits $x = c$, $y = c'$.

174. For the simple definite integral the following theorem holds: When $F(x)$ is a known continuous function whose derived function is integrable and coincides generally with a function $f(x)$, we have:

$$\int_a^x f(x)\,dx = F(x) - F(a).$$

The value of the definite integral depends therefore only on the values of the function F at the limits of integration. An analogue for the double integral is presented in the Theorem of Green (1793—1841)*): Concerning the reduction of a double integral of a unique function to simple integrals along the boundary curve.

In the plane xy let there be given a finite connected domain, bounded by one or more closed curves. In Fig. 16 ex. gr. we suppose the domain to consist of the part of the plane enclosed by the external curve omitting the two areas bounded by the ovals; in it there are therefore three closed boundary curves. Let a function $f(x, y)$ be given for all points within and on the boundaries of this domain and let it be integrable within the domain. Such a function may, as in § 167, be continuous, although it is also possible that in a "discrete" multiplicity of curves it may become discontinuous, indeterminate or

*) An Essay on the application of mathematical analysis to the theories of Electricity and Magnetism. Nottingham 1828; reprinted in Crelle Journ. f. Math., Vol. 39, 44, 47; and again in Mathematical papers of the late George Green. London 1871. — Riemann: Grundlagen für eine allgemeine Theorie etc. § 7—9.

even infinite; we call the function then "in general" continuous. In the last case if the function become determinately infinite it must be algebraically infinite at each isolated point in a lower order than the second and along entire curves in a lower order than the first.

For each value of y then $\int_a^x f(x, y)dx$ represents a function of x also in general continuous and finite, that is integrable with respect to y. Denoting:

$$\int_a^x f(x, y)dx = F(x, y) - F(a, y),$$

we have:

$$f(x, y) = \frac{\partial F(x, y)}{\partial x}.$$

Each parallel to the axis of x must cut the boundary curves in a finite even number of points. In our figure it is two or four. Denoting the values of x at the entrances and exits belonging to a definite value of y by x_1, x_2, x_3, x_4, the double integral:

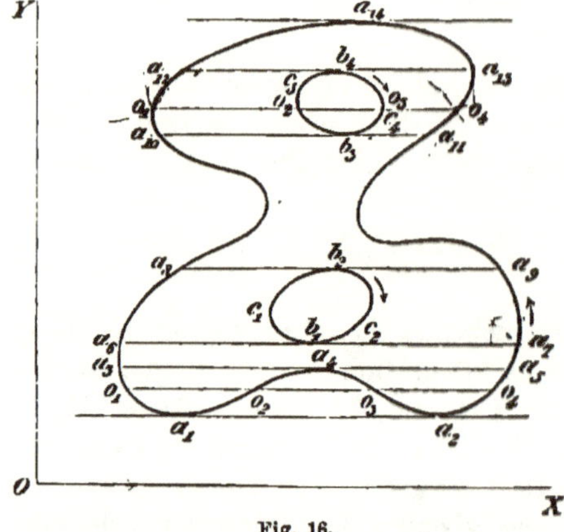

Fig. 16.

$$\iint f(x, y)\, dx\, dy$$

is equal to

$$\int dy \int f(x, y)\, dx,$$

if the integral respecting x in this successive integration be extended, for each value of y, between the limits from x_1 to x_2, and from x_3 to x_4. But in the above notation:

$$\int_{x_1}^{x_2} f(x, y)dx + \int_{x_3}^{x_4} f(x, y)dx = F(x_2, y) - F(x_1, y) + F(x_4, y) - F(x_3, y).$$

Therefore:

$$\iint f(x,y)dxdy = \int dy F(x_2,y) - \int dy F(x_1,y) + \int dy F(x_4,y) - \int dy F(x_3, y).$$

In these integrals the values of x as functions of y corresponding to the equations of the boundary curves are to be substituted and then the integrations with respect to y to be effected between the extreme values, namely those at which any entrance of a parallel to the axis of x coincides with its exit.

Constructing at each point o the normal to the boundary curve and denoting the angle its direction entering the domain makes with the positive axis of ordinates by φ_1, φ_2, φ_3, φ_4, this angle being always measured in the same direction of rotation from the positive axis of ordinates to the negative axis of abscissæ, then at the entrances o_1, o_3 the angles φ_1 and φ_3 are always in the third or fourth quadrant, but at the exits o_2, o_4, the angles φ_2, φ_4 are in the first or second quadrant. Therefore if ds denote the positive value of the element of the arc of a boundary curve, we have:

$$dy = -ds\sin\varphi_1,\quad dy = ds\sin\varphi_2,\quad dy = -ds\sin\varphi_3,\quad dy = ds\sin\varphi_4.$$

Accordingly:

$$\iint f(x,y)\,dx\,dy = \int F(x_2,y)\sin\varphi_2\,ds + \int F(x_1,y)\sin\varphi_1\,ds$$
$$+\int F(x_4,y)\sin\varphi_4\,ds + \int F(x_3,y)\sin\varphi_3\,ds.$$

All these partial integrals can be comprehended under the following single conception:

We can describe each such integral as extended along a portion of a boundary curve, inasmuch as the function that is to be integrated:

$$F(x,y)\sin\varphi\,ds$$

has always to be formed for the continuously consecutive points of the boundary curve, with positive values of ds.

Now we adopt as a convention: The length of the arc of any boundary curve is a positive increasing magnitude, when we trace the curve from any of its points so as to keep the bounded area on the left. We thus obtain the partial integrals that the points o_2 or o_4 form along the segments: a_1a_4, a_2a_5, a_5a_7, a_7a_9, a_9a_{11}, $a_{11}a_{13}$, $a_{13}a_{14}$ and from b_1 by c_1 to b_2, b_3 by c_3 to b_4; and again, those that the points o_1 or o_3 form along the segments: a_3a_1, a_4a_2, a_6a_3, a_8a_6, $a_{10}a_8$, $a_{12}a_{10}$, $a_{14}a_{12}$ and from b_2 by c_2 to b_1, b_4 by c_4 to b_3. We can therefore say: The integral

$$\int F(x,y)\sin\varphi\,ds$$

is to be formed for the points of all the boundary curves, these being traced so that the domain they have to bound is constantly on the left.

We have conversely the definition: A simple integral

$$\int F(x,y)\sin\varphi\,ds$$

extended along a closed curve in a positive direction, signifies, in terms of a single variable, the value of

$$\int F(x, y)\,dy,$$

when in the first place the entrances and exits of parallels to the axis of x:

$$x_1 = \psi_1(y) < x_2 = \psi_2(y) < x_3 = \psi_3(y) < x_4 = \psi_4(y) \dots$$

are calculated for each value of y from $\varphi(x, y) = 0$ the equation of the boundary curve, and then the sum of integrals:

$$\int F(\psi_2(y), y)dy - \int F(\psi_1(y), y)dy + \int F(\psi_4(y), y)dy - \int F(\psi_3(y), y)\,dy \dots$$

is formed for increasing values of y.

Employing this definition, the following is the statement of Green's theorem:

When the function $f(x, y)$ is integrable within a domain, the integral:

$$\int^x f(x, y)\,dx = F(x, y)$$

is in general for each value of y a continuous function of x and for each value of x an integrable function with respect to y. The double integral:

$$\iint f(x, y)\,dx\,dy$$

extended throughout the entire domain is equal to the simple integral:

$$\int F(x, y)\sin\varphi\,ds$$

extended along the boundary curves of the domain in a positive cir-

This equality still holds good when the values of the func s f and F are altered arbitrarily at infinitely many points, provided the integrals are not thereby changed.

Therefore the value of the definite double integral depends only on the values of F at the points of the boundary curves.

We can also frame the theorem thus, reversing the order of ideas: When the partial derivate $\frac{\partial F}{\partial x}$ of a function $F(x, y)$ is integrable over the entire domain, even admitting that there are points or curves at which it is discontinuous or infinite, we have always:

$$\iint \frac{\partial F(x, y)}{\partial x}\,dx\,dy = \int F(x, y)\sin\varphi\,ds.$$

Likewise, interchanging the letters x and y, and denoting the angle that the inward direction of the normal makes with the positive axis of x by ψ, so that at the entrances of parallels to the axis of y

the positive value of $dx = ds \sin\psi = ds \cos\varphi$, but at their exits dx is equal to $-ds \sin\psi = -ds \cos\varphi$; we prove the equation:

$$\int\!\!\int \frac{\partial F(x,y)}{\partial y}\, dx\, dy = -\int F(x,y) \sin\psi\, ds = -\int F(x,y)\cos\varphi\, ds.$$

By the above definition $-\int F(x,y)\cos\varphi\, ds$ is the integral $-\int F(x,y)\,dx$ formed along the boundary curves in a positive circuit.

175. Some consequences of this result claim our special attention.

1. Let a unique function of two variables $f(x, y)$ be given for a domain limited in any manner by one or more boundary curves; let its partial derived functions be:

$$\frac{\partial f(x,y)}{\partial x} = P, \quad \frac{\partial f(x,y)}{\partial y} = Q, \text{ and let: } \frac{\partial P}{\partial y} = \frac{\partial Q}{\partial x}.$$

Now when the functions $\frac{\partial P}{\partial y}$ and $\frac{\partial Q}{\partial x}$ are integrable within the domain, we have:

$$\int\!\!\int \frac{\partial P}{\partial y}\, dx\, dy = -\int P \cos\varphi\, ds, \quad \int\!\!\int \frac{\partial Q}{\partial x}\, dx\, dy = \int Q \sin\varphi\, ds,$$

therefore:

$$\int\!\!\int \left(\frac{\partial Q}{\partial x} - \frac{\partial P}{\partial y}\right) dx\, dy = \int (Q\sin\varphi + P\cos\varphi)\, ds = \int (Q\,dy + P\,dx).$$

But by hypothesis the function under the double integral sign vanishes, consequently we obtain the theorem:

If P and Q be the partial derived functions of a unique function of two variables, the value of the integral:

$$\int (P\cos\varphi + Q\sin\varphi)\, ds = \int (P\,dx + Q\,dy)$$

is zero, when it is formed in a positive circuit for all the boundary curves of a domain within which the functions $\frac{\partial P}{\partial y}$, $\frac{\partial Q}{\partial x}$ *are integrable and in general, with possibly a linear set of exceptions, satisfy the equation* $\frac{\partial P}{\partial y} = \frac{\partial Q}{\partial x}$.

2. From this theorem follows: When the domain is limited only by a single closed boundary curve (s i m p l y c o n n e c t e d) and when within it the conditions just stated are fulfilled, the value of the integral formed for this one closed boundary curve is zero.

3. If two points within such a simply connected domain, whose coordinates are $x_0 y_0$, $x_1 y_1$, be joined by arbitrary curves $s_1, s_2, s_3 \ldots$ included within the domain, the value of the integral:

$$\int (P\cos\varphi + Q\sin\varphi)\, ds = \int (P\,dx + Q\,dy)$$

is always the same for whichever curve it may be formed. For, any two of these curves ex. gr. s_1 and s_2 surround a part of the domain, consequently the sum of two integrals, taken, the first along s_1 from x_0y_0 to x_1y_1, the second along s_2 from x_1y_1 to x_0y_0, is zero; or writing this in a formula:

$$\int\limits_{x_0,y_0}^{x_1,y_1}\underset{(\text{along } s_1)}{(Pdx+Qdy)} = -\int\limits_{x_1,y_1}^{x_0,y_0}\underset{(\text{along } s_2)}{(Pdx+Qdy)} = \int\limits_{x_0,y_0}^{x_1,y_1}\underset{(\text{along } s_2)}{(Pdx+Qdy)}.$$

4. The integration to be effected in theorem 3. takes a simple form when the domain is such as to include the entire rectangle having the two points x_0y_0, x_1y_1 as vertices and its sides parallel to the coordinate axes. For simplicity we shall assume here that within and upon the boundaries of such a rectangle the functions $P, Q, \frac{\partial P}{\partial y}, \frac{\partial Q}{\partial x}$ are continuous without exception, and that the equation $\frac{\partial P}{\partial y} = \frac{\partial Q}{\partial x}$ is everywhere valid. We can then integrate along the sides of this rectangle and obtain either:

$$\int\limits_{x_0,y_0}^{x_1,y_1}(Pdx+Qdy) = \int\limits_{x_0}^{x_1}P(x,y_0)\,dx + \int\limits_{y_0}^{y_1}Q(x_1,y)\,dy,$$

or:

$$\int\limits_{x_0,y_0}^{x_1,y_1}(Pdx+Qdy) = \int\limits_{y_0}^{y_1}Q(x_0,y)\,dy + \int\limits_{x_0}^{x_1}P(x,y_1)\,dx.$$

Denoting the integral as a function of its upper limit by $F(x_1, y_1)$ — this we can easily see is a continuous function, because:

$$F(x_1+h, y_1+k) - F(x_1, y_1) = \int\limits_{x_1}^{x_1+h}P(x,y_0)\,dx + \int\limits_{y_1}^{y_1+k}Q(x_1+h,y)\,dy$$

$$+\int\limits_{y_0}^{y_1}\{Q(x_1+h,y) - Q(x_1,y)\}\,dy,$$

and $Q(x, y)$ is a continuous function — on differentiating the first formula with respect to x_1, we find the equation:

$$\frac{\partial F(x_1, y_1)}{\partial x_1} = P(x_1, y_0) + \frac{\partial \int\limits_{y_0}^{y_1} Q(x_1, y)\,dy}{\partial x_1}.$$

Now because $\frac{\partial Q(x,y)}{\partial x}$ is a continuous function, we are entitled to differentiate under the integral sign (§ 169); further we have the equation:

$$\frac{\partial Q(x,y)}{\partial x} = \frac{\partial P(x,y)}{\partial y};$$

therefore:

$$\frac{\partial\int\limits_{y_0}^{y_1} Q(x_1, y)\,dy}{\partial x_1} = \left\{\int\limits_{y_0}^{y_1} \frac{\partial Q(x,y)}{\partial x}\,dy\right\}_{x=x_1} = \left\{\int\limits_{y_0}^{y_1} \frac{\partial P(x,y)}{\partial y}\,dy\right\}_{x=x_1} = P(x_1, y_1) - P(x_1, y_0),$$

and consequently:

$$\frac{\partial F(x_1, y_1)}{\partial x_1} = P(x_1, y_1);$$

likewise:

$$\frac{\partial F(x_1, y_1)}{\partial y_1} = Q(x_1, y_1).$$

The integral function $\int\limits_{x_0, y_0}^{x_1, y_1}(Pdx + Qdy)$ can therefore be defined as that continuous function of the variables x_1, y_1, which vanishes for the values x_0, y_0, and whose partial derived functions with respect to x_1 and y_1 are the functions P and Q.

By this enunciation the function $F(x_1, y_1)$ is completely defined.

For, all continuous functions of two variables whose partial derived functions within a domain respectively coincide, can differ only by an additive constant. Suppose, in fact, that F and Φ are two distinct functions for which:

$$\frac{\partial F(x, y)}{\partial x} = \frac{\partial \Phi(x, y)}{\partial x}, \quad \frac{\partial F(x, y)}{\partial y} = \frac{\partial \Phi(x, y)}{\partial y},$$

then in consequence of this first equation we have for every value of y:

$$F(x, y) = \Phi(x, y) + C,$$

where C being a quantity independent of x, can be only a continuous function of y (§ 100); denoting this by Y, we have:

$$\frac{\partial F(x, y)}{\partial y} = \frac{\partial \Phi(x, y)}{\partial y} + \frac{\partial Y}{\partial y},$$

but in consequence of the second equation $\frac{\partial Y}{\partial y} = 0$. Accordingly (§ 100) Y is a constant also with respect to y.

The problem therefore is solved: *When two continuous functions P and Q are given which satisfy the equation $\frac{\partial P}{\partial y} = \frac{\partial Q}{\partial x}$ within a simply connected domain; it is required to find those continuous functions whose partial derived functions with respect to x and y coincide with the values P and Q. All such functions are collectively expressed by:*

$$\int\limits_{x_0, y_0}^{x, y}(Pdx + Qdy) + Const..$$

Conversely, knowing beforehand such a continuous function $F(x, y)$, the definite integral is thereby ascertained; for we have:

$$\int_{x_0, y_0}^{x, y} (P dx + Q dy) = F(x, y) + C = F(x, y) - F(x_0, y_0),$$

because the left side vanishes for $x = x_0$, $y = y_0$, therefore C must be equal to $-F(x_0, y_0)$.

Thus also the connexion between the definite and the indefinite integral in two variables is developed subject to all the hypotheses here necessary.

Note: It is to be observed, that this deduction of a continuous function $F(x, y)$ from its partial derived functions P and Q requires not only that these derived functions be continuous but also that they admit of being differentiated respectively for y and x within the domain*); while there is no analogue to this with functions of a single variable.

It is therefore not unimportant for us to realize that the condition for an exact differential must be modified in certain cases. Let us formulate the problem as follows:

Given in a rectangular domain, from $x = a$ to $x = b$, $y = \alpha$ to $y = \beta$, two continuous functions P and Q. What further conditions must they fulfil, in order that there may be a continuous function $F(x, y)$ in the domain, for which:

$$\frac{\partial F}{\partial x} = P(x, y), \qquad \frac{\partial F}{\partial y} = Q(x, y),$$

and how is this function determined?

All continuous functions whose partial derived function with respect to x coincides with P are collectively included in the form:

$$F(x, y) = \int_a^x P(x, y) dx + Y,$$

where Y is a continuous function of y only. Making $x = a$, we find:

$$F(x, y) - F(a, y) = \int_a^x P(x, y) dx.$$

---

*) There may be points at which the functions P and Q cease to be continuous and finite. But these points can be enclosed within arbitrarily small neighbourhoods; the enclosing curves are then counted among the boundaries of the domain, rendering it multiply connected. But they do not ultimately come into consideration in the formation of the simple integral, if for the interior of such a region the double integral $\iint \left(\frac{\partial P}{\partial y} - \frac{\partial Q}{\partial x}\right) dx\, dy$ must vanish. The equation $\iint \left(\frac{\partial P}{\partial y} - \frac{\partial Q}{\partial x}\right) dx\, dy = 0$, formed for every arbitrary part of the entire domain, is the necessary and sufficient hypothesis on which the derived theorems are based.

Differentiating this equation with respect to y, since we must have:

$$\frac{\partial F(x, y)}{\partial y} = Q(x, y), \qquad \frac{\partial F(a, y)}{\partial y} = Q(a, y),$$

we find the relation:

$$(1) \qquad Q(x, y) - Q(a, y) = \frac{\partial}{\partial y}\int_a^x P(x, y)\,dx.$$

We find in like manner the analogous equation:

$$(2) \qquad P(x, y) - P(x, \alpha) = \frac{\partial}{\partial x}\int_\alpha^y Q(x, y)\,dy.$$

Therefore the functions P and Q cannot be independent; they must satisfy these equations, of which one is a consequence of the other. These conditions are necessary. Does it then follow, that, since the integrals can be differentiated, the functions P and Q also can be differentiated respectively for y and x? Instead of equation (1) in which a and x signify arbitrary values we can write:

$$\frac{Q(x+h, y) - Q(x, y)}{h} = \frac{\partial}{\partial y}\frac{1}{h}\int_x^{x+h} P(x, y)\,dx = \operatorname*{Lim}_{\Delta y = 0}\frac{1}{h}\int_x^{x+h}\frac{P(x, y+\Delta y) - P(x, y)}{\Delta y}\,dx.$$

This expression on the right side has therefore for arbitrarily small values of h a limiting value for $\Delta y = 0$. But at each point at which P is a continuous function of x, the integral on the right can be replaced by its mean value, so that we obtain the equation:

$$\frac{Q(x+h, y) - Q(x, y)}{h} = \operatorname*{Lim}_{\Delta y = 0}\frac{P(x+\Theta h, y+\Delta y) - P(x+\Theta h, y)}{\Delta y}.$$

Here Θ is a function depending on Δy, and the interval Θh can be diminished arbitrarily by choice of h. Nevertheless we must not argue that the determinate limiting value which is found on the right, is the differential quotient with respect to y of the function $P(x, y)$ at a determinate point. For, that limiting value only arises by the argument Θh also varying, whereas the derived function with respect to y is defined as

$$\operatorname*{Lim}_{\Delta y = 0}\frac{P(x, y+\Delta y) - P(x, y)}{\Delta y}.$$

It is only in case the derived function $\frac{\partial P}{\partial y}$ exists for all points within the domain, and is an integrable function with respect to both the variables that we can infer by the proof in § 169 that differentiation and integration are interchangeable in equation (1), and thence the existence of $\frac{\partial Q}{\partial x}$, as well as the equality $\frac{\partial P}{\partial y} = \frac{\partial Q}{\partial x}$. Cases may be assigned in which it becomes necessary to formulate the condition generally. Let $\psi(z)$ be a function, that within a determinate interval is continuous, but has no second derived function. Substituting then $z=\varphi(x,y)$, a function $\psi(\varphi(x,y))=F(x,y)$ is obtained, for which no mixed differential coefficient $\frac{\partial^2 F}{\partial x \partial y}$ exists, although this function has a total first differential whose integral remains always independent of the path of integration.

The condition $\frac{\partial P}{\partial y}=\frac{\partial Q}{\partial x}$ was essential for the proof we have given of theorem 1. in this section, that the sum of the integrals:

$$\int (Pdx+Qdy)$$

formed in a positive circuit for all the boundary curves of a multiply connected domain is zero. Whether this condition is necessary, we have not investigated.

176. The conditions of integrability still hold, when instead of the variables x and y two new variables u and v are introduced, whose first and second derived functions exist.

Let $x=\varphi(u,v)$, $y=\psi(u,v)$, then by the equations:

$$dx=\frac{\partial\varphi}{\partial u}du+\frac{\partial\varphi}{\partial v}dv,\quad dy=\frac{\partial\psi}{\partial u}du+\frac{\partial\psi}{\partial v}dv,$$

the differential $Pdx+Qdy$ is converted into the form P_1du+Q_1dv, where:

$$P_1=P\frac{\partial\varphi}{\partial u}+Q\frac{\partial\psi}{\partial u},\quad Q_1=P\frac{\partial\varphi}{\partial v}+Q\frac{\partial\psi}{\partial v},$$

and we have:

$$\frac{\partial P_1}{\partial v}=\frac{\partial\varphi}{\partial u}\left\{\frac{\partial P}{\partial x}\frac{\partial\varphi}{\partial v}+\frac{\partial P}{\partial y}\frac{\partial\psi}{\partial v}\right\}+\frac{\partial\psi}{\partial u}\left\{\frac{\partial Q}{\partial x}\frac{\partial\varphi}{\partial v}+\frac{\partial Q}{\partial y}\frac{\partial\psi}{\partial v}\right\}$$
$$+P\frac{\partial^2\varphi}{\partial u\partial v}+Q\frac{\partial^2\psi}{\partial u\partial v},$$

$$\frac{\partial Q_1}{\partial u}=\frac{\partial\varphi}{\partial v}\left\{\frac{\partial P}{\partial x}\frac{\partial\varphi}{\partial u}+\frac{\partial P}{\partial y}\frac{\partial\psi}{\partial u}\right\}+\frac{\partial\psi}{\partial v}\left\{\frac{\partial Q}{\partial x}\frac{\partial\varphi}{\partial u}+\frac{\partial Q}{\partial y}\frac{\partial\psi}{\partial u}\right\}$$
$$+P\frac{\partial^2\varphi}{\partial u\partial v}+Q\frac{\partial^2\psi}{\partial u\partial v}.$$

Accordingly:

$$\frac{\partial P_1}{\partial v}=\frac{\partial Q_1}{\partial u}$$

and:

$$\int(P_1du+Q_1dv)=\int(Pdx+Qdy)+C.$$

In the deduction of this formula the existence of all the second differential quotients of the original function is presupposed.

Fourth Book.

Integrals of complex functions. General properties of analytic functions.

First Chapter.

The definite integral of a unique analytic function in the complex domain.

177. A function $f(z)$ of a complex variable z was defined in § 80 as a quantity which can be calculated from z by any finite or even infinite number of arithmetical operations. When such a function has throughout a connected domain, except in singular points, a determinate derivate $f'(z)$ independent of the differential $dz = dx + i\,dy$, we called it (§ 84) an analytic function in that domain. The two constituents of an analytic function $f(z) = u + iv$ are, as was then deduced, continuous functions of the two variables x and y having determinate derivates both for x and for y that satisfy the equations:

$$\frac{\partial u}{\partial x} = \frac{\partial v}{\partial y}, \quad \frac{\partial u}{\partial y} = -\frac{\partial v}{\partial x}.$$

Moreover it was shown that these equations taken along with the continuity of the derivates are also the sufficient conditions that the function $f(x + iy) = u + iv$ may have a derivate independent of the differential, and thus be an analytic function of z in the sense originally defined.

The following investigations, by which we are about to establish the integral conception in the complex domain, deal first with the function of a complex variable in general and then pass on to analytic functions, whose general properties form our ultimate object.

In the complex plane let two given points z_0 and Z be joined together by an arbitrary curve of finite length. The equation of the curve may be supposed given either in the form $\varphi(x, y) = 0$ or by means of a parameter $x = \varphi(t)$, $y = \psi(t)$. To each value of t should then correspond uniquely one value of x and one of y; further, to continuously consecutive values of the variable t, which we assume to undergo no alternation of increase or decrease between the limits t_0 and t, correspond the continuously consecutive points from z_0 to Z. Now inserting between z_0 and Z arbitrarily many points $z_1, z_2, \ldots z_{n-1}$ along the curve, let us form for any function $f(z)$ the sum:

$$f(z_0)(z_1 - z_0) + f(z_1)(z_2 - z_1) + f(z_2)(z_3 - z_2) \cdots + f(z_{n-1})(Z - z_{n-1});$$

the complex limiting value to which this sum tends, as the value of n increases arbitrarily, is called *the definite integral of the function $f(z)$, formed from z_0 to Z along the path prescribed by the equation of the curve.**)

Under what conditions is there a determinate limiting value? Assuming first that the function $f(z)$ remains finite in all points of the curve, so that its modulus does not exceed some assignable superior limit, and also that the points wherein the real and the imaginary constituents of $f(z)$ undergo finite discontinuities or infinitely many oscillations with finite fluctuations form a discrete set, the existence of a determinate limiting value is easily evident by means of the Theorems concerning the real integral. For if by the substitution:

$$z = x + iy = \varphi(t) + i\psi(t),$$

we convert $f(z)$ into $f_1(t) + if_2(t)$, and if $t_0, t_1, t_2, \ldots t_{n-1}, T$ are the values corresponding to the points $z_0, z_1, z_2, \ldots z_{n-1}, Z$, the above sum passes over into:

$$\sum_{k=0}^{k=n-1} \{f_1(t_k) + if_2(t_k)\}\{\varphi(t_{k+1}) + i\psi(t_{k+1}) - \varphi(t_k) - i\psi(t_k)\},$$

that resolves itself into a real constituent:

$$\sum_{k=0}^{k=n-1} f_1(t_k)\{\varphi(t_{k+1}) - \varphi(t_k)\} - \sum_{k=0}^{k=n-1} f_2(t_k)\{\psi(t_{k+1}) - \psi(t_k)\}$$

and an imaginary:

$$i\sum_{k=0}^{k=n-1} f_2(t_k)\{\varphi(t_{k+1}) - \varphi(t_k)\} + i\sum_{k=0}^{k=n-1} f_1(t_k)\{\psi(t_{k+1}) - \psi(t_k)\}.$$

To the first sum we can give the form:

$$\sum_{k=0}^{k=n-1} f_1(t_k)\,\frac{\varphi(t_{k+1}) - \varphi(t_k)}{t_{k+1} - t_k}\,(t_{k+1} - t_k);$$

and analogous forms to the three others.

The curve chosen as path of integration must be of finite length; it is certain therefore to have everywhere, except possibly in a

*) Cauchy: Mémoire sur les intégrales définies, prises entre des limites imaginaires. 1825. Comptes rendus, 1846. — Riemann: Grundlagen für eine allgemeine Theorie der Functionen einer veränderlichen complexen Grösse. 1851. Werke, pp. 3-47, 1876.

discrete set of points, a progressive differential quotient that is itself generally a continuous function and is identical with the regressive differential quotient; or in geometric terms: the curve has no angles except at discrete points. The intervals can therefore (§ 100) be assumed so small, that generally for each value of t:

$$\frac{\varphi(t_{k+1}) - \varphi(t_k)}{t_{k+1} - t_k} = \varphi'(t_k) + \delta,$$

where δ signifies an arbitrarily small quantity. Accordingly each of the four sums takes a form such as:

$$\sum_{k=0}^{k=n-1} f_1(t_k)\varphi'(t_k)(t_{k+1} - t_k) + \delta \sum_{k=0}^{k=n-1} f_1(t_k)(t_{k+1} - t_k),$$

that passes over into the definite integral:

$$\int_{t_0}^{T} f_1(t)\varphi'(t)\,dt$$

as the value of n increases arbitrarily. *Therefore:*

$$\text{I. } \operatorname{Lim}\sum f(z_k)(z_{k+1} - z_k) = \int_{z_0}^{Z} f(z)\,dz = \int_{t_0}^{T} \{f_1(t)\varphi'(t) - f_2(t)\psi'(t)\}\,dt$$

$$+ i\int_{t_0}^{T} \{f_2(t)\varphi'(t) + f_1(t)\psi'(t)\}\,dt = \int_{t_0}^{T} \{f_1(t) + if_2(t)\}\,\{\varphi'(t) + i\psi'(t)\}\,d$$

is a determinate finite quantity.

For instance, if the integral is to be formed in a straight path from the point $z_0 = x_0 + iy_0$ to the point $Z = X + iY$, we have:

$$x = x_0 + (X - x_0)t, \quad y = y_0 + (Y - y_0)t,$$

therefore:

$$\int_{z_0}^{Z} f(z)\,dz = \{X - x_0 + i(Y - y_0)\}\int_{t=0}^{t=1} f(x_0 + iy_0 + t(X - x_0) + it(Y - y_0))\,dt.$$

When we require to integrate along the arc of a circle of radius r whose centre is the point $x_0 + iy_0$, we have by the equations:

$$x = x_0 + r\cos t,\ y = y_0 + r\sin t,\ dx = -r\sin t\,dt,\ dy = r\cos t\,dt;$$

$$\int_{z_1}^{Z} f(z)\,dz = \int_{t_0}^{T} f(x_0 + iy_0 + re^{it})\,ire^{it}\,dt = ir\int_{t_0}^{T} f(x_0 + iy_0 + re^{it})e^{it}\,dt,$$

where t_0 and T denote the values of t belonging to the initial and terminal points of the arc.

178. It results from Equation I. that we also have for the complex integral the following theorems:

1) $$\int_{z_0}^{Z} f(z)\,dz = -\int_{Z}^{z_0} f(z)\,dz,$$

only that both integrals must refer to the same curve. Again:

2) $$\int_{z_0}^{Z} f(z)\,dz = \int_{z_0}^{z_k} f(z)\,dz + \int_{z_k}^{Z} f(z)\,dz,$$

only that the paths of integration from z_0 by z_k to Z or from z_0 by Z to z_k must be the same in all the integrals.

There is likewise here a Theorem of the Mean Value; although it has not as simple a form as for the real integral.

In fact, from the equation:

$$\int_{z_0}^{Z} f(z)\,dz = \int_{t_0}^{T} f_1(t)\{\varphi'(t) + i\psi'(t)\}\,dt + i\int_{t_0}^{T} f_2(t)\{\varphi'(t) + i\psi'(t)\}\,dt,$$

follows:

$$\int_{z_0}^{Z} f(z)\,dz = M_1(Z - z_0) + iM_2(Z - z_0),$$

where M_1 signifies a mean value of $f_1(t)$ and M_2 a mean value of $f_2(t)$. When these functions are continuous along the path of integration, this equation can be given the form:

3) $$\int_{z_0}^{Z} f(z)\,dz = \{f_1(t_0 + \Theta(T - t_0)) + if_2(t_0 + \Theta'(T - t_0))\}\ (Z - z_0).$$

The values of Θ and Θ', however, will in general be different.

It can further be deduced from this equation, that the integral in every point of the path of integration presents a function of the complex variable, and moreover generally an analytic function. For if z denote an arbitrary point on the path of integration, the differential quotient in an arbitrary direction of the integral:

$$\int_{z_0}^{z} f(z)\,dz$$

with respect to its upper limit is derived by forming the quotient:

$$\frac{1}{h}\int_{z}^{z+h} f(z)\,dz,$$

where h signifies any quantity converging to zero and the integration is along any path between the points z and $z + h$. If the values t and $t + k$ correspond to these points in such a way that when $h = 0$

along the path of integration k also converges to zero, we have by equation 3):

4)
$$\frac{1}{h}\int_{z}^{z+h} f(z)\,dz = f_1(t+\Theta k) + i f_2(t+\Theta' k),$$

therefore:

$$\operatorname*{Lim}_{h=0} \frac{1}{h}\int_{z}^{z+h} f(z)\,dz = f_1(t) + i f_2(t) = f(z).$$

The definite integral, regarded as a function of its upper limit, has therefore generally in the points of the path of integration the derived function $f(z)$. In particular also:

$$\frac{\partial \int_{z_0}^{z} f(z)\,dz}{\partial x} = f(z), \qquad \frac{\partial \int_{z_0}^{z} f(z)\,dz}{\partial y} = i f(z),$$

by which is expressed, in conformity with § 84, that the integral is a function of the complex variable $x + iy = z$.

When along the path of integration a unique analytic function $F(z)$ is known, whose derivate is equal to $f(z)$, we have likewise:

5)
$$\int_{z_0}^{Z} f(z)\,dz = F(Z) - F(z_0).$$

For if by the substitution $z = \varphi(t) + i\psi(t)$, $f(z)$ passes into $f_1(t) + i f_2(t)$, and $F(z)$ into $F_1(t) + i F_2(t)$; we have:

$$F'(z) = f(z) = \{F_1'(t) + i F_2'(t)\}\frac{dt}{dz},$$

therefore:

$$\{f_1(t) + i f_2(t)\}\{\varphi'(t) + i\psi'(t)\} = F_1'(t) + i F_2'(t),$$

consequently:

$$\int_{z_0}^{Z} f(z)\,dz = \int_{t_0}^{T}\{f_1(t) + i f_2(t)\}\{\varphi'(t) + i\psi'(t)\}\,dt = \int_{t_0}^{T}\{F_1'(t) + i F_2'(t)\}\,dt$$
$$= \{F_1(T) + i F_2(T)\} - \{F_1(t_0) + i F_2(t_0)\} = F(Z) - F(z_0).$$

The definition of the definite integral, furthermore, can be extended as it was before:

When the function $f(z)$ becomes infinite in discrete points c_1, c_2, etc., the equation of definition takes the form:

$$\int_{z_0}^{c_1} f(z)\,dz = \operatorname*{Lim}_{\delta=0} \int_{z_0}^{c_1-\delta} f(z)\,dz.$$

Here also, as for the real integral, the following criterion holds:

This right hand integral has a determinate limiting value for $\delta = 0$, provided the amount of the product $(c - z)^\nu f(z)$ remains finite when $z = c$, for ν a positive proper fraction, in whatever way the value of z converges to c. (Cf. § 181, 3.)

When the path of integration proceeds in a determinate manner to infinity, we have:

$$\int_{z_0}^{\infty} f(z)\,dz = \operatorname{Lim} \int_{z_0}^{z} f(z)\,dz, \qquad \text{for } z = \infty,$$

when the point z passes to infinity along the prescribed line.

The integral is certain to have a determinate limiting value, when in this process the amount of $z^\nu f(z)$ remains finite and ν is a number greater than 1. (See § 181, 4.)

179. One essential difference there always is, between the integral of a complex function and that of a real function, notwithstanding the similarity of the Theorems of last Section with those we had before: A complex integral can be taken along very different paths between two determinate limits z_0 and Z, while for the real integral the mere requirement that it shall be real prescribes always one path only between the limits. The complex integral of one and the same function between the same limits can therefore assume various values according to the path of integration.

When we integrate ex. gr. $\int \frac{dz}{z}$ from the point $+1$ to the point -1 along the upper semicircle round the origin, putting:

$$z = \cos(t) + i\sin(t), \quad dz = \{-\sin(t) + i\cos(t)\}\,dt,$$

we have:

$$\int_{+1}^{-1} \frac{dz}{z} = i \int_0^{\pi} dt = i\pi.$$

But along the lower semicircle its value is:

$$\int_{+1}^{-1} \frac{dz}{z} = i \int_0^{-\pi} dt = -i\pi.$$

The question therefore arises: *Under what conditions is the integral of a complex function a unique function of its upper limit, independent of the path of integration?* This question, as Riemann has shown, is answered by means of Green's Theorem and the Corollaries that follow from it (§ 174—176).

Let the unique function $f(z) = u + iv$ be defined for a given simply connected domain. Let the functions u and v be continuous throughout this domain. Should this not be the case in certain points, let us enclose them within curves and count these on to the boundary curves of the domain which thereby becomes multiply connected. Integration within such a multiply connected domain will occupy us in what follows presently. Further let us put:

$$\int_{z_0}^{Z} f(z)\,dz = F(Z) - F(z_0) = (U + iV) - (U_0 + iV_0) = \int_{x_0, y_0}^{x, y} (u + iv)(dx + idy),$$

where U and V similarly signify real and continuous functions of x and y.

The function $U + iV$ is required to have the property, that in each point of the domain it is independent of the path of integration and satisfies the equations:

$$\frac{\partial U}{\partial x} + i\frac{\partial V}{\partial x} = f(z) = u + iv, \quad \frac{\partial U}{\partial y} + i\frac{\partial V}{\partial y} = if(z) = i(u + iv),$$

so that we must have:

$$\frac{\partial U}{\partial x} = u, \quad \frac{\partial U}{\partial y} = -v, \quad \frac{\partial V}{\partial x} = v, \quad \frac{\partial V}{\partial y} = u.$$

We are accordingly led back to the problem previously treated, whose solution (§ 175, p. 317) informs us:

In the simply connected domain if U is to be a continuous function, whose partial derivates for x and y are everywhere respectively u and $-v$, and if V is likewise to be a continuous function for which they are respectively v and u, these functions u and v having determinate partial differential coefficients with respect both to x and to y that satisfy the relations:

$$\frac{\partial u}{\partial y} = -\frac{\partial v}{\partial x}, \quad \frac{\partial v}{\partial y} = \frac{\partial u}{\partial x};$$

then the functions U and V are obtained, formed by the integrals:

$$\int_{x_0, y_0}^{x, y} (u\,dx - v\,dy) \quad \text{and} \quad \int_{x_0, y_0}^{x, y} (v\,dx + u\,dy),$$

along any arbitrary path. But the above relations inform us that $f(z)$ is an analytic function with the determinate derivate:

$$f'(z) = \frac{\partial u}{\partial x} + i\frac{\partial v}{\partial x} = \frac{1}{i}\left\{\frac{\partial u}{\partial y} + i\frac{\partial v}{\partial y}\right\}.$$

It is accordingly proved: *When in a simply connected finite domain $f(z)$ is without exception an analytic function, the integral $\int_{z_0}^{z} f(z)\,dz$ is an analytic function of z completely independent of the path of integration within the entire domain, and its derivate is $f(z)$.*

180. If a new variable be introduced into the integral $\int f(z)\,dz$ by substituting $z = \psi(z')$, where $\psi(z')$ is an analytic function in the entire domain within which the integral is to be formed, then by § 176 the property of integrability holds undisturbed with respect to the new variable z', and thus:

$$\int f(z)\,dz = \int f(\psi(z'))\,\psi'(z')\,dz'$$

is an analytic function of z' with the derivate $f(\psi(z')) \,.\, \psi'(z')$.

When the integral has to be extended over a domain reaching to infinity, we can transform it by Inversion (§ 79) into an integral that is to be investigated within a finite region. From the substitution:

$$z = \frac{1}{z'}\,,\quad dz = -\frac{dz'}{z'^2}\,,\ \text{we find:}\ \int f(z)\,dz = -\int f\left(\frac{1}{z'}\right)\frac{dz'}{z'^2}\,,$$

and to arbitrarily increasing values of z correspond values of z' with arbitrarily small modulus. If then we have to integrate $\int f(z)\,dz$ along a curve upon which z becomes infinite, we have:

$$\operatorname{Lim}_{\text{for } z=\infty} \int^{z} f(z)\,dz = \operatorname{Lim}_{\text{for } z'=0} -\int^{z'} f\left(\frac{1}{z'}\right)\frac{dz'}{z'^2}\,.$$

181. A number of corollaries depend on the Theorem of § 179:

1. *When we form the integral along a closed curve that lies within the simply connected domain, its value is zero.*

For, when the path from z_0 by z_1 to z is one part of this curve, and the path from z_0 by z_2 to z the other, the integrals along these paths are equal; and because along one and the same path:

$$\int_{z_0}^{z} f(z)\,dz = -\int_{z}^{z_0} f(z)\,dz, \qquad \S\ 178,\ 1),$$

the sum of the two integrals for the closed path from z_0 to z and from z to z_0 is zero.

2. When the domain is not simply connected, both the integrals:

$$\int (u + iv)\,(dx + i\,dy) = \int (u\,dx - v\,dy) + i\int (v\,dx + u\,dy)$$

formed for all the boundary curves of the domain in a positive circuit are zero (§ 175, 1.). *If therefore we form* $\int f(z)\,dz$ *in a positive circuit for all the boundary curves of a multiply connected domain wherein* $f(z)$ *is an analytic function, the value the integral assumes is zero.*

Along each separate boundary curve it has a determinate value.

When within the domain a closed curve is drawn, (α, fig. 17), which by itself encloses a simply connected domain, the integral along this curve vanishes; but when a curve is drawn which together with a boundary curve encloses a domain, ex. gr. β, the integral along this curve is equal to the value that it assumes for the associated boundary curve.

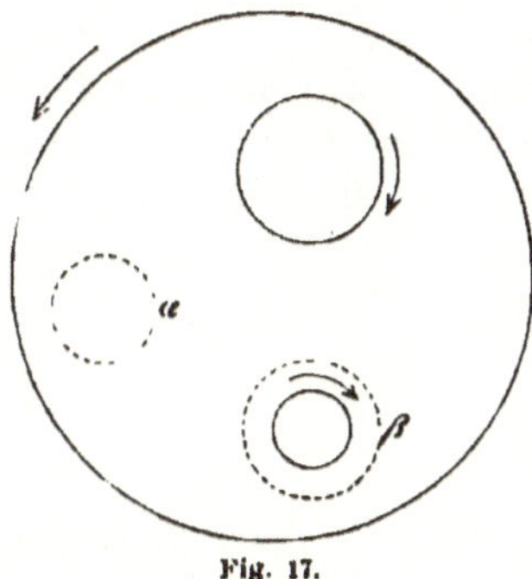

Fig. 17.

From this we form the rule:

3. When $f(z)$, the function that is to be integrated, loses the property of an analytic function in a discrete set of points of a simply connected finite domain, by either ceasing to be continuous, or to be finite, or to have a derivate, let us enclose each such isolated point within a curve arbitrarily near it, and, counting all these among the boundary curves of the domain, let us carry on the investigation in the multiply connected domain. When the integral for the boundary curve round any such point vanishes, the boundary of this point can be dispensed with, although it is always possible that the integral function up to such a point may likewise lose the property of an analytic function. For, it was only proved for points at which $f(z)$ remains continuous, that the integral function also is analytic.

There are therefore two things to be examined: first, whether the integral round a singular point vanishes; second, whether the integral up to the singular point continues finite or analytic in general.

We can at once see that: if the integral round the singular point do not vanish, the integral function is an ambiguous function, and the singular point is one of its branching points; along a branching section starting from this point the values of the integral differ by a constant quantity.

For if the value of the integral round one boundary curve of the point α be A, it has the same value for every boundary curve round the point α, because two boundary curves determine a ring surface in which the function is analytic without exception. At the two sides of a curve starting from the singular point the values of the integral differ by A; the integral is therefore ambiguous.

A determinate value of the integral up to the singular point is in this case quite out of the question; the value of it then depends on the path by which the singular point is reached; how often ex. gr. the branching section is crossed on the way.

But next: if the integral round the boundary of a point α vanish, the integral remains a unique function within a domain however small round α, but this point itself may possibly be a singular point for the integral.

A necessary condition in order that the integral up to the point α may remain finite, in which case it is certainly continuous, is: that for any number δ however small, it shall be possible to assign a circle with α as centre and with radius r, such that for all values of

$$z - \alpha = \varrho(\cos\psi + i\sin\psi)$$

for which $\varrho \leq r$, the amount of the integral

$$\int_r^0 f(\alpha + \varrho e^{i\psi})e^{i\psi}d\varrho$$

shall independently of ψ be smaller than δ. This will certainly be the case ex. gr. when the function $f(z)$ is so constituted that: for $\varrho < r$ and $0 < \nu < 1$,

$$\text{abs}\,[f(\alpha + \varrho e^{i\psi}) \cdot \varrho^\nu]$$

is constantly smaller than a finite quantity G. For we have:

$$\text{abs}\int_r^0 f(\alpha + \varrho e^{i\psi})e^{i\psi}d\varrho < \int_0^r \text{abs}\,[f(\alpha + \varrho e^{i\psi})]d\varrho < G\int_0^r \frac{d\varrho}{\varrho^\nu} = G\frac{r^{1-\nu}}{1-\nu},$$

and r can be chosen so as to make this quantity smaller than δ.

We can also assign a condition which is sufficient in order that the integral taken round the boundary of a point may vanish. This condition is: that

$$\lim_{z=\alpha}\{(z-\alpha)f(z)\} = 0,$$

and therefore that a domain can be bounded off round the point α within which the amount of the product $(z - \alpha)f(z)$ is equal to or smaller than an arbitrarily small number δ.

For then, integrating round the circle with radius r, we have:

$$\text{abs}\int f(z)dz < \int_0^{2\pi} \text{abs}\,[f(\alpha + re^{i\psi})\, ire^{i\psi}]\,d\psi < 2\pi\delta,$$

and this value is arbitrarily small; it is therefore impossible that the integral round the boundary should have a finite value, so that, since its value is determinate and is the same for each boundary curve, it must be zero.

We can now estimate the bearing upon integration of whatever singularities there may be.

a) When the function $f(z)$ loses the property of an analytic function in a point by becoming discontinuous or by ceasing to satisfy the equation $\frac{\partial f}{\partial x} + i\frac{\partial f}{\partial y} = 0$, while still remaining finite, this point has no influence whatever upon the integral. In fact, for a positive ν either equal to or less than 1 we have

$$\lim_{z=\alpha} \{(z-\alpha)^{\nu} f(z)\} = 0.$$

The integral is still an analytic function in the point α, its derived function is $\operatorname{Lim} f(z)$ for $z = \alpha$. (See moreover Remark c.)

b) But when the function $f(z)$ becomes infinite at a point, whether an essential or a non-essential singular point, this is always an infinity point for the integral also. In fact, here

$$\lim_{z=\alpha} \{(z-\alpha)^{\nu} f(z)\}$$

is not finite for ν less than 1; because in a non-essential singular point

$$\operatorname{Lim} \{(z-\alpha)^{m} f(z)\}$$

is finite only for m equal to or greater than 1, but in an essential singular point it is finite for no assignable m whatever. In the examples of next Section we shall discuss this as well as the question whether such a singular point is a branching point or not.

c) It is necessary also to consider whether possibly the function $f(z)$ may lose the analytic property all along a curve c situated within the domain, by no longer satisfying the equation

$$\frac{\partial f}{\partial x} + i\frac{\partial f}{\partial y} = 0$$

in any point of this curve, although continuous in its neighbourhood; while the derived functions remain finite. This assumption however, as we shall prove, involves a contradiction.

In fact if we surround the curve c by a boundary l arbitrarily close to it, the integral round this boundary will vanish, because the values of the function on the part of l to the right of c differ from those to the left arbitrarily little when l closes in arbitrarily up to c; therefore the sum of the integrals, being taken in opposite directions, becomes arbitrarily small. Moreover the integral remains finite for each point of the curve c. Therefore such a curve would not be singular for the integral. Thence it follows that even in the points of the curve we have:

$$U + iV = \int^{x,y}(u+iv)(dx+idy),$$

or the equations:

$$U = \int^{x,y}(u\,dx - v\,dy), \qquad V = \int^{x,y}(v\,dx + u\,dy).$$

Therefore the integral $U + iV$ is an analytic function, whose derivate is $f(z) = u + iv$. It will be shown subsequently, that the derivate of an analytic function has also a derivate independent of the quotient

$\frac{dy}{dx}$; consequently the equations

$$\frac{\partial u}{\partial y} = -\frac{\partial v}{\partial x}, \quad \frac{\partial v}{\partial y} = \frac{\partial u}{\partial x}$$

must also hold for the points of the curve, that is to say, the original assumption is impossible.

d) When the function $f(z)$ is discontinuous at the two sides of a curve lying within the domain, the paths of integration at the two sides of such a curve also lead to different values; i. e. the integral also is no longer an analytic function in the domain.

4. When the domain of the integral contains the point infinity, we convert it into a finite domain by the substitution $z = \frac{1}{z'}$. To the integral

$$-\int f\left(\frac{1}{z'}\right)\frac{dz'}{z'^2}$$

taken arbitrarily closely round the origin in a positive circuit will correspond an integral in z along a curve likewise surrounding the origin but arbitrarily remote and in a circuit keeping the infinite on the left: such a curve is said to surround the point infinity. The value of this integral is certainly zero when

$$\lim_{z'=0}\left\{z' f\left(\frac{1}{z'}\right)\frac{1}{z'^2}\right\} = \lim_{z=\infty}\{f(z)\,.\,z\} = 0.$$

The integral function will remain finite even for the point infinity, according to the theorem just proved, when we have:

$$\lim_{z'=0}\left\{z'^{v} f\left(\frac{1}{z'}\right)\ \frac{1}{z'^2}\right\} = \lim_{z'=0}\left\{f\left(\frac{1}{z'}\right)\cdot\frac{1}{z'^{2-v}}\right\} = \lim_{z=\infty}\left\{f(z)\,z^{2-v}\right\} = G$$

some finite quantity, where the exponent $2 - v > 1$ or $v < 1$.

182. Of unique analytic functions presented in an explicit form we have only become acquainted with rational algebraic functions and with the infinite series of ascending positive integer powers within its circle of convergence. These we have now to integrate, taking their singular points specially into account.

1. The integral of an integer rational function.

When n is a positive integer, the power z^n is an analytic function without exception in the entire plane. Therefore along any path of integration:

$$\int^{z} z^n\,dz = \frac{z^{n+1}}{n+1} + C.$$

When the integration is extended to infinity, the integral function also becomes infinite. The point infinity is a non-essential singular point.

It is easy to prove the theorem: that the complex integral of a sum of functions is equal to the sum of the integrals formed for the several summands; by its help we obtain from this the formula for the integration of any integer rational function:

$$\int^{z}(a_0+a_1z+a_2z^2+\cdots a_nz^n)dz=a_0z+\frac{1}{2}a_1z^2+\frac{1}{3}a_2z^3+\cdots\frac{1}{n+1}a_nz^{n+1}+C.$$

2. The integral of a fractional rational function.

Every fractional rational function can be broken up into an integer function and into partial fractions each having a constant numerator and its denominator an integer power; for, the identities developed in §§ 111, 113 hold also for the complex variable. Accordingly the integration of a fractional function requires us only to investigate:

$$\int\frac{dz}{z-a}\text{ and more generally }\int\frac{dz}{(z-a)^n};$$

in place of these, writing z instead of $z-a$, adopting therefore the point a as zero or origin, we can deal with the simpler integrals:

$$\int\frac{dz}{z}\text{ and }\int\frac{dz}{z^n}.$$

Considering first the case of n a positive integer greater than 1; it is evident that the function $\frac{1}{z^n}$ loses the character of an analytic function at the singular point zero; in all the rest of the plane, the point infinity included, it is regular. Surround the singular point with a circle of radius r. Along the circumference of this circle, since

$$z=r(\cos\varphi+i\sin\varphi),\quad dz=r(-\sin\varphi+i\cos\varphi)d\varphi,$$

the value of the definite integral is:

$$\int\frac{dz}{z^n}=\frac{1}{r^{n-1}}\int\frac{-\sin\varphi+i\cos\varphi}{\cos n\varphi+i\sin n\varphi}d\varphi=\frac{i}{r^{n-1}}\int_0^{2\pi}(\cos\overline{n-1}\varphi-i\sin\overline{n-1}\varphi)d\varphi=0,$$

since $n>1$. Thus the result of integrating round the boundary of the origin vanishes, so that its boundary curve has not to be taken into account. Along every path in the plane, even such as pass through the point infinity, since for it

$$\operatorname*{Lim}_{z=\infty}\left\{z\frac{1}{z^n}\right\}=0,$$

we have the integral:

$$\int^{z}\frac{dz}{z^n}=\frac{z^{-n+1}}{-n+1}+C,$$

and it vanishes along every closed path. But the point zero or origin is a non-essential singular point of the integral.

The case $n = 1$, that already served as an example in § 179, requires a particular treatment; for here round a circle with the point zero or origin as centre we have:

$$\int \frac{dz}{z} = i \int_0^{2\pi} d\varphi.$$

Therefore independently of the radius r the integral is equal to $2i\pi$. Around the point infinity the integral has likewise a finite value; for, integrating along a circle with an arbitrarily great radius R, keeping the origin on the right, the integral is converted by the substitution

$$z = \frac{1}{z'}, \quad dz = -\frac{dz'}{z'^2},$$

into: $-\int \frac{dz'}{z'}$ to be integrated along a circle of radius $\frac{1}{R}$ keeping the origin on the left; thus its value is $-2i\pi$. The value of the integral will accordingly depend on the path of integration; along a finite closed curve its value will be $2i\pi k$ or zero, according as the closed curve goes round the origin k times or not at all. The integral $\int_{z_0}^{z} \frac{dz}{z}$ is a unique function of its upper limit z, when the paths that lead from the lower limit z_0 to z do not cross a section leading from the origin to the point infinity; it is only in the plane perforated along this section that the values of the integral calculated along such curves are continuous; on opposite banks arbitrarily near the branching section they differ by the constant quantity $2i\pi$.

But the many-valued function which is presented by the integral is the logarithm treated in § 82, 5; for, $l(z)$ is that function whose derivate is $\frac{1}{z}\cdot$ The method there employed of rendering it a unique function by means of arbitrarily many leaves, is valid also for the integral. For each path not crossing the branching section we have:

$$\int_{z_0}^{z} \frac{dz}{z} = l(z) - l(z_0),$$

and thus in particular $\int_1^z \frac{dz}{z} = l(z)$. In writing this equation, however, we must observe that the quantity on the left is now determinate, while that on the right is still many-valued. What is the value on the right belonging to a determinate path? Before we can characterise a definite kind of path, we must adopt a definite branching section, ex. gr. the positive axis of ordinates. We reject the positive axis of abscissæ, to avoid having the lower limit $+ 1$ on the branching section, for in that case we should have to distinguish between its two banks.

To proceed from 1 to z in the same leaf, means to take a path that, while otherwise arbitrary, does not cross this section. To proceed from 1 to z and at the same time pass from the first into the $(k+1)^{\text{th}}$ leaf, means to take an arbitrary path which crosses this line $(k+k')$ times in the direction from right to left and k' times in the opposite direction; to pass from the first into the $-(k+1)^{\text{th}}$ leaf, means to cross the section $(k'+k)$ times from left to right and k' times from right to left.

Remaining in the first leaf, we can as a particular case integrate from 1 to $z = x + iy$, provided z is not in the second quadrant $(x < 0,\ y > 0)$, by proceeding first from 1 parallel to ordinates to y and then parallel to abscissæ to the value x.

Putting:

$$\int_1^z \frac{dz}{z} = \int_1^{x_0+iy_0} \frac{dx+idy}{x+iy} = \int \frac{(x-iy)\,dx}{x^2+y^2} + i\int \frac{(x-iy)\,dy}{x^2+y^2},$$

then (as in § 175, 4.) the first integral on the right is zero along the first part of the path of integration, along the second it is either:

$$\int_1^{x_0} \frac{(x-iy_0)\,dx}{x^2+y_0^2} = \frac{1}{2}\,l(x_0^2+y_0^2) - \frac{1}{2}\,l(1+y_0^2) + i\left\{\tan^{-1}\frac{y_0}{x_0} - \tan^{-1}y_0\right\},$$

or:

$$= \frac{1}{2}\,l(x_0^2+y_0^2) - \frac{1}{2}\,l(1+y_0^2) + i\left\{\tan^{-1}\frac{y_0}{x_0} - \tan^{-1}y_0\right\} - i\pi,$$

according as x_0 is positive or negative; the circular function here always signifies a value between $-\frac{1}{2}\pi$ and $+\frac{1}{2}\pi$. For in the second case, as x passes through zero, $\tan^{-1}\frac{y_0}{x}$ is led over continuously into the value $\tan^{-1}\frac{y_0}{x} - \pi$.

We find for the second integral along the path parallel to ordinates:

$$i\int_0^{y_0} \frac{(1-iy)\,dy}{1+y^2} = \frac{1}{2}\,l(1+y_0^2) + i\tan^{-1}y_0.$$

Here the circular function signifies always a value between $-\frac{1}{2}\pi$ and $+\frac{1}{2}\pi$; therefore in the first leaf, (excluding $y_0 > 0$, $x_0 < 0$), we have:

$$\int_1^{z=x_0+iy_0} \frac{dz}{z} = \frac{1}{2}\,l(x_0^2+y_0^2) + i\tan^{-1}\frac{y_0}{x_0} \cdot + 0, \text{ or, } - i\pi;$$

according as $x_0 > 0$, or, < 0.

For a point of the second quadrant let us first proceed from $x = 1$ parallel to ordinates to the point $1 - iy_0$, thence parallel to abscissæ to the point $x_0 - iy_0$ and lastly again parallel to ordinates to $x_0 + iy_0$; thus:

$$\int_1^z \frac{dz}{z} = \int_1^{x_0 - iy_0} \frac{dz}{z} + \int_{x_0 - iy_0}^{x_0 + iy_0} \frac{dz}{z} = \frac{1}{2} l(x_0^2 + y_0^2) - i \tan^{-1} \frac{y_0}{x_0} - i\pi + \int_{x_0 - iy_0}^{x_0 + iy_0} \frac{dz}{z}.$$

But the last integral is equal to:

$$\int_{x_0 - iy_0}^{x_0 + iy_0} \frac{i\,dy}{x + iy} = i \int_{-y_0}^{+y_0} \frac{(x_0 - iy)\,dy}{x_0^2 + y^2} = 2i \tan^{-1} \frac{y_0}{x_0},$$

therefore for the points of the second quadrant:

$$\int_1^z \frac{dz}{z} = \frac{1}{2} l(x_0^2 + y_0^2) + i \tan^{-1} \frac{y_0}{x_0} - i\pi.$$

Along the right side of the positive axis of ordinates the values of the integral are $\frac{1}{2} l(y_0^2) + \frac{1}{2} i\pi$, but along the left they are $\frac{1}{2} l(y_0^2) - \frac{3}{2} i\pi$; they differ by $2i\pi$.

3. The integral of an infinite series of powers.

Within the circle round the centre α, in which the complex series of ascending positive integer powers:

$$a_0 + a_1(z - \alpha) + a_2(z - \alpha)^2 + \cdot \quad \cdot a_n(z - \alpha)^n + \cdots$$

converges, it expresses a unique and continuous function $f(z)$ without singular points; for every point $z = Z$ within the circle of convergence the series converges absolutely; and therefore it also converges uniformly; i. e. a value of n can be assigned from which onwards every remainder is of smaller amount than an arbitrarily small number δ, for every value of z such that abs $[z - \alpha] <$ abs $[Z - \alpha]$. For,

$$\text{abs}[a_n(Z - \alpha)^n + a_{n+1}(Z - \alpha)^{n+1} + \cdots] < A_n R^n + A_{n+1} R^{n+1} + \cdots,$$

when A is written for abs a, and R for abs $[Z - \alpha]$. But since the series:

$$A_0 + A_1 R + A_2 R^2 + \cdots$$

converges, a value can be assigned to n, for which the right side of the above inequality will be always smaller than δ, *à fortiori* we shall have:

$$\text{abs}[a_n(z - \alpha)^n + a_{n+1}(z - \alpha)^{n+1} \cdots] < A_n r^n + A_{n+1} r^{n+1} + \cdots < \delta,$$

when abs $[z - \alpha] = r < R$.

From this results the following theorem:

When the complex series of powers is integrated from the point z_0 up to the point Z, that both lie within its convergency, its integral is an analytic function and presents itself as the difference between the values for the arguments Z and z_0 of the series of powers that arises on integrating the given series term by term:

$$\int_{z_0}^{Z} f(z)dz = \left\{a_0(Z-\alpha) + \frac{a_1}{2}(Z-\alpha)^2 + \frac{a_2}{3}(Z-\alpha)^3 + \cdots\right\}$$
$$- \left\{a_0(z_0-\alpha) + \frac{a_1}{2}(z_0-\alpha)^2 + \frac{a_2}{3}(z_0-\alpha)^3 + \cdots\right\}.$$

As a particular case we have:

$$\int_{\alpha}^{Z} f(z)dz = a_0(Z-\alpha) + \frac{a_1}{2}(Z-\alpha)^2 + \frac{a_2}{3}(Z-\alpha)^3 + \cdots \frac{a_n}{n+1}(Z-\alpha)^{n+1} + \text{etc.}.$$

The circle of convergence of this new series can be neither smaller nor larger than that of the original. It is however possible that this second series may still converge (conditionally or unconditionally) upon the circle itself, while the original series diverges in the points of the circumference. In this case the series expresses the integral of the function $f(z)$ for the point on the circle of convergence also. For, when Z denotes such a point, we have:

$$\int_{\alpha}^{Z} f(z)dz = \operatorname*{Lim}_{\delta=0} \int_{\alpha}^{Z-\delta} f(z)dz$$
$$= \operatorname*{Lim}_{\delta=0} \left\{a_0(Z-\delta-\alpha) + \frac{a_1}{2}(Z-\delta-\alpha)^2 + \frac{a_2}{3}(Z-\delta-\alpha)^3 + \cdots\right\}.$$

Since a series of powers remains continuous provided it converges in the points of the limiting circle (§ 83), this right hand limiting value passes over continuously into the series:

$$a_0(Z-\alpha) + \frac{a_1}{2}(Z-\alpha)^2 + \frac{a_2}{3}(Z-\alpha)^3 + \cdots + \text{etc.}.$$

When the infinite plane is the convergency of the series, the value of the integral is expressed by the series of powers for values of Z however great in amount. For the point $Z = \infty$, the integral, meaning thereby the limiting value along a determinate path, may possibly remain finite, only its value can no longer be expressed by a series of powers.

This is exemplified in the case of the integral $\int e^z dz$, whose value for every finite value of Z is:

$$\int_0^Z e^z dz = \int_0^Z \left\{1 + \frac{z}{\lfloor 1} + \frac{z^2}{\lfloor 2} + \frac{z^3}{\lfloor 3} + \cdots \frac{z^n}{\lfloor n} + \cdots\right\} dz$$
$$= Z + \frac{Z^2}{\lfloor 2} + \frac{Z^3}{\lfloor 3} + \cdots \frac{Z^{n+1}}{\lfloor n+1} + \cdots = e^Z - 1$$

When we require this integral along a path to infinity along which the abscissa x increases arbitrarily negatively and the ordinate y positively, we can proceed by describing the path of integration first up to $-a$ along the axis of abscissæ, then the path b parallel to ordinates, and ultimately causing a and b to increase arbitrarily. Conceiving ex. gr. the parabola $y = x^2$ as path of integration, let us consider the point

$$z = -a + ia^2$$

upon it. The value of the integral extended to this point is:

$$\int_0^z e^z dz = \int_0^{-a} e^x dx + i\int_0^{a^2} e^{-a+iy} dy$$

$$= (e^{-a} - 1) + e^{-a}(e^{ia^2} - 1) = -1 + e^{-a+ia^2}.$$

When a converges to infinity, the value of the right side passes over into -1. But it is only along a determinate path of integration that a determinate value of the integral is generated.

In order to study in this example the behaviour of the essential singular point in integration generally, we transfer it to the origin or point zero by the substitution $z = \frac{1}{z'}$ and consider the integral

$$-\int e^{\frac{1}{z'}} \frac{dz'}{z'^2} = -\int \left\{\frac{1}{z'^2} + \frac{1}{z'^3 \lfloor 1} + \frac{1}{z'^4 \lfloor 2} + \frac{1}{z'^5 \lfloor 3} + \cdots + \frac{1}{z'^{n+2} \lfloor n} + \cdots\right\} dz'$$

taken in a negative circuit round the point $z' = 0$.

We have here a series advancing by powers of the variable $\frac{1}{z'}$; we shall therefore first prove in general the theorem (see § 131):

When $F(z)$ is a series which advances by powers of a function $f(z)$, its integral within a domain wherein the series converges and $f(z)$ is an analytic function, is obtained by integrating its several terms.

With a view to this proof, we have to demonstrate that the series:

$$F(z) = a_0 + a_1\{f(z)\} + a_2\{f(z)\}^2 + \cdots + a_n\{f(z)\}^n + \cdots$$

converges uniformly within its convergency. It is obvious that if the series converge for a determinate value of Z, it is absolutely convergent for every value of z for which abs $f(z) <$ abs $f(Z)$. For, inasmuch as the series converges for $f(Z)$, its terms must decrease in amount in such a way that, calling $A_n =$ abs $[a_n]$, $R =$ abs $[f(Z)]$, we shall have:

$$A_{n+1} R^{n+1} < A_n R^n, \text{ and that, } \lim_{n=\infty} \frac{A_{n+1}}{A_n} R$$

shall at most $= 1$.

Accordingly

$$F(z) = a_0 + a_1 \frac{f(z)}{F(z)} F(z) + a_2 \left\{\frac{f(z)}{F(z)}\right\}^2 \{F(z)\}^2 + \cdots a_n \left\{\frac{f(z)}{F(z)}\right\}^n \{F(z)\}^n + \cdots$$

is an absolutely convergent series; for, its series of moduli:

$$A_0 + A_1 \frac{r}{R} \cdot R + A_2 \left(\frac{r}{R}\right)^2 R^2 + \cdots A_n \left(\frac{r}{R}\right)^n R^n + \cdots$$

converges, because while $r < R$:

$$\operatorname{Lim} \frac{A_{n+1}}{A_n} \cdot \frac{r}{R} \cdot R < 1.$$

But as long as the series converges absolutely, it converges also uniformly; in fact then, for every value of z, the amount of:

$$a_n \{f(z)\}^n + a_{n+1} \{f(z)\}^{n+1} + \cdots$$

is smaller than δ, provided we choose n large enough to make

$A_n R^n + A_{n+1} R^{n+1} + \cdots < \delta$, remembering that $R >$ abs $f(z)$.

We have consequently:

$$\int_{z_0}^{z_1} F(z) dz = a_0 \int_{z_0}^{z_1} dz + a_1 \int_{z_0}^{z_1} f(z) dz + a_2 \int_{z_0}^{z_1} \{f(z)\}^2 dz + \cdots a_n \int_{z_0}^{z_1} \{f(z)\}^n dz + \cdots$$

Since $f(z)$ is a unique analytic function, the integrals on the right are also analytic functions independent of the path of integration.

This series converges uniformly within its convergency.

Employing this theorem as a Lemma, we can find the value of:

$$-\int \left\{ \frac{1}{z'^2} + \frac{1}{z'^3 \lfloor 1} + \frac{1}{z'^4 \lfloor 2} + \cdots \frac{1}{z'^{n+2} \lfloor n} + \cdots \right\} dz'$$

integrated round the point zero in a negative circuit, since the series converges for every finite value of z' but zero. The value of this integral is found by the substitution:

$$z' = \varrho(\cos\varphi + i \sin\varphi),$$

to be:

$$\int_0^{2\pi} \left\{ \frac{1}{\varrho^2} e^{-2i\varphi} + \frac{1}{\varrho^3 \lfloor 1} e^{-3i\varphi} + \frac{1}{\varrho^4 \lfloor 2} e^{-4i\varphi} + \cdots \frac{1}{\varrho^{n+2} \lfloor n} e^{-(n+2)i\varphi} + \cdots \right\} \varrho i e^{+i\varphi} d\varphi = 0.$$

The essential singular point of the exponential function is therefore not a branching point of the integral function; but it is an essential singular point; for the integral:

$$-\int_a^z e^{\frac{1}{z'}} \frac{dz'}{z'^2} = \frac{1}{z} + \frac{1}{z^2 \lfloor 2} + \frac{1}{z^3 \lfloor 3} + \cdots \frac{1}{z^n \lfloor n} + \cdots + C$$

is an infinite series of powers and has the singular point $z = 0$.

But the functions $e^{\frac{1}{z}}$ and $e^{\frac{1}{z}} \cdot \frac{1}{z}$ are instances in which the essential singular point $z = 0$ is at the same time a branching point of the integral. Moreover when a non-essential singular point of the function $f(z)$ is considered in the general form:

$$f(z) = \frac{1}{(z - \alpha)^m} \{a_0 + a_1(z - \alpha) + a_2(z - \alpha)^2 + \cdots\},$$

where the series of powers must be convergent in the neighbourhood of the point α, we see that (if $m > 1$) the point α will always remain a non-essential singular point for the integral $\int f(z)\,dz$, and that it will at the same time be a branching point, provided the coefficient a_{m-1} in the term $a_{m-1}(z - \alpha)^{m-1}$ does not vanish.

These examples therefore teach us that the character of an infinity point of the function can only alter for the integral function in so far as it may possibly become at the same time a branching point.

Second Chapter.

Expansion of unique analytic functions in series of powers. General properties.

183. We have more than once pointed out that, except in the case of rational algebraic functions, the definition of functions by means of arithmetical operations left the problem of their calculation still unsolved; for real functions this problem found its solution in Taylor's Theorem: Given the value of an arbitrarily defined function and of each of its derived functions at one point, and knowing that these are all continuous within an interval, the value of the function and of each of its derived functions can be calculated for every point of the interval by means of Taylor's series, provided this series converges.

We are now however concerned with the calculation of a function in a complex domain; our present investigations will show that this is accomplished by means of the following simple theorem:

Being given the value of an arbitrarily defined function and of each of its derived functions at one point, and knowing that the function is analytic within a finite connected domain, i. e. is unique, continuous, and has a determinate finite first derived function, its Taylor's series converges round this point within a circle that is included in the domain. Around each point in the domain such an expansion is possible, and by means of it the function and all its derived functions can be calculated for every value of the variable.

Whereas therefore under the restriction to the real interval the continuity of the derived functions and the convergence of Taylor's series still belonged to the hypotheses that had to be established before the problem could be proved to admit of solution, here simply the character of the analytic function is seen to be hypothesis sufficient, so that we can enunciate the theorem also in the pregnant form:

Every function that is analytic without exception in a connected domain can be expressed for the neighbourhood of each point in it by an ascending series of positive integer powers.

To this is owing the fundamental importance of these series, which also in our previous investigations always had our special attention.

Not only do they present the simplest method of calculating any function other than a rational algebraic function, but they define in general all continuous functions of a complex variable whose first derivate is determinate.

The way of arriving at this theorem was pointed out by Cauchy.

184. When $f(z)$ is an analytic function without exception in a simply connected domain inclusive of the boundary, and $t = u + iv$ denotes an internal point, the quotient $\frac{f(z)}{z-t}$ is in the same domain likewise an analytic function, only that the one point $z = t$ is a non-essential singularity. An arbitrarily small circle drawn with radius ϱ round t as centre will be the second boundary curve of a domain in which the above quotient is analytic without exception. Integrating the quotient along both the external boundary curve and this circle ϱ, keeping the ring surface on the left, the sum of these two integrals is zero. (§ 181.)

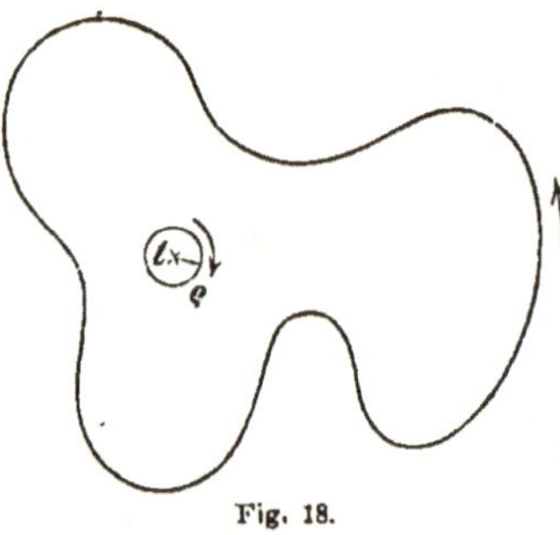

Fig. 18.

We find the result of integrating round the arbitrarily small circle of radius ϱ negatively, i. e. keeping its interior on the right, as follows: For any point on the circumference of this circle we have:

$$z - t = \varrho(\cos\varphi + i\sin\varphi) = \varrho e^{i\varphi}, \quad dz = \varrho i e^{i\varphi} d\varphi = \varrho i(\cos\varphi + i\sin\varphi)d\varphi,$$

therefore:

$$\int \frac{f(z)}{z-t}\, dz = -i \int_{\varphi=0}^{\varphi=2\pi} f(t + \varrho e^{i\varphi})d\varphi.$$

But since the function f is continuous in the neighbourhood of the point $z = t$, we can choose ϱ so small that the difference between the values $f(t)$ and $f(t + \varrho e^{i\varphi})$, for all values of φ, shall be smaller than a number δ whose modulus is arbitrarily small. Accordingly

$$i\int_0^{2\pi} f(t + \varrho e^{i\varphi})d\varphi \quad \text{differs from} \quad i\int_0^{2\pi} f(t)d\varphi = 2i\pi f(t)$$

by a quantity whose modulus is smaller than that of the arbitrarily small number $2\pi\delta$; i. e. independently of the value ϱ we have the integral:

$$i\int_{\varphi=0}^{\varphi=2\pi} f(t + \varrho e^{i\varphi})d\varphi \quad \text{equal to } 2i\pi f(t).$$

Therefore the following equality is established for the integral taken positively round the outer boundary, and in general along any included curve surrounding the point t:

I. $$\int \frac{f(z)}{z-t}\,dz - 2i\pi f(t) = 0 \quad \text{or:} \quad f(t) = \frac{1}{2i\pi}\int \frac{f(z)}{z-t}\,dz.$$

This equation asserts that: If we are given the values along a closed curve of a function $f(z)$ that is known to be analytic without exception in the enclosed space, we can find the value of $f(z)$ for each internal point t by a definite integral.*)

185. We began with the hypothesis that the function $f(z)$ has a first differential coefficient everywhere within and along the boundary; we are now in fact able easily to express this first derived function by means of a definite integral. For we have:

$$\frac{f(t+\Delta t)-f(t)}{\Delta t} = \frac{1}{2i\pi}\int f(z)\cdot\frac{dz}{(z-t-\Delta t)(z-t)}.$$

When Δt converges to zero, since the quantity on the right side of this equation is also a continuous function of Δt, we obtain:

II. $$f'(t) = \frac{1}{2i\pi}\int \frac{f(z)}{(z-t)^2}\,dz,$$

where this integration round t may be along any path that remains within the domain originally defined. But a further consequence is: that the derived function $f'(t)$ itself is analytic everywhere in the interior; for we have:

$$\frac{f'(t+\Delta t)-f'(t)}{\Delta t} = \frac{1}{2i\pi}\int f(z)\,\frac{2(z-t)-\Delta t}{(z-t-\Delta t)^2(z-t)^2}\,dz,$$

therefore:

III. $$f''(t) = \frac{\underline{|2}}{2i\pi}\int \frac{f(z)}{(z-t)^3}\,dz.$$

In like manner all the successive higher derived functions are found for each point situated within the domain and we have:

IV. $$f^n(t) = \frac{\underline{|n}}{2i\pi}\int \frac{f(z)}{(z-t)^{n+1}}\,dz.$$

A function that is analytic in a simply connected domain therefore is not only itself finite and continuous, as is likewise its first derivate,

---

*) This theorem must not be taken to mean that we may arbitrarily assume the values of the function $f(z)$ along the boundary and then calculate the value for each internal point by equation I. On the contrary the values of the function at the boundary, as we can easily conceive, must satisfy antecedent conditions in order that they may give rise to an analytic function for the interior. The investigations concerning this question — definition of a function by conditions of boundary and of continuity — are collectively comprehended under the title: the Principle of Dirichlet.

but it has also for each point within the domain higher derivates, and these all, as many as may be formed, are analytic functions.

186. Equation I. leads also to the expansion of the function $f(t)$ in an infinite series of ascending positive integer powers.

Let any point a be selected subject only to the condition: that the greatest circle with a as centre which does not go outside the boundary curve of the domain shall include the point t for which the values of the function are to be calculated. Let this circle be adopted as curve of integration of Equation I., then for every point z on its circumference we have:

$$\text{abs}[t-a] < \text{abs}[z-a].$$

Accordingly

$$\frac{1}{z-t} = \frac{1}{z-a} \cdot \frac{1}{1-\frac{t-a}{z-a}} = \frac{1}{z-a}\left\{1 + \frac{t-a}{z-a} + \frac{(t-a)^2}{(z-a)^2} + \cdots \frac{(t-a)^n}{(z-a)^n} \cdots\right\}$$

is a convergent series. By the Lemma proved in § 182, 3, therefore:

$$\int \frac{f(z)}{z-t}\,dz = \int \frac{f(z)}{z-a}\,dz + (t-a)\int \frac{f(z)}{(z-a)^2}\,dz$$

$$+ (t-a)^2 \int \frac{f(z)}{(z-a)^3}\,dz + \cdots (t-a)^n \int \frac{f(z)}{(z-a)^{n+1}}\,dz \cdots,$$

and we obtain the equation:

$$f(t) = \frac{1}{2i\pi}\left\{\int \frac{f(z)}{z-a}\,dz + (t-a)\int \frac{f(z)}{(z-a)^2}\,dz\right.$$

$$\left. + (t-a)^2 \int \frac{f(z)}{(z-a)^3}\,dz \cdots + (t-a)^n \int \frac{f(z)}{(z-a)^{n+1}}\,dz \cdots\right\},$$

that can also in consequence of Equations I.—IV. be written in the form:

$$\text{V. } f(t) = f(a) + (t-a)f'(a) + \frac{(t-a)^2}{\underline{|2}} f''(a) + \cdots \frac{(t-a)^n}{\underline{|n}} f^n(a) \cdots + \text{etc.}.$$

This series of powers certainly converges absolutely for every value of t situated within the greatest circle round the centre a that does not go outside of the boundary curves of the original domain. This is Taylor's expansion for a complex function.

The contents of equation V. may be stated in the following words:

Given the value of a function and of each of its successive derivates at a point a, if we know that the function is analytic within a circle round the centre a, the value of the function will be calculated for every point within this circle by means of the infinite series of ascending positive integer powers V.;

or: The value of a function and of each of its successive derivates being given at a point a, it is an analytic function in the neighbourhood of this point, only, when a circle of arbitrarily small finite radius can be assigned round a as centre, within which the series V. is convergent.

The series V. will generally not converge for every value t within the region originally assumed, wherein $f(z)$ was supposed analytic. But we can vary the centre from a so as to arrive at a circle, and with it at an expansion, that will include the required point. For, if this point t be at a finite distance however small beyond the boundary of the circle, let us draw from the point a to t any curve that keeps always at a finite distance from the boundary curves. The circle whose centre is a will meet this curve in a point a' between a and t. The function f and arbitrarily many of its successive derivates can be calculated from series V. for a point on the curve and arbitrarily near a' within the circle; this point can then be made the centre of a new expansion all whose coefficients are known. The point a'' in which the new circle of convergence crosses the curve is of course nearer to t; and the continuation of this process must ultimately lead to a circle that includes the point t, since the radii cannot become infinitely small, because the path $a\,a'a'' \ldots t$ is always at a finite distance from the boundary curves.

This process of continuation can also be employed in the case of an analytic function defined only by a series of powers, in order to extend it out beyond its convergency into any domain not including a singular point. (§ 87.)

Each such series of powers defines the function for a determinate circular convergency and is called an element of the function. According to the centre chosen for the expansion, different elements of the function are obtained, moreover, one and the same value of the argument belongs to different elements. But when the function is unique, its different elements must lead to the same value for the same point. When in this extension of the function by its elements we do not pass out beyond some limited connected domain, the function exists only for this domain.

Any pair of analytic functions of the complex variable defined arbitrarily within given domains are to be considered as belonging to the same function, only, when the elements of one of the functions can be derived from those of the other.

In this sense an analytic function is completely determined when its value and the values of each of its derived functions at a point are known, or in other words, when the values of the function are given along a line however short. For then its derived functions can be determined. Functions that are not thus connected, are to be regarded as independent of each other.*)

*) Hankel: Untersuchungen über die unendlich oft oscillirenden und unstetigen Functionen, p. 44 etc., 49 etc.; reprinted, Math. Annalen, XX, p. 104 etc., p. 109 etc.. Weierstrass, Monatsberichte der Berliner Akademie. 1870 August.

187. If we know that a function loses the property of a unique analytic function at certain points $c_1, c_2, \ldots$ of the plane, and we desire to expand this function in a series of powers in the neighbourhood of a point a, the circle of convergence may only be so large as to pass through one or more of the points c, but must include none of them. For, the series of powers is a unique analytic function in the entire circle, while by hypothesis the function loses this property in each point c.

As soon therefore as we know the properties of a function, we are in a position to foretell the extent of the convergencies of its elements. This we proceed to illustrate by the functions already studied.

I. It is known regarding the exponential function

$$e^{x+iy} = e^x (\cos y + i \sin y),$$

that it is analytic in the entire plane, for every finite value of $x + iy$. There exists round each point a an expansion with an arbitrarily great convergency, the series is:

$$e^z = e^a \left\{1 + \frac{z-a}{1} + \frac{(z-a)^2}{1 \cdot 2} + \cdots\right\}.$$

II. The function $l(z)$, which we have defined as the inverse of the exponential function, is a many-valued function of z. Its values can be uniquely coordinated to the points of a winding surface with infinitely many leaves that are connected along branching sections from the point 0 to ∞. When we describe with a centre a in any leaf a circle including the point zero, the logarithm is not a unique analytic function within this domain; moreover the domain itself is not closed according to the idea we formed of the winding surface, for its boundary circle crosses the branching section an odd number of times and does not lead back to the original point, but into a different leaf. But when we describe round a as centre a circle that at the utmost passes through the branching point $z = 0$, the function is everywhere analytic within this circle, and the circle itself is closed, although parts of it may possibly be in different leaves. The function $l(z)$ must therefore admit of expansion by powers of $(z - a)$, and because its successive derived functions are $\frac{1}{z}, \frac{-1}{z^2}, \frac{\underline{|2}}{z^3}, \frac{-\underline{|3}}{z^4}$, etc., the series is, for abs $[z - a] <$ abs a:

$$l(z) = l(a) + \frac{z-a}{a} - \frac{(z-a)^2}{2a^2} + \frac{(z-a)^3}{3a^3} - \frac{(z-a)^4}{4a^4} + \cdots \text{ etc.}.$$

This series expresses that value of the logarithm into which the value assumed for $l(a)$ changes continuously along any path within the circle of convergence. When we extend the function $l(z)$ out beyond this circle of convergence, by adopting some new point a' within it as centre of the expansion:

$$l(z) = l(a') + \frac{z-a'}{a'} - \frac{(z-a')^2}{2a'^2} + \frac{(z-a')^3}{3a'^3} - \cdots, \text{ abs } [z - a'] < \text{abs } a',$$

we can calculate the corresponding logarithm for each point of the plane, and the value thus obtained is that actually lying in the same leaf as $l(a)$ when we join the points a and z by a curve which does not cross the branching section, and then choose successive points of this curve as centres of expansion until we arrive at a circle which includes the point z, and in which the curve drawn from its centre to z is not intersected by the branching section.

III. The binomial $(1+z)^n$, when the number n is a rational fraction $p:q$, is a q-valued function, whose branching points are $z=-1$ and $z=\infty$. This is presented as a unique function of z by means of a surface with q leaves that are connected along a branching section from -1 to ∞. Taking any leaf and selecting any point a in it as centre of expansion, its convergency will be a circle that at the utmost passes through, but does not include, the branching point $z=-1$; the series is:

$$(1+z)^n=(1+a)^n+\frac{n}{1}(z-a)(1+a)^{n-1}+\frac{n(n-1)}{\underline{|2}}(z-a)^2(1+a)^{n-2}$$
$$+\frac{n(n-1)(n-2)}{\underline{|3}}(z-a)^3(1+a)^{n-3}+\cdots,\ \text{abs}[z-a]<\text{abs}[-1-a].$$

This series expresses that value of the root at the point z, which proceeds by continuous change from the value assumed for $(1+a)^n$ along any path within the circle of convergence. In particular, if a be the origin of coordinates and we choose for 1^n the simple value 1, we have:

$$(1+z)^n=1+\frac{n}{1}z+\frac{n(n-1)}{\underline{|2}}z^2+\frac{n(n-1)(n-2)}{\underline{|3}}z^3+\cdots,\ (\text{abs}\,z<1).$$

When the exponent n is a complex number, this series still holds unchanged; we have then

$$(1+z)^n=e^{n\,l(1+z)}$$

an infinitely many-valued function; selecting for $z=0$ the value 1, the above series presents that value of $e^{n\,l(1+z)}$ which proceeds from the number 0 by continuous change when the point z moves ex. gr. along a radius from the origin. We can fix this value uniquely in the form:

$$e^{n\left\{\frac{1}{2}l(\overline{1+x}^2+y^2)+i\tan^{-1}\frac{y}{1+x}\right\}};$$

restricting, as usual, the angle whose tangent is named, to mean a quantity between $-\frac{1}{2}\pi$ and $+\frac{1}{2}\pi$.

188. The implicit algebraic function w, which is defined by the equation $f(z^m, w^n)=0$, was proved in Book II, Chapter IV, to be an n-valued function with a certain finite number of critical points and non-essential singular points. By the word critical we designate all those points z for which $\frac{\partial f(z,w)}{\partial w}$ is also zero, because the value of w

belonging to each is a multiple root. In the neighbourhood of any other (regular) point each branch of w is an analytic function, and consequently with any regular point a as centre must admit of an expansion whose convergency extends at least as far as it can without including any non-essential singular or critical point. Denoting by w_a' one of the values, necessarily simple, belonging to the point a, the expansion is:

$$w = w_a' + (z - a)\left(\frac{dw}{dz}\right)_{a,w_a'} + \frac{(z-a)^2}{\underline{|2}}\left(\frac{d^2w}{dz^2}\right)_{a,w_a'} + \cdots \frac{(z-a)^n}{\underline{|n}}\left(\frac{d^nw}{dz^n}\right)_{a,w_a'} + \cdots,$$

where $\left(\frac{d^nw}{dz^n}\right)_{a,w_a'}$ denotes the value of the nth derived function formed at the point $z = a$, $w = w_a'$. We know a circle of convergence of this series, when we have previously determined all such singular and critical points of the function.

It has now become necessary to find the nth derived function. We showed in § 94 how the first derivate of the implicit function is calculated:

$$\frac{dw}{dz} = -\frac{\partial f}{\partial z} : \frac{\partial f}{\partial w},$$

and that it has a finite determinate value for every regular point because $\frac{\partial f}{\partial w}$ is not zero. Every higher derived function is found from this by successive differentiation: the partial differential coefficients $\frac{\partial f}{\partial z}$, $\frac{\partial f}{\partial w}$, alike with f, are algebraic expressions in z and w; consequently, as long as w is a unique analytic function of z, they are likewise unique analytic functions of z. Therefore also their quotient is an analytic function of z, and by the same rule of differentiation we find:

$$\frac{d^2w}{dz^2} = -\frac{\frac{\partial f}{\partial w}\left(\frac{\partial^2 f}{\partial z^2} + \frac{\partial^2 f}{\partial w\,\partial z}\frac{dw}{dz}\right) - \frac{\partial f}{\partial z}\left(\frac{\partial^2 f}{\partial z\,\partial w} + \frac{\partial^2 f}{\partial w^2}\frac{dw}{dz}\right)}{\left(\frac{\partial f}{\partial w}\right)^2}.$$

The Theorem of the interchange of the order of differentiations holds here, for, the algebraic functions $\frac{\partial^2 f}{\partial z\,\partial w}$, $\frac{\partial^2 f}{\partial w\,\partial z}$ are continuous functions of the variables z and w; therefore, inserting its value for $\frac{dw}{dz}$, we have:

$$\frac{d^2w}{dz^2} = -\frac{\frac{\partial^2 f}{\partial z^2}\left(\frac{\partial f}{\partial w}\right)^2 - 2\frac{\partial^2 f}{\partial z\,\partial w}\frac{\partial f}{\partial z}\frac{\partial f}{\partial w} + \frac{\partial^2 f}{\partial w^2}\left(\frac{\partial f}{\partial z}\right)^2}{\left(\frac{\partial f}{\partial w}\right)^3}.$$

We have then in this expression on the right to substitute for z the value a and for w the value w_a'.

Successive differentiation of this quotient, a calculation without difficulty but leading to long formulas, presents as many derived functions as we please. It is essential that in all of them the only quantity occurring in the denominator is $\frac{\partial f}{\partial w}$, which does not vanish as long as we keep in a domain of regular points, so that the values of all the derived functions are finite and determinate, since the points are excluded in which w becomes infinite.

Although we have thus indicated that which is essential in the equations whereby the derived functions are calculated, it is still important for a subsequent application of the formulas, to introduce some abridgments of notation which give us a comprehensive view of the equations that are to be formed.*) Let us denote the derivate $\frac{d^k w}{dz^k}$ by w_k, the partial derived function $\frac{\partial^k f(z,w)}{\partial z^{k-p}\partial w^p}$ by $f_{k-p,p}$, and let us write:

$$\frac{1}{\underline{|k}}\sum_{p=0}^{p=k} k_p f_{k-p,p} w_1^p = \Phi_k;$$

then, for the ratio $\frac{w-w_a'}{z-a}=w_1$, which for $z=a$ becomes the derivate $\frac{dw}{dz}$, we have the equation:

$$0=\Phi_1+(z-a)\Phi_2+(z-a)^2\Phi_3+\cdots,$$

ascending by powers of z to the term $(z-a)^m$ or $(z-a)^n$ according as m is greater or is less than n.

From the equation $w-w_a'=w_1(z-a)$ we find that:

$$\frac{d^k w}{dz^k}=(z-a)\frac{d^k w_1}{dz^k}+k\frac{d^{k-1}w_1}{dz^{k-1}};$$

so that for $z=a$ we have the relation

$$\left(\frac{d^k w}{dz^k}\right)_a=k\frac{d^{k-1}w_1}{dz^{k-1}}.$$

Accordingly the higher derived functions w_2, w_3, etc., are found by differentiating the above equation totally with respect to z, and putting $z=a$, $\frac{d^{k-1}w_1}{dz^{k-1}}=\frac{1}{k}\cdot w_k$ in the derived equations. Only the terms up to the power $(z-a)^{k-1}$ in the equation are required in establishing the recurring formula for the kth derived function. For the lowest values of k we find the following equations:

*) Plücker (1801—68): Theorie der algebraischen Curven, p. 156. Bonn 1839. Liouville (1809—82): Mémoire sur quelques propositions générales de géométrie et sur la théorie de l'élimination dans les équations algébriques. J. de math., T. VI.

$$\Phi_1 = 0,$$

$$\frac{1}{2}\frac{\partial\Phi_1}{\partial w_1} w_2 + \Phi_2 = 0,$$

$$\frac{1}{\underline{|3}}\frac{\partial\Phi_1}{\partial w_1} w_3 + \frac{1}{2}\frac{\partial\Phi_2}{\partial w_1} w_2 + \Phi_3 = 0,$$

$$\frac{1}{\underline{|4}}\frac{\partial\Phi_1}{\partial w_1} w_4 + \frac{1}{\underline{|3}}\frac{\partial\Phi_2}{\partial w_1} w_3 + \frac{1}{2}\frac{\partial^2\Phi_2}{\partial w_1^2}\left(\frac{w_2}{2}\right)^2 + \frac{1}{2}\frac{\partial\Phi_3}{\partial w_1} w_2 + \Phi_4 = 0,$$

$$\frac{1}{\underline{|5}}\frac{\partial\Phi_1}{\partial w_1} w_5 + \frac{1}{\underline{|4}}\frac{\partial\Phi_2}{\partial w_1} w_4 + \frac{1}{2\underline{|3}}\frac{\partial^2\Phi_2}{\partial w_1^2} w_2 w_3 + \frac{1}{\underline{|3}}\frac{\partial\Phi_3}{\partial w_1} w_3$$
$$+ \frac{1}{2}\frac{\partial^2\Phi_3}{\partial w_1^2}\left(\frac{w_2}{2}\right)^2 + \frac{1}{2}\frac{\partial\Phi_4}{\partial w_1} w_2 + \Phi_5 = 0.$$

. . . .

It is evident in this form also, that, for the regular point, in which

$$\frac{\partial\Phi}{\partial w_1} = f_{0,1} = \frac{\partial f}{\partial w}$$

does not vanish, all the differential coefficients as many as we may form, continue finite.

Returning to the main result of the investigation of the algebraic function in a regular point, we formulate it in the theorem:

Knowing a single value w_a' at a regular point $z = a$ of an algebraic function w given by $f(z^m, w^n) = 0$, we can calculate the value of the function belonging to any other point z within the regular domain round a, on expanding arbitrarily many terms of the series of powers by a determinate succession of possible arithmetical operations.

We may frame the theorem still more completely: *Knowing every critical and singular point of an algebraic function given by $f(z^m, w^n) = 0$, and furthermore one of its values at a regular point, we can by a determinate succession of finite arithmetical operations resolve the n-valued function by means of branching sections into n branches, each of which is discontinuous only along the branching sections and becomes infinite only in non-essential singular points; moreover, for any arbitrary point every branch can have its value calculated by means of arbitrarily many terms of an infinite series of ascending positive integer powers.*

We arrive at this theorem by means of the method of continuous extension of a function out beyond its circle of convergence repeatedly described in the foregoing examples. Let the value of the function given for a be denoted as belonging to the first leaf. From the point a let us draw curves, each until it is arbitrarily near to one critical point, but keeping at a finite distance from every other, and let us surround that critical point by an arbitrarily small circle. Such a curve along with the small circle is called a loop. By expansion we can establish whether the circuit of the entire loop introduces a change of the value of the function belonging to a; if it do not, the point is not a branching point for this first leaf; if it do, let us establish by repeated expansion, employing the newly found value for the point $z = a$, how many leaves are connected with the first in the point. Let these be called $1, 2, \dots p$, then for each of these leaves the significance of any other critical point can be determined by means of the loops. In this way too we shall become acquainted in another branching point with the value for $z = a$ belonging to a new leaf that did not occur among the values $1, 2, \dots p$.

If the case were to occur, that the circuit of all the loops does not lead beyond the leaves 1 to p, or, in general, that all the n values which belong to $z = a$ are not obtained in this process, this signifies, as we shall subsequently prove (§ 199), that the algebraic function $f(z^m, w^n)$ may be resolved into factors which are rational in z and w. In the case of an irreducible function this does not occur, the theorem holds good for it as above enunciated.

189. Having established that analytic functions can be expressed by series of ascending positive integer powers, it is still necessary that we should discuss their singular points, in order that we may ascertain how analytic functions admit of expansion in the neighbourhood of any singular point according to its kind.

We must first of all premise that in a domain wherein a function is generally a unique analytic function, it can have no other singularities than the non-essential and essential points in which it becomes infinite.

For, if an analytic function $f(z)$ have finite discontinuities in separate points α of a domain, the product $(z - \alpha)f(z)$ is an analytic function in the neighbourhood of and including the point α at which it is zero, and its first derived function is equal to $\operatorname{Lim} f(z)$ for $z = \alpha$. Accordingly there is an expansion of the form:

$$(z - \alpha)f(z) = \operatorname*{Lim}_{z=\alpha} f(z) \cdot (z - \alpha) + a_2(z - \alpha)^2 + a_3(z - \alpha)^3 + \cdots \text{etc.}.$$

Therefore the value of the function $f(z)$ at the point α must be $\operatorname*{Lim}_{z=\alpha} f(z)$, thus this point is regular.

It is likewise impossible (see § 181. 3. c) that an analytic function while itself finite and continuous in a domain should become singular in separate points or along an entire curve in consequence of any violation of the equation $\frac{\partial f}{\partial x} + i\frac{\partial f}{\partial y} = 0$ between its partial derivates. The integral $\int f(z)dz$ even up to any point α of the irregular curve would then be an analytic function; for the integral is continuous and its derived function is $\operatorname{Lim} f(z)$ for $z = \alpha$. But an analytic function, as was proved, has arbitrarily many successive derivates all of which are continuous functions. Therefore $f(z)$ must have a determinate derived function even at the point α, and accordingly at that point the equation $\frac{\partial f}{\partial x} + i\frac{\partial f}{\partial y} = 0$ must be satisfied.

In the unique analytic function, therefore, only essential and non-essential singular points have to be considered, and in these all the successive derived functions will be seen also to become infinite.

Let a circle described round the centre α with the radius R be given. Suppose we know that a function $f(z)$ is regular in the point α and is analytic within the entire domain inclusive of the boundaries, except at the points $c_1, c_2, \ldots c_n$ within this circle which are to be singular points, essential or non-essential. When a concentric circle is described round α that excludes all the points c, a series can be assigned within this smaller circle, ascending by positive integer powers of $z - \alpha$, whose coefficients are the derived functions at the point α. We are now, however, about to show that there also exists an expansion valid for the entire domain of radius R, no longer containing only positive integer powers of z with the derived functions at the point α as their coefficients, but enabling us to calculate the function for every point z within the circle R, so that therefore there is no need of extensions of the series of powers, as in the case of the previous expansion.

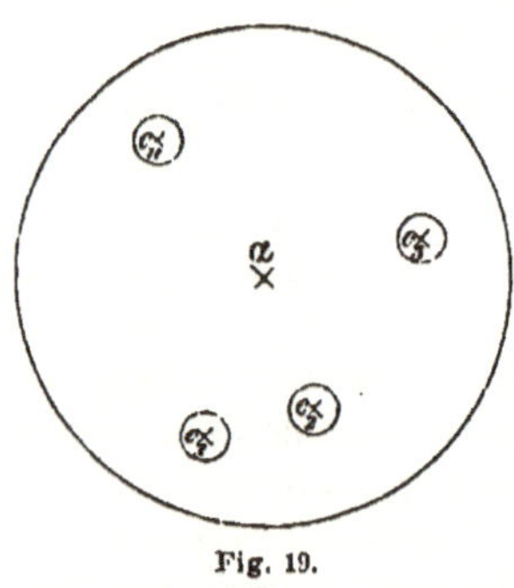

Fig. 19.

Let us enclose the points $c_1, c_2, \ldots c_n$ in circles of arbitrarily small radius ϱ; then within the one-leaved but multiply connected surface, $f(z)$ is an analytic function; and the sum of the integrals taken round the circle α and round the circles $c_1, c_2, \ldots c_n$, keeping the surface on the left, is zero. We can express this by the equation:

$$\int_{(\alpha)} f(z)\,dz = \int_{(c_1)} f(z)\,dz + \int_{(c_2)} f(z)\,dz + \cdots + \int_{(c_n)} f(z)\,dz,$$

integrating now round each circle c so as to keep its area on the left.

When t signifies an arbitrary point within the multiply connected surface, $\frac{f(z)}{z-t}$ is in this surface an analytic function alike with $f(z)$, only that it has the point $z = t$ as a non-essential singular point. Hence we have also:

$$\int_{(\alpha)} \frac{f(z)}{z-t}\,dz = \int_{(t)} \frac{f(z)}{z-t}\,dz + \int_{(c_1)} \frac{f(z)}{z-t}\,dz + \int_{(c_2)} \frac{f(z)}{z-t}\,dz \cdots + \int_{(c_n)} \frac{f(z)}{z-t}\,dz,$$

or:

$$\int\limits_{(t)}\frac{f(z)}{z-t}\,dz=\int\limits_{(\alpha)}\frac{f(z)}{z-t}\,dz-\int\limits_{(c_1)}\frac{f(z)}{z-t}\,dz-\int\limits_{(c_2)}\frac{f(z)}{z-t}\,dz\cdots-\int\limits_{(c_n)}\frac{f(z)}{z-t}\,dz.$$

Now on the left side of this equation we have the integral:

$$\int\limits_{(t)}\frac{f(z)}{z-t}\,dz=2i\pi f(t).$$

But the integrals on its right can be expressed by series. Since for the first of them z signifies a point upon the circle R, we have $\operatorname{mod}[z-\alpha]>\operatorname{mod}[t-\alpha]$; therefore:

$$\frac{1}{z-t}=\frac{1}{z-\alpha}\cdot\frac{1}{1-\dfrac{t-\alpha}{z-\alpha}}=\frac{1}{z-\alpha}\left\{1+\frac{t-\alpha}{z-\alpha}+\frac{(t-\alpha)^2}{(z-\alpha)^2}+\frac{(t-\alpha)^3}{(z-\alpha)^3}+\cdots\right\}$$

is an absolutely convergent series. Accordingly:

$$\text{VI.}\int\limits_{(\alpha)}\frac{f(z)}{z-t}\,dz=\int\limits_{(\alpha)}\frac{f(z)}{z-\alpha}\,dz+(t-\alpha)\int\limits_{(\alpha)}\frac{f(z)}{(z-\alpha)^2}\,dz+(t-\alpha)^2\int\limits_{(\alpha)}\frac{f(z)}{(z-\alpha)^3}\,dz+\cdots;$$

but the coefficients of this series are no longer the derived functions at the point α as before, although they have the same form; for the circle of radius R, round which these integrals are formed, can no longer be arbitrarily contracted about the point α.

For each integral round a point c: $\operatorname{mod}[t-c]>\operatorname{mod}[z-c]$, because t is a point outside its circle ϱ; consequently:

$$\frac{1}{z-t}=\frac{-1}{t-c}\cdot\frac{1}{1-\dfrac{z-c}{t-c}}=-\frac{1}{t-c}\left\{1+\frac{z-c}{t-c}+\frac{(z-c)^2}{(t-c)^2}+\frac{(z-c)^3}{(t-c)^3}+\cdots\right\}$$

is an absolutely convergent series; and we have: $-\int\limits_{(c)}\frac{f(z)}{z-t}\,dz=$

$$\text{VII.}\quad+\frac{1}{t-c}\int\limits_{(c)}f(z)dz+\frac{1}{(t-c)^2}\int\limits_{(c)}f(z)(z-c)dz+\frac{1}{(t-c)^3}\int\limits_{(c)}f(z)(z-c)^2dz+\text{etc.}.$$

Denoting briefly the coefficients of series VI. divided by $2i\pi$ by $A_0, A_1, A_2, \ldots$; likewise those of series VII., referring to the point c_k, by $C_1^{(k)}, C_2^{(k)}, C_3^{(k)}, \ldots$, and in this the definite integrals can be taken along a circle of arbitrarily small radius; the result is: For all points t within the circle R, in which $c_1, c_2, \ldots c_n$ are singular points for the unique analytic function $f(z)$, we have the expansion:

$$\begin{aligned}\text{VIII.}\quad f(t)&=A_0+A_1(t-\alpha)+A_2(t-\alpha)^2+A_3(t-\alpha)^3+\cdots\\&+\frac{C_1'}{t-c_1}+\frac{C_2'}{(t-c_1)^2}+\frac{C_3'}{(t-c_1)^3}+\frac{C_4'}{(t-c_1)^4}+\cdots\\&\quad\cdots\cdots\cdots\cdots\cdots\\&+\frac{C_1^{(n)}}{t-c_n}+\frac{C_2^{(n)}}{(t-c_n)^2}+\frac{C_3^{(n)}}{(t-c_n)^3}+\frac{C_4^{(n)}}{(t-c_n)^4}+\cdots\end{aligned}$$

in series proceeding both by positive and negative integer powers of t.

From this expansion moreover we realise how the two kinds of singular points differ in the nature of their expansions in powers.

The non-essential was characterised by the possibility of assigning some positive integer m, for which $f(z)(z-c)^m$ at the point c is equal to a finite quantity. Hence follows: When c is a non-essential point, of infinitude m, all the coefficients higher than C_m vanish for it. In fact, because the radius ϱ can be chosen so small that $f(z)(z-c)^m$ shall differ inappreciably from a finite number G, the integral

$$\int\limits_{(c)} f(z)(z-c)^{m+k}dz$$

will differ in value inappreciably from

$$G\int\limits_{(c)}(z-c)^k dz,$$

and the value of this integral is zero. Conversely, if for a point c all the higher coefficients $C_{m+1}, C_{m+2}, \ldots$ vanish, it must evidently be a non-essential singular point, because then $(z-c)^m f(z)$ remains finite for $z=c$.

Every other point in which $f(z)$ becomes infinite is an essential singular point, and we now see that the essential singular point in any analytic function behaves as it did in the exponential series where we first took cognisance of it: there exists for its neighbourhood an infinite series proceeding by negative powers.

The value of the reciprocal function $\frac{1}{f(z)}$ will be zero at every non-essential singular point of $f(z)$. For, when $(z-c)^m f(z)=G$ for $z=c$, we have $\frac{1}{f(z)}=\frac{(z-c)^m}{G}=0$ for $z=c$. But an essential singular point remains such also for the reciprocal function. For if in the region of an essential singular point the reciprocal function $\frac{1}{f(z)}$ were regular, the same should be the case also with $f(z)$; and if the singularity of $\frac{1}{f(z)}$ were only non-essential, $f(z)$ should be regular. Any essential singular point of $f(z)$ is also essentially singular for the function $\frac{1}{f(z)-C}$, C denoting any arbitrary constant; we conclude hence: Round the essential singular point a circle of radius ϱ can be assigned, that certainly includes points at which $f(z)$, $\frac{1}{f(z)}$, $\frac{1}{f(z)-C}$ each becomes greater in amount than any arbitrarily prescribed number K however great; i. e. points at which the amount of $f(z)$ exceeds K, as well as points at which it differs from zero, or from any number C, by less than the arbitrarily small quantity $\frac{1}{K}$. In other words: The value of a function at any essential singular point is completely indeterminate between infinite limits, in the vicinity of the point it converges to every arbitrary value. In the case of the exponential function this property was already pointed out. § 82, 4.

The integral of $f(z)$ taken round the arbitrarily close boundary of an essential or non-essential singular point c is zero, only when, in the expansion relative to the point c, the coefficient of the term $\frac{1}{z-c}$ is zero. The integral taken up to the singular point always becomes infinite; its infinitude is $m-1$ for a non-essential point of $f(z)$ of the order m, and for a non-essential point of the order 1 the integral becomes logarithmically infinite.

190. We have now to examine how the domain of $f(z)$ admits of extension out beyond the domain R: Let $R' > R$ be the radius of a circle round the centre a, in it let $f(z)$ be generally an analytic function, only with the additional singular points $c_{n+1}, \ldots c_{n+k}$; then in Expansion VIII. the coefficients $A_0, A_1, A_2 \ldots$ change, while all the coefficients C remain as they were, and new terms arise by the points $c_{n+1}, \ldots c_{n+k}$.

Assuming first that the function $f(z)$ is not singular at infinity, but converges for $z=\infty$ to a determinate finite value G, and that it has in all cases a finite number m of singular points in the entire plane; then the integrals in Series VI. that determine the coefficients A all converge to zero, for R can be chosen so great that $f(z)$ shall differ inappreciably from the finite number G, accordingly we have also:

$$\operatorname{mod}\int\frac{f(z)\,dz}{(z-a)^n} < \int_0^{2\pi}\frac{\operatorname{mod} G}{R^{n-1}}\,d\varphi$$

arbitrarily small. Only, the value of the first term:

$$A_0 = \frac{1}{2i\pi}\int\frac{f(z)}{z-a}\,dz$$

is equal to G.

Therefore a unique analytic function which has m singular points in the entire plane and behaves regularly at infinity is of the form:

$$f(t) = \frac{C_1'}{t-c_1} + \frac{C_2'}{(t-c_1)^2} + \frac{C_3'}{(t-c_1)^3} + \cdots + \frac{C_1^{(m)}}{t-c_m} + \frac{C_2^{(m)}}{(t-c_m)^2} + \frac{C_3^{(m)}}{(t-c_m)^3} + \cdots + G.$$

A function which has only finite non-essential singular points and is regular at infinity, differs from a proper fractional rational function only by an additive constant. For then the terms that refer to the various points c all come to an end, and when we reduce them to a common denominator, the order of the numerator is at least one less than that of the denominator.

When the function $f(z)$ has a non-essential singular point at infinity, and when moreover the number of its singular points in the entire plane is finite, let us convert it by the substitution $z - \beta = \frac{1}{z'}$, or by $z = \frac{1}{z'}$ if the origin $z = 0$ be not singular, into a function of z' that behaves regularly at infinity. For this function then, since the new origin is now a non-essential singular point, the expansion:

$$f\left(\frac{1}{t'}\right) = \frac{C_1'}{\frac{1}{t'} - c_1} + \frac{C_2'}{\left(\frac{1}{t'} - c_1\right)^2} + \frac{C_3'}{\left(\frac{1}{t'} - c_1\right)^3} + \cdots + \frac{C_1^{(m)}}{\frac{1}{t'} - c_m}$$

$$+ \frac{C_2^{(m)}}{\left(\frac{1}{t'} - c_m\right)^2} + \frac{C_3^{(m)}}{\left(\frac{1}{t'} - c_m\right)^3} + \cdots + G + \frac{K_1}{t'} + \frac{K_2}{t'^2} + \cdots \frac{K_n}{t'^n},$$

is valid in the entire plane, i. e.:

$$f(z) = \frac{C_1'}{z - c_1} + \frac{C_2'}{(z - c_1)^2} + \frac{C_3'}{(z - c_1)^3} + \cdots + \frac{C_1^{(m)}}{z - c_m} + \frac{C_2^{(m)}}{(z - c_m)^2}$$

$$+ \frac{C_3^{(m)}}{(z - c_m)^3} + \cdots + G + K_1 z + K_2 z^2 + \cdots K_n z^n.$$

Hence follows: *A function that has a non-essential singular point at infinity, and a finite number of finite non-essential singular points, is an improper fractional rational function.* When it has no finite singular point, it is an integer function.

Lastly, when the point infinity is an essential singular point, the expansion:

$$\frac{K_1}{t'} + \frac{K_2}{t'^2} + \cdots \frac{K_n}{t'^n} + \cdots$$

in the last series does not come to an end; therefore:

$$K_1 z + K_2 z^2 + \cdots K_n z^n + \cdots$$

also is an infinite series that converges for every finite value of z. We have thus attained the most general form of the expansion:

When $f(z)$ is an analytic function that has the non-essential or essential singular points $c_1, c_2, \ldots c_m$ in the finite plane, and also an essential singular point at infinity, the expansion:

$$f(z) = \frac{C_1'}{z - c_1} + \frac{C_2'}{(z - c_1)^2} + \frac{C_3'}{(z - c_1)^3} + \cdots + \frac{C_1^{(m)}}{z - c_m} + \frac{C_2^{(m)}}{(z - c_m)^2}$$

$$+ \frac{C_3^{(m)}}{(z - c_m)^3} + \cdots + G + K_1 z + K_2 z^2 + K_3 z^3 + \cdots,$$

is valid for every finite value of z, and its coefficients C and K are to be calculated by means of definite integrals, K in particular from the form:

$$K_n = \frac{1}{2i\pi}\int_{\text{round } z'=0} f\left(\frac{1}{z'}\right) z'^{n-1}\,dz' = \frac{1}{2\pi}\,\text{Lim}\int f\left(\frac{1}{r}\,e^{-i\varphi}\right) r^n e^{in\varphi}\,d\varphi, \text{ for } r=0,$$

or also

$$= \frac{1}{2i\pi}\int \frac{f(z)}{z^{n+1}}\,dz, \text{ round the point } z=\infty.$$

When there are no finite singular points, $f(z)$ can be expressed by an infinite series of ascending positive integer powers:

$$f(z) = G + K_1 z + K_2 z^2 + K_3 z^3 + \cdots$$

which converges in the entire plane. *In this case $f(z)$ is said to be an integer transcendental function.**) To this class the exponential function belongs. The calculation of the coefficients K can moreover then be reduced to any curve round the origin, and as no singular points are included, we have:

$$K_n = \frac{1}{2i\pi}\int \frac{f(z)}{z^{n+1}}\,dz = \frac{1}{\underline{|n}}\,f^n(0).$$

The following statements supplement the theorems resulting from the expansion of the analytic function in a series of the Form VIII.:

An analytic function that has no singular point, and therefore nowhere becomes infinite, is a constant.

It cannot be an integer function either rational or transcendental, because either would have a singular point at infinity, non-essential in the former case and essential in the latter.

Further: *An analytic function without any essential singular point, that nowhere vanishes, is a constant.*

(The exponential function is zero at no finite point, its vanishing point coincides with its essential singularity.)

An analytic function must assume every arbitrary value at least once; otherwise it is a constant.

The value may possibly belong to the essential singular point.

An analytic function is determined when its values are given along a finite portion of a curve however short.

For then all the successive derivates of the function at a point can be calculated; therefore the series of powers in the neighbourhood of this point is obtained, and this can be extended out beyond its convergency (§ 186).

An analytic function is constant when it is constant along a finite portion of a curve however short.

For then all its derived functions at a point are zero.

---

*) The classification of transcendental analytic functions into integer and fractional with the subdivision of these according to the number of their essential singular points was given by Weierstrass in the Memoir cited in the foot-note p. 130. (Also p. 148.) In it he also showed how the functions may be developed analytically when the number of their non-essential singular points is unrestricted.

191. Inversion of the unique analytic function.

When the analytic function $f(z) = w$ has no singular point in a finite domain, its inverse function $z = \psi(w)$ is also analytic. For, since there is a determinate derivate:

$$\frac{dw}{dz} = f'(z),$$

there exists also a derived function of z with respect to w: namely

$$\frac{dz}{dw} = \frac{1}{f'(z)},$$

at each point within the region, and, putting $w = u + iv$, we have

$$\frac{\partial z}{\partial u} + i\frac{\partial z}{\partial v} = 0.$$

This derivate of z with respect to w can become infinite only at separate points at which $f'(z) = 0$, and these will be found to be branching points for the function $z = \psi(w)$.

We may examine this in detail as follows, in order to assure ourselves that the derived function $\frac{dz}{dw}$ depends uniquely on w. To a determinate value $z = \alpha$ belongs a determinate value $w = \beta$. On inversion, a finite number of different values of z can correspond to the single value $w = \beta$. Assuming then a determinate value for w and considering one of the values $z = \alpha$ belonging to it, we may enquire how z varies when the value of β is changed infinitesimally. We shall show that this variation is continuous and unique, by proving that the value of the derivate $\frac{dz}{dw}$ also is determinate as long as w does not pass through a point at which the corresponding value of z belongs to the equation $f'(z) = 0$. Again it is to be remarked, that this equation possesses a finite number of solutions in any closed domain; that to each solution belongs a determinate value of w, but that inversely to each such value of w can belong different values of z of which however in general only one will satisfy the equation $f'(z) = 0$.

The unique and continuous variation of z is perceived as follows. Let α be a point for which $f'(\alpha)$ does not vanish. Then denoting a neighbouring value of w by $\beta + \Delta w$, let the value $\alpha + \Delta z$ correspond to it, thus when the right side of the equation

$$\beta + \Delta w = f(\alpha + \Delta z)$$

is arranged by powers of Δz, we have:

$$\Delta w = \Delta z f'(\alpha) + \frac{\Delta z^2}{\underline{|2}} f''(\alpha) + \frac{\Delta z^3}{\underline{|3}} f'''(\alpha) + \cdots \text{etc.}.$$

For a given value of Δw within a finite domain for Δz, different values of Δz may possibly satisfy this equation; but since $f'(\alpha)$ is not zero, only one of these values of Δz can become zero when Δw converges to zero. The others converge to the values that satisfy:

$$f'(\alpha) + \frac{\Delta z}{\lfloor 2} f''(\alpha) + \frac{\Delta z^2}{\lfloor 3} f'''(\alpha) + \cdots = 0,$$

and the roots of this equation each increased by α are the remaining values of z that belong to the point $w = \beta$.

This one value of Δz depends continuously on Δw, inasmuch as the quotient $\frac{\Delta z}{\Delta w}$ tends to a determinate magnitude, for we have:

$$1 = \frac{\Delta z}{\Delta w}\left\{f'(\alpha) + \frac{\Delta z}{\lfloor 2} f''(\alpha) + \frac{\Delta z^2}{\lfloor 3} f'''(\alpha) + \cdots\right\},$$

therefore when Δz converges to zero, the limiting value is:

$$\frac{dz}{dw} = \frac{1}{f'(\alpha)}.$$

This holds for every point at which $f'(\alpha)$ is not zero. The function z can be continued in a determinate unique manner.

It is otherwise when we come to a point α for which $f'(\alpha) = 0$; we have then:

$$\Delta w = \frac{\Delta z^2}{\lfloor 2} f''(\alpha) + \frac{\Delta z^3}{\lfloor 3} f'''(\alpha) + \cdots,$$

or more generally

$$= \frac{\Delta z^m}{\lfloor m} f^m(\alpha) + \frac{\Delta z^{m+1}}{\lfloor m+1} f^{m+1}(\alpha) + \cdots,$$

and to the point $w = \beta$ belongs a value $z = \alpha$ to be counted doubly (or m-ly) as there are at the neighbouring point $\beta + \Delta w$ two values $z = \alpha + \Delta_1 z$ and $z = \alpha + \Delta_2 z$ (or m values) which converge to zero with Δw. Therefore when w arrives at a point for which $f'(z) = 0$, a branching point is reached and the function can no longer be continued uniquely.

In fact the point $w = \beta$, in whose neighbourhood we have:

$$\Delta w = \frac{\Delta z^m}{\lfloor m} f^m(\alpha) + \frac{\Delta z^{m+1}}{\lfloor m+1} f^{m+1}(\alpha) + \cdots,$$

is a branching point for the required function $z = \psi(w)$ in which m of its leaves are cyclically connected; or expressed otherwise: The point being enclosed by an arbitrarily small circle let Δw go through the values corresponding to the points of this circle, then by a single circuit, Δz will change from whichever of the m values it may have begun with into some other; when Δw repeats the circuit, Δz passes from this second value into a third; after $m - 1$ circuits, Δz assumes an mth value and only when m circuits are completed does it resume the original value. For, from the equation:

$$\frac{\Delta w}{\Delta z^m} = \frac{1}{\underline{|m}} f^m(\alpha) + \frac{\Delta z}{\underline{|m+1}} f^{m+1}(\alpha) + \frac{\Delta z^2}{\underline{|m+2}} f^{m+2}(\alpha) + \cdots,$$

which for a vanishing value of Δz passes over continuously into:

$$\frac{dw}{dz^m} = \frac{1}{\underline{|m}} f^m(\alpha),$$

it follows, that with the centre β a circle can be described so small that for each of its points the value of

$$\frac{\Delta z}{\underline{|m+1}} f^{m+1}(\alpha) + \frac{\Delta z^2}{\underline{|m+2}} f^{m+2}(\alpha) \cdots$$

shall be smaller than a quantity δ whose amount is arbitrarily small; therefore we have also

$$\frac{\Delta w}{(\Delta z)^m} = \frac{1}{\underline{|m}} f^m(\alpha) + (< \delta) = C + (< \delta).$$

When we put $\Delta w = re^{i\varphi}$ and write this last equation in the form:

$$\frac{e^{i\varphi}}{\left(\frac{\Delta z}{r^{\frac{1}{m}}}\right)^m} = C + (< \delta),$$

we see that the values of Δz divided by the positive value of the root $r^{\frac{1}{m}}$ are expressed with arbitrarily close approximation by:

$$e^{\frac{i\varphi}{m}} \cdot C^{-\frac{1}{m}}, \quad e^{\frac{i\varphi+2i\pi}{m}} \cdot C^{-\frac{1}{m}}, \ldots, e^{\frac{i\varphi+2(m-1)i\pi}{m}} \cdot C^{-\frac{1}{m}},$$

where $C^{-\frac{1}{m}}$ means one of its m possible values, the same throughout. When φ changes by 2π, i. e. when Δw completes a circuit, each of these values passes over into another, consequently the different values of $\Delta z : r^{\frac{1}{m}}$, which come arbitrarily near this value, must also interchange cyclically.

Accordingly: Considering a point α within the circle in which the function $f(z)$ is convergent, if we denote those points for which $f'(z) = 0$ by $\alpha_1, \alpha_2, \ldots \alpha_n$ (α itself must not be one of these), then round the centre α a circle can be described with radius r excluding these n points. Let us interpret the values of w in a second plane. To the centre α corresponds the point β. To all points of the circumference must correspond points of a determinate finite closed curve in the plane w surrounding the point β; moreover this curve cannot intersect itself, for then two different points z would be coordinated to the intersection w. To each point in the circle r corresponds a determinate point in the domain round β, but also inversely to each point in the domain round β only a single determinate point in the circle r. There must exist an expansion for z of the form:

$$z = \alpha + a_1(w-\beta) + a_2(w-\beta)^2 + a_3(w-\beta)^3 + \cdots = \psi(w);$$

this is convergent in a circle with the centre β that is altogether included within the bounded region round β, hence the radius of the circle of convergence is to be specified analytically as follows:

As z travels through all the points $\alpha + re^{i\varphi}$ let us determine the least modulus among the values $(w-\beta) = f(z) - f(\alpha)$ for these points; this minimum is the radius of convergence of the series ascending by powers of w.

The coefficients of the series are the successive derivates of z with respect to w formed for the point α; they are obtained by successively differentiating the equation

$$\frac{dz}{dw} f'(z) = 1$$

relatively to w, or in the form of the definite integrals ($k > 1$):

$$a_k = \frac{1}{2i\pi}\int \frac{\psi(w)\,dw}{(w-\beta)^{k+1}} = \frac{1}{2i\pi}\int \frac{(z-\alpha)f'(z)\,dz}{(f(z)-f(\alpha))^{k+1}},$$

taken round the point $z = \alpha$. In the numerator of the second integral we have been able to write $(z-\alpha)$ instead of z, because the integral of the part multiplied by α vanishes.

Since the successive differentiation of z with respect to w as independent variable leads to complicated recurring formulas, it is important that we should know how these coefficients may be otherwise expressed. Forming the difference $w - \beta$ by means of the equations:

$$w = f(z) = b_0 + b_1 z + b_2 z^2 + \cdots \quad \text{and} \quad \beta = b_0 + b_1\alpha + b_2\alpha^2 + \cdots,$$

we can arrange it by powers of $(z-\alpha)$, and are thus led to the problem of inverting the series of ascending positive integer powers:

$$w - \beta = (z-\alpha)\{c_0 + c_1(z-\alpha) + c_2(z-\alpha)^2 + \cdots\} = (z-\alpha)F(z-\alpha).$$

192. The function $F(z-\alpha) = \frac{f(z)-f(\alpha)}{z-\alpha}$ does not vanish at the point $z = \alpha$, for there its value is

$$\lim_{z=\alpha} \frac{f(z)-f(\alpha)}{z-\alpha} = f'(\alpha),$$

and it does not vanish anywhere in the convergency, because $w = \beta$ only for $z = \alpha$. Denoting the value of $\frac{1}{F(z-\alpha)}$ by $\varphi(z-\alpha)$, this is a function which can be expanded in a series in the neighbourhood of the point α. Our problem accordingly is reduced to the special form of one first solved by Lagrange: From the equation:

$$(w-\beta)\varphi(z-\alpha) = z - \alpha, \text{ or: } w'\varphi(z') = z'$$

to calculate z' as a function of w'. This problem leads to the former series:

$$z' = a_1 w' + a_2 w'^2 + a_3 w'^3 + \cdots,$$

and the coefficients a_k are to be determined from the integral:

$$a_k = \frac{1}{2i\pi}\int^{\bullet}\frac{(z-\alpha)f'(z)}{(f(z)-f(\alpha))^{k+1}}\,dz.$$

Now putting:

$$\frac{f(z)-f(\alpha)}{z-\alpha} = F(z-\alpha) = \frac{1}{\varphi(z-\alpha)},$$

whence:

$$f'(z) = \frac{1}{\varphi(z-\alpha)} - \frac{(z-\alpha)\varphi'(z-\alpha)}{(\varphi(z-\alpha))^2},$$

we see that the integral breaks up into the following two parts:

$$a_k = \frac{1}{2i\pi}\int^{\bullet}\frac{(\varphi(z-\alpha))^k\,dz}{(z-\alpha)^k} - \frac{1}{2i\pi}\int^{\bullet}\frac{(\varphi(z-\alpha))^{k-1}\varphi'(z-\alpha)}{(z-\alpha)^{k-1}}\,dz.$$

The first part by § 185 is equal to

$$\frac{1}{\underline{|k-1}}\;\frac{d^{k-1}(\varphi(z-\alpha))^k}{dz^{k-1}}$$

at the point $z=\alpha$. The second part is equal to

$$\frac{1}{\underline{|k-2}}\;\frac{d^{k-2}\{(\varphi(z-\alpha))^{k-1}\varphi'(z-\alpha)\}}{dz^{k-2}} = \frac{1}{\underline{|k-2}}\;\frac{1}{k}\;\frac{d^{k-1}(\varphi(z-\alpha))^k}{dz^{k-1}},\ \text{for } z=\alpha;$$

so that:

$$a_k = \frac{1}{\underline{|k-2}}\left(\frac{1}{k-1}-\frac{1}{k}\right)\frac{d^{k-1}(\varphi(z-\alpha))^k}{dz^{k-1}} = \frac{1}{\underline{|k}}\;\frac{d^{k-1}(\varphi(z-\alpha))^k}{dz^{k-1}},\ \text{for } z=\alpha.$$

Accordingly we have the above solution stated in Lagrange's form:

From the equation: $w'\varphi(z') = z'$, in which φ signifies a unique analytic function, the following expansion is found:

$$z' = w'\varphi(0) + \frac{w'^2}{\underline{|2}}\left\{\frac{d(\varphi(z'))^2}{dz'}\right\}_{z'=0} + \frac{w'^3}{\underline{|3}}\left\{\frac{d^2(\varphi(z'))^3}{dz'^2}\right\}_{z'=0} + \frac{w'^4}{\underline{|4}}\left\{\frac{d^3(\varphi(z'))^4}{dz'^3}\right\}_{z'=0} + \cdots \text{etc.}.$$

Since:

$$f(z) - f(\alpha) = \frac{z-\alpha}{\varphi(z-\alpha)},\quad \frac{1}{f'(z)} = \frac{(\varphi(z-\alpha))^2}{\varphi(z-\alpha)-(z-\alpha)\varphi'(z-\alpha)},$$

the manner in which the preceding condition of convergence is brought into relation with the function $\varphi(z')$ is as follows:

When r signifies the radius of the greatest circle round $z'=0$ as centre within which $\varphi(z')-z'\varphi'(z')$ does not vanish, let us find the smallest value assumed by $w'=\frac{z'}{\varphi(z')}$ upon this circle, this minimum is the radius of convergence for w'.

The domain of this circle in the plane w' can by inversion be uniquely transformed into a domain of the plane z' included in the circle of radius r, wherein also $\varphi(z')$ is not zero.

Here the point $w'=0$ corresponds to the point $z'=0$.

Third Chapter.

Expansion of ambiguous analytic functions, specially of the algebraic function.

193. Both the problem: of solving for w the implicit algebraic function defined by $f(z, w) = 0$, and that just considered: of forming the inverse function of a series of powers, led to ambiguous functions. In each case we showed how Taylor's expansion is applicable within restricted domains in which the function remains regular. The algebraic function (§ 188) presented two kinds of points in which the function w ceases to be regular: first, the points that were called non-essential singular points, in which w becomes infinite, although its product by a rational algebraic function remains finite; second, all the points that were described as critical, in which the value of w counts as a multiple root, for which therefore $\frac{\partial f(z, w)}{\partial w} = 0$. It may be, that both specialities concur in the same point. In our last problem the ramifications formed the irregular points. Generalising this problem to the inversion of any unique analytic function in an arbitrary domain that also contains essential and non-essential singular points, such singularities will also occur in the inverse function.

Accordingly in what follows we shall investigate the properties of ambiguous analytic functions in general, and specially those of the algebraic function.

By an n-valued analytic function of z, defined for the entire plane or for a finite domain, is meant a function which generally for each value of z has n different values. Each of these values must satisfy the differential equation $\frac{\partial f}{\partial x} + i\frac{\partial f}{\partial y} = 0$ — except in non-essential or essential singular points. But furthermore the function must have branching points. By a branching point or ramification is meant a point z in which two at least of the corresponding values of the function become equal, and in which moreover the values of the function have changed when z has travelled round a curve enclosing the point (§ 191). The ramification may happen to be also a singular point. Now as we have already seen that in any domain that is regular without exception the function can be expressed by Taylor's series, and that the expansion is generalised by the method developed in § 189 in case singular points occur in the domain, it still remains to be shown that an expansion is possible also in the neighbourhood of a branching point and how it proceeds. Then the values of the derived functions in the branching point must be determined. Lastly it has to be shown how the analytic expression of the function is modified when the ramification is also an essential or non-essential singular point.

In transferring these theorems to the ambiguous algebraic function, the further question specially requires an answer: whether each of its critical points is a ramification. As it will appear that this is not always the case, but that the critical point may be merely a multiple point without branching, the question arises: what are then the values of the successive derivates; a question that for real values of the function was already (§ 60) answered in the simplest cases.

194. In our investigation of an ambiguous function $f(z)$ in the neighbourhood of a branching point α, at present supposed not to be also singular, in which m branches of the function are connected so as to form a cycle, we shall set out from the consideration of a definite integral, in accordance with the method we have always employed hitherto. The analytical operations can be geometrically elucidated by constructing round the point α, instead of the single plane of z, a Riemann's winding surface of the order $m-1$. This consists of m leaves cyclically connected along an arbitrarily drawn branching section; superincumbent points in these leaves represent the same value of z, but to each of these points is always uniquely coordinated only one of the m values of the function. A closed curve, ex. gr. a circle, required to surround the point α, must be constructed so that starting from a point z of the first leaf it is drawn round the point α in that leaf, but then having reached the branching section it enters into the second leaf, thence after a complete circuit round the point α it enters into the third leaf, and so on, lastly into the mth. From this at the branching section it returns into the first leaf and completes its circuit there where it began.

Substituting in the function $f(z)$ for z the new variable ζ connected with z by the equation:

$$\zeta^m = z - \alpha, \text{ and therefore, extracting the root: } \zeta = (z-\alpha)^{\frac{1}{m}},$$

let us begin with some one of the m values of this root belonging to a determinate z, then while z describes a circle round the point α, the value of ζ will change continuously with it; ζ however will not have resumed its former value on z completing a circuit, but will have changed continuously into another root differing from the original in amplitude by $2\pi : m$. The first return of ζ to its initial value occurs when z has completed the entire m circuits. Interpreting the values of ζ in a plane, ζ first completes the circle round the point $\zeta = 0$ when z has completed its circuit on all the m leaves of the winding surface. To the points z of a single leaf within a circle of radius r, correspond only the points ζ within a circular sector of radius $r^{\frac{1}{m}}$ and central angle $2\pi : m$. Having fixed which value of the radical $(z - \alpha)^{\frac{1}{m}}$ shall be chosen initially, we have established a definite relation between the consecutive circular sectors and the various leaves.

The values which the function $f(z)$ assumes in the various leaves can be coordinated uniquely and continuously to the points of the m circular sectors, so that we can say, the function

$$f(z) = f(\zeta^m + \alpha) = \varphi(\zeta)$$

is a unique and continuous function for the interior of the circle round the point $\zeta = 0$. But it is also an analytic function. For, the function $f(z)$ must have a determinate finite derivate at every point, except the branching point; therefore we have:

$$f'(z) = \varphi'(\zeta) \cdot \frac{d\zeta}{dz} = \varphi'(\zeta) \cdot \frac{1}{m}(z - \alpha)^{\frac{1}{m} - 1};$$

i. e. the quantity:

$$\varphi'(\zeta) = mf'(z) \cdot (z - \alpha)^{\frac{m-1}{m}}$$

is everywhere determinate and finite. In the point $z = \alpha$ itself, $f(z)$ and therefore also $\varphi(\zeta)$ must remain finite and continuous; hence this cannot be a singular point for the otherwise unique analytic function $\varphi(\zeta)$; consequently (§ 189) $\varphi'(\zeta)$ also has a determinate value. When this value is finite and different from zero, the equation:

$$f'(z) = \frac{\varphi'(\zeta)}{m(z - \alpha)^{\frac{m-1}{m}}},$$

shows that the derived function $f'(z)$ becomes infinite of the order $\frac{m-1}{m}$ in the branching point. When $\varphi'(\zeta)$ vanishes, $f'(z)$ may be finite or even zero, as shall be more strictly determined hereafter.

To the integral $\int f(z)dz$ formed for the complete circular circuit in all the m leaves, beginning in the first leaf with one of the m possible values of $f(z)$ and changing this continuously with z, corresponds thus the integral

$$m\int \varphi(\zeta)\zeta^{m-1}d\zeta$$

taken along the circle with radius $r^{\frac{1}{m}}$ round the point zero; since

$$dz = m\zeta^{m-1}d\zeta.$$

According to the fundamental Theorem proved concerning the unique analytic function, this integral, and in general every integral taken along a curve enclosing the point $\zeta = 0$ and containing no singular point, has the value zero. It follows similarly, because $\int\varphi(\zeta)d\zeta$ integrated along the same path is zero, that the corresponding integral:

$$\frac{1}{m}\int\frac{f(z)}{(z-\alpha)^{\frac{m-1}{m}}}dz$$

also vanishes. Hence for the ambiguous function the analogous theorem is:

When the branching point of an ambiguous function, in which m leaves are cyclically connected, is surrounded by a closed curve that necessarily winds m times round the point, if there be no non-essential or essential singular point within this curve, both the integrals

$$\int f(z)dz \text{ and } \int\frac{f(z)}{(z-\alpha)^{\frac{m-1}{m}}}dz\text{*)}$$

along this closed path are zero.

The above Theorem led to the analytical expression of the unique function in the regular domain (§ 184); its counterpart does the same for ambiguous functions, only it must also first be generalised for a domain with a multipartite boundary.

When there are singular points in the circle round $\zeta = 0$, supposing first that none of them coincides with this point zero, let us surround each such point $\zeta = c$ by an arbitrary closed curve, ex. gr. a circle with radius ϱ. The curve corresponding to this curve (c) on the winding surface by reason of the equation:

$$z = \zeta^m + \alpha = (c + \varrho e^{i\varphi})^m + \alpha,$$

is likewise closed; it is contained altogether in one leaf, when the corresponding curve (c) in the circle ζ lies altogether within one of the sectors corresponding to each leaf; when this curve (c) enters into different sectors, the other is also found in different leaves. But always the theorem holds for each unique analytic function $\varphi(\zeta)$ and $\varphi(\zeta)\,.\,\zeta^{m-1}$, that the sum of all the integrals taken in a positive

*) In general $\int\frac{f(z)}{(z-\alpha)^{\frac{m-1-k}{m}}}dz = 0$ for $k \geqq 0$.

circuit along the boundary curves of a domain in which $\varphi(\zeta)$ has no singular points, is zero. Accordingly the generalisation of the above theorem is as follows:

When we form either integral:

$$\int f(z)dz \text{ or: } \int \frac{f(z)}{(z-\alpha)^{\frac{m-1}{m}}}dz$$

in a positive circuit for all the curves constituting the multipartite boundary of a domain, which consists of m leaves that are cyclically connected along a branching section starting from the point $z = \alpha$, if there be no singular point within such a domain, the sum of all the integrals in each case is zero.

195. We can now proceed in analogy with the earlier development as follows. First, let $f(z)$ and therefore also $\varphi(\zeta)$ have no singular points within the winding surface round the point α bounded by the radius r. With t any value within the circle of radius $r^{\frac{1}{m}}$ round the point zero let us form the function $\frac{\varphi(\zeta)}{\zeta - t}$. Enclosing also the point t by an arbitrarily small circle, we have a domain with bipartite boundary without any singular point. Then (§ 184):

I. $$\varphi(t) = \frac{1}{2i\pi}\int \frac{\varphi(\zeta)}{\zeta - t}d\zeta.$$

The integration is to be along the circle with radius $r^{\frac{1}{m}}$ enclosing the point zero.

Hence, denoting $t^m + \alpha$ by u, so that u is a point on a determinate leaf of the winding surface within the circle of radius r, it follows that:

Ia. $$f(u) = \frac{1}{2i\pi m}\int \frac{f(z)}{(z-\alpha)^{\frac{m-1}{m}}} \frac{dz}{(z-\alpha)^{\frac{1}{m}} - (u-\alpha)^{\frac{1}{m}}}.$$

Since the values of z signify the points of the bounding curve, while u represents any point w i t h i n it, we have

$$\text{mod }[u-\alpha]^{\frac{1}{m}} < \text{mod }[z-\alpha]^{\frac{1}{m}}.$$

Accordingly:

$$\frac{1}{(z-\alpha)^{\frac{1}{m}} - (u-\alpha)^{\frac{1}{m}}} = \frac{1}{(z-\alpha)^{\frac{1}{m}}}\left\{1 + \left(\frac{u-\alpha}{z-\alpha}\right)^{\frac{1}{m}} + \left(\frac{u-\alpha}{z-\alpha}\right)^{\frac{2}{m}} + \left(\frac{u-\alpha}{z-\alpha}\right)^{\frac{3}{m}} + \cdots\right\},$$

and we have (cf. § 186):

II. $$f(u) = \frac{1}{2i\pi m}\left\{\int \frac{f(z)\,dz}{(z-\alpha)^{\frac{m}{m}}} + (u-\alpha)^{\frac{1}{m}} \int \frac{f(z)\,dz}{(z-\alpha)^{\frac{m+1}{m}}} + (u-\alpha)^{\frac{2}{m}} \int \frac{f(z)\,dz}{(z-\alpha)^{\frac{m+2}{m}}} + \cdots\right\}.$$

Each integral in this series refers to the closed curve taken positively in all the leaves round the point α; in the various leaves $f(z)$ and $(z-\alpha)^{\frac{m+1}{m}}$ assume their prescribed values.

The root $(u-\alpha)^{\frac{1}{m}}$ and its powers assume their different values according as the value of $f(u)$ is to be determined in one or other of the m leaves.

The statement of equation II. in words is:

When the branching point in which m values of the function are cyclically connected is not also a singular point, each of the m values of the function in its neighourhood can be expanded in an ascending series of positive integer powers of $(u-\alpha)^{\frac{1}{m}}$. This neighbourhood is coextensive with a domain wherein there is neither a singular point nor another branching point of the function.

The significance of the coefficients in II. can be shown otherwise. Putting $u=\alpha$, we have:

$$f(\alpha) = \frac{1}{2i\pi m}\int \frac{f(z)\,dz}{z-\alpha}.$$

If further we differentiate the equation with respect to u, which in case of a series of powers is done by differentiating term by term, we obtain, on multiplying both sides by $(u-\alpha)^{\frac{m-1}{m}}$ and putting $u=\alpha$:

$$\lim_{u=\alpha}\left\{f'(u)\,(u-\alpha)^{\frac{m-1}{m}}\right\} = \frac{1}{2i\pi m^2}\int \frac{f(z)\,dz}{(z-\alpha)^{\frac{m+1}{m}}}.$$

By the same process is found:

$$\lim_{u=\alpha}\left\{f''(u)\,(u-\alpha)^{\frac{2(m-1)}{m}} + \frac{m-1}{m} f'(u)\,(u-\alpha)^{\frac{m-2}{m}}\right\} = \frac{2(2-m)}{m^2}\cdot\frac{1}{2i\pi m}\int \frac{f(z)\,dz}{(z-\alpha)^{\frac{m+2}{m}}}, \ \ldots \text{ etc.}.$$

Writing series II. briefly in the form:

$$f(u) = a_0 + (u-\alpha)^{\frac{1}{m}} a_1 + (u-\alpha)^{\frac{2}{m}} a_2 + \cdots,$$

we see that the derived functions $f'(u)$, $f''(u)$, ... become infinite at α in the respective orders $\frac{m-1}{m}$, $\frac{2m-1}{m}$, ..., provided a_1 is not zero.

But when some of the coefficients following a_0 vanish, so that the series presents the form:

$$f(u) = a_0 + (u-\alpha)^{\frac{\mu}{m}} a_\mu + (u-\alpha)^{\frac{\mu+1}{m}} a_{\mu+1} + (u-\alpha)^{\frac{\mu+2}{m}} a_{\mu+2} + \cdots,$$

the expansion of $f'(u)$ begins with the term $\frac{\mu}{m}(u-\alpha)^{\frac{\mu-m}{m}} a_\mu$, and for $u = \alpha$ this expression becomes either infinitely great of the order $\frac{m-\mu}{m}$, or infinitely small of the order $\frac{\mu-m}{m}$, according as m is greater or less than μ. For $m = \mu$ the term is finite. The first derivate of a function in a branching point can therefore also be zero or finite. But, whether it be infinite or not, a number k can always be assigned such that every derivate of order equal to or higher than k shall become infinite in the branching point. For, the kth derived function begins with the term $(u-\alpha)^{\frac{\mu-km}{m}}$. We have accordingly the theorem:

When the lowest power occurring in the expansion of the m-valued function is $(u-\alpha)^{\frac{\mu}{m}}$:

The kth derived function becomes infinite of the order $\frac{km-\mu}{m}$ in the branching point α, when k is chosen $> \frac{\mu}{m}$;

When $k = \frac{\mu}{m}$ is an integer, the kth derived function is finite at α;

All derived functions of order $k < \frac{\mu}{m}$, vanish in the branching point α.

196. According to the process shown in § 189, the expansion can also be generalised to a domain wherein there are essential or non-essential singular points of the function $f(z)$ and therefore also of the function $\varphi(\zeta)$; on the hypothesis, that none of these points is also a ramification.

Let $c_1, c_2, \ldots c_k$ be the singular points of $f(z)$, the corresponding singular points of $\varphi(\zeta)$ being respectively

$$\gamma_1 = (c_1-\alpha)^{\frac{1}{m}},\ \gamma_2 = (c_2-\alpha)^{\frac{1}{m}},\ \cdots \gamma_k = (c_k-\alpha)^{\frac{1}{m}},$$

where in each case it needs only one of the m possible values of the radical to determine a point γ that is an infinity point for the function $\varphi(\zeta)$; thus we have the equation:

$$\text{III.}\quad \varphi(t) = \frac{1}{2i\pi}\left\{\int\limits_{(0)} \frac{\varphi(\zeta)\,d\zeta}{\zeta-t} - \int\limits_{(\gamma_1)} \frac{\varphi(\zeta)\,d\zeta}{\zeta-t} - \int\limits_{(\gamma_2)} \frac{\varphi(\zeta)\,d\zeta}{\zeta-t} - \cdots \int\limits_{(\gamma_k)} \frac{\varphi(\zeta)\,d\zeta}{\zeta-t}\right\}.$$

The integrals are to be taken round the point zero and round the points γ, so as to keep each of them on the left.

From this equation, substituting $\zeta = (z-\alpha)^{\frac{1}{m}}$, $t = (u-\alpha)^{\frac{1}{m}}$, we find for the function $f(u)$ the relation:

III a. $$f(u) = \frac{1}{2i\pi m}\left\{\int_{(\alpha)} \frac{f(z)}{(z-\alpha)^{\frac{m-1}{m}}} \frac{dz}{(z-\alpha)^{\frac{1}{m}} - (u-\alpha)^{\frac{1}{m}}} - \int_{(c_1)} \frac{f(z)}{(z-\alpha)^{\frac{m-1}{m}}} \frac{dz}{(z-\alpha)^{\frac{1}{m}} - (u-\alpha)^{\frac{1}{m}}} - \cdots - \int_{(c_k)}\right\}.$$

The first integral is to be formed for the external boundary of the winding surface; the others refer severally to each curve that encloses a (non-branching) singular point c. The first integral can consequently be expanded by powers of $\left(\frac{u-\alpha}{z-\alpha}\right)^{\frac{1}{m}}$, since u signifies a point within the curve to which the values of z refer. In any of the other integrals, z signifies a point on the curve round the point c, and u lies outside this curve; therefore we have

$$\text{mod}\left[\frac{\zeta-\gamma}{t-\gamma}\right] = \text{mod}\,\frac{(z-\alpha)^{\frac{1}{m}} - (c-\alpha)^{\frac{1}{m}}}{(u-\alpha)^{\frac{1}{m}} - (c-\alpha)^{\frac{1}{m}}} < 1,$$

and the quotient

$$1 : \left\{(z-\alpha)^{\frac{1}{m}} - (u-\alpha)^{\frac{1}{m}}\right\}$$

can be expressed by a series beginning with the terms:

$$\frac{-1}{(u-\alpha)^{\frac{1}{m}} - (c-\alpha)^{\frac{1}{m}}}\left\{1 + \frac{(z-\alpha)^{\frac{1}{m}} - (c-\alpha)^{\frac{1}{m}}}{(u-\alpha)^{\frac{1}{m}} - (c-\alpha)^{\frac{1}{m}}} + \left\{\frac{(z-\alpha)^{\frac{1}{m}} - (c-\alpha)^{\frac{1}{m}}}{(u-\alpha)^{\frac{1}{m}} - (c-\alpha)^{\frac{1}{m}}}\right\}^2 + \cdots\right\}.$$

Accordingly, from equation III a. we have the following expansion:

IV. $$f(u) = \frac{1}{2i\pi m}\left\{\int_{(\alpha)} \frac{f(z)\,dz}{(z-\alpha)^{\frac{m}{m}}} + (u-\alpha)^{\frac{1}{m}}\int_{(\alpha)} \frac{f(z)\,dz}{(z-\alpha)^{\frac{m+1}{m}}} + (u-\alpha)^{\frac{2}{m}}\int_{(\alpha)} \frac{f(z)\,dz}{(z-\alpha)^{\frac{m+2}{m}}} + \cdots\right\}$$

$$+ \frac{1}{2i\pi m}\left\{\frac{1}{(u-\alpha)^{\frac{1}{m}} - (c_1-\alpha)^{\frac{1}{m}}}\int_{(c_1)} \frac{f(z)\,dz}{(z-\alpha)^{\frac{m-1}{m}}}\right.$$

$$+ \frac{1}{\left\{(u-\alpha)^{\frac{1}{m}} - (c_1-\alpha)^{\frac{1}{m}}\right\}^2}\int_{(c_1)} \frac{f(z)}{(z-\alpha)^{\frac{m-1}{m}}}\left\{(z-\alpha)^{\frac{1}{m}} - (c_1-\alpha)^{\frac{1}{m}}\right\} dz$$

$$+\frac{1}{\left\{(u-\alpha)^{\frac{1}{m}}-(c_1-\alpha)^{\frac{1}{m}}\right\}^3}\int_{(c_1)}\frac{f(z)}{(z-\alpha)^{\frac{m-1}{m}}}\left\{(z-\alpha)^{\frac{1}{m}}-(c_1-\alpha)^{\frac{1}{m}}\right\}^2 dz+\cdots\Bigg\}$$

$$+\cdots\cdots\cdots\cdots$$

$$+\frac{1}{2i\pi m}\Bigg\{\frac{1}{(u-\alpha)^{\frac{1}{m}}-(c_k-\alpha)^{\frac{1}{m}}}\int_{(c_k)}\frac{f(z)\,dz}{(z-\alpha)^{\frac{m-1}{m}}}$$

$$+\frac{1}{\left\{(u-\alpha)^{\frac{1}{m}}-(c_k-\alpha)^{\frac{1}{m}}\right\}^2}\int_{(c_k)}\frac{f(z)}{(z-\alpha)^{\frac{m-1}{m}}}\left\{(z-\alpha)^{\frac{1}{m}}-(c_k-\alpha)^{\frac{1}{m}}\right\}dz+\cdots\Bigg\}.$$

The singular point c_i is non-essential or essential, according as the expansion relative to it is finite or not. In the former case an integer n can be assigned such that the product:

$$f(u)\left(u-\alpha^{\frac{1}{m}}-c_i-\alpha^{\frac{1}{m}}\right)^n$$

is finite for $u=c_i$.

197. When the ramification α in which m leaves of the function are cyclically connected is also a singular point, the point $\zeta=0$ is likewise singular in the function $\varphi(\zeta)$ that results from $f(z)$ by the substitution $\zeta=(z-\alpha)^{\frac{1}{m}}$. Hence the expansion (§ 189) within a domain in which there are no further singular points, is:

$$\varphi(t)=\frac{1}{2i\pi}\left\{\int_{(0)}\frac{\varphi(\zeta)}{\zeta}d\zeta+t\int_{(0)}\frac{\varphi(\zeta)}{\zeta^2}d\zeta+t^2\int_{(0)}\frac{\varphi(\zeta)}{\zeta^3}d\zeta+\cdots\right\}$$

$$+\frac{1}{2i\pi}\left\{\frac{1}{t}\int_{(0)}\varphi(\zeta)\,d\zeta+\frac{1}{t^2}\int_{(0)}\varphi(\zeta)\zeta\,d\zeta+\frac{1}{t^3}\int_{(0)}\varphi(\zeta)\zeta^2\,d\zeta+\cdots\right\}.$$

To this corresponds for $f(u)$ the formula:

$$\text{V.}\quad f(u)=\frac{1}{2i\pi m}\Bigg\{\int_{(u)}\frac{f(z)}{z-\alpha}dz+(u-\alpha)^{\frac{1}{m}}\int_{(\alpha)}\frac{f(z)}{(z-\alpha)^{\frac{m+1}{m}}}dz$$

$$+(u-\alpha)^{\frac{2}{m}}\int_{(\alpha)}\frac{f(z)}{(z-\alpha)^{\frac{m+2}{m}}}dz+\cdots\Bigg\}$$

$$+\frac{1}{2i\pi m}\Bigg\{\frac{1}{(u-\alpha)^{\frac{1}{m}}}\int_{(u)}\frac{f(z)}{(z-\alpha)^{\frac{m-1}{m}}}dz+\frac{1}{(u-\alpha)^{\frac{2}{m}}}\int_{(\alpha)}\frac{f(z)}{(z-\alpha)^{\frac{m-2}{m}}}dz$$

$$+\frac{1}{(u-\alpha)^{\frac{3}{m}}}\int_{(\alpha)}\frac{f(z)}{(z-\alpha)^{\frac{m-3}{m}}}dz+\cdots\Bigg\}.$$

Here the curve of integration must surround the point α and be closed by completing the circuit in all the leaves, it can proceed arbitrarily near the branching point.

This singular point α is non-essential or essential according as the number of terms in the second part of the expansion is finite or not. When finite, an integer n can be assigned such that the product

$$f(u)\overline{u-\alpha}^{\frac{n}{m}}$$

continues finite for $u = \alpha$. In the branching point, $f(u)$ then becomes infinite of the order $n:m$.

The integral $\int f(z)\,dz$ taken in a closed circuit round the branching point is zero, only when the coefficient of the term $\frac{1}{u-\alpha}$ is zero; for this coefficient is the integral itself. When this condition is fulfilled, the branching point, notwithstanding that it is also singular, is not a logarithmic branching point for the ambiguous integral function.

The value of the integral up to the branching point is finite, only when $\int \varphi(\zeta)\zeta^{m-1}d\zeta$ also continues finite. This requires (§ 181)

$$\{\varphi(\zeta)\zeta^m\}_{\zeta=0} = \{f(z)(z-\alpha)\}_{z=\alpha}$$

to vanish. Therefore in the second part of the above expansion every term, whose denominator contains a power of $(u-\alpha)$ with exponent equal to or greater than unity, must vanish; therefore in the branching point, which can only be a non-essential singular point, the infinitude of the function must be less than unity.

198. If the point infinity $z = \infty$ be a branching point in which m values of the function are cyclically connected, and if all the singular points of the function lie within a finite domain, then by the substitution $z = \frac{1}{z'}$, $f(z)$ becomes $f\left(\frac{1}{z'}\right) = \psi(z')$, and this function has the origin $z' = 0$ as its branching point and it has no singular points within a finite domain round the origin.

Accordingly (§ 195 II.) for an arbitrary point u' of this domain:

$$\psi(u') = \frac{1}{2i\pi m}\left\{\int\limits_{(0)}\frac{\psi(z')\,dz'}{z'^{\frac{m}{m}}} + u'^{\frac{1}{m}}\int\limits_{(0)}\frac{\psi(z')\,dz'}{z'^{\frac{m+1}{m}}} + u'^{\frac{2}{m}}\int\limits_{(0)}\frac{\psi(z')\,dz'}{z'^{\frac{m+2}{m}}} + \cdots\right\}.$$

The integrals are to be taken in a positive circuit round the origin. Hence, the following must be the expansion for the original function:

$$f(u) = \frac{1}{2i\pi m}\left\{\int\limits_{-(\infty)}\frac{f(z)dz}{z} + \frac{1}{u^{\frac{1}{m}}}\int\limits_{-(\infty)}\frac{f(z)dz}{z^{1-\frac{1}{m}}} + \frac{1}{u^{\frac{2}{m}}}\int\limits_{-(\infty)}\frac{f(z)dz}{z^{1-\frac{2}{m}}} + \cdots\right\}.$$

The integrations refer to the point infinity, i. e. they are to be taken along an arbitrarily remote curve enclosing the point zero, and in the direction that keeps the finite surface likewise on the left.

When the point $z = \infty$ is at the same time a singular point, we obtain by the same substitution from Formula V. § 197 the expansion:

$$f(u) = \frac{1}{2i\pi m}\left\{\int\limits_{-(\infty)} \frac{f(z)dz}{z} + \frac{1}{u^{\frac{1}{m}}}\int\limits_{-(\infty)} \frac{f(z)dz}{z^{1-\frac{1}{m}}} + \frac{1}{u^{\frac{2}{m}}}\int\limits_{-(\infty)} \frac{f(z)dz}{z^{1-\frac{2}{m}}} + \cdots\right\}$$

$$+ \frac{1}{2i\pi m}\left\{u^{\frac{1}{m}}\int\limits_{-(\infty)} \frac{f(z)dz}{z^{1+\frac{1}{m}}} + u^{\frac{2}{m}}\int\limits_{-(\infty)} \frac{f(z)dz}{z^{1+\frac{2}{m}}} + u^{\frac{3}{m}}\int\limits_{-(\infty)} \frac{f(z)dz}{z^{1+\frac{3}{m}}} + \cdots\right\}.$$

When infinity is a non-essential singular point, an integer n can be assigned for which the value of

$$\left\{f(u) : u^{\frac{n}{m}}\right\}_{u=\infty}$$

is finite. The function then becomes infinite of the order $n : m$ in the branching point, and the second part of the above expansion contains only the powers from $u^{\frac{1}{m}}$ to $u^{\frac{n}{m}}$.

The value of $\int f(z)dz$ integrated round the point infinity is zero when the coefficient of the term $\frac{1}{u}$ vanishes.

The same integral taken up to the point infinity is finite when

$$\left\{f\left(\frac{1}{z'}\right)\frac{1}{z'}\right\}$$

vanishes for $z' = 0$; therefore also $\{f(z)z\}_{z=\infty} = 0$. In the $(m-1)$-branching point $z = \infty$ therefore the function must vanish in a higher order than the first, i. e. this cannot be a singular point, and the first part of the expansion must begin with the term $1 : u^{\frac{m+1}{m}}$.

199. The investigations of the constitution of ambiguous functions in the neighbourhood of a branching point are necessary in order that we may attain a definitive insight into the theory of algebraic functions.

As a culmination to these investigations we may establish a theorem by which algebraic functions are completely characterised as a special class among ambiguous functions.

For, in a similar manner as among unique analytic functions rational algebraic functions admitted of being defined as those which have only non-essential singular points, either finite or at infinity, we have the following theorem respecting ambiguous functions:*)

When a function w has n values for each value of z, and in the entire infinite plane its only irregular points are non-essential singular points and branching points such as have been above discussed, the function must be root of an algebraic equation:

$$f(z^m, w^n) = 0,$$

of the nth degree in w, and of a degree m in z that is equal to the sum of the orders (infinitudes) of the infinity points.

Let the n values of the function be denoted by $w_1, w_2, \ldots w_n$; and further let $\alpha_1, \alpha_2, \ldots \alpha_\mu$ be the finite singular points that are not branching points; in each of these points one branch of the function becomes infinite, let the orders of becoming infinite be denoted respectively by $i_1, i_2, \ldots i_\mu$, so that therefore the products:

$$w(z-\alpha_1)^{i_1};\ w(z-\alpha_2)^{i_2};\ \ldots w(z-\alpha_\mu)^{i_\mu}$$

remain finite; where w in each signifies that branch which becomes infinite in the point α involved. Moreover points α may be coincident.

Further let $\beta_1, \beta_2, \ldots \beta_\nu$ be the finite branching points that are also infinity points; in these several points let $k_1, k_2, \ldots k_\nu$ leaves of the function respectively be connected; and let the respective infinitudes (cf. § 197) be denoted by $l_1, l_2, \ldots l_\nu$; the products:

$$w(z-\beta_1)^{\frac{l_1}{k_1}};\ w(z-\beta_2)^{\frac{l_2}{k_2}};\ \ldots w(z-\beta_\nu)^{\frac{l_\nu}{k_\nu}}$$

are therefore finite. Lastly let the point $z = \infty$ be a branching point in which k leaves are connected and let its infinitude be l, so that

$$w\left(\frac{1}{z}\right)^{\frac{l}{k}}$$

is finite. Let us put

$$(i_1 + i_2 + \cdots i_\mu) + (l_1 + l_2 + \cdots l_\nu) + l = m.$$

Forming now the symmetric function of the values of w:

$$S = (\sigma - w_1)(\sigma - w_2)\ldots(\sigma - w_n),$$

this, as in general every symmetric function of the quantities w, is a single-valued function of z; for, even the paths along which certain values of w interchange cyclically lead always to the same value of the function S. S becomes infinite of the order i_μ in each point α_μ, of the order l_ν in each point β_ν, and lastly of the order l in the point $z = \infty$. Thus as the single-valued function S has only non-essential singularities, it must be a rational fractional function of z (§ 190) that can be set down in the form:

---

*) Riemann: Theorie der Abel'schen Functionen. Werke (pp. 81—135), p. 101. Briot († 1882) et Bouquet († 1885): Théorie des fonctions elliptiques, 2. éd., p. 216.

$$S = \frac{f(z, \sigma)}{\varphi(z)},$$

where f is some function of the order m with respect to z, and

$$\varphi(z) = (z - \alpha_1)^{l_1} (z - \alpha_2)^{l_2} \ldots (z - \alpha_\mu)^{l_\mu} (z - \beta_1)^{l'_1} \ldots (z - \beta_\nu)^{l'_\nu}$$

Accordingly

$$S \cdot \varphi(z) = f(z, \sigma)$$

is an integer function of the mth degree in z and an integer function of the nth degree in σ.

Hence, inasmuch as this polynomial $f(z^m, \sigma^n)$ of the nth degree in σ vanishes whenever σ assumes one of the values $w_1, w_2, \ldots w_n$; w must satisfy, or, as the enunciation asserts, be root of the algebraic equation:

$$f(z^m, w^n) = 0.$$

This algebraic expression is **irreducible**, i. e. it cannot be resolved into rational factors of a lower degree in w, provided the n generally different values of the function w are connected in such a way that, by suitable choice of paths which enclose the branching points, any value w_i of the function can be carried over continuously into any other w_k; in other words: when the n-leaved surface requisite for exhibiting the function w uniquely is connected not only in separate points but along entire branching sections.

For, if $f(z^m, w^n)$ break up into the product $g(z, w) \cdot h(z, w)$, since each of these factors is of a lower degree than the nth in w, neither of them can vanish for all the n values of w, they must therefore both vanish for every value of z, ex. gr.

$$g(z, w_i) = 0, \quad h(z, w_k) = 0.$$

Now since w is determined by any algebraic equation as a continuous function of z, round each point can be assigned a finite region throughout which $g(z, w_i)$, and likewise $h(z, w_k)$, each regarded as a function of z, has the value zero. But hence follows that each of these functions must be zero in the entire connected n-leaved surface; for, the function can be extended out from the finite domain into each leaf by means of the expansion in series of positive integer powers. But since by hypothesis w_i can be carried over continuously into every other value of the function, we have therefore:

$$g(z, w_1) = 0, \; g(z, w_2) = 0, \ldots g(z, w_n) = 0,$$

which is only possible when g is of the nth degree with respect to w. When the sum of the infinitudes of the function w is m, g must also be of the mth degree in z, i. e. the factor $h(z, w_k)$ is of the order 0 with respect to z also. We have therefore:

$$f(z^m, w^n) = \text{Const.}\; g(z^m, w^n). \quad \text{Q. E. D.}$$

It is evident conversely: If the algebraic equation $f(z, w) = 0$ be such that we cannot pass from any one arbitrary initial value w by arbitrary circuits of the branching points to every other value of the root, the algebraic form must break up into rational factors; for, each connected cycle satisfies an irreducible equation of lower degree with rational coefficients.

200. Supposing now the irreducible equation $f(z^m, w^n) = 0$ given; it is required to establish criteria for estimating the properties of each critical point and also a method of obtaining the expansions valid in its neighbourhood.

In this investigation we may restrict ourselves to points in which the values of z and w are finite; for, those points in which they are not finite can be transformed by substituting $z = \frac{1}{z'}$ and $w = \frac{1}{w'}$ respectively, into points with finite values.

We shall first show how certain simple cases can be settled without recourse to special methods. From this will emerge the most general statement of the problem under any conditions whatever, and its solution is presented in § 202.

Suppose the function w is known to assume the value b for a determinate value $z = a$, then the algebraic form $f(z, w)$ can be expanded by powers of $(z - a)$ and $(w - b)$ (§ 94). The coefficients of this expansion are the partial derived functions of f with respect to z and w formed for the point $z = a$, $w = b$; we shall briefly denote them by:

$$\left(\frac{\partial^k f(z, w)}{\partial z^{k-p}\, \partial w^p}\right)_{a,b} = f_{k-p,p}.$$

Then we have:

$$\begin{aligned} f(z^m, w^n) = &\left\{f_{1,0}(z-a) + f_{0,1}(w-b)\right\} \\ &+ \frac{1}{\underline{|2}}\left\{f_{2,0}(z-a)^2 + 2f_{1,1}(z-a)(w-b) + f_{0,2}(w-b)^2\right\} \\ &+ \cdots + \frac{1}{\underline{|k}}\left\{f_{k,0}(z-a)^k + k_1 f_{k-1,1}(z-a)^{k-1}(w-b) + \cdots \right. \\ &\qquad \left. + k_p f_{k-p,p}(z-a)^{k-p}(w-b)^p + \cdots + f_{0,k}(w-b)^k\right\} \\ &+ \cdots\cdots\cdots\cdots, \end{aligned}$$

a sum concluding with terms in which $(z - a)$ rises to the mth and $(w - b)$ to the nth power.

When the system of values $z = a$, $w = b$, is a **regular** point for the function w, $f_{0,1}$ is not zero; in this case an expansion in a series of ascending positive integer powers:

$$w - b = (z-a)\left(\frac{dw}{dz}\right)_{a,b} + \frac{1}{\underline{|2}}(z-a)^2\left(\frac{d^2w}{dz^2}\right)_{a,b} + \frac{1}{\underline{|3}}(z-a)^3\left(\frac{d^3w}{dz^3}\right)_{a,b} + \cdots$$

is valid for the neighbourhood of the point, as was developed in § 188.

The derivates $\frac{d^k w}{dz^k}$ are obtained from the formula $\frac{dw}{dz} = -\frac{f_{1,0}}{f_{0,1}}$ by successive differentiation. In particular, if $f_{1,0}, f_{2,0}, \ldots f_{k-1,0}$ all vanish at the point $z = a$, $w = b$, the expansion begins with the term:

$$w - b = \frac{1}{\underline{|k}}(z-a)^k\left(\frac{d^k w}{dz^k}\right)_{a,b} + \cdots, \qquad \frac{d^k w}{dz^k} = -\frac{f_{k,0}}{f_{0,1}}.$$

If for illustration we consider the relation as one between an ordinate (w) and an abscissa (z), the tangent to the algebraic curve at the point $z = a$, $w = b$, is then parallel to abscissæ and has a contact of the order $k - 1$; it meets the curve in k consecutive points.

The point $z = a$, $w = b$ is a critical point, when for it $f_{0,1} = 0$. We proceed to examine what expansions are then valid.

As a first case we have to consider the critical point when $f_{1,0}$ is not zero. Then z can be expanded by integer powers of w in the manner just described, and assuming, in order to mention at once the most general eventuality, that all the partial derived functions $f_{0,2}, f_{0,3}, \ldots f_{0,k-1}$ also vanish for the critical point, we obtain the expansion:

$$z - a = \frac{1}{\underline{|k}}(w-b)^k\left(\frac{d^k z}{dw^k}\right)_{a,b} + \frac{1}{\underline{|k+1}}(w-b)^{k+1}\left(\frac{d^{k+1}z}{dw^{k+1}}\right)_{a,b} + \cdots,$$

which we may write:

$$(z-a) = (w-b)^k \alpha\{1 + \alpha_1(w-b) + \alpha_2(w-b)^2 + \cdots R_n(w-b)^n\},$$
$$(\text{Lim } R_n = 0).$$

Extracting the kth root on both sides, and arranging the right in powers of $(w - b)$ by means of the Polynomial Theorem, we find:

$$\left(\frac{z-a}{\alpha}\right)^{\frac{1}{k}} = (w-b)\{1 + \alpha_1'(w-b) + \alpha_2'(w-b)^2 + \cdots R_n'(w-b)^n\}, (\text{Lim } R_n' = 0).$$

Denoting $\left(\frac{z-a}{\alpha}\right)^{\frac{1}{k}}$ by t, we have now to solve the problem discussed in § 192: to invert the series

$$t = (w-b)\{1 + \alpha_1'(w-b) + \alpha_2'(w-b)^2 + \cdots\}.$$

By this means we find an expansion of the form:

$$w - b = t + \beta_2 t^2 + \beta_3 t^3 + \cdots,$$

therefore:

$$w-b=\left(\frac{z-a}{\alpha}\right)^{\frac{1}{k}}+\beta_2\left(\frac{z-a}{\alpha}\right)^{\frac{2}{k}}+\beta_3\left(\frac{z-a}{\alpha}\right)^{\frac{3}{k}}+\cdots,\quad \alpha=-\frac{1}{\underline{|k}}\frac{f_{0,k}}{f_{1,0}}.$$

This critical point is accordingly a branching point of the order $k-1$; in it k of the n leaves required to exhibit the function w uniquely are cyclically connected. When the point in question is a real point of an algebraic curve with real coefficients, the real figure of the curve has here a tangent with contact of the order $k-1$ as before, but parallel to ordinates. The curve also crosses the tangent at the point (of inflexion) when k is an odd number.

201. The critical point has next to be investigated for the case that $f_{0,1}$ and $f_{1,0}$ simultaneously vanish. We briefly denote henceforth $z-a$ simply by z, and $w-b$ by w, merely expressing thereby that the origin of coordinates replaces the point $z=a$, $w=b$; further we shall write the value of $\frac{d^r w}{dz^r}$ at that point $=w_r$. When all the partial derived functions of the 2nd, 3rd, ... $(k-1)$th orders vanish besides the first two, the expanded equation is:

$$0=f(z+a,w+b)=\frac{1}{\underline{|k}}\{f_{k,0}z^k+k_1 f_{k-1,1}z^{k-1}w+\cdots k_p f_{k-p,p}z^{k-p}w^p+\cdots+f_{0,k}w^k\}$$
$$+\frac{1}{\underline{|k+1}}\{f_{k+1,0}z^{k+1}+\cdots\}+\cdots$$

In this case the point considered is called a k-elementary point*), inasmuch as the system of values, $z=0$, $w=0$, along with the several series of systems of values which satisfy $f=0$ in its neighbourhood, form k elements of the function defined by $f=0$ at that point. In fact putting $z=0$ in this equation, we obtain an equation of the nth degree for w; assuming for the present that $f_{0,k}$ does not vanish, k of its roots are zero. For, w^k can be taken out as a factor. Therefore k leaves now meet in the critical point, and the question arises whether they branch in it.

This requires us to investigate the quotient $w:z$ near the k-elementary point, in order to establish what values the first derived function $\frac{dw}{dz}=\mathrm{Lim}\,\frac{w}{z}$, as well as the higher derived functions assume in that point. Dividing by z^k let us form the equation:

$$0=\frac{1}{\underline{|k}}\left\{f_{k,0}+k_1 f_{k-1,1}\left(\frac{w}{z}\right)+\cdots k_p f_{k-p,p}\left(\frac{w}{z}\right)^p+\cdots+f_{0,k}\left(\frac{w}{z}\right)^k\right\}$$
$$+\frac{z}{\underline{|k+1}}\left\{f_{k+1,0}+\cdots\right\}+\cdots,$$

which is of the nth degree in the quotient, but gives only k finite values of it when $z=0$ as roots of the equation:

*) Nöther, Math. Annal., Vol. IX, p. 160.

$$\frac{1}{\underline{|k}}\left(f_{k,0} + k_1 f_{k-1,1} w_1 + \cdots k_p f_{k-p,p} w_1^p + \cdots + f_{0,k} w_1^k\right) = \Phi_k = 0.$$

By our hypothesis the k roots of this equation are determinate quantities not infinite; we shall further assume, that they are also all different.

In the k-elementary point there is then no branching; for, all the higher derived functions, belonging respectively to the various values of w, remain finite. These are to be deduced from a system of equations which are obtained, in analogy with those established in § 188, by successive total differentiation.

Denoting the quotient $\frac{w}{z}$ by w_1, because $\lim_{z=0}\left(\frac{w}{z}\right) = w_1 = \frac{dw}{dz}$, we should have in general:

$$\frac{d^n w}{dz^n} = z\frac{d^n w_1}{dz^n} + n\frac{d^{n-1} w_1}{dz^{n-1}},\ \text{therefore when } z=0:\ \frac{d^{n-1} w_1}{dz^{n-1}} = \frac{1}{n}\frac{d^n w}{dz^n}.$$

Now the equation for w_1 arranged by powers of z is:

$$\Phi_k + z\Phi_{k+1} + z^2\Phi_{k+2} + \cdots = 0,$$

and from this we obtain for determining the successive derivates at the point $z=0$ the equations:

$$\frac{w_2}{\underline{|2}}\frac{\partial\Phi_k}{\partial w_1} + \Phi_{k+1} = 0,$$

$$\frac{w_3}{\underline{|3}}\frac{\partial\Phi_k}{\partial w_1} + \frac{1}{2}\left(\frac{w_2}{2}\right)^2\frac{\partial^2\Phi_k}{\partial w_1^2} + \frac{w_2}{2}\frac{\partial\Phi_{k+1}}{\partial w_1} + \Phi_{k+2} = 0,$$

$$\frac{w_4}{\underline{|4}}\frac{\partial\Phi_k}{\partial w_1} + \frac{w_3}{\underline{|3}}\left(\frac{\partial\Phi_{k+1}}{\partial w_1} + \frac{w_2}{2}\frac{\partial^2\Phi_k}{\partial w_1^2}\right)$$
$$+\left(\frac{\partial\Phi_{k+2}}{\partial w_1}\frac{w_2}{2} + \frac{1}{2}\frac{\partial^2\Phi_{k+1}}{\partial w_1^2}\left(\frac{w_2}{2}\right)^2 + \frac{1}{\underline{|3}}\frac{\partial^3\Phi_k}{\partial w_1^3}\left(\frac{w_2}{2}\right)^3\right) + \Phi_{k+3} = 0.$$

. . . .

These equations present successively finite determinate values for $w_2, w_3, \ldots$, since the factor $\frac{\partial\Phi_k}{\partial w_1}$ does not vanish. Corresponding to each different value of w_1 they give uniquely a different value of each higher derived function, and accordingly k different expansions.

The result is formulated in the theorem:

When in the k-elementary point all the values of the first derived function are finite and different, k elements of the function meet in this point and each of these is only a simple element, i. e. each can be expressed by a series of positive integer powers.

When the point considered is real, and the k values of the first derivate likewise real, k branches of the algebraic curve with distinct directions of tangents pass through the k-elementary point, and each right line through it $w = \alpha z$ has in it at least k points common with the curve; it is then called a multiple point of the order k without ramification.

When one root of the equation $\Phi_k = 0$ becomes infinite, this means that $f_{0,k} = 0$, and the degree of Φ_k reduces to $k - 1$. In order to obtain that element of the function which belongs to the one infinite root, let us first expand z as a function of w, as at the end of § 200. The series begins with the term w^2, or with some higher power of w when further consecutive values of the derivates of z with respect to w also vanish. When $w^{k'}$ is the first term, we obtain by inversion a k'-branched element of the function, having the initial term $z^{\frac{1}{k'}}$. The geometric statement of this case is: In the multiple point one of the k branches of the curve has a tangent parallel to ordinates having k' consecutive points on it, or having a contact of the order $k' - 1$ with that branch of the curve.

The theorem still holds for each simple root of $\Phi_k = 0$ even when there are multiple roots besides. But, for a multiple root it does not hold; for, because $\frac{\partial \Phi_k}{\partial w_1}$ vanishes for such a root, the values of the higher derived functions for it are no longer generally finite. Thus the question finally outstanding is: What is the form of the expansion for a multiple value of w_1, finite or infinite?

202. Although this question can also be solved by successive substitutions*), a process having the preference: that it employs only the Theorem for the possibility of the expansion in a regular point of an algebraic function in order to deduce from it the existence and nature of the expansions in the singular point, still, since the general investigations in this chapter establish the existence of the expansion, it appears suitable that we should go back to the method given by Newton, which has been elaborated by Puiseux.**)

Suppose the equation $f(z + a,\ w + b) = 0$, which defines the function that is to be investigated in the neighbourhood of its k-elementary point $z = 0$, $w = 0$, arranged by powers of z and w. Since we must henceforth assume that some partial derivates of the kth order (specially $f_{0,k}$) also may vanish, let us conceive the terms of the equation arranged as follows. Take first the term independent of w in which z occurs in the lowest power; there must be such a term, for otherwise the factor w could be separated and the equation would

---

*) See Hamburger: Ueber die Entwickelung algebraischer Functionen in Reihen. Zeitschrift f. Math. u. Physik, Vol. XVI. Nöther: Ueber die singulären Werthsysteme einer algebraischen Function. Math. Annal., Vol. IX, pp. 166—182.

**) Newton in the letters to Oldenburg, June 13 and October 24, 1676. Newton's method was explained and proved by Stirling, who says of it: "quae est omnium quam quis excogitare potest, generalissima et elegantissima", Lineae tertii ordinis, 1717; Cramer: Analyse des lignes courbes, 1750; Puiseux, see reference in foot-note § 91, p. 154. Compare also the exposition given in Briot et Bouquet.

not be irreducible. Let this term be z^l, its coefficient $A_{l,0}$. Let us then take the term independent of z in which w occurs in the lowest power. Let it be $A_{0,h}w^h$. Between these two let us arrange all those terms of the form $A_{\alpha,\beta}z^\alpha w^\beta$ whose exponents α and β are respectively lower than l and h, while of terms with the same power z^α (or w^β) we always take only that one in which the exponent of w (or z) is lowest. The terms thus selected can then be so arranged that the series of numbers α shall decrease from l to 0, and that the series of numbers β shall increase from 0 to h. All the rest of the terms, in case there be any over, may be denoted by $\varphi(z, w)$ so that we shall write down the aggregate in the form:*)

$$f(z+a, w+b) = \left\{A_{l,0}z^l + \sum A_{\alpha,\beta}z^\alpha w^\beta + A_{0,h}w^h\right\} + \varphi(z, w) = 0.$$

Since the point is by hypothesis k-elementary, a term of the kth dimension is certain to occur among the bracketed terms. For $z = 0$, h values of w vanish, what is required therefore is to obtain expansions of these h roots in series of ascending powers of z.

Such an expansion begins with the term: $w = vz^\mu \ldots$, where μ must be a positive number, integer or fractional. If the series were known, all the powers of w could be expressed by series valid within the same circle of convergence, the expansion for w^β beginning with the term: $v^\beta z^{\beta\mu}$. Substituting then these series in the above form, the expression should vanish identically: i. e. the coefficients of its various powers of z should be separately zero. But when we substitute

$$w = vz^\mu, \quad w^\beta = v^\beta z^{\beta\mu},$$

the form of the above expression becomes:

$$\left\{A_{l,0}z^l + \sum A_{\alpha,\beta}v^\beta z^\alpha z^{\beta\mu} + A_{0,h}v^h z^{h\mu}\right\} + \varphi(z, vz^\mu);$$

the substitution of further terms of w only introduces terms whose dimensions exceed those written down. But even among those written down, in consequence of our selection, the dimensions of the terms in $\varphi(z, vz^\mu)$ are certainly higher than of those bracketed; for, if there be a term $z^\alpha w^\beta$ in φ, there is a term inside the brackets in which at least one exponent is less than the corresponding α or β. If the value of μ were known, those terms within the brackets for which the exponent is lowest could easily be pointed out; but as μ has first to be found, the inverse questions arise: What assumption as to terms of equal

*) Example: From the equation: $A_{1,5}zw^5 + A_{5,2}z^5w^2 + A_{1,6}zw^6 + A_{0,7}w^7 + A_{7,1}z^7w + A_{4,4}z^4w^4 + A_{0,8}w^8 + A_{9,0}z^9 + A_{5,4}z^5w^4 + A_{7,2}z^7w^2 + A_{10,0}z^{10} = 0$, in which $z = 0$, $w = 0$ is a 6-elementary point, the terms:

$$A_{9,0}z^9 + A_{7,1}z^7w + A_{5,2}z^5w^2 + A_{4,4}z^4w^4 + A_{1,5}zw^5 + A_{0,7}w^7$$

have to be singled out; all the rest belong to the aggregate $\varphi(z, w)$.

lowest dimensions within the brackets leads to a rational determination of μ? and must a value of μ found in this way necessarily be the exponent of the initial term of an expansion?

The first question can be solved graphically: Draw the rectangular system of coordinates having OX and OY as axes, and, adopting any unit of length, mark the values of α and β which occur as exponents in each of the bracketed terms by a point with the coordinates $x = \alpha$, $y = \beta$. Thus the first term gives the point on the axis of abscissæ: $x = l$, $y = 0$; then from those that follow we have certain points with decreasing values of x and increasing values of y; lastly the point upon the axis of ordinates $x = 0$, $y = h$. Fig. 20 records the points belonging to the example in the last foot-note.

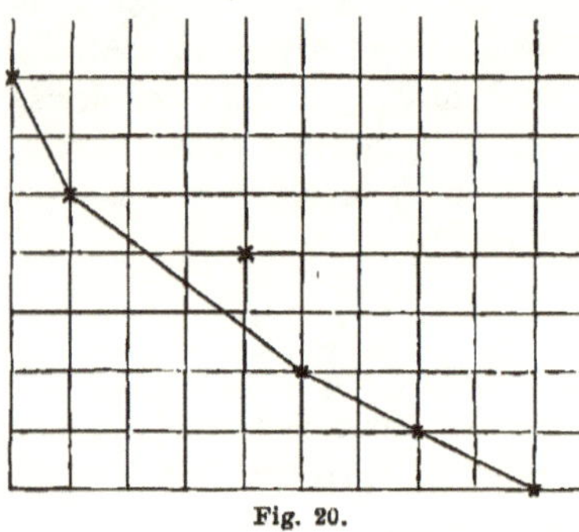
Fig. 20.

If now two terms ex. gr. $A_{\alpha,\beta} z^{\alpha} w^{\beta}$ and $A_{\alpha',\beta'} z^{\alpha'} w^{\beta'}$ are to become of equal dimensions when z^{μ} is put for w, remembering that only positive values of μ are considered, $\alpha + \beta\mu$ must be $= \alpha' + \beta'\mu$, i. e.

$$\mu = \frac{\alpha - \alpha'}{\beta' - \beta}.$$

But from this relation the following theorem results: When the points α, β, and α', β', are joined by a right line, whose equation therefore is:

$$\frac{x - \alpha}{\alpha' - \alpha} - \frac{y - \beta}{\beta' - \beta} = 0, \quad \text{or:} \quad (x - \alpha) + \mu(y - \beta) = 0;$$

points α'', β'', belonging to terms whose dimensions become the same, lie upon this right line; while points whose terms become of lower dimensions lie on the side of it towards the origin, and the rest, whose dimensions become greater, lie on the other side of the right line. For, if $\alpha'' + \beta''\mu = \alpha + \beta\mu = \alpha' + \beta'\mu$ for a point α'', β'', its coordinates satisfy the equation of the right line. When $\alpha'' + \beta''\mu < \alpha + \beta\mu$, the result of substituting $x = \alpha''$, $y = \beta''$ in the expression

$$x - \alpha + \mu(y - \beta)$$

is negative, as it is also for the origin; when

$$\alpha'' + \beta''\mu > \alpha + \beta\mu,$$

the result is positive. Hence the rule, on any assumption as to μ: Only those of the bracketed terms can be of equal lowest dimensions whose corresponding points are joined by a right line that leaves all the other points on the side remote from the origin. Accordingly the solution depends on drawing through the given points only such connecting lines from the point $x = l$, $y = 0$ to the point $x = 0$, $y = h$ as shall form a polygon convex to the origin ($\mu > 0$) but concave to all the remaining points. Hence: turn a right line clockwise round the point $x = l$, $y = 0$, from being along the axis of abscissæ, until it first meets one or more of the points marked down. Let α_1, β_1 be the most distant of these from the turning point, then the terms of lowest dimensions are:

$$A_{l,0}\, z^l + \sum A_{\alpha,\beta} z^\alpha w^\beta + A_{\alpha_1,\beta_1} z^{\alpha_1} w^{\beta_1},$$

and the corresponding value of μ, found from the equation

$$l = \alpha_1 + \mu \beta_1: \quad \mu = \frac{l - \alpha_1}{\beta_1}$$

is a rational number; to indicate that it is not necessarily an integer, let it be denoted by

$$\mu = \frac{l - \alpha_1}{\beta_1} = \frac{p_1}{q_1},$$

where p_1 and q_1 are relatively prime. Substituting the value $w = v z^{\frac{p_1}{q_1}}$, we find an expression from which the factor z^l can be separated, and since the term independent of z must vanish of itself, we obtain for the determination of v the equation of the degree β_1:

$$A_{l,0} + \sum A_{\alpha,\beta} v^\beta + A_{\alpha_1,\beta_1} v^{\beta_1} = 0.$$

This presents β_1 values for v, finite but not all necessarily distinct; so that from this first assumption we should obtain the initial terms of β_1 series.

Let us now consider a second side of the polygon, rotating the right line further from left to right round the point α_1, β_1, till it meets one or more of the points marked down. Let the furthest of these from α_1, β_1, be α_2, β_2. The corresponding terms:

$$A_{\alpha_1,\beta_1} z^{\alpha_1} w^{\beta_1} + \sum A_{\alpha,\beta} z^\alpha w^\beta + A_{\alpha_2,\beta_2} z^{\alpha_2} w^{\beta_2}$$

are then of equal dimensions when we put:

$$\alpha_1 + \mu\beta_1 = \alpha_2 + \mu\beta_2, \text{ or: } \mu = \frac{\alpha_1 - \alpha_2}{\beta_2 - \beta_1} = \frac{p_2}{q_2};$$

and for determining the coefficient v in the expansion $w = v z^{\frac{p_2}{q_2}}$, on substituting this value for w, we find the equation of the degree $\beta_2 - \beta_1$:

$$A_{\alpha_1,\beta_1} + \sum A_{\alpha,\beta} v^{\beta - \beta_1} + A_{\alpha_2,\beta_2} v^{\beta_2 - \beta_1} = 0.$$

Continuing this process, we ultimately obtain from the last (i^{th}) side of the polygon, which must pass through the point $x = 0$, $y = h$, the combination:

$$A_{\alpha_{i-1},\beta_{i-1}} z^{\alpha_{i-1}} w^{\beta_{i-1}} + \sum A_{\alpha,\beta} z^\alpha w^\beta + A_{0,h} w^h, \quad \mu = \frac{\alpha_{i-1}}{h - \beta_{i-1}} = \frac{p_i}{q_i},$$

and from this for determining the coefficient v in the expansion $w = v . z^{\frac{p_i}{q_i}}$ we find the equation of the degree $h - \beta_{i-1}$:

$$A_{\alpha_{i-1},\beta_{i-1}} + \sum A_{\alpha,\beta} v^{\beta-\beta_{i-1}} + A_{0,h} v^{h-\beta_{i-1}} = 0.$$

The result is therefore stated: When the polygon consists of i sides, i different initial terms of expansions $w = v . z^{\frac{p}{q}}$ are possible; and if the coordinates of the vertices of the polygon are called:

$$l, 0; \ \alpha_1, \beta_1; \ \alpha_2, \beta_2; \ \ldots \ \alpha_{i-1}, \beta_{i-1}; \ 0, h,$$

the numbers of such possible expansions belonging to the values:

$$\frac{p_1}{q_1} = \frac{l - \alpha_1}{\beta_1}; \ \frac{p_2}{q_2} = \frac{\alpha_1 - \alpha_2}{\beta_2 - \beta_1}; \ \ldots \ \frac{p_{i-1}}{q_{i-1}} = \frac{\alpha_{i-2} - \alpha_{i-1}}{\beta_{i-1} - \beta_{i-2}}; \ \frac{p_i}{q_i} = \frac{\alpha_{i-1}}{h - \beta_{i-1}},$$

are respectively:

$$\beta_1; \quad \beta_2 - \beta_1; \quad \ldots \quad \beta_{i-1} - \beta_{i-2}; \quad h - \beta_{i-1},$$

on the whole therefore h expansions of the h values of w that vanish for $z = 0$ have been proved possible.

The quotients $p_1 : q_1, p_2 : q_2, \ldots p_i : q_i$ form a decreasing series of numbers, as a glance at the figure shows, because the tangents of the angles between the sides of the polygon and the positive axis of x are the negative values of these numbers.

It must also be shown that these expansions are all actually necessary in order to obtain the h expansions of w.

Let us consider the β_1 possible expansions belonging to the first side of the polygon, and so to the ratio $p_1 : q_1$; this number β_1 is either equal to q_1 or is a multiple of it, suppose $\beta_1 = k_1 q_1$. Now since for each point α, β that lies upon this side of the polygon,

$$\alpha + \mu\beta = \alpha_1 + \mu\beta_1 = l,$$

we have:

$$\mu = \frac{l - \alpha}{\beta} = \frac{p_1}{q_1};$$

therefore also each such β is equal to q_1 or to some multiple of q_1, suppose $\beta = k q_1$.

In the equation from which the corresponding value of v is calculated let us substitute $\lambda^{\frac{1}{q_1}}$ for v, thus it becomes:

$$A_{l,0} + \Sigma A_{\alpha,\beta} \lambda^k + A_{\alpha_1,\beta_1} \lambda^{k_1} = 0.$$

This presents k_1 finite values for λ, some of them moreover may be equal. Then to each simple root of such an equation corresponds a cycle of q_1 values; the expansion begins with the term:

$$w = \lambda^{\frac{1}{q_1}} . z^{\frac{p_1}{q_1}},$$

in which we can retain for the root $\lambda^{\frac{1}{q_1}}$ some one of its q_1 values, while $z^{\frac{p_1}{q_1}}$ assumes all its q_1 different values.

Accordingly, when all the roots of the equations for λ are simple, the $\beta_1, \beta_2 - \beta_1, \ldots h - \beta_{i-1}$ expansions which belong to the i sides of the polygon resolve respectively into k_1 cycles each having q_1 values, into k_2 cycles each having q_2 values, . . ., into k_i cycles each having q_i values; so that in fact all these expansions are necessary in order to exhibit the h values of w that vanish for $z = 0$.

Relatively to each simple root λ, the next term in the expansion:

$$w = vz^{\frac{p}{q}} + v_1 z^{\frac{p+1}{q}} + \cdots = vz^{\frac{p}{q}}\left(1 + \frac{v_1}{v} z^{\frac{1}{q}} + \cdots\right),$$

is obtained by attending to the term of the next dimension of z in substituting this series for w. The coefficient of this term equated to zero, presehts a linear equation for determining v_1.

To investigate the signification of a multiple root of the equation:

$$A_{l,0} + \Sigma A_{\alpha,\beta}\lambda^k + A_{\alpha_1,\beta_1}\lambda^{k_1} = 0,$$

we suppose it to have j roots equal to λ; then each of the forms:

$$w = \lambda^{\frac{1}{q_1}} z^{\frac{p_1}{q_1}},$$

must be initial term in $q_1 j$ expansions; but these again will resolve into certain cycles. In order to perceive that they do, let us substitute

$$z = z'^{q_1}, \quad w = \left(\lambda^{\frac{1}{q_1}} + w'\right) z'^{p_1},$$

in the original algebraic equation. This is thereby converted into an equation between z' and w', which, developed by powers of these quantities, must have j roots w' vanishing along with $z' = 0$. In fact

$$\{A_{l,0}z^l + \Sigma A_{\alpha,\beta} z^\alpha w^\beta + A_{\alpha_1,\beta_1} z^{\alpha_1} w^{\beta_1}\} + \{\Sigma A_{\alpha',\beta'} z^{\alpha'} w^{\beta'} + A_{0,h} w^h + \varphi(z,w)\},$$

in which the first brackets contain all the terms of lowest dimensions, passes over, as regards these terms, into:

$$A_{l,0} z'^{lq_1} + \Sigma A_{\alpha,\beta} z'^{\alpha q_1}\left(\lambda^{\frac{1}{q_1}} + w'\right)^\beta z'^{\beta p_1} + A_{\alpha_1,\beta_1} z'^{\alpha_1 q_1}\left(\lambda^{\frac{1}{q_1}} + w'\right)^{\beta_1} z'^{\beta_1 p_1},$$

or, dividing by the lowest power of z', the value of whose exponent is $lq_1 = \alpha q_1 + \beta p_1 = \alpha_1 q_1 + \beta_1 p_1$, into:

$$A_{l,0} + \Sigma A_{\alpha,\beta}\left(\lambda^{\frac{1}{q_1}} + w'\right)^\beta + A_{\alpha_1,\beta_1}\left(\lambda^{\frac{1}{q_1}} + w'\right)^{\beta_1},$$

and of these, the term independent of w' vanishes, because λ is a root of the equation:

$$A_{l,0} + \Sigma A_{\alpha,\beta}\lambda^{\frac{\beta}{q_1}} + A_{\alpha_1,\beta_1}\lambda^{\frac{\beta_1}{q_1}} = A_{l,0} + \Sigma A_{\alpha,\beta}\lambda^k + A_{\alpha_1,\beta_1}\lambda^{k_1} = 0.$$

But because j roots are to be equal to λ in this equation, its first $j-1$ derivates with respect to λ also vanish.

Now let the equation between z' and w' be investigated, as was the original equation between z and w, by constructing the polygon corresponding to the dimensions of its terms. Each side of the polygon leads to a selection of terms of equal lowest dimensions, and each simple root corresponding to this selection leads to the initial term

$$w' = \lambda'^{\frac{1}{q'}} z'^{\frac{p'}{q'}}$$

of an expansion, for which the denominator q' specifies the number in the cycle. Now since

$$z' = z^{\frac{1}{q_1}}, \text{ and } w = \left(\lambda^{\frac{1}{q_1}} + w'\right) z'^{p_1},$$

the first two terms of the expansion of w in powers of z are:

$$w = \lambda^{\frac{1}{q_1}} z^{\frac{p_1}{q_1}} + \lambda'^{\frac{1}{q'}} z^{\frac{p'+q'p_1}{q'q_1}}.$$

Here we can retain for each root $\lambda^{\frac{1}{q_1}}$ and $\lambda'^{\frac{1}{q'}}$ some one of its possible values, while the roots of z assume all possible $q_1 q'$ values. Thus a cycle of $q_1 q'$ branches arises from this simple root; the next term in the expansion is to be found by substituting:

$$w = \lambda^{\frac{1}{q_1}} z^{\frac{p_1}{q_1}} + \lambda'^{\frac{1}{q'}} z^{\frac{p'+q'p_1}{q_1q'}} + vz^{\frac{p'+q'p_1+1}{q_1q'}} + \cdots \text{ etc.}.$$

But when there is a multiple root in the equation between w' and z', let us similarly introduce the variables z'' and w''; then as result of substituting

$$w = \lambda^{\frac{1}{q_1}} z^{\frac{p_1}{q_1}}, \quad z = z'^{q_1}, \quad w = \left(\lambda^{\frac{1}{q_1}} + w'\right) z'^{p_1}, \quad z' = z''^{q'}, \quad w' = \left(\lambda'^{\frac{1}{q'}} + w''\right) z''^{p'},$$

provided this new substitution leads to a cycle between w'' and z'', we shall have:

$$w'' = \lambda''^{\frac{1}{q''}} z''^{\frac{p''}{q''}},$$

therefore:

$$w' = \lambda'^{\frac{1}{q'}} z'^{\frac{p'}{q'}} + \lambda''^{\frac{1}{q''}} z'^{\frac{p''+q''p'}{q''q'}},$$

$$w = \lambda^{\frac{1}{q_1}} z^{\frac{p_1}{q_1}} + \lambda'^{\frac{1}{q'}} z^{\frac{p'+q'p_1}{q'q_1}} + \lambda''^{\frac{1}{q''}} z^{\frac{p''+q''p'+p_1q'q''}{q_1q'q''}} + \cdots,$$

a cycle of $q_1 q' q''$ values arises.

Now several such substitutions may be necessary, but ultimately a finite number of them must lead to a simple root.

This we see as follows: Since λ is a multiple root of the equation of degree k_1, j will be less than or at most equal to k_1. Therefore there is generally a decrease of multiplicity, and as the process is continued we generally reach a simple point that presents a root also only simple. The process would only fail to lead to the desired end, if after a certain stage the multiplicity of the root remained always equal to the degree of the equation in question. But this supposition contradicts the hypothesis that the form considered is **irreducible**.

For, assuming, as we may without any restriction of generality, that from the outset the equation:

$$A_{l,0} + \Sigma A_{\alpha,\beta}\lambda^k + A_{\alpha_1,\beta_1}\lambda^{k_1} = 0$$

has k_1 roots equal, and is therefore:

$$A_{\alpha_1,\beta_1}(\lambda - \lambda_0)^{k_1} = 0,$$

the form of the original equation must be

$$z^{\alpha_1} A_{\alpha_1,\beta_1}(w^{q_1} - \lambda_0 z^{p_1})^{k_1} + \varphi(z, w) = 0.$$

By substituting: $z = z'^{q_1}$, $w = \left(\lambda_0^{\frac{1}{q_1}} + w'\right)z'^{p_1}$, and dividing by the factor $z'^{q_1(\alpha_1+p_1k_1)} = z'^{q_1 l}$, we obtain the equation between z' and w':

$$A_{\alpha_1,\beta_1}\left\{\left(\lambda_0^{\frac{1}{q_1}} + w'\right)^{q_1} - \lambda_0\right\}^{k_1} + \psi(z', w') = 0,$$

where $\psi(z', w')$ denotes the terms proceeding from φ; the lowest power of w' is w'^{k_1}. By hypothesis this equation is to give only a single root, to be counted k_1 times, for the expansion of w'. But this requires that the corresponding polygon should reduce to a single right line and that therefore the corresponding equation should be:

$$A(w' - \lambda' z'^{p'})^{k_1} + \varphi_1(z', w') = 0.$$

Now supposing the further substitutions carried out, and that we found every time an equation of degree k_1 between w and z, there should exist an expansion:

$$w = \lambda^{\frac{1}{q_1}} z^{\frac{p_1}{q_1}} + \lambda' z^{\frac{p_1+p'}{q_1}} + \lambda'' z^{\frac{p_1+p'+p''}{q_1}} + \cdots,$$

that must be valid however great the exponents of z become. By this expansion k_1 roots of the equation would be expressed. The algebraic form would therefore have k_1 equal roots within a certain domain of z and consequently in general. Since, starting from a value w, we can establish an irreducible form (§ 199) whose root is w, it follows that the original algebraic form on our hypothesis must have contained as a factor the k_1^{th} power of an irreducible factor or must have been equal to the k_1^{th} power of such a form.

203. The example in the foot-note to § 202, p. 381:

$$A_{9,0}z^9 + A_{7,1}z^7 w + A_{5,2}z^5 w^2 + A_{4,4}z^4 w^4 + A_{1,5}z w^5 + A_{0,7}w^7 + \varphi(z, w) = 0,$$

where

$$\varphi(z, w) = A_{1,6}z w^6 + A_{7,2}z^7 w^2 + A_{5,4}z^5 w^4 + A_{0,8}w^8 + A_{10,0}z^{10},$$

in which seven values of w vanish for $z = 0$, leads, as the polygon fig. 20 shows, to the three aggregates of terms of equal dimensions:

$$\text{I.}\quad A_{9,0}z^9 + A_{7,1}z^7w + A_{5,2}z^5w^2 = 0,$$

$$\text{II.}\quad A_{5,2}z^5w^2 + A_{1,5}zw^5 = 0,$$

$$\text{III.}\quad A_{1,5}zw^5 + A_{0,7}w^7 = 0.$$

To I. belongs the value:

$$\mu = \frac{9-7}{1} = \frac{9-5}{2} = 2;$$

accordingly putting $w = vz^2$, we obtain the quadratic for v:

$$A_{9,0} + A_{7,1}v + A_{5,2}v^2 = 0.$$

When this has unequal roots, v_1 and v_2, there are two expansions of w by integer powers of z beginning respectively with the terms

$$w = v_1z^2 \text{ and } w = v_2z^2.$$

These are not branched here. The next term in each is found by substituting:

$$w = v_1z^2\left(1 + \frac{v_3}{v_1}z\right) \text{ and } w = v_2z^2\left(1 + \frac{v_3'}{v_2}z\right),$$

and equating to zero the sum of the terms found to be of the tenth dimension; thus:

$$v_3 = -\frac{A_{10,0}}{2A_{5,2}v_1 + A_{7,1}},\quad v_3' = -\frac{A_{10.0}}{2A_{5,2}v_2 + A_{7,1}}.$$

When the roots of the quadratic are equal, $v_1 = v_2 = v$, and thus:

$$A_{7,1}^2 - 4A_{9,0}A_{5,2} = 0 \text{ and } 2A_{5,2}v + A_{7,1} = 0,$$

let us put $w = (v + w')z^2$; in the equation between z and w' resulting from division by z^9, two values of w' vanish along with $z = 0$. This equation is:

$$A_{9,0} + A_{7,1}(v+w') + A_{5,2}(v+w')^2 + A_{4,4}z^3(v+w')^4 + A_{1,5}z^2(v+w')^5$$
$$+ A_{0,7}z^5(v+w')^7 + \cdots + A_{10,0}z = 0,$$

or, as it may be arranged in consequence of the condition for v:

$$A_{10,0}z + A_{5,2}w'^2 + \cdots = 0.$$

Hence results:

$$w' = \pm\sqrt{\frac{-A_{10,0}}{A_{5,2}}}\cdot z^{\frac{1}{2}},$$

and consequently we obtain:

$$w = vz^2 + \sqrt{\frac{-A_{10,0}}{A_{5,2}}}\cdot z^{\frac{5}{2}} + \cdots,$$

that is to say, a series that proceeds by powers of $z^{\frac{1}{2}}$; two values of the function ramify cyclically.

But when $A_{10,0} = 0$, the initial terms are: $A_{1,5}z^2v^5 + A_{5,2}w'^2 = 0$, accordingly:

$$w' = \pm\sqrt{\frac{-A_{1,5}v^5}{A_{5,2}}}\,z,$$

and we obtain the following two separate expansions that coincide only in the initial term:

$$w = v z^2 + \sqrt{\frac{-A_{1,5} v^5}{A_{5,2}}}\, z^3 + \cdots, \quad w = v z^2 - \sqrt{\frac{-A_{1,5} v^5}{A_{5,2}}}\, z^3 - \cdots \text{etc.}.$$

To II. belongs the value:

$$\mu = \frac{5-1}{5-2} = \frac{4}{3},$$

and for $w = v z^{\frac{4}{3}}$ we find the equation:

$$A_{5,2} v^2 + A_{1,5} v^5 = 0 \text{ or: } A_{5,2} + A_{1,5}\lambda = 0, \quad \lambda^{\frac{1}{3}} = v.$$

There is accordingly a cycle of three branches; the expansion pertaining to it is:

$$w = \sqrt[3]{\frac{-A_{5,2}}{A_{1,5}}}\, z^{\frac{4}{3}} + v z^{\frac{5}{3}} + \cdots \text{ etc.}.$$

To III. belongs the value:

$$\mu = \frac{1}{7-5} = \frac{1}{2} \text{ and } A_{1,5} + A_{0,7}\lambda = 0;$$

therefore:

$$w = \sqrt{\frac{-A_{1,5}}{A_{0,7}}} \cdot z^{\frac{1}{2}} + v z^{\frac{2}{2}} + \cdots \text{ etc.}.$$

We have thus indicated the explicit forms of the seven values of w which vanish along with $z = 0$ in the 6-elementary point.

CORRECTIONS.

p. 8, l. 20-19 up, read "the limiting value of: 0.3; 0.33; etc.".
„ 13 „ 11 up, read "$B^x = A = B^{x'}$".
„ 33 „ 2 down, read "$x > a$".
„ 39 „ 4-5 down, read "(1) we write $\Theta\Delta x$ for Δx and then".
„ 60 „ 7 up, read "of any portion of a body cut off by a plane:"
„ 80 „ 14-3 up, read "ultimately smaller than that preceding it;"
„ 89 „ 5 up, read "$f(x_1 - \varepsilon, y_1 - \varepsilon')$".
„ 180 „ 5-7 down, read "f has finite discontinuities at any finite number of points $c_1, c_2, \ldots c_m$."
„ 239 „ 9 up, read "value may or may not".
„ 243 „ 8 up, read "throughout continuous;"
„ 261 „ 6-5 up, read "(In other words,"
„ 309 „ 7 down, read "values".

INDEX.

www.ingramcontent.com/pod-product-compliance
Lightning Source LLC
Chambersburg PA
CBHW030823110726
47900CB00006B/1725